Evolutionary Biogeography of the Marine Algae of the North Atlantic

NATO ASI Series

Advanced Science Institutes Series

A series presenting the results of activities sponsored by the NATO Science Committee, which aims at the dissemination of advanced scientific and technological knowledge, with a view to strengthening links between scientific communities.

The Series is published by an international board of publishers in conjunction with the NATO Scientific Affairs Division

A Life Sciences **B Physics**	Plenum Publishing Corporation London and New York
C Mathematical and Physical Sciences **D Behavioural and Social Sciences** **E Applied Sciences**	Kluwer Academic Publishers Dordrecht, Boston and London
F Computer and Systems Sciences **G Ecological Sciences** **H Cell Biology**	Springer-Verlag Berlin Heidelberg New York London Paris Tokyo Hong Kong Barcelona

Series G: Ecological Sciences Vol. 22

Evolutionary Biogeography of the Marine Algae of the North Atlantic

Edited by

David J. Garbary

Department of Biology
St. Francis Xavier University
Antigonish, Nova Scotia
Canada B2G 1C0

G. Robin South

Huntsman Marine Science Centre
St. Andrews, New Brunswick
Canada EOG 2X0

Springer-Verlag Berlin Heidelberg New York
London Paris Tokyo Hong Kong Barcelona
Published in cooperation with NATO Scientific Affairs Division

Proceedings of the NATO Advanced Research Workshop on Evolutionary Biogeography of the Marine Algae of the North Atlantic held in St. Andrews, N.B., Canada, September 18–22, 1989

Library of Congress Cataloging-in-Publication Data
NATO Advanced Research Workshop on Evolutionary Biography of Marine Algae of the North Atlantic (1989 : St. Andrews, N. B.) Evolutionary biogeography of the marine algae of the North Atlantic / edited by David J. Garbary, G. Robin South. p. cm.—(NATO ASI series. Series G, Ecological sciences ; vol. 22) "Proceedings of the NATO Advanced Research Workshop on Evolutionary Biogeography of the Marine Algae of the North Atlantic, held in St. Andrews, N. B., Canada, September 18–22, 1989"—T.p. verso. "Published in cooperation with NATO Scientific Affairs Division."

ISBN-13: 978-3-642-75117-2 e-ISBN-13: 978-3-642-75115-8
DOI: 10.1007/978-3-642-75115-8

1. Marine algae—North Atlantic Ocean—Geographical distribution—Congresses. 2. Marine algae—North Atlantic Ocean—Evolution—Congresses. I. Garbary, David J. II. South, G. Robin. III. North Atlantic Treaty Organization. Scientific Affairs Division. IV. Title. V. Series: NATO ASI series. Series G, Ecological sciences ; no. 22. QK572.5.N67N38 1990 589.392'11—dc20 90-43825

Softcover reprint of the hardcover 1st edition 1990

2131/3140-543210 – Printed on acid-free-paper

PREFACE

Evolutionary biogeography of the marine algae of the North Atlantic was the primary theme of the Fourth Seaweed Biogeography Workshop. This five day symposium was held in St. Andrews, New Brunswick, in September 1989, and was organized under the auspices of the International Working Group on Seaweed Biogeography. Previous workshops in this series were held at Hamburg (1982), Groningen (1984) and Helgoland (1986), and were published in Helgoländer Meeresunters (38: 225-417 and 41: 233-383).

The Organizing Committee for the conference comprised the two of us, and C. van den Hoek, who formulated the scientific program, and identified the primary participants.

The primary financial support for the meeting was provided by a grant from NATO (ARW 931/88), under their Advanced Research Workshop program.

The Primary Theme

This meeting was based on the notion that the current status of algal biogeography, and its advance, could be best addressed by a multidisciplinary approach to a single theme: THE EVOLUTIONARY BIOGEOGRAPHY OF THE MARINE ALGAE OF THE NORTH ATLANTIC. Participants were asked to focus their presentations on this theme, or to provide philosophical and/or methodological discussions of biogeography.

In addition to strictly algal biogeographers, people in related fields of geology, theoretical biogeography and marine ecology were invited, so that their theoretical and philosophical concerns, as well as their methods, could be better evaluated by the more traditional 'phycologists', many of whom undertake biogeography as a supplement to their primary systematic, physiological, or ecological interests. Biogeography, is by its nature, interdisciplinary, and the St. Andrews meeting was a reflection of this.

The St. Andrews Symposium: A NATO Advanced Research Workshop

The meeting was attended by about 30 scientists from eight countries, and included a series of 20 formal paper presentations, of which 17 are included in this volume. The presentations not in this volume are:
1) M. Ragan, C.J. Bird and E.L. Rice "Nuclear small-subunit ribosomal ribonucleic acid sequences and molecular relationships within the Gracilariales";
2) S. Srivastava "Evolution of the North Atlantic - Jurassic to present"; and
3) J. Cabioch "Adaptive strategies and distribution of Corallinales".
Two presentations of computer sofware for biogeography and phylogenetic analysis were also made (P. Legendre, D. Garbary). At the conclusion of the meeting, a panel discussion was used as a venue for discussing theoretical issues, progress that had been made in understanding the overall problem that was the theme of the meeting, and how algal biogeographers might redirect their approach so that biogeographic hypotheses might be more rigorously evaluated.

Throughout the meeting, informal discussions were held in Anderson House, before, during and after meals, and long into the evenings. These discussions were suitably lubricated from the fridge in the basement of Anderson House. The overall tone of the five day meeting was, at the same time, both intense and informal, and highly productive. This volume identifies the formal contributions of the symposium; however, it cannot capture the excitement and general atmosphere of the symposium.

Organization of Entries

The papers in this volume have been put into a particular order to reflect some of the dominant themes of the conference, as well as seaweed biogeography in general. None of the sections outlined below are mutually exclusive, and some papers might easily fit into other sections.

The first section treats conceptual and methodological issues and includes two papers. The opening paper by I. Ball is a discussion of some of the conceptual/philosophical issues in contemporary biogeographic theory, and his paper provides a brief introduction to this literature for a phycological audience. This is followed by a complementary paper by P. Legendre describing quantitative methods in biogeography, providing a list of the appropriate computer programs for implementing these methods.

The next section includes two papers, by D. Schnitker & J. Jorgensen and C. van den Hoek & A. M. Breeman, who examine the effects of Pleistocene glaciation on algal distributions and the changes that occurred after retreat of the ice. This is examined on a local scale (phytoplankton in the Bay of Fundy), and on a wider scale (seaweeds in the North Atlantic). The next section treats concepts of dispersal and adaptation in an ecological and evolutionary context; in the first instance dealing with barnacle evolution in eastern Canada (E. Bourget *et al.*), and the second dealing with evolution of herbivory on coralline algae in the North Atlantic and North Pacific (R. Steneck). Processes of speciation were a major theme of the meeting and the next three papers by P.V.M. Bot & C. van den Hoek, B. Egan *et al.*, and B. Stache provide insight into seaweed speciation and genetics. They use a variety of different approaches, including hybridization studies, morphometric analysis and DNA/DNA hybridization. Two papers dealing with evolution of Laminariales were presented, by K. Lüning & I. tom Dieck and L. Druehl, which applied morphological and physiological features, as well as molecular biology to the problem of kelp phylogeny. These papers illustrate, on the one hand, the extent to which phylogenetic relationships can be hypothesized using more traditional features; and on the other, the exciting potential for resolving phylogenetic problems using molecular techniques.

The relationship between systematics and biogeography in Rhodophyta is discussed in three papers by A. Athanasiadis, C. Maggs, and M.D. Guiry & D. Garbary studying a diversity of red algal taxa. These papers emphasize the importance of taxonomy in dealing with historical biogeography.

The division of the North Atlantic into biogeographic regions is an ongoing theme in seaweed biogegraphy; this is treated by Tittley *et al.* using a variety of quantitative methods, including an innovative use of cladistic analysis.

The final section of these proceedings includes two highly synthetic papers discussing the evolutionary biogeography the of entire seaweed flora of the Arctic (Wilce) and the red algae of the North Atlantic (M. Hommersand). Both of these papers emphasize the importance of earth history and systematics in developing an understanding of evolutionary biogeography of marine algae.

Preparation of articles for publication

Authors were asked to provide copies of manuscripts and computer disk versions of the papers at the St. Andrews meeting. Manuscripts were reviewed, generally by two referees, at least one of whom was not a participant at the meeting. The editors are especially grateful to the reviewers who read and critically evaluated the various manuscripts. The individuals who did not participate in the conference, but who provided reviews included: M. Alvarez, M.

Blouw, M. Cormaci, P.W. Gabrielson, R.G. Hooper, S. Lindstrom, J.L. Olsen, C. Plowright, G. Russell, E. Theriot, and A. Whittick.

After review and revisions, manuscripts were copy edited, checked for style, and were then formatted using Microsoft Word (version 4) (Copyright Microsoft Corp.), implemented on a Macintosh SE30 computer. Manuscripts were set in Times, and camera-ready copy was produced using an Apple laser printer.

Dedication to "Bud" Brinkhuis and "Barry" Egan

After this conference was organized, and during the preparation of this volume, two of the contributing authors, P.F. "Barry" Egan and B. "Bud" Brinkhuis, passed away. It is with considerable sadness that we mark the deaths of these friends and fellow phycologists. The deaths, of an established colleague, and a young scientist of considerable promise, make this a particular tragedy. This volume is dedicated, in part, to these two phycologists.

Acknowledgments

This volume, and the symposium on which it is based, would not have come to fruition without the help and dedication of a wide variety of people and organizations. First, we thank the Staff (especially Irene Pohle)and Directors of the Huntsman Marine Science Centre, and the Department of Fisheries and Oceans, Biological Station, St. Andrews for the high standard of general organization of the meeting, and of the facilities that were made available. All of the participants were housed in Anderson House, whose Staff are complimented for the quality of meals, and the fine surroundings in which many of the discussions were held. Barbara South is to be congratulated for the excellent banquet that she hosted.

We are most grateful to the external and internal reviewers who evaluated the manuscripts. Mr. J. Hendra and Mr. A. M. Garbary provided major assistance in editing and preparation of the camera-ready copy. We would also like to thank the authors for their patience and cooperation during the production of these proceedings.

Finally we would like to thank NATO who provided the primary financial assistance towards travel and accomodation for the symposium contributors, and the Natural Sciences and Engineering Research Council of Canada who provided additional funds, through a Conference Grant, for support of particular individuals.

David J. Garbary
G. Robin South
May, 1990

TABLE OF CONTENTS

Part 5. Kelp Evolution

Part 6. Taxonomy and Biogeography

Part 7. Biogeographic Provinces

Part 8. Evolution of Floras in the Arctic and Atlantic Oceans

THE FRAMING OF BIOGEOGRAPHIC HYPOTHESES

Ian R. Ball
Department of Biology
Memorial University of Newfoundland
St John's Newfoundland, Canada A1B 3X9

Abstract - Current discussions of the philosophical and logical bases of biogeography are summarised. The relationships between systematics and biogeography are discussed and the importance of proper assessment of the ontological status of the units of biogeography is addressed. Criteria for the framing of biogeographical hypotheses, invoking clarity, comparability and rigour, are suggested.

INTRODUCTION

In the past decade there has been much discussion concerning the philosophical and logical bases of biogeography (Rosen 1978; Nelson & Platnick 1981; Nelson & Rosen 1981; Ball 1983; Craw 1988; Sober 1988; Wiley 1988). Much of this has stemmed directly from the discursive, yet still instructive, writings of Croizat (1964). A recent and thorough review is that by Rosen (1988). There can be little doubt that this discipline is undergoing revolutionary change and discussion of its bases is essential during that process. Yet, in agreeing with Ghiselin (1980) in his cautioning against the misuse, or naive use, of philosophy in biogeography, I also agree that the constant questioning of its logical and philosophical foundations is essential to its healthy growth. This is true, both for its theoretical development and its practical application, as I shall demonstrate, albeit briefly, in this essay.

THE NATURE OF BIOGEOGRAPHY

The pattern of geographical distribution exhibited by any group of organisms is a result of historical and ecological constraints; in other words, it is a product of historical and proximal causes. One may take the historical constraints as given,and validly study only the proximal causes, but this is unlikely to yield explanations of high generality that will contribute to a unified theory of biogeography. On the other hand, it will contribute to

NATO ASI Series, Vol. G 22
Evolutionary Biogeography of the
Marine Algae of the North Atlantic
Edited by D. J. Garbary and G. R. South

our knowledge of the biology of the organisms themselves, and of their community structures. It is the unravelling, but not necessarily the decoupling, of these constraints that has been the focus of much of the recent writing on biogeographical theory (refs in Myers & Giller 1988; see also special issues of Systematic Zoology, 37(3/4), 1988).

An earlier distinction between descriptive, narrative and analytical biogeography (Ball 1975) has persisted, and seems to have proved useful (Cracraft 1988; Rosen 1988). From the viewpoint of history specialists, in particular, animal and plant groups have generally emphasized the ecological constraints, often to the abandonment of all historical considerations other than those of comparatively recent dispersal events. Biogeography consisted primarily of the description of individual distributions and the delimitation, often statistically based, of biotic regions. Appended to these were narratives based on experience, and on knowledge of the organism's ecology and biology, usually invoking centres of origin and dispersal away from them. These narratives pertained solely to the particular organisms or were used to form generalizations about the origins, usually from an ecological/climatological viewpoint, of the regions themselves. Competition, predation, habitat selection and colonization were the catch words of the day. Debate was centred on the narrative domains of different workers, usually with appeals to 'reason' or 'authority', and a further analytical approach was deemed unnecessary. This history, as it relates to algal biogeography, has been recounted by Garbary (1987), and need not be repeated here.

It is now commonplace to state that the distributions of living organisms are non-random; that there are discernible patterns of occurrences. The presence of non-random events implies causality, and inspires the search for testable, explanations (eg Cracraft 1988); mere description is not enough. What is more, since the evolution of life and the history of the earth are inextricably bound together (Croizat 1964), it follows that we should be seeking unified explanations, not ones applicable only to the taxon of interest to the particular specialist. Such explanations must invoke history. In short, it is historical factors that influence the raw material available for ecological principles to act upon. The historical constraints are paramount.

This symposium has been concerned with evolutionary biogeography and therefore it is appropriate to confine the subject matter of this essay to historical approaches. Brundin (1988) has pointed out there is a great asymmetry between philosophical and methodological discussions based on abstract groups, and actual applications to real groups. Yet, if methods and philosophies are to develop effectively, they must be continually tested in the real world. It is the specialist who can best do this. My aim, then, is to draw attention to these approaches, and their literature, in the hope that algal biogeography might thereby be advanced, and that phycologists will provide rigorous tests of newer theoretical developments.

THE UNITS OF STUDY

The question of the logical status of the units to be studied biogeographically is of paramount importance; that is because we must deal with logical 'individuals'. The debate concerning the roles of logical individuals and logical classes in systematic biology arose largely because of the development of phylogenetic systematics. In brief, phylogeneticists accept only strictly monophyletic groups (clades) as having any evolutionary status or reality. The species is regarded as an individual for the simple reason that classes do not evolve, cannot speciate; individuals can. A moment's reflection will reveal that all higher taxa under the phylogenetic system, as monophyla, must be individuals as well. They are recognised, but not defined (individuals are, they are not defined entities), by their possession of unique evolutionary novelties, or synapomorphies. Taxa delimited solely on the basis of shared primitive features, or symplesiomorphies, represent grades, not clades, and as such have the logical status of classes. The conclusion, then, must be that systematics is concerned not with the recognition or delimitation of classes, but rather with the recognition of chunks of the genealogical nexus (see Ghiselin, 1980). Biogeography is concerned with the distributions of these parts, i.e. with the distributions of strict monophyla.

What constitutes an individual in biogeography has been subject to less consideration and is open to debate. As illustration, I would suggest that Kentucky is an individual, as is Madagascar. Kentucky plus California is a class, much as are statistically defined floral provinces. Areas of endemism (see below) can be regarded as evolutionary units, and also may be regarded as individuals, although we have few rules for determining their boundaries.

BIOGEOGRAPHY AND SYSTEMATICS

Phylogenetics, or cladistics, demands that we deal with monophyla; these result from individual speciation events. If we regard most speciation events as allopatric, it can be seen that the spatial distribution of closely related species forming a monophylum, may be regarded as a series of successive speciation events, or a succession of vicariant divisions, through time, of an ancestral biota (Croizat 1964). Dispersal from a common centre need not be invoked except to account for the prior expansion of range, to fixed boundaries, of the original protospecies. This forms the basis of Croizat's Panbiogeography, which sought to establish areal relationships by investigating large numbers of distributions of 'consanguinous' taxa. An application of this method to marine littoral organisms is that of Sluys (1989); I am concerned about some of the taxonomic decisions of that work, but it provides a secure introduction to the literature. To the original version of Panbiogeography, various computational sophistications have now been added (see Legendre 1990, this volume).

A development of Panbiogeography, and one disavowed by Croizat, is also recognised as a part of vicariance biogeography (Nelson & Platnick 1981; Humphries *et al.* 1988; Wiley 1988). If a species responds, by speciating, to all the vicariant events postulated above, then it follows that each of the final areas will have its own endemic species. Reconstructing the phylogenetic relationships of these endemics will give the history of the relationships between the areas; that is, the sequence of the vicariant events that resulted in the disruption of the original area.

We see then, that such areas of endemism reflect lineages produced by fragmentation through time. Compare this with the evolutionary lineages, or clades, produced by phylogenesis, and it is apparent that there is a logical cohesiveness between systematics and biogeography, in that each is dealing with the historical (genealogical) relationships between individual historical entities, and not with the relationships between classes. Those who deal only with faunal and floral provinces, however, are paralleling the pheneticists, and deal primarily with arbitrary classes.

There are other parallels; in establishing biogeographical relationships, the cladistic biogeographer treats areas as if they were taxa, and species in those areas as if they were the characters of the taxa. Endemic distributions are analagous to the uniquely derived characters, or apomorphies, of the phylogenetic systematist; and, just as systematists must cope with convergence, secondary loss, and the like, so the biogeograher must deal with 'non-homologies' in the form of widespread distributions and extinctions.

It should also be pointed out that Sober (1988) has questioned, without resolving the issue, the analogy between systematics and biogeography just outlined. He points out that vertical transmission of characters is assumed in phylogenetic inferences, whereas competing biogeographical analyses may use either vertical or horizontal transmission (eg dispersal or fusion of biotas). The point is well taken, and should be examined carefully.

Ideally, cladistic biogeography is the analysis of the relationships between areas of endemism as exhibited by several, often distantly related, groups. If there is but one areal history, then clearly it must be reflected in the evolutionary histories of all groups inhabiting the total area. In practice, there is a great deal of noise and non-congruence caused by dispersal, extinction or by the failure of some particular group to respond, by speciation, to some actual vicariant event; after all, what is a barrier to gene flow in one group, is not necessarily such a barrier for another group.

Nevertheless, for the reasons outlined, there must be some congruence or compatibility in the relative distributions of all groups for a specified geographical area which must have had a unique history; earth and life do, indeed, evolve in concert. The resolution of the anomalies afforded by dispersal, extinction and evolutionary stability, in the face of specific vicariant events has been the subject of much study by means of consensus and compatibility techniques involving parsimony criteria. These are discussed by Legendre (1990).

FRAMING BIOGEOGRAPHIC HYPOTHESES

The increased precision and rigour provided by vicariant approaches to biogeography gives the possibility of removing biogeographical theories from a non-testable narrative status, to one in which testable hypotheses of general significance may be generated; at least this should be our aim. Even so, there remain biogeographers who take a most pessimistic view of the attainability of these ideals (Ghiold & Hoffman 1984). I remain hopeful.

The framing of biogeographical hypotheses should, I suggest, comprise three facets, though these are not without difficulties:

1. Clarity. - The units of study should be clearly denoted and be of the logical status of individuals, whether they be taxa or areas. The history of biotas is to be discovered by the analysis of the distributions and relationships of endemic species. Areas of endemism, specified by several groups, reflect the individual geographical units for study. For any endemic species discovered, the first question must be: where does its sister species occur? Traditional specialized faunal and floral biogeographers are dealing with biotic provinces, based on some measure of overall similarity, which are arbitrary classes, likely to have unique, individual histories. They may be of immediate ecological interest, but it is dangerous to use such concepts to infer the general history of the areas as a whole. Debate concerning the validity of a specialised picture must take place within the framework of a general theory involving the area in question.

2. Comparability. - The hypothesis should be presented in such a way that direct comparison with hypotheses derived for other organisms in the same area can be made. If there is a unified story it must apply to other groups irrespective of their ecological properties. This implies consensus and compatibility analyses (Legendre 1990).

3. Rigour. - Hypotheses should be presented in such a way that potential observations that would serve to refute the hypothesis may be assessed. These may include additional phylogenetic, distributional, or geological data. Narratives are rarely so testable, and usually rely on 'expert authority'for their survival; usually they die with their protagonists.

Yet the taxonomic specialist is the person best placed to formulate and carry out specific tests of properly formulated hypotheses.

It must be added that most theoreticians in biogeography are dealing with land based faunas and floras. Much less is known concerning barriers in the sea (see Valentine & Jablonski 1983). Some marine biogeographers remain faithful to the centre of origin/dispersal paradigm as applied to individual groups (see Briggs 1987). This is why it is so important for debate to be continued from a multidisciplinary approach.

REFERENCES

Ball IR (1975) Nature and formulation of biogeographical hypotheses. Syst Zool 24: 407-430

Ball IR (1983) Planarians, plurality and biogeographical explanations. In: Sims RW, Price JH & Whalley PES (eds), Evolution, Time and Space: The emergence of the biosphere, Academic Press, London & New York, p 409-430.

Briggs JC (1987) Biogeography and Plate Tectonics. Elsevier, Amsterdam & New York

Brundin L (1988) Phylogenetic Biogeography. In: Myers AA & Giller PS (eds), Analytical Biogeography, Chapman & Hall, London & New York, p 343-369

Cracraft J (1988) Deep-history biogeography; retrieving the historical pattern of evolving continental biotas. Syst Zool 37: 221-236

Craw R (1988) Panbiogeography:method and synthesis in biogeography. In: Myers AA & Giller PS (eds), Analytical Biogeography, Chapman & Hall, London & New York, p 405-435

Croizat, L. (1964). Space, Time and Form: The Biological Synthesis. Published by the author, Caracas, Venezuela

Garbary, D. (1987) A critique of traditional approaches to seaweed distribution in light of the development of vicariance biogeography. Helgol Meeresunters 41: 235-244

Ghiold, J. & A. Hoffman (1984) Clypeasteroid echinoids and historical biogeography. N Jb Geol Palaeon Mh 9: 529-538

Ghiselin, M.T. (1980) Biogeographical units; more on radical solutions. Syst Zool 29: 80-85

Humphries CJ, Ladiges PY, Roos M & Zandee M (1988) Cladistic Biogeography. In: Myers AA & Giller PS (eds), Analytical Biogeography. Chapman & Hall, London & New York, p 371-404.

Legendre P (1990) Quantitative methods and biogeographic analysis. In: Garbary DJ, South GR (eds), Evolutionary Biogeography of the Marine Algae of the North Atlantic. Springer-Verlag, Berlin, p 9-34

Myers AA, Giller PS (eds) (1988), Analytical Biogeography. Chapman & Hall, London & New York

Nelson, G. & N. Platnick (1981) Systematics and Biogeography; cladistics and vicariance. Columbia University Press, New York

Nelson, G. & D.E. Rosen (eds) (1981) Vicariance Biogeography: a critique. Columbia University Press, New York

Rosen, B: R. (1988) Biogeographic patterns: a perceptual overview. In: Myers AA, Giller PS (eds), Analytical Biogeography. Chapman & Hall, London & New York, p 23-55

Rosen, D.E. (1978). Vicariant patterns and historical explanation in biogeography. Syst Zool 27: 159-188

Sluys, R. (1989) A Monograph of the Marine Triclads. A.A. Bolkema, Rotterdam & Brookfield, Ct

Sober, E. (1988) The conceptual relationship of cladistic phylogenetics and vicariance biogeography. Syst Zool 37: 245-253.

Valentine, J.W. & D. Jablonski (1983). Speciation in the shallow sea: general patterns and biogeographic controls. In: Sims RW, Price JH, Whalley PES (eds), Evolution, Time and Space: The emergence of the biosphere. Academic Press, London & New York, p 201-226

Wiley, E.O. (1988) Vicariance Biogeography. Ann Rev Ecol Syst 19: 513-542

QUANTITATIVE METHODS AND BIOGEOGRAPHIC ANALYSIS

Pierre Legendre
Département de sciences biologiques
Université de Montréal
C.P. 6128, Succursale A
Montréal, Québec
Canada H3C 3J7

Abstract - Numerical methods are presented that represent three different approaches to biogeographic problems. The first approach is multivariate data analysis. The delineation of biogeographic "provinces" or areas is a type of descriptive analysis that can be accomplished by clustering faunal data (with or without spatial contiguity constraint) and drawing the resulting choropleth map. On the other hand, ecological biogeographers like to use ordinations of sampling localities and interpret the main axes of variation in terms of environmental gradients; canonical ordination, where a species presence or abundance data table and an environmental data matrix are both analyzed simultaneously, can be used with profit in this context. Secondly, the analysis of spatial patterns can help identify the type of spatial distribution of the biological material, both at the population and at the community level, while Mantel tests and other derived analyses make it possible to test hypotheses concerning causal factors possibly responsible for the observed spatial structures. Finally, phylogenetic-tree reconstruction methods, as well as other techniques, can be used for historical biogeographic studies; these include the study of taxa cladograms and of area cladograms.

INTRODUCTION

Myers & Giller (1988) have clearly demonstrated that historical and ecological biogeography can both contribute to explaining the distributions of organisms over the surface of the Earth. In both of these approaches, the hypotheses to be tested are developed by studying the non-randomness of the distribution patterns of living organisms in nature. While the distinction between a purely historical and a purely ecological perspective often depends on the spatial or temporal scale of the study, the two approaches must rely on sound numerical methods to support as efficiently as possible their respective hypotheses, using observational data.

This paper will focus on the most important numerical methods needed to demonstrate the non-randomness of the patterns found in nature and to approach biogeographic problems from an ecological and a historical perspective. [Asterisks (*) indicate specialized methods, not available in general-purpose statistical packages, for which computer programs are suggested in Table 5.]

NATO ASI Series, Vol. G 22
Evolutionary Biogeography of the
Marine Algae of the North Atlantic
Edited by D. J. Garbary and G. R. South

MULTIVARIATE DATA ANALYSIS

A large fraction of the currently published biogeographic studies concerns present-day faunas and floras. Typical problems involve the delineation of biogeographic "provinces" or areas, or the exploration of ecological relationships by correlating geographic distributions of organisms to environmental conditions. Methods of univariate or multivariate data analysis, coupled with classical geographic analysis, are the appropriate tools for such problems.

A quantitative study of a spatially structured phenomenon, like the geographic distribution of a species or a group of species, and of their controlling factors, should start with the study of distribution maps of the variables of interest: the species, and the environmental variables. Biogeographers usually satisfy themselves with rather unsophisticated kinds of maps. The two most common types are the following (both types can be produced by computer software).

1- Isoline maps are similar to the contoured maps used to represent altitudes on topographic maps, where the nested isolines represent different intensities of some continuous variable. Since these maps are in most cases derived from point samples obtained from the study area, intermediate values have to be estimated by interpolation. The sampling design is of prime importance for the accuracy of such reconstructions; this problem has been discussed by Burgess *et al.* (1981), McBratney *et al.* (1981) and Legendre *et al.* (1989), among others. Some of the common interpolation methods, including trend surface analysis* and kriging*, have recently been described for ecologists by Burrough (1987) and by Legendre & Fortin (1989). This type of mapping places the emphasis on the continuity properties of the variable to be mapped. The main trends in multivariate data can also be mapped by these methods, after reducing the data tables to synthetic variables, using ordination methods (examples below).

2- In so-called "choropleth maps" on the other hand, the study area is divided into nonoverlapping regions (or polygons), and boundaries are drawn around the regions on the basis that all points included in any given polygon allegedly possess some common property. This type of mapping is used to emphasize discontinuities in data. For multivariate data, the division of the area into regions can be obtained from clustering methods.

Clustering and ordination are two common tools of multivariate data analysis; they have been used by ecologists and biogeographers for a long time (Orlóci, 1978; Legendre & Legendre, 1983, 1984a; Jongman *et al.*, 1987; etc.) and are currently available in easy-to-use computer packages, both on mainframes and on microcomputers.

Clustering

The basic idea behind clustering is to partition a collection of objects into classes on the basis of their overall similarity or dissimilarity. (We will consider here only the clustering methods that produce nonoverlapping clusters, although methods exist that find overlapping clusters.) Many different clustering methods have been developed, based on mathematical models that are all adequate for producing discontinuous groups: geometric, graph theoretic, probabilistic, information theoretic, etc. Users also have to know that clustering can be obtained in two basically very different ways:

1- Many of the commonly used methods proceed in two steps: first an overall measure of resemblance (similarities or distances, reviewed in most textbooks of multivariate data analysis) is computed among all pairs of objects, which are sampling localities in most biogeographic studies; then clustering is performed, based upon this resemblance matrix. This family of methods includes linkage clustering (single, complete, intermediate); the average clustering methods (UPGMA, WPGMA, UPGMC, WPGMC), Ward's minimum-variance clustering, McQuitty's similarity analysis, flexible clustering, as well as the other methods that can be computed with the general agglomerative algorithm of Lance & Williams; and many of the divisive methods. "Eye-balling" methods, such as an ordination (principal components or correspondence analysis, for instance) followed by approximate delineation of clusters on the ordination diagram, can be included in this category. The choice of a resemblance measure appropriate to the data at hand is critical to obtaining meaningful results; guidance is provided by Legendre & Legendre (1983, 1984a) and by Gower & Legendre (1986).

2- In other methods, the assessment of similarity is done as part of the clustering method itself. These include the well-known *k*-means method, the ecologically-oriented method of information analysis (Williams *et al.*, 1966), as well as Lefkovitch's (1980, 1982, 1987) conditional clustering.

It is important to understand that clustering is a collection of heuristic procedures. This means that, in most cases, several methods and algorithms are equally appropriate, but none of them may also include assumptions related to the ecological or biogeographic theory of interest in the study. As a consequence, results also have to be judged heuristically. In practice, this means that among the methods that are appropriate to the data, users of clustering are encouraged to apply several methods to their data. Decision, as to the most interesting result among those that are technically appropriate, is often based on one of the following considerations, which are external to the clustering analyses:

- retain for interpretation only the groups that are evidenced by all methods used (strict consensus: see below);

- keep the partition or hierarchy that optimizes some objective criterion, such as the "cophenetic" or matrix correlation with the starting resemblance matrix;
- consider only the partition or hierarchy that best fits some other criterion of interest, such as geographic contiguity of the data points in each group (see also the methods of constrained clustering, below);
- consider only the partition or hierarchy that best fits some other data set of interest, for instance a matrix of environmental data (see also the methods of canonical ordination, below).

Let us remember that biogeographers often use clustering methods to delineate homogeneous geographic areas, based upon faunistic or floristic data. In this respect, clustering may be looked at as a family of methods — and indeed, as the only family of methods presently available — for mapping multivariate data. It is important at this point to realize that two different objectives may be pursued, and that they lead to two different alternatives in clustering.

In the first case, biogeographers are interested to locate all the areas where each given type of fauna or flora is found. Typically, one may want to test hypotheses of environmental control of species assemblages. In that case, geographically discontinuous ("disconnected") groups are just as interesting as continuous ("connected") areas, so that the results obtained by clustering areas on the basis of the species data can be mapped directly, and subsequently interpreted, using the environmental data.

On the other hand, the prime purpose of the study may be to outline connected geographic areas, based on the species data. Typical studies include the delineation of biogeographic "provinces", and the study of geographic dispersal of species assemblages after some event. In such studies, there is an implicit constraint that the geographic areas based on the species data will be of interest only if they are internally connected; that is, if none of them is split into geographically disconnected sub-areas. It may happen, of course, that the usual clustering methods will produce connected areas; an example is the study of Bock & Root (1981) who delineated biogeographic regions for landbird winter abundances in southern Canada and the United States. In other cases, however, some of the regions obtained by simple clustering may come out scattered over the study area. This is why several authors (Ray & Berry 1966; Webster & Burrough 1972; Lefkovitch 1978, 1980; Legendre & Legendre, 1984b; and others) have suggested including a constraint of spatial contiguity in the clustering algorithm itself*, in order to guarantee that the clustering results will also be geographically meaningful. Technically, this means that after the resemblance matrix has been computed among sites from the species data, one proceeds with any of the usual clustering methods, with the difference that the algorithm is constrained to cluster only those sites or site groups that are geographically contiguous. The constraint is provided to the program in the form of a list of connections, or spatial links, among neighbouring localities. Clustering with constraint has interesting properties. On the one hand, it reduces the set of mathematically possible solutions to those that are geographically meaningful. This avoids the well-known problem of clustering methods, where different solutions may obtain from different clustering algorithms applied to the same data set; constraining all these algorithms to produce results that are geographically consistent forces them to converge towards very similar solutions. On the other hand, the partitions obtained in this way reproduce a larger fraction of the structure's information than equivalent partitions obtained without constraint (Legendre, 1987). Finally, constrained agglomerative clustering is faster with large data sets than the unconstrained equivalent, because the search for "the next pair to join" is limited to adjacent groups only (Openshaw 1974; Lebart 1978). Examples of the use of constrained clustering for biogeographic analysis are found in Legendre & Legendre (1984b) on freshwater fishes, and in Legendre & Fortin (1989) on terrestrial vegetation.

Ordination

An ordination is a plot of object-points along an axis corresponding to a quantitative or (at least) an ordered relationship, or on a system of two or more such axes. Scatter diagrams, which are two-dimensional ordinations, are familiar to all scientists. The purpose of ordinations is to exhibit the salient features of a data set, such as gradients, groupings, or the presence of outliers. When the data contain more than two dimensions, instead of plotting all pairwise comparisons of variables, methods exist to find the "best" projection of the objects in one, two, or more dimensions; these graphs are called ordinations in a space of reduced dimensionality ("reduced space"). The best-known of these methods are principal components analysis, correspondence analysis, and principal coordinates analysis; several more specialized methods, often referred to as *scaling* methods, also exist; Gower (1987) gives a clear overview of these methods, and illustrates each one with vegetation data.

Ordination must be seen as the geometric problem of projecting points from a multidimensional space onto a smaller number of axes. The main ordination techniques differ from one another, mostly in the distance relationship they allow to preserve through that projection.

In *principal components analysis*, the axes on which the points are plotted in the original, multidimensional space are the original variables, so that the distance between any two points in that space can easily be calculated using the *Euclidean distance* formula. We say, then, that principal components analysis preserves the Euclidean distance, although distances are somewhat distorted when an n-dimensional data set is projected into 2 or 3 dimensions. Solutions of the so-called canonical equation give the various eigenvalues; to each one is associated an eigenvector, from which the principal components are computed. The principal components give the coordinates of the data points in the new space; that is, their positions on the successive principal axes. The method guarantees that the first principal axis passes through the direction of greatest variance of the scatter of points in the original, multidimensional space; the variance of the points along this first axis is given by its eigenvalue. Similarly, the succeeding principal axes, with variance equal to their respective eigenvalue, pass through the greatest successive directions of the *n*-dimensional scatter of points. The principal axes are orthogonal to one another, which means that their correlation to one another is zero. Since the first few principal axes account for the largest fractions of the data set variance, they are the most efficient axes to summarize, in a few important dimensions, the greatest part of the variability of a larger number of original variables.

*Correspondence analysis** differs from principal components analysis in the fact that the distance among points is the so-called *chi-square distance*, instead of the Euclidean distance. While the Euclidean distance is appropriate to represent relationships among points as measured, for instance, by environmental variables, or by other variables that can take positive, null, or negative values, the chi-square distance is appropriate to analyze either a species presence/absence or abundance table, or else a contingency table comparing two qualitative (categorical) variables (Legendre & Legendre, 1983, 1984a; Gower, 1987).

In *principal coordinates analysis**, as well as in the other forms of metric or non-metric scaling, the resemblance matrix is the user's choice, so that scientists can produce ordinations using any one of the metric or semimetric distance measures available in the literature (reviewed in most texts of numerical ecology), or from the corresponding similarity functions. In these ordination methods, the programs read in a previously computed resemblance matrix, and find the best projection of the data points in a space of reduced dimensionality. "Best" is defined by different criteria (depending on the method), the basic concept always being to distort, as little as possible, the original interpoint distances during the process of projecting the points onto the reduced space.

The chief role of ordinations in biogeography is to help identify gradients or other structures in the species abundance data obtained at the sampling localities, and to explain the spatial arrangement of the community structure using other environmental variables by *a posteriori* correlation or regression. This mode of analysis is called "indirect gradient analysis" (Whittaker, 1967; ter Braak, 1987b). The converse can also be done, where the ordination is done on the environmental variables, and the interpretation made using the species abundance data, either one at a time (single species biogeography) or all together (community biogeography). Reduced-space ordination of species abundance data is usually done by correspondence analysis or by principal coordinates analysis (or other form of scaling), while ordination of environmental variables is done either by principal components or by principal coordinates analysis (or other form of scaling). Ordinations can also be used in conjunction with clustering, and in particular with hierarchical clustering, to help decide what is the most informative partition (cutting level) in a dendrogram to produce a choropleth map. Finally, since ordination axes do represent new synthetic quantitative variables, they can be mapped directly (Fig. 1).

There are ways of jointly analyzing directly a species abundance data table and an environmental data matrix ("direct gradient analysis": ter Braak, 1987b), instead of relying on *a posteriori* correlations to associate environmental characteristics to species ordination axes. These methods are called canonical ordination (or constrained multivariate regression), and they pertain to two families: like principal components analysis, *canonical correlation analysis* and *redundancy analysis** preserve the Euclidean distance among points, while *canonical correspondence analysis** preserves the chi-square distance, just like standard correspondence analysis. I will briefly discuss the new method of canonical correspondence analysis (ter Braak, 1986, 1987a, 1987b) which was designed to interpret species assemblage variations by environmental variables, and should prove particularly useful for biogeographic analysis.

Direct and indirect gradient analyses will lead to the same results, if the major species gradients can readily be accounted for by the set of available environmental variables. It often

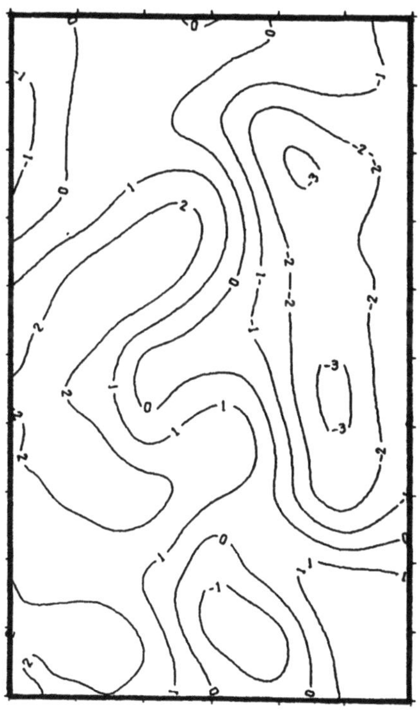

Fig. 1. Terrestrial vegetation data: 200 sampling stations, 28 tree species. Isoline map obtained by kriging from the first principal coordinate of a Steinhaus similarity matrix among sampling localities. Figure from Fortin (1985), with permission. The data set is also described in Legendre & Fortin (1989). The mapped area is 600 m wide and 1 km high. The negative areas of the map are the zones of highest density of the maple sugar *Acer saccharum*

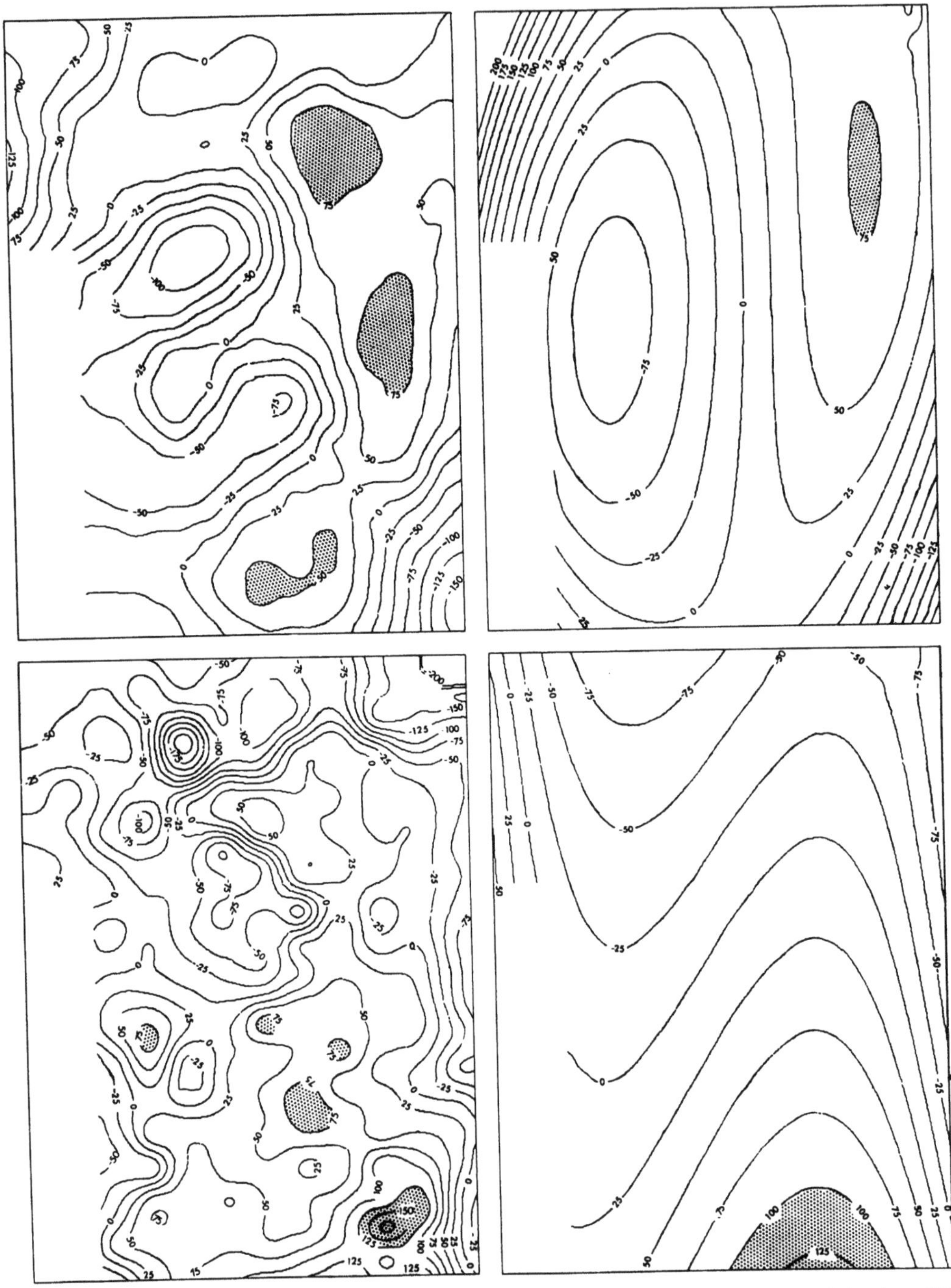

Fig. 2. Maps obtained by block kriging for the sample scores, on canonical axes I (top) and II (bottom), in the species space (left) and in the trend-surface geographic space (right); values multiplied by 100 for mapping. Peaks are shadowed. No samples had been taken from the blanked area on the left

happens, however, that a large part of the species variability is not directly controlled by environmental variables (bottom-up process), but, instead, by other factors, such as historical events, or by top-down ecological processes. In such cases, the larger part of the species variability, that occupies the first principal axes of the ordination, has nothing to do with environmental variables, so that their influence will not be brought out. Even though the environmental variables cannot explain a large part of the species variability, they may be useful to explain some of the remaining variation, and that fraction may still be important. The role of canonical ordination is to bring out that part of the species variability (first data table) that can be accounted for by the environmental variables (second data table). So, canonical ordination is related to multiple regression as well as to ordination. More precisely, the technique of canonical correspondence analysis selects as the first ordination axis the linear combination of environmental variables that maximizes the dispersion of the species scores; the species scores remain weighted averages of the site scores, as in the normal form of correspondence analysis, and the first eigenvalue measures the dispersion of the species scores along the axis. The second and further axes also select linear combinations of environmental variables that are maximally related to the species data, subject to the constraint that the successive fractions of the environmental variables' variability have to be uncorrelated with one another. Canonical correspondence analysis then produces a "constrained correspondence analysis" solution, since the site scores in the species space are still weighted averages of the species scores, which are also constrained, to be expressed as linear combinations of the environmental variables (ter Braak, 1987b).

In biogeographic analysis, the second data table can be made to contain the geographic coordinates (x, y) of the data points instead of environmental variables, as suggested by ter Braak (1987a). In that case, an ordination of the species data can be obtained that will be constrained to be consistent with the geographic distribution of the sampling localities. Columns can also be included in the geographic data table that contain values of x^2, y^2, xy, and possibly higher powers of the basic coordinates, to allow the species data to fit to a higher-degree polynomial of the basic geographic coordinates, in the manner of trend surface analysis (Burrough, 1987; Legendre & Fortin, 1989), which can be extended in this way to multivariate data. On the other hand, a biplot of species and geographic variables should indicate what species have the most important spatial structures. An example is presented for the 200 points of vegetation data of Figure 1. The geographic data matrix contains all the terms for a cubic trend surface regression of the form

$$z = b_1x + b_2y + b_3x^2 + b_4xy + b_5y^2 + b_6x^3 + b_7x^2y + b_8xy^2 + b_9y^3 \qquad (1)$$

Table 1 shows the weights of the species and of the geographic variables in the formation of the first two canonical axes. The first two eigenvalues are 0.23237 and 0.17912 respectively; they measure the amount of variation of the species data that is explained by the first and the second canonical axes and, hence, by the geographic variables; these first two canonical axes together account for 50% of the 28-species-geography biplot, and for 10% of the total 28-species variability, at the spatial scale determined by the sampling grid. Figure 2 shows maps obtained by kriging the sample scores: on the left, the sample scores are computed in the species space; and on the right in the geographic coordinates space. In other words, the sample data values from which the maps on the left-hand side have been interpolated are weighted averages of the species (weights given in Table 1), while the data values for the maps on the right-hand side are trend-surface maps based on equation (1) above, with weights also given in the Table. For canonical axis I (top portion of the Figure), the first community structure surface (left) is very

Table 1. Species and standardized geographic variable weights (canonical coefficients) for canonical axes I and II of the vegetation data study

Species	Axis I	Axis II	Geographic var.	Axis I	Axis II
1. *Abies balsamea*	-1.98	-1.55	1. x	-34.23	-20.02
2. *Acer pensylvanicum*	1.50	-3.17	2. y	-0.82	3.05
3. *Acer rubrum*	-0.22	0.93	3. x^2	71.88	44.81
4. *Acer saccharum*	1.29	0.43	4. xy	-5.29	-6.73
5. *Acer saccharinum*	-1.00	-3.87	5. y^2	5.76	-0.29
6. *Betula alleghaniensis*	-0.22	-1.14	6. x^3	-38.43	-24.73
7. *Betula papyrifera*	-0.97	-0.24	7. x^2y	6.91	3.34
8. *Betula populifolia*	-0.86	1.30	8. xy^2	-6.04	0.65
9. *Carpinus caroliniana*	-0.11	-1.21	9. y^3	0.41	-0.17
10. *Carya cordiformis*	0.96	-2.64			
11. *Fagus grandifolia*	2.66	-0.23			
12. *Fraxinus americana*	-0.86	-0.88			
13. *Fraxinus nigra*	-0.98	-1.72			
14. *Fraxinus pennsylvanica*	0.65	-2.45			
15. *Juglans cinerea*	0.65	0.17			
16. *Malus pumila*	-2.28	0.78			
17. *Ostrya virginiana*	1.11	-0.51			
18. *Pinus strobus*	-0.62	3.93			
19. *Populus balsamifera*	-2.10	2.28			
20. *Populus grandidentata*	2.18	-1.25			
21. *Populus tremuloides*	-1.07	-0.31			
22. *Prunus pensylvanica*	0.94	-0.64			
23. *Prunus serotina*	0.26	-0.60			
24. *Thuja occidentalis*	-2.58	0.09			
25. *Tilia americana*	1.13	-2.09			
26. *Tsuga canadensis*	-0.56	-0.75			
27. *Ulmus americana*	-1.35	-0.26			
28. *Ulmus rubra*	-2.07	0.17			

similar to the map in Figure 1, that represents the first principal coordinate obtained by unconstrained ordination of the vegetation data; the signs are inverted, a result of arbitrary decisions made by the computer programs and has no ecological significance. The species loadings in Table 1 show that the peaks of this map correspond to high densities of species 11, 20, 2 and 4, while the troughs correspond to high densities of species 24, 16, 19 and 28. The interesting fact is that this first component of the community structure can be pretty well approximated (correlation = 69%) by the trend surface map on the right, with high values in the upper right and the extreme upper left regions, and low values present in the center-left and in the lower right-hand regions; in other words, this equation predicts pretty well the community structure, from the simple knowledge of the geographic position of the samples. The same goes for axis II (lower part of Figure 2), where the second community structure surface (left) is well approximated by the trend surface on the right, to which it is correlated at 71%. This hints to a control of this community structure by mechanisms that make it strongly spatially structured. Leduc *et al.* (submitted MS) have investigated the hypothesis that while some of the tree species is this assemblage seem to be controlled by spatially autocorrelated environmental factors, other species' distributions are controlled by patch or gap dynamics.

Geographic variables can be used in yet another way by biogeographers: large-scale gradients in both the species and the environmental data can be removed by regression on the geographic coordinates (trend surface analysis). The study of the relations between species and environmental variables is then carried out on the regression residuals, either by direct or by indirect gradient analysis; this is an elegant way of removing this confounding factor, which is a common cause for spurious correlations in biogeographic analyses (Legendre & Troussellier, 1988). When removing geographic trends, however, one must always keep in mind that one may be discarding an important component of the information, if the community structure is indeed determined by a spatial process.

ANALYSIS OF SPATIAL PATTERNS

Much can be said on the importance of spatial heterogeneity in ecological and biogeographic theory. Several theories and models implicitly or explicitly assume that elements of an ecosystem that are close to one another in space are more likely to be influenced by the same generating processes. Organisms are not distributed evenly or at random in nature, for the good reason that spatial structures are functional in ecosystems, and not the result of some random process; a spatially homogeneous ecosystem simply could not function. This is why present-day biogeographers are interested in detecting the spatial arrangements of populations and communities. As mentioned in the previous section, mapping is a good way of grasping the salient features of geographic distributions. Qualitative examination of maps, when coupled with wishful thinking, may, however, play tricks on the best-intentioned of biogeographers. This is why it is important to be able to rely on some form of statistical analysis to confirm or infirm hypotheses about the observed spatial distributions of organisms and assemblages. This is the first role of spatial pattern analysis in biogeography. We will see that spatial analysis can go further and make it possible to test simple causal models that involve spatial variables as predictors.

This section will be brief, since the statistical analysis of spatial patterns for community analysis has been reviewed recently (Legendre & Fortin, 1989). I will limit the exposé to a guide to the most appropriate methods (with references), classified by the ecological and biogeographic questions they allow to be answered. Readers should refer to the above-mentioned paper for a more complete presentation of the methods, including formulas and ecological examples.

In the exploratory or hypothesis-generating phase of a research program, biogeographers should spend some effort testing for the statistical significance of the spatial structure they have detected or hypothesized. This can be done either in order to establish that there is a significant spatial structuring, and to determine the kind or shape of the pattern, or, to establish that there is no significant spatial pattern in the data, in order to validate the use of parametric statistical tests. Indeed, the presence of autocorrelation in data — spatial or temporal — invalidates the use of the classical tests, since the assumption of independence of the observations is not met. Testing procedures that are valid in the presence of spatial autocorrelation are found in Cliff & Ord (1981), Edgington (1987), Griffith (1987), Odland (1988) and Legendre *et al.* (1990).

There are various ways of testing for the presence of spatial autocorrelation in data:

1- A common structure encountered by biogeographers is the presence of a large-scale gradient in species composition. Such gradients may reflect a change in major environmental conditions, such as harshness of climate in north-south gradients, or dryness as one moves inland from an ocean margin. The Mantel test* (1967) of matrix correspondence is an easy and elegant way of verifying such an hypothesis. First, distance matrices are computed among the sampling localities, in two different ways. The distance matrix **X** is computed for the species composition data, using one of the many ecological resemblance coefficients adapted to species abundance data; an often used coefficient is the Odum (1950) or Bray & Curtis (1957) *percentage difference*. The second matrix **Y** contains the geographic distances among sampling localities, i.e., their distance in m or in km on the Earth surface. The null hypothesis to be tested is that the distances among points in matrix **X** are not linearly related to the corresponding distances in matrix **Y**; or in other words, that the species composition difference does not increase as the geographic distance among sampling stations increases. The normalized form of the Mantel statistic has the same formula as the Pearson product-moment correlation coefficient, computed for pairs of distance values in the off-diagonal of the two distance matrices. This statistic cannot be tested as a regular correlation coefficient, however, because distances in a matrix are not independent of one another; instead, it is tested either by a permutation procedure, or using a normal approximation which becomes asymptotically valid as the size of the matrices being compared increases. Schnell *et al.* (1986) have used the Mantel test to study patterns of geographic variation in dolphin morphology in the Pacific Ocean. The Mantel test has also been used to test the agreement of ecological data with a model (Burgman, 1987; Hudon & Lamarche, 1989; Legendre & Fortin, 1989), as well as hypotheses in panbiogeographic analysis (Page, 1987: see below).

2- In addition to gradients, it is possible to explore other shapes of spatial structures using *structure functions*. The most commonly used structure functions for univariate data (one species at a time) are the variogram* and the spatial correlograms*, and the Mantel correlogram* for multivariate data (whole species assemblages, as in the Mantel test). In particular, a *correlogram* is a graph where autocorrelation values are plotted in ordinate, against distances *d* among localities in abscissa. Figure 3 is an example of a correlogram for variable NO_3^- in the Thau marine lagoon, in southern France. In such an analysis, the problem is subdivided into a number of discrete distance classes, and an autocorrelation statistic is computed for each distance class. One must assume that a single dominant spatial structure exists over the study area; or, in other words, that the main large-scale structure is the same everywhere in that area. In correlograms, each statistic (each point in the correlogram) can be tested for statistical significance, and an overall test can be performed on the whole correlogram as well. Visual analysis of the shape of correlograms indicates whether the underlying spatial structure is aggregated, or has the form of a gradient, a wave, and so on. The correlogram in Figure 3 is characteristic of an aggregated structure; the second positive peak of the correlogram, at distance class 7 km, indicates the average distance between successive peaks or successive troughs of the variable, in the lagoon (see Legendre *et al.* (1989) for a description of

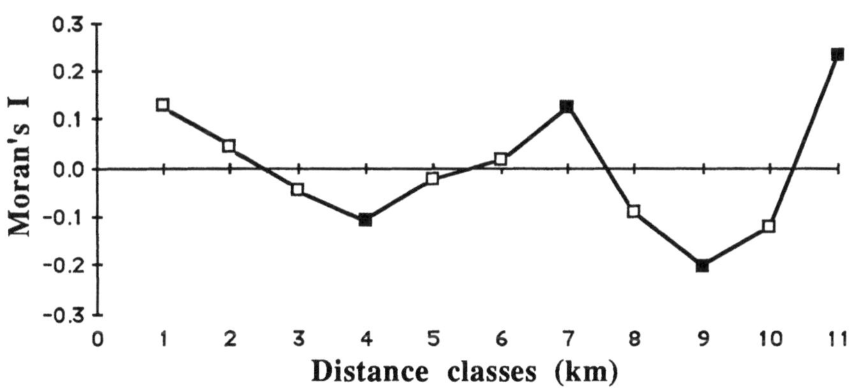

Fig. 3. Spatial correlogram for variable NO_3^- in the Thau lagoon on June 17, 1986. Black squares represent significant values of the Moran's I statistic at the 5% level; white squares are non-significant values

the lagoon and a mapping of this variable). Fig. 1 of Legendre & Fortin (1989) presents the correlograms of nine common types of spatial structures. The absence of a significant spatial structure, for the given observation scale, is indicated by a non-significant correlogram. For multivariate data, which in biogeography are often multi-species assemblages, the Mantel correlogram technique can be used (Sokal, 1986; Oden & Sokal, 1986); the end result is also a correlogram, where the statistics displayed in ordinate are Mantel statistics computed for the various distance classes, instead of Moran's I values. Structure functions allow biogeographers to describe the shape of their spatial structures at the same time as performing tests of statistical significance. In this way, they complement the visual exploration of data that can be done by mapping for univariate data, or by constrained clustering, and by normal or constrained ordination in the case of multivariate data (as explained in the previous section).

In the hypothesis-testing (model testing) phase of a research program, biogeographers often want to test models of assumed causal relationships among two or several variables or groups of variables. Legendre & Troussellier (1988) have shown that even highly significant correlations between biological and environmental variables can be spurious, because of the confounding effect of spatial autocorrelation. In other words, two variables or groups of variables may appear correlated simply because both obey the same spatial structure, which may be controlled by other factors that are not explicitly included in the model of assumed causal relationships, such as currents in the marine or estuarine environments. How is it possible to remove the effect of a common spatial structure from such studies of causal relationships? Smouse *et al.* (1986) have found a solution to this problem, suggesting that the variations in the two variables or groups of variables be expressed by two distance matrices, **A** and **B**, and that the effect of a third distance matrix **C**, that represents geographic relationships among sampling stations be removed, by a partial Mantel test procedure. Legendre & Troussellier (1988) have used this technique in an ecological modelling framework and have shown a case where the wrong ecological conclusion would have been reached using classical modelling techniques, had the confounding spatial structure not been studied.

I mentioned above that statistical tests adapted to the presence of spatial autocorrelation are becoming available. Among them, analysis of variance is of special interest to biogeographers, who are often interested in testing for significant differences among geographic areas on a map, for variables that are spatially autocorrelated. A statistically valid form of analysis of variance has been developed by Legendre *et al.* (1990) for these situations, where there is spatial autocorrelation in the variable being tested, as well as in the classification criterion, which is made of the geographic areas into which the map is subdivided. The problem with classical analysis of variance is that positive autocorrelation reduces within-group variability, thus artificially increasing the relative amount of among-group variance. Negative autocorrelation may produce the opposite effect. This difficulty can be viewed as a loss of an unknown number of degrees of freedom. Besides biogeography, such problems can be found in population genetics, in ecology and in other branches of biology, as well as in economics, epidemiology, geography, geology, marketing, political science, and sociology. A computer-intensive method has been developed to overcome this problem in certain cases. It is based on repeated permutations of the geographic areas on the map, while keeping them internally connected, followed by the computation of a pooled within-group sum of squares statistic after each permutation to produce a reference distribution to which the actual value, computed for the real map, can then be compared and its statistical significance assessed.

The paper of Legendre *et al.* (1990) presents a biogeographic example. The ecological hypothesis of interest was that the number of habitats available for freshwater fishes, in the James Bay territory of northwestern Québec (Canada), varied with the nature of the surface deposits; significant differences in the number of fish species were hypothesized among three contiguous geographic areas, dividing the territory under study according to geomorphology. A large-scale, well-known north-south gradient in fish numbers had first been eliminated by regression (trend surface analysis) from the 64, one-degree-square territory units, each about 7000 km^2 in area; this still left the data autocorrelated. Classical parametric ANOVA declared the differences among the three groups to be very highly significant ($p = 0.0009$). Since we know that spatial autocorrelation makes such a test too liberal — that is, differences among groups are too often declared significant when in fact they may not be — the analysis was repeated using the permutational procedure. After permuting 250 times at random the division of the territory into 3 regions, the probability of the null hypothesis was found to be 0.064, which is insufficient to reject the null hypothesis at the 0.05 level. Thus, we find no convincing evidence that fish species diversity, as measured by the number of species, varies in this territory according to the nature of the surface deposits. The modified test prevented a statistically erroneous decision from being made, which could have led to needless management decisions.

HISTORICAL BIOGEOGRAPHY

Historical (or phylogenetic) biogeography is interested in the simultaneous evolution of taxa, species assemblages (communities), and geographic areas. It is divided into a number of schools of thought, each one associated with the names of one or a few major proponents, without whom the vitality of the field would not be what it is now. I will briefly summarize these approaches, to set the stage for the presentation of the most commonly used numerical methods of data analysis, which is the aim of this paper.

The major division of the field is between *dispersal* and *vicariance* biogeography. The dispersal approach tries to follow the movements of organisms through geographic areas, while the vicariance approach studies how the creation of barriers, or other geologic processes, may have split up (vicariated) biota; leading to allopatric speciation. Vicariance is the property of closely related taxa that occupy similar biota located in distinct and often widely separated geographic areas. Dispersal and vicariance are probably constantly alternating processes in nature, although it may seem more appropriate to use the dispersal mode of analysis when analyzing intermediate time and space processes like postglacial reinvasion of a territory, for instance; on the contrary, the vicariance approach is certainly the main one of interest in studying long temporal and large spatial scale problems, like continental drift and its effects upon populations and communities, since it is at large scales that biogeographic patterns may involve related species groups that may have become separated by vicariance events (Brooks & McLennan, 1990). Important papers on vicariance biogeography are found in the two special issues recently devoted by *Systematic Zoology* to this question (Vol. 37 Nos. 3 and 4, 1988).

Vicariance biogeography is further divided into *panbiogeography* on the one hand, and *cladistic biogeography* on the other. Panbiogeography (Croizat 1952, 1958, 1981) uses "tracks" (lines on a map, each representing the distribution of one or more taxa), and "nodes" (regions with large numbers of tracks) to express phylogenetic relationships among areas. Tracks, represented by minimum-spanning trees, can in turn be translated into connectivity matrices, and Page (1987) has suggested using the Mantel test of matrix correspondence (see above) to assess the statistical significance of track concordance for different species.

Cladistic biogeography, on the other hand, uses cladograms as its main analytic tool. Cladograms can be used either to analyze the evolution of species (species cladograms, in phylogenetic systematics) or the biohistorical relationships of geographic areas (area cladograms). Among the methods of cladistic biogeography, Wiley (1988a) recognizes:

1- the reduced area cladogram of Rosen (1978, 1979), a method of analysis performed by hand, and now abandoned.

2- the component analysis of Platnick & Nelson (1978) and Nelson & Platnick (1981) — which has nothing to do with principal component analysis mentioned in the first section of this paper. Exposés of the method can be found in the above-mentioned papers, in Humphries *et al.* (1988), as well as in Wiley (1988a, b). Wiley finds major flaws in this method because it relies on consensus trees for the final solution. While they are good representations of the congruence of a set of cladograms, consensus trees may, when they contain unresolved polychotomies, turn out to be less parsimonious than the data allow when analyzed by parsimony methods.

3- Brooks *et al.* (1981) parsimony analysis of area relationships (below); and

4- Zandee & Roos (1987) compatibility analysis of area relationships (below), which is, in biogeographic analysis, another form of parsimony analysis and has nothing to do with the compatibility analysis used in phylogenetic analysis.

The relative merits of these methods have been thoroughly reviewed recently (Wiley, 1988a, b; Humphries *et al.*, 1988), and I will limit my exposé to a presentation of the principles of parsimony analysis in cladistic reconstruction, and then to a few statements about the way parsimony is applied to biogeographic analysis. Theriot (1989) reviews cladistic analysis in algal systematics.

Some basic principles of cladistic analyis are as follows: the first axiom of cladistic phylogenetics is that only derived states (called *apomorphic* states) contain information as to phylogenetic affinity (Mitchell 1901; Le Gros Clark & Sonntag 1926). If one considers a nominal variable observed over five species belonging to a common higher taxonomic category, then the first task of cladistic analysis is to formulate hypotheses as to the transformation series of the states of this character. Hennig (1950, 1966) suggests a number of methods for establishing the ancestral state of a character, as well as the transformation series; the best known, and most commonly applied, of these methods is outgroup analysis. In the case of a binary (two-state) character, the derived state, often called "1", identifies all the species that ought to be grouped together, under the assumptions that: a) the mutation that led to the derived state is likely to have occurred only once in the lineage, and, b) no reversal ("1"→ "0") is likely to have occurred during the course of evolution. Thus, a single binary character leads to an incompletely resolved tree, in which all species sharing state "0" form a polytomy (or "bush") at the base of the tree, while higher up the tree, all species sharing state "1" form another polytomy that represents the monophyletic group. Consider now several binary characters: if the derived state is known in each case, then each derived state is the basis for recognizing a monophyletic group of species. If the characters are compatible, drawing all these monophyletic groups will lead to a more completely resolved tree. In the artificial example of Table 2, proposed by Felsenstein (1982), the first four characters are compatible and lead to a completely resolved tree for the five species (Fig. 4), with monophyletic groups **ac, ace, bd** and **d**. What happens now if we consider also characters 5 and 6? They show the existence of a new phylogenetic group made of species **b**, **c** and **e**, which is incompatible with the tree derived from characters 1 to 4. One can of course try to revise the ancestral state hypotheses; if the incompatibilities persist, two main solutions are offered:

1) allow parallel evolution (recurrent appearance of the derived state in various branches of the tree), as well as reversals towards the ancestral state ("1"→"0"), to occur within the reconstructed tree. In that case, one will be likely to find the tree that *minimizes* the number of such extra evolutionary events, that contradict the assumptions above. This is the principle of *parsimony*, or minimum-evolution in phylogenetic reconstruction (Edwards & Cavalli-Sforza, 1963, 1964); it states that the most likely phylogenetic reconstruction is that which minimizes the number of extra evolutionary events, due to reversals, parallelism, convergence or inadequate coding of characters.

2) identify the largest subset of characters that are compatible with one another, in the sense that they lead to a phylogenetic tree without extra evolutionary steps, and derive the cladogram using these characters only. This is the method of *compatibility* in phylogenetic reconstruction (LeQuesne, 1969; Estabrook, 1972; McMorris, 1975; etc.). It has nothing to do with the

biogeographic method of component compatibility briefly described below; phylogenetic trees used as the basis for cladistic biogeographic analysis may have been derived either by phylogenetic parsimony or by compatibility analysis.

A third solution, actually used in biogeographic analysis, consists of finding a consensus among the various single-character cladograms or the taxon-area cladograms. Notice that the very idea of consensus underlies a variety of different concepts (Adams (1972) consensus, strict consensus, majority rule, and so on. See the review paper by McMorris *et al.* (1982), as well as the special issue devoted by the *Journal of Classification* to the subject (Day, 1986)). In historical biogeographic analysis, this solution was tried in the context of component analysis and has been criticized by various authors, including Wiley (1988a) and Humphries *et al.* (1988), as mentioned above.

The parsimony method most often used in vicariance biogeography is to construct a Wagner tree (Kluge & Farris, 1969). Actually, the Wagner method first produces a network — that is, an undirected, or unrooted tree — that simply minimizes the number of evolutionary events, or, in other words, the number of mutations from "0" to "1" or from "1" to "0"; minimizing the total number of evolutionary events is actually equivalent to minimizing the

Table 2. Hypothetical binary data table

Taxa	Characters					
	1	2	3	4	5	6
a	1	1	0	0	0	0
b	0	0	1	0	1	1
c	1	1	0	0	1	1
d	0	0	1	1	0	0
e	0	1	0	0	1	1

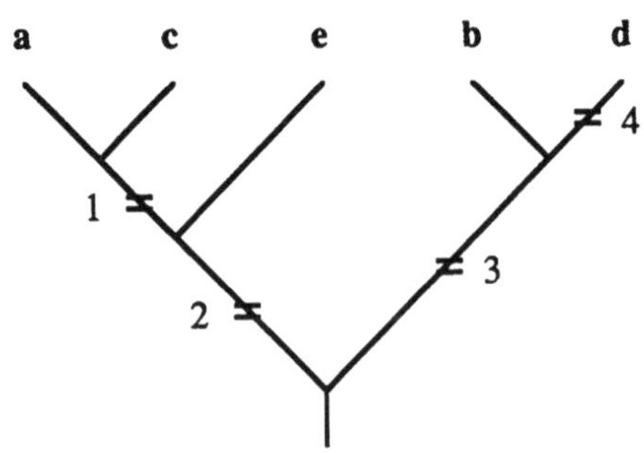

Fig. 4. Cladogram for the first 4 characters of Table 2

number of extra evolutionary steps. The tree is then rooted at the node that attaches the outgroup, real or hypothetical, that has been included in the analysis. The Wagner method, like many of the other methods of phylogenetic tree reconstruction (Table 3), was defined for binary characters. In phylogenetic analysis, multi-state characters whose transformation sequences are known, can be transformed into series of binary characters using the method of Kluge & Farris (1969), for instance. For quantitative characters, other methods have been proposed, based on the analysis of distance matrices (Fitch & Margoliash, 1967; Farris, 1972). See Felsenstein (1982) for a review of the phylogenetic-tree reconstruction methods, and Legendre (1986) for an application of several of these methods to community dispersal analysis from (species ¥ area) data matrices (presence/absence, and frequencies of occurrence). Widely distributed computer packages to carry out these analyses are HENNIG 86*, MacCLADE*, PAUP* and PHYLIP*; Wagner trees can also be reconstructed using the CLADAREA* package. Among programs that perform parsimony analysis, and algorithms within the same program, differences may exist in their ability to find quickly the most parsimonious solution, or to find at all a solution that is close to the most parsimonious one; the reason is that the search for the most parsimonious solution represents a NP-complete problem, a computationally difficult problem whose solution can only be guaranteed by enumerating all possible trees; complete enumeration rapidly becomes impossible as the number of OTUs increases.

Table 3. Differences among the main methods of phylogenetic tree reconstruction for binary characters. See Felsenstein (1982) for details

Method	Minimize	No more than once per character	Forbidden
Parsimony methods:			
Wagner	0 → 1, 1 → 0	-----	-----
Camin-Sokal	0 → 1	-----	1 → 0
Dollo	1 → 0	0 → 1	(01)[1]
Polymorphism	(01)[1]	0 → (01)[1]	1 → 0
Compatibility:	eliminate incompatible characters	-----	0 → 1 → 0 1 → 0 → 1

[1] (01) represents the polymorphic state. The polymorphism parsimony method minimizes the duration of retention of that state.

How does one go from species dendrograms to biogeographic analysis? The most simple way of obtaining an area cladogram based on phylogenetic information, used already in Rosen's reduced area cladogram method, consists of replacing species names by the names of the areas where these species occur (example in Fig. 5). The four methods of cladistic biogeographic analysis mentioned above differ in the way they assemble such taxon-area cladograms based on different taxonomic groups, and in the way they deal with areas that are found at the top of more than one branch of a species cladogram.

I will summarize in a few statements the differences between the methods based on parsimony; that is, Brooks *et al.* (1981) parsimony analysis, and Zandee & Roos (1987) component compatibility analysis. Contrary to the dispersal study of Legendre (1986) mentioned above, cladistic vicariance analysis is not carried out on a simple (species ¥ area) presence/absence data matrix. Instead, phylogenetic information is incorporated into the parsimony analysis by coding it into the data matrix, and this can be done in various ways (Humphries *et al.* (1988) for detailed examples). The most simple form of coding is based upon the so-called "assumption 0" of Zandee and Roos (1987), which states that the original taxon cladograms are considered to be the best available estimates of the phylogenies, so that they are used directly for coding the (character ¥ area) data matrix. In this matrix, the lines are the various geographic areas under consideration. Each taxon-area cladogram, constructed for a

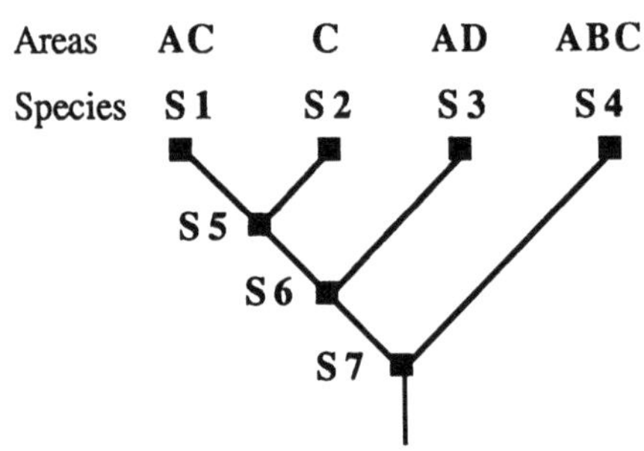

Fig. 5. A taxon-area cladogram for 7 species (S1-S7) and 4 areas (A-D)

Table 4. Binary data matrix scored from Figure 5

Areas	Species (characters)						
	S1	S2	S3	S4	S5	S6	S7
A	1	0	1	1	1	1	1
B	0	0	0	1	0	0	1
C	1	1	0	1	1	1	1
D	0	0	1	0	0	1	1

given taxonomic group, is considered in turn. Each node is considered as a character, including both the ancestral and the terminal nodes (S1 to S7 in Fig. 4); an area is coded "1" for a species (character) when it lies in a branch of the taxon-area cladogram derived from that node. An example of coding is provided in Table 4 for the single taxon-area cladogram of Figure 5; Table 4 can be computed using program CLADAREA*.

In the parsimony method of Brooks *et al.* (1981), this matrix is then analyzed by a Wagner parsimony algorithm, and the tree that minimizes the total number of steps is retained (packages HENNIG 86*, PAUP*, PHYLIP*, CLADAREA*). In the Zandee & Ross (1987) method of component compatibility, on the contrary, the tree is retained that minimizes both the number of extra steps and the number of characters requiring multiple origins. These characters receive extra weight in the measure of parsimony to be minimized: double weight if the species originates twice, etc. Details of the computation method are clearly explained in Humphries *et al.* (1988). A third method of parsimony analysis, proposed by Mickevich (1981), is based on transformation series analysis, however, this method has not been fully evaluated by biogeographers (Humphries *et al.*, 1988). In all of these methods, extinction is the biogeographically analogous to phylogenetic reversals, while widespread primitive distributions are analogous to parallelism.

These analyses may, of course, be carried out for a single taxonomic group at a time; the result differs from the simple taxon-area cladograms in that there is now a single area at the tip of each branch. However, since a cladogram represents a summary of the histories of the areas as interpreted from species-distribution data, it makes a stronger summary statement when it is based upon a large number of different taxonomic groups.

Page (1988) has designed methods of comparing area cladograms derived from independent data, and of testing statistical hypotheses of biogeographic congruence (program COMPONENT*). In a similar vein, Lapointe & Legendre (1990) have developed a method to compare dendrograms (taxa or area) using a permutation procedure.

CONCLUSION

This paper devoted to methods of data analysis presents but a brief outline of three major methodological fields of interest to biogeographers, each of which can be, and has been in recent years, expanded into books and lengthy review articles. Instead of limiting the discussion to the more sophisticated techniques, i.e. those related to vicariance biogeography, I tried to present a broader panorama of methods, many of which may be neglected by biogeographers for lack of acquaintance with them, that allow them to answer different kinds of questions of biogeographic interest, some pertaining to the field of ecological biogeography, others to historical biogeography. These include: mapping univariate and multivariate data;

Table 5. The following programs are available to compute the various methods of biogeographic analysis identified by asterisks (*) in the text. Only those methods are listed that are not available in general-purpose statistical packages. Footnotes indicate addresses for programs

PROGRAM	METHODS
CANOCO[1]	Simple and canonical correspondence analysis, redundancy analysis.
CLADAREA[2]	Area cladistic analysis. Wagner tree reconstruction (parsimony).
COMPONENT[3]	Construction and comparison of area cladograms.
CORR2D[4]	Two-dimensional correlogram.
GEO-EAS[5]	Variogram, kriging.
GEOLIN[6]	Variogram, kriging. Also GEOSREC for non-linear and BLUEPACK for non-stationary kriging.
GEOSTAT[7]	Variogram, kriging.
HENNIG 86[8]	Wagner tree reconstruction (parsimony).
Kellogg's[9]	Variogram, kriging.
MacCLADE[10]	Interactive analysis of phylogenies and character evolution (parsimony).
NTSYS-PC[11]	Simple Mantel test; correspondence analysis.
PAUP[12]	Phylogenetic tree reconstruction (parsimony).
PHYLIP[13]	Phylogenetic tree reconstruction (parsimony, compatibility).
"R"[14]	Spatial autocorrelation, simple and partial Mantel tests, Mantel correlogram, clustering with space and time contiguity constraint. Connecting networks.
SAAP[15]	Spatial autocorrelograms (Moran's I and Geary's c).
SASP[16]	Two-dimensional spectral analysis.
SURFER[17]	Kriging from linear variogram only; other interpolation methods.
SYMAP[18]	Trend surface analysis; other interpolation methods.
UNIMAP[19]	Variogram, kriging; other interpolation methods.

[1]Cajo J. F. ter Braak, Agricultural Mathematics Group, TNO Institute for Applied Computer Science, Box 100, NL-6700 AC Wageningen, Netherlands.

[2]Willem N. Ellis, Instituut voor Taxonomische Zoologie, afd. Entomologie, Plantage Middenlaan 64, NL-1018 DH Amsterdam, Netherlands.

[3]Roderic D. M. Page, Department of Zoology, University of Auckland, Private Bag, Auckland, New Zealand.

[4]Applied Biostatistics Inc., 100 North Country Road, Bldg. B, Setauket, New York 11733, USA. (written by G. M. Jacquez)

[5]Evan J. Englund (Geo-EAS), USEPA EMSL-LV, EAD, P.O. Box 93478, Las Vegas, Nevada 89193-3478, USA. (developed by EPA)

[6]Geovariances International, 1 rue Charles-Meunier, F-77210 Avon-Fontainebleau, France.

[7]Geostat Systems International Inc., 4385 rue Saint-Hubert, Suite 1, Montréal, Québec, Canada H2J 2X1.

[8]James S. Farris, 41 Admiral Street, Port Jefferson Station, New York, NY 11776, USA.

[9]Computer Laboratory, W.K. Kellogg Biological Station, Michigan State University, Hickory Corners, Michigan 49060, USA.

Table 5. cont.

[10]Wayne P. Maddison and David R. Maddison, Museum of Comparative Zoology, Harvard University, Cambridge, Massachusetts 02138, USA.
[11]PC version from Applied Biostatistics Inc., 100 North Country Road, Bldg. B, Setauket, New York 11733, USA. (developed by F.J. Rohlf)
[12]David L. Swofford, Illinois Natural History Survey, University of Illinois, Urbana, Illinois 61801, USA.
[13]Joseph Felsenstein, Department of Genetics, SK-50, University of Washington, Seattle, Washington 98125, USA.
[14]developed by Alain Vaudor (P. Legendre's lab: see title page), is available for Macintosh microcomputers, VAX, and IBM mainframes.
[15]Daniel Wartenberg, Department of Environmental and Community Medicine, Robert Wood Johnson Medical School, 675 Hoes Lane, Piscataway, New Jersey 08854, USA.
[16]E. Renshaw, Department of Statistics, University of Edinburgh, King's Buildings, Mayfield Road, Edinburgh EH9 3JZ, United Kingdom.
[17]Golden Software Inc., P. O. Box 281, Golden, Colorado 80402, USA
[18]No longer distributed by Laboratory for Computer Graphics and Spatial Analysis, Harvard University, USA. It is however still available at many computing centers.
[19]European Software Contractors A/S, Nørregade, DK-2800 Lyngby, Denmark.

relating distributional and environmental data through ordination; the statistical analysis of spatial patterns, and biohistorical reconstruction.

This presentation of numerical methods would be of little utility to field biogeographers if they were not given access to computer software capable of carrying out these calculations. Accordingly, computer programs, for methods that are not readily available in general-purpose statistical packages, are listed in Table 5.

ACKNOWLEDGEMENTS

This is publication No. 369 from the Groupe d'Écologie des Eaux douces, Université de Montréal. I am indebted to Pierre Drapeau, Ph.D. student at Université de Montréal, who directed the sampling program that produced the data used in Figures 1 and 2. Fruitful discussions with Dr. Cajo J.F. ter Braak during the international symposium on *Progress in theoretical vegetation science* held in Vienna in July 1988 led to the formulation of the application of canonical correspondence analysis to produce a space-constrained ordination, as illustrated in Figure 2. This study was supported by NSERC grant No. A7738 to P. Legendre.

REFERENCES

Adams EN III (1972) Consensus techniques and the comparison of taxonomic trees. Syst Zool 21: 390-397
Bock CE, Root TL (1981) Winter abundance patterns of landbirds in the United States and southern Canada. Amer Birds 35: 891-897
Bray RJ, Curtis JT (1957) An ordination of the upland forest communities of southern Wisconsin. Ecol Monogr 27: 325-349
Brooks DR, Thorson TB, Mayes MA (1981) Freshwater stingrays (Potamotrygonidae) and their helminth parasites; testing hypotheses of evolution and coevolution. pp. 147-175. In: Funk VA, Brooks DR (eds) Advances in cladistics, 1. Columbia Univ. Press, New York
Brooks DR, McLennan DA (1990) Historical ecology as a research program. In: Mayden RL (ed) Historical ecology and phylogeny of North American freshwater fishes. Stanford Univ. Press, Stanford
Burgess TM, Webster R, McBratney AB (1981) Optimal interpolation and isarithmic mapping of soil properties. IV. Sampling strategy. J Soil Sci 32: 643-659
Burgman MA (1987) An analysis of the distribution of plants on granite outcrops in southern Western Australia using Mantel tests. Vegetatio 71: 79-86
Burrough PA (1987) Spatial aspects of ecological data. Chapter 7, pp. 213-251. In: Jongman RHG, ter Braak CJF, van Tongeren OFR (eds) Data analysis in community and landscape ecology. Pudoc, Wageningen
Cliff AD, Ord JK (1981) Spatial processes: models and applications. Pion Limited, London
Croizat L (1952) Manual of phytogeography. Junk, The Hague
Croizat L (1958) Panbiogeography. Published by the author, Caracas, Venezuela
Croizat L (1981) Biogeography: past, present, and future. pp. 501-523. In: Nelson G, Rosen DE (eds) Vicariance biogeography, a critique. Columbia Univ. Press, New York
Day WHE (1986) Foreword: comparison and consensus of classifications. J Classif 3: 183-185
Edgington ES (1987) Randomization tests, 2nd ed. Marcel Dekker Inc., New York
Edwards AWF, Cavalli-Sforza LL (1963) The reconstruction of evolution. Ann Hum Genet 27: 105
Edwards AWF, Cavalli-Sforza LL (1964) Reconstruction of evolutionary trees. pp. 67-76. In: Heywood VH, McNeill J (eds) Phenetic and phylogenetic classification. Systematic Association Publ No. 6, London
Estabrook GF (1972) Cladistic methodology: a discussion of the theoretical basis for the induction of evolutionary history. Ann Rev Ecol Syst 3: 427-456
Farris JS (1972) Estimating phylogenetic trees from distance matrices. Amer Nat 106: 645-668
Felsenstein J (1982) Numerical methods for inferring evolutionary trees. Quart Rev Biol 57: 379-404
Fitch WM, Margoliash E (1967) Construction of phylogenetic trees. Science 155: 279-284
Fortin M-J (1985) Analyse spatiale de la répartition des phénomènes écologiques: méthodes d'analyse spatiale, théorie de l'échantillonnage. Mémoire de Maîtrise, Université de Montréal
Gower JC (1987) Introduction to ordination techniques. pp. 3-64. In: Legendre P, Legendre L (eds) Developments in Numerical Ecology. NATO ASI Series, Vol. G 14. Springer-Verlag, Berlin
Gower JC, Legendre P (1986) Metric and Euclidean properties of dissimilarity coefficients. J Classif 3: 5-48
Griffith DA (1987) Spatial autocorrelation — A primer. Resource Publications in Geography, Association of American Geographers, Washington, D.C.
Hennig W (1950) Grundzüge einer Theorie der phylogenetischen Systematik. Deutscher Zentralverlag, Berlin
Hennig W (1966) Phylogenetic systematics. University of Illinois Press, Urbana, Illinois
Hudon C, Lamarche G (1989) Niche segregation between American lobster *Homarus americanus* and rock crab *Cancer irroratus*. Mar Ecol Prog Ser 52: 155-168
Humphries CJ, Ladiges PY, Roos M, Zandee M (1988) Cladistic biogeography. pp. 371-404. In: Myers AA, Giller PS (eds) Analytical biogeography — An integrated approach to the study of animal and plant distributions. Chapman and Hall, London

Jongman RHG, ter Braak CJF, van Tongeren OFR (eds) (1987) Data analysis in community and landscape ecology. Pudoc, Wageningen
Kluge AG, Farris JS (1969) Quantitative phyletics and the evolution of anurans. Syst Zool 18: 1-32
Lapointe F-J, Legendre P — A statistical framework to test the consensus of two nested classifications. Syst Zool 39: 1-14
Lebart L (1978) Programme d'agrégation avec contraintes (C.A.H. contiguïté). Cah Anal Données 3: 275-287
Leduc A, Drapeau P, Bergeron Y, Legendre P (submitted) Importance of spatial phenomena in the explanation of forest cover variations.
Le Gros Clark WE, Sonntag CF (1926) A monograph of *Orycteropus afer*. III. The skull. Proc Gen Meet Sci Bus, Zool Soc London 1926: 445-485
Lefkovitch LP (1978) Cluster generation and grouping using mathematical programming. Math Biosci 41: 91-110
Lefkovitch LP (1980) Conditional clustering. Biometrics 36: 43-58
Lefkovitch LP (1982) Conditional clusters, musters, and probability. Math Biosci 60: 207-234
Lefkovitch LP (1987) Species associations and conditional clustering: clustering with or without pairwise resemblances. pp. 309-331. In: Legendre P, Legendre L (eds) Developments in Numerical Ecology. NATO ASI Series, Vol. G 14. Springer-Verlag, Berlin
Legendre L, Legendre P (1983) Numerical ecology. Developments in environmental modelling, 3. Elsevier, Amsterdam
Legendre L, Legendre P (1984a) Écologie numérique, 2ième éd. Tome 2: La structure des données écologiques. Masson, Paris et les Presses de l'Université du Québec
Legendre P (1986) Reconstructing biogeographic history using phylogenetic-tree analysis of community structure. Syst Zool 35: 68-80
Legendre P (1987) Constrained clustering. pp. 289-307. In: Legendre P, Legendre L (eds) Developments in Numerical Ecology. NATO ASI Series, Vol. G 14. Springer-Verlag, Berlin
Legendre P, Fortin M-J (1989) Spatial pattern and ecological analysis. Vegetatio 80: 107-138
Legendre P, Legendre V (1984b) Postglacial dispersal of freshwater fishes in the Québec peninsula. Can J Fish Aquat Sci 41: 1781-1802
Legendre P, Oden NL, Sokal RR, Vaudor A, Kim J (1990) Approximate analysis of variance of spatially autocorrelated regional data. J Classif 7: 53-75
Legendre P, Troussellier M (1988) Aquatic heterotrophic bacteria: modeling in the presence of spatial autocorrelation. Limnol Oceanogr 33: 1055-1067
Legendre P, Troussellier M, Jarry V, Fortin M-J (1989) Design for simultaneous sampling of ecological variables: from concepts to numerical solutions. Oikos 55: 30-42
LeQuesne WJ (1969) A method of selection of characters in numerical taxonomy. Syst Zool 18: 201-205
Mantel, N. 1967. The detection of disease clustering and a generalized regression approach. Cancer Res 27: 209-220.
McBratney AB, Webster R, Burgess TM (1981) The design of optimal sampling schemes for local estimation and mapping of regionalized variables. I. Theory and method. Computers & Geosciences 7: 331-334
McMorris FR (1975) Compatibility criteria for cladistic and qualitative taxonomic characters. pp. 399-415. In: Estabrook GF (ed) Proceedings of the Eight International Conference on Numerical Taxonomy. Freeman, San Francisco
McMorris FR, Meronk DB, Neumann DA (1982) A view of some consensus methods for trees. pp. 122-126. In: Felsenstein J (ed) Numerical taxonomy. NATO ASI Series, Vol. G 1. Springer-Verlag, Berlin
Mickevich MF (1981) Quantitative phylogenetic biogeography. pp. 209-222. In: Funk VA, Brooks DR (eds) Advances in cladistics, 1. Columbia Univ. Press, New York
Mitchell PG (1901) On the intestinal tract of birds, with remarks on the valuation and nomenclature of zoological characters. Trans Linnean Soc London (Zool Ser 2) 8: 173-275
Myers AA, Giller PS (1988) Process, pattern and scale in biogeography. pp. 3-12. In: Myers AA, Giller PS (eds) Analytical biogeography — An integrated approach to the study of animal and plant distributions. Chapman and Hall, London

Nelson G, Platnick NI (1981) Systematics and biogeography: cladistics and vicariance. Columbia Univ. Press, New York
Oden NL, Sokal RR (1986) Directional autocorrelation: an extension of spatial correlograms to two dimensions. Syst Zool 35: 608-617
Odland J (1988) Spatial autocorrelation. Scientific Geography Series, Vol. 9. Sage Publications Inc., Newbury Park, California
Odum EP (1950) bird populations of the Highlands (North Carolina) plateau in relation to plant succession and avian invasion. Ecology 31: 587-605
Openshaw S (1974) A regionalisation program for large data sets. Computer Appl 3-4: 136-160
Orlóci L (1978) Multivariate analysis in vegetation research, 2nd ed. Dr. W. Junk B.V., The Hague
Page RDM (1987) Graphs and generalized tracks: quantifying Croizat's panbiogeography. Syst Zool 36: 1-17
Page RDM (1988) Quantitative cladistic biogeography: constructing and comparing area cladograms. Syst Zool 37: 254-270
Platnick NI, Nelson G (1978) A method of analysis for historical biogeography. Syst Zool 27: 1-16
Ray DM, Berry BJL (1966) Multivariate socioeconomic regionalization: a pilot study in central Canada. pp. 75-130. In: Ostry S, Rymes T (eds) Papers on Regional Statistical Studies. Univ. of Toronto Press, Toronto
Rosen DE (1978) Vicariant patterns and historical explanation in biogeography. Syst Zool 27: 159-188
Rosen DE (1979) Fishes from the uplands and intermontane basins of Guatemala: revisionary studies and comparative geography. Bull Am Mus Nat Hist 162: 267-376
Schnell GD, Douglas ME, Hough DJ (1986) Geographic patterns of variation in offshore spotted dolphins (*Stenella attenuata*) of the eastern tropical Pacific Ocean. Marine Mammal Science 2: 186-213
Smouse PE, Long JC, Sokal RR (1986) Multiple regression and correlation extensions of the Mantel test of matrix correspondence. Syst Zool 35: 627-632
Sokal RR (1986) Spatial data analysis and historical processes. pp. 29-43. In: Diday E et al. (eds) Data analysis and informatics, IV. Proceedings of the Fourth International Symposium on Data Analysis and Informatics, Versailles, France, 1985. North-Holland, Amsterdam
ter Braak CJF (1986) Canonical correspondence analysis: a new eigenvector technique for multivariate direct gradient analysis. Ecology 67: 1167-1179
ter Braak CJF (1987a) The analysis of vegetation-environment relationships by canonical correspondence analysis. Vegetatio 69: 69-77
ter Braak CJF (1987b) Ordination. Chapter 5, pp. 91-173. In: Jongman RHG, ter Braak CJF, van Tongeren OFR (eds) Data analysis in community and landscape ecology. Pudoc, Wageningen
Theriot E (1989) Phylogenetic systematics for phycology. J Phycol 25: 407-411
Webster R, Burrough PA (1972) Computer-based soil mapping of small areas from sample data. I. Multivariate classification and ordination. II. Classification smoothing. J Soil Sci 23: 210-221, 222-234
Whittaker RH (1967) Gradient analysis of vegetation. Biol Rev 49: 207-264
Wiley EO (1988a) Vicariance biogeography. Ann Rev Ecol Syst 19: 513-542
Wiley EO (1988b) Parsimony analysis and vicariance biogeography. Syst Zool 37: 271-290
Williams WT, Lambert JM, Lance GN (1966) Multivariate methods in plant ecology. V. Similarity analyses and information-analysis. J Ecol 54: 427-445
Zandee M, Roos MC (1987) Component-compatibility in historical biogeography. Cladistics 3: 305-332

LATE GLACIAL AND HOLOCENE DIATOM SUCCESSIONS IN THE GULF OF MAINE: RESPONSE TO CLIMATOLOGIC AND OCEANOGRAPHIC CHANGE

Detmar Schnitker and Julie B. Jorgensen
University of Maine
Darling Marine Center
Walpole
Maine, 04573
U.S.A.

Abstract - The diatom floras of a 16 meter long piston core from the Wilkinson Basin in the Gulf of Maine were analyzed quantitatively and their composition compared with diatom floras contained in surface sediment samples from the Canadian Arctic to Cape Hatteras.

The very few and corroded specimens found in the lower portion of the core indicate that initial deposition from about 17,600 yrs to 13,500 yrs BP took place below a floating ice shelf. A *Thalassiosira gravida* - dominated flora occurs in two successive sharp peaks, about 12,500 and 10,500 yrs ago, indicating the presence of an ice-shelf edge or seasonally open sea ice, with nearly full glacial conditions in-between. This two-step deglaciation is reminiscent of the Younger Dryas and was probably brought about by the influx of glacial meltwater from the St. Lawrence. The end of glacial conditions is marked by a brief 50-fold increase in the abundance of diatom frustules and the dominance of *Chaetoceros diadema*, a species indicative of high biological productivity in coastal upwelling systems. During the remainder of the early Holocene the diatom floras were affected by dissolution and the residual flora (dominated by *Paralia sulcata*) is not suitable for paleoenvironmental interpretation. Beginning about 5,000 yrs ago, the diatom flora approached the composition of the modern Gulf of Maine flora, with the rising dominance of *Thalassiosira eccentrica* and accessory species, and the rise in frustule abundance. This 'modernization' is thought to be the result of the onset of tidal resonance within the Gulf of Maine.

INTRODUCTION

The southeastern edge of the late Pleistocene North American ice sheet met the ocean in the Gulf of Maine; and glacial history and paleoceanography are thus intertwined in this region. Late Wisconsin Laurentide ice overrode the Maritime Provinces, New England and the Gulf of Maine, terminating on the Scotian Shelf, Browns Bank and Georges Bank, with an outlet glacier reaching the open ocean through the deep Northeast Channel (Mayewski *et al.* 1981). The Gulf of Maine area evolved from these full glacial late Pleistocene conditions, when the ocean was excluded from it, to the fully marine Holocene interglacial environment, a biologically highly productive body of water. The record of this development is contained

NATO ASI Series, Vol. G 22
Evolutionary Biogeography of the
Marine Algae of the North Atlantic
Edited by D. J. Garbary and G. R. South

within the sediments that have accumulated on the bottom of the Gulf of Maine. These sediments have, until now, received relatively little attention.

PHYSIOGRAPHIC AND OCEANOGRAPHIC SETTING

The Gulf of Maine is a deep marginal shelf basin of about 90,000 km^2. It is partially enclosed by Nova Scotia and New Brunswick to the northeast, Maine to the northwest, and Massachusetts to the southwest. It is open to the North Atlantic Ocean to the southeast, but much of this marine connection is across the shallow (<50 m) Georges and Browns Banks. The only deep connection (232 m sill depth) is through the Northeast Channel. Within the Gulf of Maine, there are many prominent bedrock ledges, rises, and deep intervening silled basins (Fig. 1). The present floor of the Gulf of Maine has been shaped by Pleistocene glacial scouring and deepening of a Late Tertiary fluvial system of Paleozoic through Early Tertiary deposits, overlain by late Pleistocene and early Holocene sedimentary units, which consist of

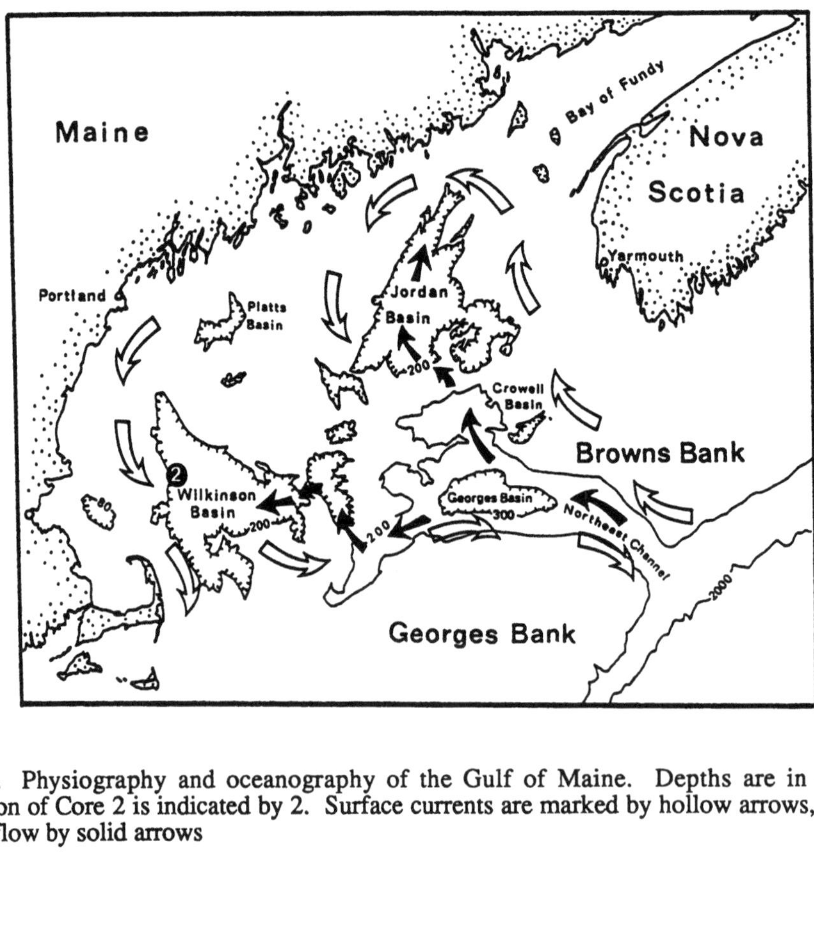

Fig. 1. Physiography and oceanography of the Gulf of Maine. Depths are in meters. Location of Core 2 is indicated by 2. Surface currents are marked by hollow arrows, bottom water flow by solid arrows

morainal deposits, glacifluvial and glacimarine units and an acoustically transparent marine unit (Oldale *et al.* 1973; Ballard & Uchupi 1975). Sediments accumulated primarily within the basins and troughs, while ridges and swells were sites of non-deposition or erosion. King & Fader (1986) applied their Nova Scotian terminology to the Gulf of Maine sedimentary sequence (from bottom to top): 1) a basal glacial deposit, the Scotian Shelf Drift; 2) a glacimarine, poorly sorted clayey and sandy silt, the Emerald Silt, (which is further subdivided into Facies A, B, and C); and 3) a postglacial, soft, silty-clay unit, the LaHave Clay.

The water in the Gulf of Maine is derived from three principal sources: 1) from an influx of warm (6-8°C) and saline (>$34^{0}/_{00}$ S) Continental Slope water which enters through the Northeast Channel; 2) from cold (2-4°C) but less saline (<$32^{0}/_{00}$ S) Nova Scotian Shelf water; and 3) from fresh water runoff. Within the Gulf of Maine, a counterclockwise circulation results from the contrast between shallow, cool, fresh waters of nearshore origin and deep, comparatively warm and salty slope water which enters as an intermittent deep flow (Ramp et al., 1985). In some years winter cooling produces cold (2–3°C) and less saline water, which accumulates in those deep basins that are not reached by the comparatively warm slope water. Strong tidal mixing equalizes the seasonal contrast in the northeastern Gulf of Maine, where slope water is mixed upwards. In the southwestern Gulf of Maine, less tidal mixing, and a smaller influx of Slope Water permit a strong seasonal stratification to develop (Bigelow 1927; Brooks 1985; Mountain & Jessen, 1987).

MATERIALS AND METHODS

In 1979 a suite of long piston cores was taken from several of the deep basins. Core E-2F-79-C2 was chosen for study, based on its length of 16.8 m, its lithologic character, its undisturbed appearance, and its apparently continuous record of late glacial to Holocene sediments. The core is from a depth of 200 m in Wilkinson Basin (42° 26.4' N, 69° 51.8' W) (Fig. 1). The core was sampled at 50 cm intervals for diatom analyses and at 20 cm intervals for foraminiferal and sediment analyses. The sediment exposed to the core liner was removed to avoid possible contamination from peripheral drag. Age determinations were based on three ß-counting radiocarbon dates of disseminated organic matter, five accelerator mass spectrometry (AMS) radiocarbon dates on selected benthic foraminifers, and three amino acid racemization dates on single-species benthic foraminifers.

In order to interpret the environmental significance of fossil floras, the diatom composition of nineteen continental shelf surface sediment samples from the Canadian Arctic, off eastern Labrador and eastern Newfoundland, and from several locations between Nova Scotia and Cape Hatteras, were analyzed for comparison. The station locations and depths are presented in Table 1.

Samples were processed according to a modification of the method described by Schrader (1974), starting with a 0.5 g aliquot of sediment.

For identification and counting of diatoms a phase contrast microscope, with a 10x eyepiece and 100x oil immersion objective, was used. Counts of 200 to 300 specimens were made when possible. Valves more than 50% intact were counted as one. Long pennate diatoms, such as *Thalassiothrix longissima* and *Thalassionema nitzschioides*, were rarely found complete; end fragments were counted, totaled, and divided by two to get an abundance count. Benthic littoral pennates were grouped under the heading "Pennates". The number of traverses made across the slide was used as a rough measure of absolute diatom abundances. A subjective record was kept of valve conditions; valves in good to fair condition were in their entirety, with such features as areolae and valve processes readily visible. Poor condition was indicated by dissolution of these features. Fragmented valves were those recorded with cracks across valves or with portions missing. The relative abundance of the robust, heavily silicified diatom *Paralia sulcata* was plotted for each sample, and used as an additional measure of the degree of dissolution.

A more complete presentation of methods and results of the diatom analyses are found in Jorgensen (1984).

Table 1. List of recent sediment–surface samples utilized for comparison with fossil floras

STATION	LATITUDE	LONGITUDE	DEPTH
1433	35°30.5' N	75°15.2' W	30 m
1420	37°39.2' N	74°41.9' W	49 m
1343	39°10.9' N	73°45.0' W	38 m
1007	41°26.5' N	74°42.3' W	37 m
PC2	42°26.4' N	69°51.8' W	200 m
PC3	42°27.6' N	69°43.5' W	270 m
PC1	42°53.5' N	67°23.1' W	239 m
1147	42°50.0' N	64°00.0' W	116 m
PC13	43°26.1' N	67°50.3' W	248 m
1246	43°30.0' N	65°00.0' W	112 m
1175	44°19.5' N	66°59.5' W	128 m
7966	44°26.3' N	63°31.2' W	111 m
8130	47°53.0' N	52°50.3' W	183 m
7740	58°12.9' N	61°41.0' W	240 m
7743	58°16.7' N	61°43.8' W	224 m
6610	76°25.0' N	112°30.0' W	349 m
6611	46°70' N	112°35' W	381 m
6613	77°10' N	111°55' W	385 m
6614	77°17' N	110°35' W	440 m

RESULTS

Sediments

Core 2 comprises two distinct sedimentary units. The lower unit, from 15.11 to 7.40 m, is variable in both color and texture. From 15.11 to 13.20 m is a light to moderate brown layer of very coarse silty clay with localized patches of well-sorted sand. At 13.19 m there is a sharp transition to an olive gray, silty clay layer. The color and texture remain the same up to 8.19 m, with the exception of a gravelly layer between 10.10 to 10.50 m. From 8.19 to 7.40 m the sediment is again moderate brown in color. The unit as a whole is virtually devoid of mollusc shell fragments, and has numerous large angular and subangular pebbles scattered throughout. Varying degrees of mottling are present. The upper unit, extending from 7.40 m to the surface, is a homogeneous, olive gray, silty clay, although darker gray layers are found from 7.40 to 5.40 m. Shells and shell fragments are present in the top 7 m, along with occasional worm tubes and mottling, but pebbles and mineral sand are absent. The coarse fraction content of the sediments is plotted in Fig. 2. The coarse material below 7.40 m is composed almost entirely of clastic ice-rafted detritus, while the minor coarse fraction above 7.40 m is composed mostly of biogenic debris.

The results of the different dating methods and the sedimentation rate derived from them are illustrated in Fig. 2. Amino acid racemization and AMS ^{14}C dating indicate that conventional ^{14}C dating of disseminated organic matter, in the lower portion of the core, yields ages that are systematically too old. The presence of sub-millimeter sized softwood fragments in some samples from these lower core sections suggests contamination with reworked fossil organic matter. We assume that the top of the core represents contemporaneous sediment. The initial, late glacial sedimentation of coarse material was rapid, occurring at nearly two meters per 1000 yrs. The Holocene sedimentation of fine-grained muds slowed considerably to 66 cm/kyr between 12.5 and 6 ky ago, and to 32 cm/kyr for the last 6 ky.

The $^{18}O/^{16}O$ ratios of benthic foraminiferal calcite and their glacial/interglacial isotopic differential (Fig. 3), are similar to those obtained from contemporaneous North Atlantic deep-sea cores, indicating that the deep water within the Gulf of Maine has been derived from Continental Slope water throughout the late glacial and Holocene period, undiluted by glacial meltwater and land run-off.

The amounts of organic carbon contained within the sediment (Fig. 3) indicate that the late glacial sediments are nearly barren, while organic matter became significant during the Holocene. Late glacial carbon/nitrogen ratios of 15 to 20 indicate the presence of much terrestrial and degraded organic matter, while Holocene C/N ratios of 6 to 8 indicate a normal marine origin for the organic matter.

Diatoms

The abundance and preservation of diatoms vary strongly within the core (Fig. 4). Diatoms are very rare within the coarse-grained glacimarine sediments, increase slightly in

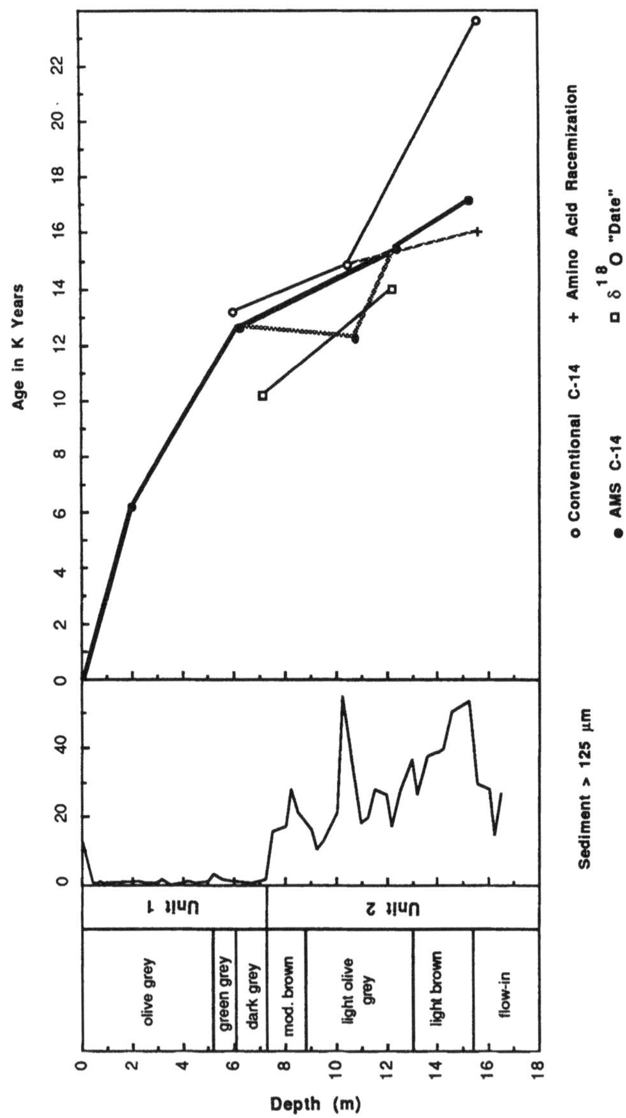

Fig. 2. Simplified log of sediment colors and lithologic units in Core 2 (left panel); occurrence of coarse-grained sediment (center panel); sedimentation rate as determined by various dating methods (right panel). Unit 1 corresponds to the Holocene, Unit 2 to the late glacial

abundance in the transition from glacial to Holocene conditions, and reach their absolute abundance peak very early in the Holocene, at 6.24 m. Between 5 and 3 m diatoms become rare, and reach moderate abundances again only above 2 m. The few specimens contained in the glacimarine sediments are relatively well preserved, heavily silicified, robust, and dissolution resistant. The abundance fluctuations in the transition zone, and the Holocene section, are roughly paralleled by the proportion of well-preserved specimens; zones of few specimens are also characterized by poor preservation.

The diatoms can be sorted into four distinct groups:

1) Species with abundance maxima in the glacimarine sediments,

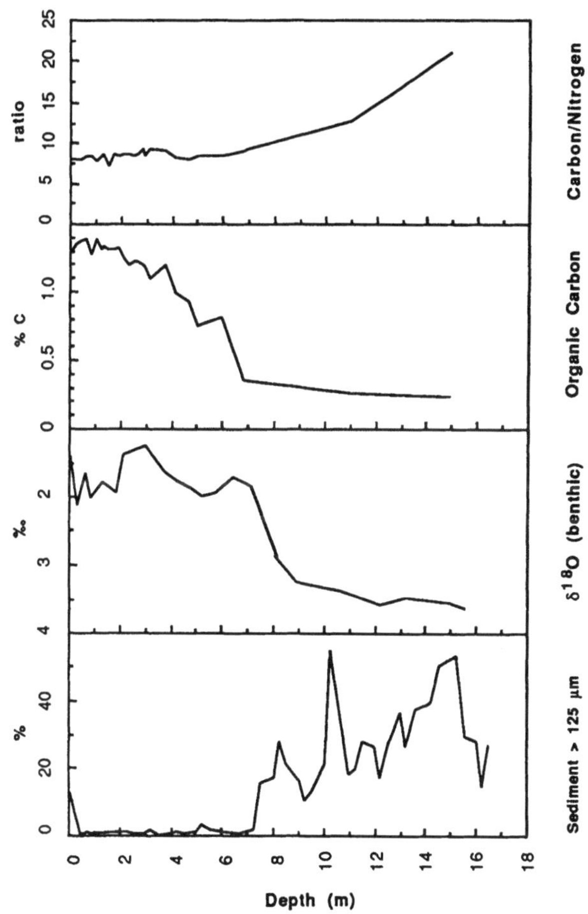

Fig. 3. Oxygen isotope ratios, indicators of deep water characteristics and correlation tool to oceanic time scales. Abundance of organic carbon, indicator of productivity. Carbon/Nitrogen ratio, indicator of provenance of organic matter: low ratios indicate marine source. Sediment grain-size distribution is presented for stratigraphic comparison

2) Species with abundance maxima during the glacial-Holocene transition,
3) Species with abundance maxima during the very early Holocene, and
4) Species with abundance maxima in the mid and late Holocene sediments.

The glacimarine flora is dominated by the relatively few valves of *Paralia sulcata*, with *Stephanopyxis turris*, and the freshwater species *Melosira granulata*, as secondary species (Fig. 5). Only robust valves were found. Valves of this type can survive reworking, and should be selectively preserved in reworked sediments (Truesdale 1977).

The transitional flora is made up of only one significant species, *Thalassiosira gravida*, which occurs in two sharp abundance peaks at about 9 and 7 meters (Fig. 5).

The very early Holocene flora is dominated by a brief but very high dominance between 6.25 and 5.57 m of *Chaetoceros diadema*, accompanied by *Thalassiosira anguste-lineata*, *Coscinodiscus curvatulus* and *Actinocyclus curvatulus* (Fig. 6). This group represents the most abundant diatom flora of the entire core (Fig. 4).

The interval from 5 to 3 m is characterized by low diatom abundances and a resurgence in the relative abundance of *Paralia sulcata*, accompanied by *Coscinodiscus radiatus*. The diatoms of this interval are very poorly preserved.

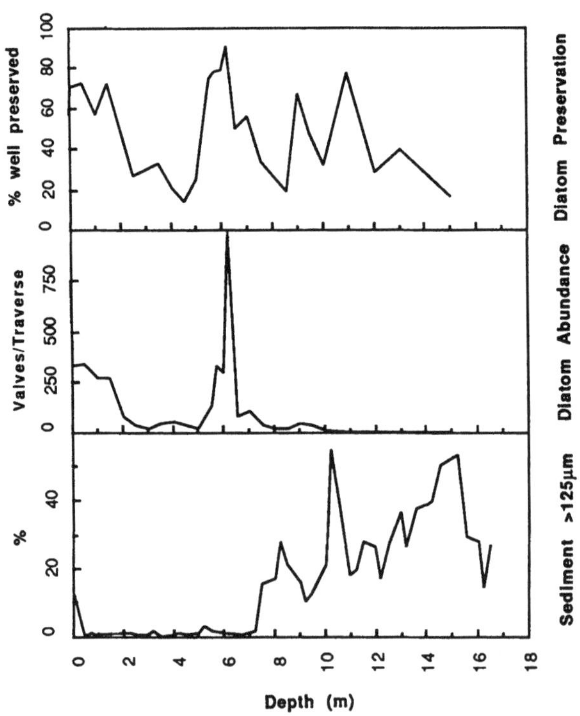

Fig. 4. Absolute abundance and state of preservation of diatoms within Core 2. Sediment grain-size distribution is presented for stratigraphic comparison

The relatively abundant and progressively better preserved diatom floras from 4 m to the top of the core, the mid- and late Holocene, are first dominated by *Cyclotella striata*, then by *Thalassiosira eccentrica*, accompanied by *Actinocyclus octonarius* (Fig. 7).

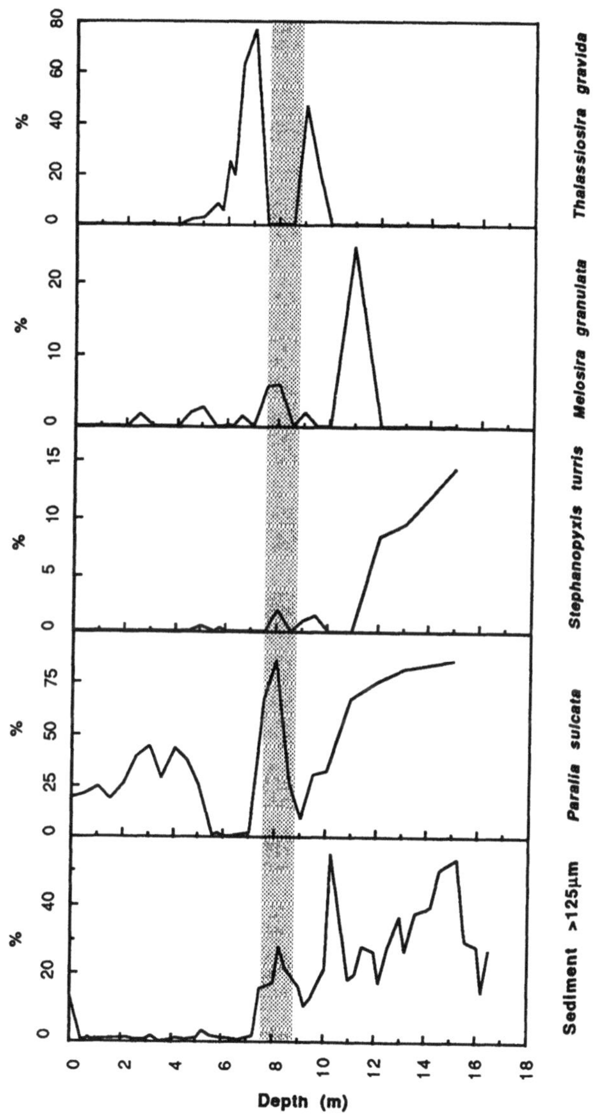

Fig. 5. Diatoms dominating the late glacial stage and the glacial/interglacial transition (right panel). The shading indicates zone of recurrence of glacial conditions - possibly the Younger Dryas. Note that the dominance of late glacial species is obtained by the presence of only a few specimens

DISCUSSION

The transition of the Gulf of Maine, from a glacial environment at the end of the last ice age to its modern boreal-temperate conditions, with high tidal ranges and high biological productivity, is tracked by the sequence of sediments and the diatom floras they contain. The four-fold floral succession outlined above is a good representation of the four principal environmental stages through which the Gulf of Maine has progressed since the last Pleistocene glaciation.

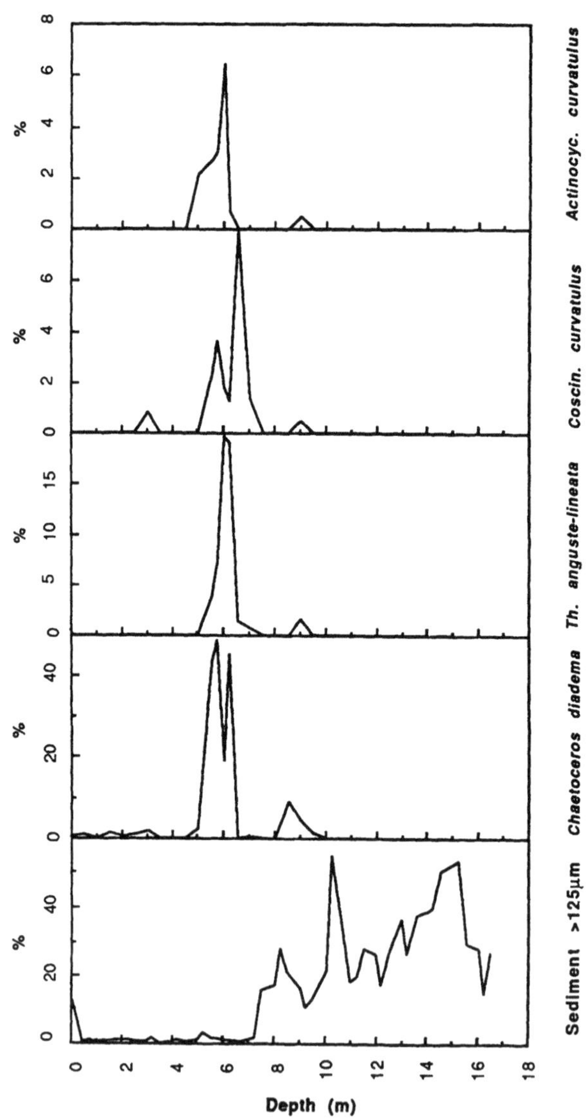

Fig. 6. Early postglacial flora which suggests high productivity (compare abundances shown on Fig. 4)

The deglacial stage, ~18,000 to 14,000 years ago

Seawater entered the Gulf of Maine about 18,000 yrs ago. As indicated by the stable oxygen isotope ratio (Fig. 3), this water was Continental Slope water, probably warm and salty. Coarse-grained sediments were deposited in widespread continuous units, evenly draped over bottom irregularities (Belknap *et al.* in press; Schnitker in prep.). These sediments are

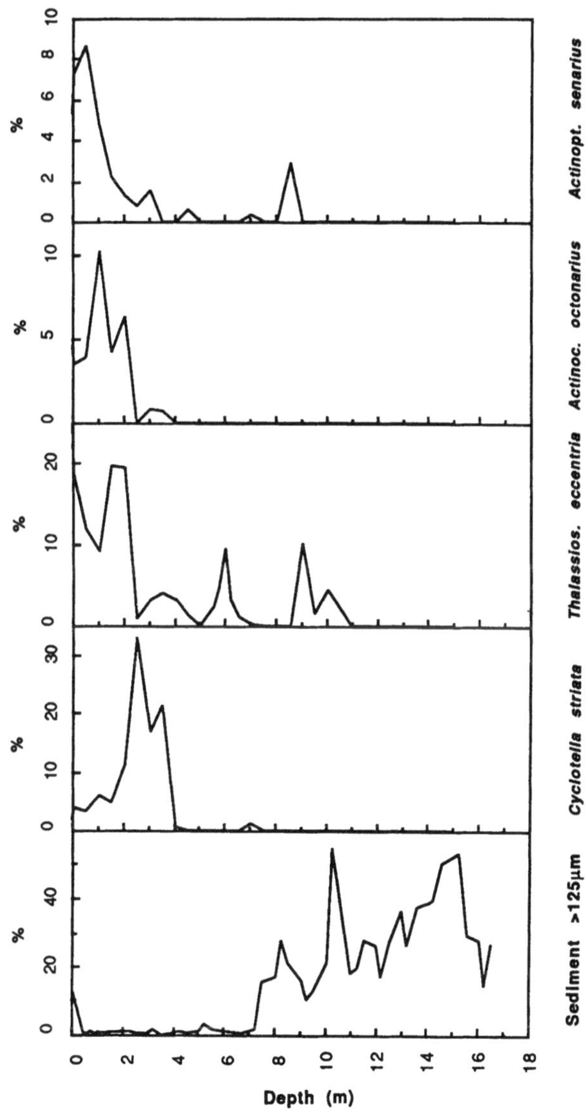

Fig. 7. Mid- and late Holocene flora, indicating the attainment of modern oceanographic and climatic conditions

mottled with infaunal burrows, contain few macrofossils, and a sparse but diverse fauna of microfossils. The grain-size composition indicates that these sediments were delivered to the Gulf of Maine by floating ice; the draping of the sediment units indicates that bottom conditions were relatively tranquil. The scarcity of diatoms in these sediments (Fig. 4) can, in part, be explained by dilution with rapidly accumulating sediments (Fig. 2). However, the late glacial sediments accumulated only three to six times as fast as Holocene sediments, while late glacial diatom abundances are one to two orders of magnitude lower than Holocene abundances. The poor preservation of the few diatoms that do occur cannot be explained by bottom turbulence, since the late glacial Gulf of Maine deep waters were relatively tranquil. The most likely origin of these diatoms is glacial reworking from older, pre- late Wisconsin deposits. The dominance of robust, heavily silicified *Paralia sulcata* and of the freshwater species *Melosira granulata* in these sediments is probably also an indicator of selective 'survival' of such frustules in the process of reworking (Truesdale 1977). Similarly reworked foraminiferal microfossils from an earlier cycle of deposition have been noted (Schnitker in prep.). The scarcity of diatoms in general, and, in particular, the notable absence of epontic diatoms, strongly suggests that phytoplankton growth was inhibited by thick floating ice (probably an ice shelf), as suggested by Tucholke & Hollister (1973) and Schnitker (1975, 1986). A stylized representation of the late glacial Gulf of Maine environment is presented in Fig. 8. The ice shelf could have persisted for nearly 4,000 years only because it was anchored and buttressed by the many ledges and sills within the Gulf of Maine, and by Georges and Browns Banks along its periphery.

The transitional stage, 14,000 to 10,500 years ago

An increase in diatom abundance, and the sudden appearance of *Thalassiosira gravida* (first at 9, then at 7 m), indicate that the ice shelf had disappeared and the Gulf of Maine was, at least seasonally, ice-free. *T. gravida* is a common, well-known, cold-water high-latitude species, but had not been observed in such high abundances in any of the samples that served as modern comparisons. The MIZEX experiment (Smith *et al.* 1987) demonstrated that this species reaches maximum abundances within a very narrow zone along the edge of the pack ice of Fram Strait. However, as shown by the shading on Fig. 5, glacial conditions recurred for a brief period; ice-rafting of sediments resumed, and the decline of *Thalassiosira gravida* allowed heavy-shelled *Paralia sulcata* to regain its relative dominance. Although dated at 12,150 to 10,880 years BP, this recurrence of glacial conditions is reminiscent of similar re-establishment of glacial conditions in Europe and the North Atlantic, termed the Younger Dryas (Rind *et al.* 1986; Broecker *et al.* 1988). The Younger Dryas was dated there to have occurred between 11,000 and 10,200 years BP.

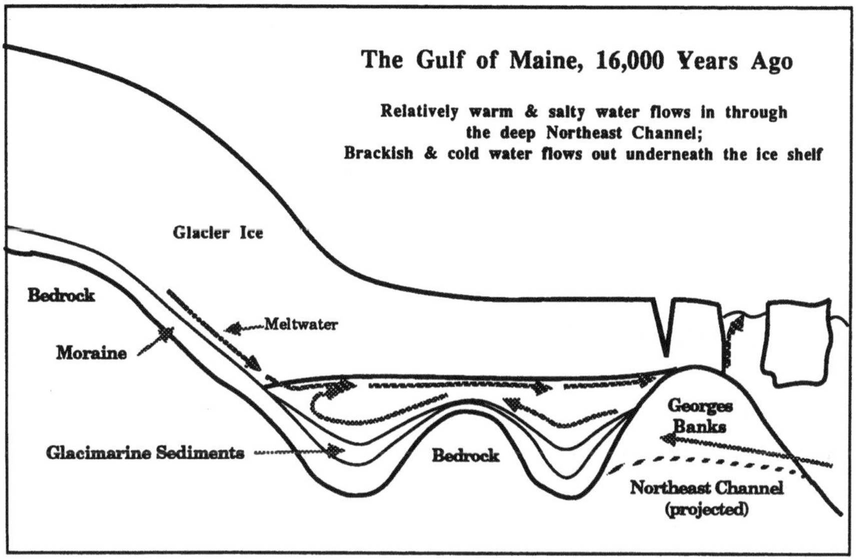

Fig. 8. Diagram of the late glacial oceanography of the Gulf of Maine

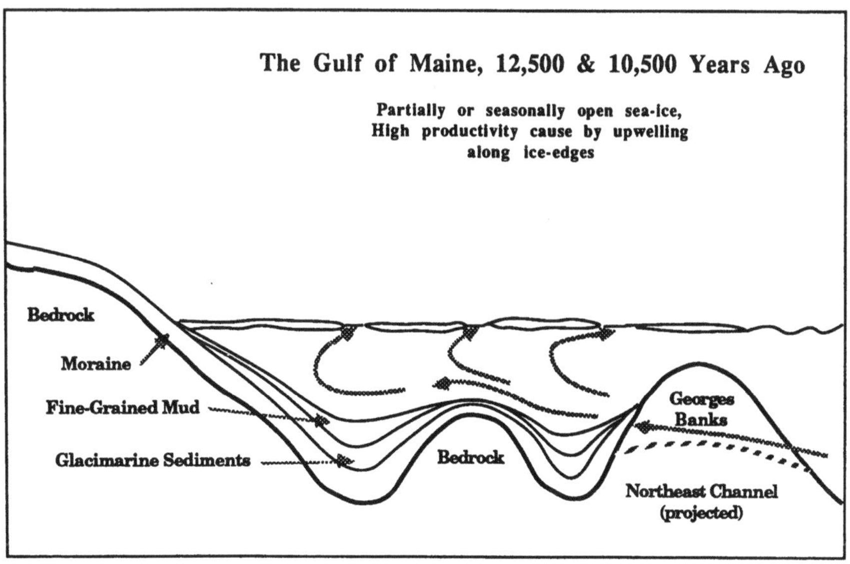

Fig. 9. Diagram of the glacial/interglacial transition with presence of sea-ice

A stylized representation of the Gulf of Maine, in its transition from a glacial to a postglacial environment, is presented in Fig. 9. With the thinning and disappearance of the ice shelf, the style of sedimentation also changed. Seismic profiles show that the last glacimarine sediments were no longer draped over the bottom topography, but became ponded within depressions (Belknap *et al.* in press; Schnitker, in prep.). Bottom currents had become more vigorous, redistributing the fine fraction of the sediment.

The early postglacial stage, ~10,000 years ago

With the disappearance of ice, as indicated by the end of deposition of coarse-grained sediment, and the sharp decline of *Thalassiosira gravida*, a very brief period of extraordinarily high primary productivity occurred. This short interval is characterized by the highest diatom abundance within the sediment (Fig. 4), and an almost equally brief peak of abundance for a few species (Fig. 6). Of these, *Chaetoceros diadema* is a coastal boreal species (Hargraves 1972), and *Thalassiosira anguste-lineata* currently occurs from the Canadian Arctic to the Gulf of Maine (Fryxell & Hasle 1977). *Chaetoceros* spores are abundant (61% of the total flora) in sediments below the cold upwelling waters off South West Africa (Schuette & Schrader 1980) and under the relatively cold upwelling regions off Peru (DeVries & Schrader 1980). A generalized view of this brief time of high productivity is presented in Fig. 10. Unfortunately, there are no additional indications that might suggest the mechanism that was driving this high productivity: whether it was caused by high nutrient content of the upwelling deep water, by increased intensity of upwelling driven by stronger winds, by intense mixing in a then shallower coastal zone, or by land runoff that was draining a recently deglaciated landscape.

The low abundances and the poor preservation of diatoms in the following section of core (Fig. 4) are probably due to silica dissolution, rather than to specific ecological conditions affecting productivity. Contemporaneous calcareous microfossils are abundant, their faunas diverse; primary productivity was high, as indicated by the amount of organic carbon present and by the carbon to nitrogen ratio (Fig. 3). A qualitative examination of smear slides of two other cores (one also from the Wilkinson Basin, the other from the Jordan Basin), found indications of silica dissolution at the same stratigraphic levels. Severe dissolution occurred in the Jordan Basin core, while dissolution was less severe, and confined to a narrower interval, in the Wilkinson Basin core (Fig. 11). The dissolution event was apparently throughout the Gulf of Maine, but its severity was modulated by presently unknown local factors.

Starting at the top of the dissolution zone, and continuing into the base of the following zone, is an interval characterized by the abundant (> 30% of total fauna) occurrence of deep-water (bathyal) benthic foraminifers (Schnitker 1986, in prep.). They indicate that Continental Slope water occupied Wilkinson Basin continually, and in little altered form. Neither the along-

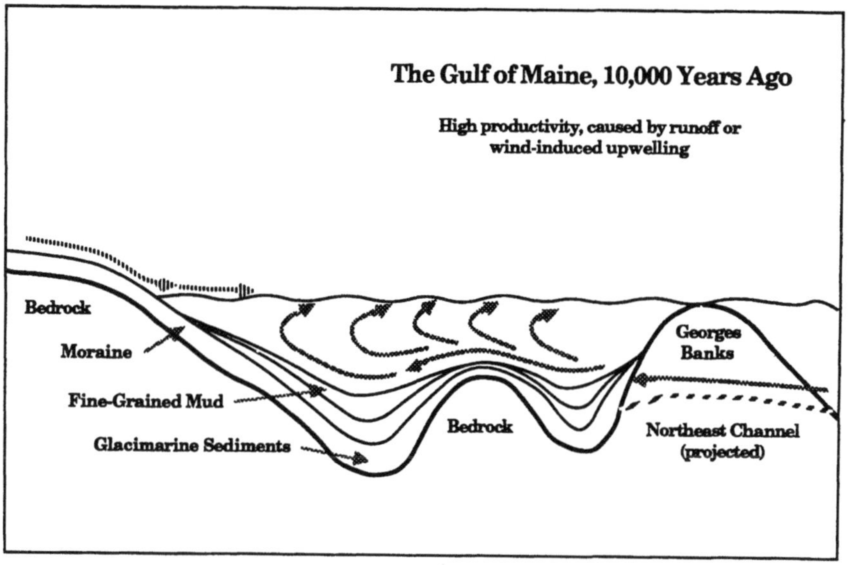

Fig. 10. Diagram of the coastal upwelling conditions of the earliest Holocene

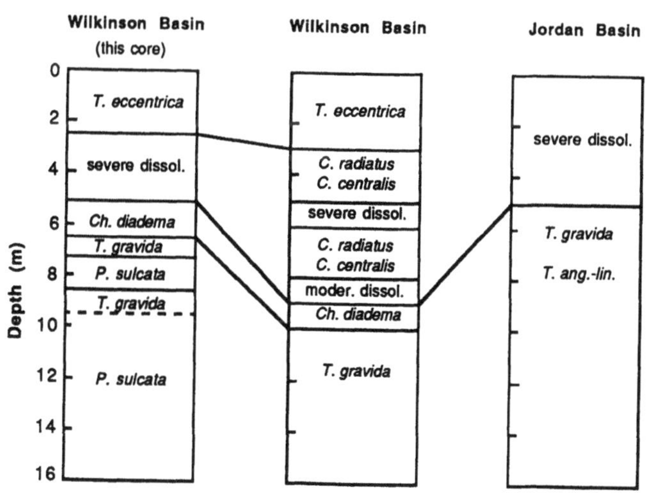

Fig. 11. Comparison of faunal sequences and silica dissolution history in different Gulf of Maine cores

flow entrainment of shallower water, nor the occasional replacement of Slope water with locally derived, winter-cooled Gulf of Maine surface water occurred, as documented for the present-day by Mountain & Jessen (1987). The qualitative examination of the second Wilkinson Basin core revealed the presence of such oceanic species as C. *centralis*, *C. marginatus*, *C. oculus-iridis/asteromphalus*, as well as numerous *Thalassiothrix longissima* fragments. These seem to indicate that the surface waters of the Gulf of Maine were influenced by this increased influx of North Atlantic ocean water (Fig. 12).

The modern stage

The *Thalassiosira eccentrica*-dominated flora, which becomes abundant at the 2 m level (Fig. 7), about 6,000 years ago, essentially represents the modern-day Gulf of Maine diatom flora. At about the same time, the modern benthic foraminiferal faunas became established. It is noteworthy that these faunas are composed of species which are indicators of a high influx of marine organic matter (Schnitker 1986, in prep.). The present-day conditions of water masses, of local oceanic climate, and of nutrient distribution are to a large extent the result of strong tidal mixing. The configuration of the Gulf of Maine is close to the size that causes resonance with the external forcing of the tides (Scott & Greenberg 1983). Lower relative sea-level, and emergent or very shallow Browns and Georges Banks in early Holocene times, resulted in a smaller, shallower Gulf of Maine in which little resonance occurred. Only with wider connections to the Atlantic, through isostatic subsidence of the peripheral banks, and an enlargement of the Gulf of Maine, through isostatic subsidence of the coastal zones and concurrent eustatic sea-level rise (Schnitker 1974; Belknap *et al.* 1986), did the dimensions of the Gulf of Maine system allow for the tides to be amplified through resonance. A stylized view of the configuration that led to strong mixing and high productivity is presented in Fig. 13.

CONCLUSION

Diatoms preserved in sediments of the Gulf of Maine suffice to trace its evolution from a glacial embayment, covered by thick shelf ice, to a two-step deglaciation with repeated covers of periodically open sea-ice. They point to a brief period of strongly enhanced upwelling and productivity at the very beginning of the Holocene. Silica dissolution destroyed much evidence for the conditions in the early Holocene, but suggestions of a strongly 'oceanic' environment are present. Modern conditions of cold water and high productivity became established only

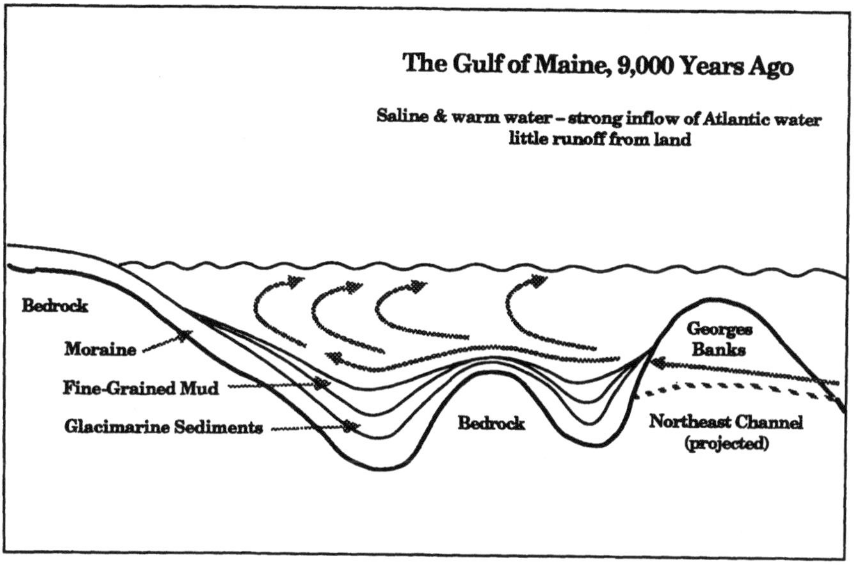

Fig. 12. Diagram of the 'oceanic stage' of the Gulf of Maine, a time of minimum land run-off and maximum ingress of Slope Water

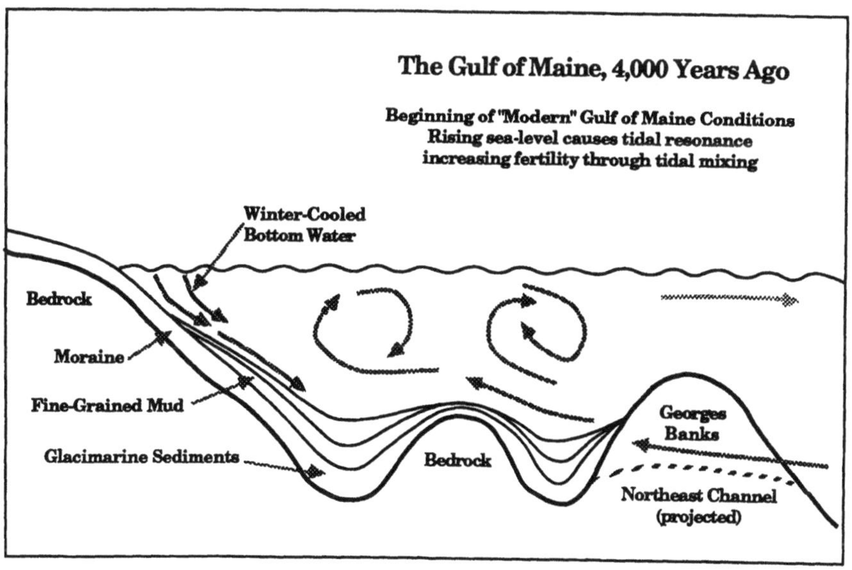

Fig. 13. Diagram of the modern oceanographic conditions of the Gulf of Maine. Strong tidal mixing brings cool, nutrient-rich deep water to the surface for high biological productivity

6,000 years ago, probably because relative sea-level rise caused the tides to reach resonance amplification.

Although only a small proportion of diatom frustules produced in the surface waters ever reach the ocean floor, and many, if not most, of these will be destroyed by ocean-floor processes of dissolution, mechanical abrasion and breakage, a surprisingly large amount of information on past environmental conditions can be retrieved from fossil assemblages. The confidence one can place in the veracity of such postdictions increases significantly when they are corroborated by other paleoenvironmental indicators that are not subject to the same vagaries of preservation.

The history of the Gulf of Maine that is presented here is only a crude approximation of its real paleoceanographic evolution. A detailed look at the succession of the less abundant species (that were largely ignored here) could add more detail to the account, but a search for additional cores with less disturbed records seems likely to be the most productive path to follow. Such experience of past environmental conditions can become valuable when the need arises to make predictions of environmental change brought about by climatic change, or by man-made interference with the physical system.

ACKNOWLEDGEMENTS

We owe gratitude to the many people, too numerous to mention here, who have assisted us with this study, in the field, in the laboratory and in reviewing this manuscript. Special thanks go to Drs D.F. Belknap, R. Fairbanks, D. Kellogg, T. Kellogg, B. McAlice, C. Sancetta, and to Joanne Levasseur. This study was supported by a grant from the National Science Foundation. Thanks are also due to the crew of the R.V. EASTWARD of Duke University for their enthusiastic help in obtaining these cores.

REFERENCES

Ballard RD, Uchupi E (1975) Triassic structure in Gulf of Maine. Bull Am Assoc Pet Geol 59: 1041-1072

Belknap DF, Shipp RC, Kelley JT, Schnitker D (in press) Quaternary geologic history of Sheepscot Bay, west-central Maine coast. In: Studies in Maine Geology, V. 3, Quaternary Geology. Tucker RD, Marvinney RG (eds). Maine Geological Survey, Augusta

Belknap DF, Shipp RC, Kelley JT (1986) Depositional setting and Quaternary stratigraphy of the Sheepscot estuary, Maine: a preliminary report. Geogr Phys Quat 40: 55-69

Bigelow HB (1927) Physical oceanography of the Gulf of Maine. Fisheries Bull 40: 51-1027

Broecker WS, Oppo D, Peng TH, Curry W, Andree M, Wolfli W, Bonani G (1988) Radiocarbon based chronology for the $^{18}O/^{16}O$ record for the last deglaciation. Paleoceanography 3: 509-515

Brooks DA (1985) Vernal circulation in the Gulf of Maine. J Geophys Res 90: 4687-4705
DeVries TJ, Schrader HJ (1980) Variation of upwelling/oceanic conditions during the latest Pleistocene through Holocene off the central Peruvian coast: a diatom record. Mar Micropal 6: 157-167
Fryxell GA, Hasle GR (1977) *Thalassiosira eccentrica* (Ehrenb.), *T. symmetrica* sp. nov., and some related centric diatoms. J Phycol 8:297-317
Hargraves PD (1972) Studies on marine plankton diatoms. I. *Chaetoceros diadema* (Ehr.) Gran: life cycle, structural morphology, and regional distribution. Phycologia 11:247-257
Jorgensen JB (1984) Diatom evidence concerning the post-glacial history of the Gulf of Maine. M.S. thesis, University of Maine
King LH, Fader GBJ (1986) Wisconsinan glaciation of the Atlantic continental shelf of southeast Canada. Geol Surv Canada Bull 363: 1-72
Mayewski PA, Denton GH, Hughes T (1981) Late Wisconsin ice sheets in North America. In: The last great ice sheets. Denton GH, Hughes T (eds). Wiley Interscience, New York, p. 67-178
Mountain DG, Jessen PF (1987) Bottom waters of the Gulf of Maine, 1978-1983. J Mar Res 45: 319-345
Oldale RN, Uchupi E, Prada KE (1973) Sedimentary framework of the western Gulf of Maine and the southeastern Massachusetts offshore area. U S Geol Surv Prof Paper 757: 1-10
Ramp SR, Schlitz RJ, Wright WR (1985) The deep flow through the Northeast Channel, Gulf of Maine. J Phys Oceanogr 15:1790-1808
Rind D, Peteet D, Broecker W, Ruddiman W (1986) The impact of cold North Atlantic sea surface temperatures on climate: implications for the Younger Dryas cooling (11-10k). Climate Dynamics 1: 3-33
Schnitker D (1974) Postglacial emergence of the Gulf of Maine. Geol Soc Am Bull 85: 491-494
Schnitker D (1975) Late glacial to recent paleoceanography of the Gulf of Maine. 1st Internat. Symp. Cont. Marg. Benthic Forams., Pt. B.. Paleoecol Biostrat; Maritime Sed Spec Pub 1: 385-392
Schnitker D (1986) Ocean basin with a past. Explorations (J. Res. Univ. Maine) 3: 29-37
Schrader HJ (1974) Proposal for a standardized method of cleaning diatom bearing deep-sea and land-exposed marine sediments. Nova Hedwigia Beih 45: 403-409
Schuette G, Schrader HJ (1980) Diatom taphocoenoses in the coastal upwelling area off south west Africa. Mar Micropal 6: 131-155
Scott DB, Greenberg DA (1983) Relative sea-level rise and tidal development in the Fundy tidal system. Can J Earth Sci 20: 1554-1564
Smith WO, Baumann MEM, Wilson DL, Aletsee L (1987) Phytoplankton biomass and productivity in the marginal ice zone of Fram Strait during the summer 1984. J Geophys Res 92(C7): 6777-6786
Truesdale RS (1977) Diatoms in the sediments of the Ross Sea, Antarctica. MS thesis, University of Maine
Tucholke BE, Hollister CD (1973) Late Wisconsin glaciation of the southwestern Gulf of Maine: new evidence from the marine environment. Geol Soc Am Bull 84:3 279-3296

SEAWEED BIOGEOGRAPHY OF THE NORTH ATLANTIC: WHERE ARE WE NOW?

C. van den Hoek and A.M. Breeman
Department of Marine Biology
Biological Centre, University of Groningen
P.O. Box 14
9750 AA Haren
The Netherlands

Abstract - Reconstructions of the glacial distribution areas of individual species belonging to the various seaweed regions, lead to the conclusion that species belonging to the Arctic-cold temperate N Atlantic flora could have been faced with such an extreme reduction of their distribution area (especially in the NW Atlantic) that they were threatened with extinction. Such severely adverse conditions did not exist in the N Pacific, and this may explain the much greater richness of the cold-temperate N Pacific flora. *Laminaria hyperborea* possibly exemplifies species that became extinct in the NW Atlantic during the glaciations, but survived in the NE Atlantic. Species in other seaweed regions than those in the Arctic-cold temperate N Atlantic were probably much less affected by the Pleistocene glaciations. Species endemic to the NW Atlantic coast, such as *Grinnellia americana,* are eurythermous and could potentially inhabit a broad area along NW African coasts and in the Mediterranean. Its absence there indicates that the central Atlantic Ocean can function as a barrier to long distance dispersal.

The rich Warm-temperate NE Atlantic region is characterized by stenothermous endemic species such as *Polyneura hilliae* and *Cladophora hutchinsiae* which, because of their narrow thermal survival range, are excluded from the Warm-temperate NW Atlantic region, with its pronounced interglacial seasonality. Some species in this category can probably be interpreted as belonging to pairs of sister species, each pair consisting of a stenothermous NE Atlantic species with a "cooler" warm-temperate range, and a stenothermous NW Atlantic species, with a "warmer" tropical range. The evolutionary advantage of the "cooler" species would be its capacity to occupy the vast rocky coasts of the Warm-temperate NE Atlantic region, and that of the "warmer" species to occupy the vast Caribbean coasts. An example of such a species pair is that of *Cladophora pellucida* (NE Atlantic) and *C. pellucidoidea* (NW Atlantic). The steepening of the latitudinal temperature gradient since the Middle Miocene, could have created the central Atlantic barrier, after which the two sister species could have diverged, both genetically and in their temperature responses. In other examples, the populations on both sides of the ocean diverged only in their temperature responses, but not in their morphologies (e.g. *Cladophora coelothrix*).

The thermal isolation of warm-water, tropical-to-subtropical species in the Mediterranean, by cold upwelling in the Atlantic entrance to this sea, was much enhanced, and affected more species during the glaciations. After the glaciations, populations of cool water species (e.g. *Laminaria ochroleuca*) could also have been isolated as disjunct populations in cooler parts of the Mediterranean. Possibly this thermal isolation may have promoted speciation in the Mediterranean and the origin of endemic species.

The reconstructed glacial distributions of tropical and tropical-to-warm temperate species do not support the hypothesis of massive extinctions during the glaciations of tropical species along Atlantic and Pacific east coasts. This hypothesis was invoked in earlier discussions to explain the paucity of the seaweed floras on these coasts. An alternative explanation for the absence, on these coasts, of tropical seaweeds with a W Atlantic-Indo-W Pacific distribution (such as *Dictyosphaeria cavernosa*) could be the progressive extinction of tropical seaweeds by the Miocene reduction in the complexity and length of the E Atlantic (and E Pacific) tropical coasts, and the concomitant steep cooling trend. This same temperature lowering could also

NATO ASI Series, Vol. G 22
Evolutionary Biogeography of the
Marine Algae of the North Atlantic
Edited by D. J. Garbary and G. R. South

explain the passage of amphipolar species such as *Rhodochorton purpureum* across the equator. The reconstructed glacial distribution of the latter species indicates that it could not have survived glacial equatorial temperatures, which hardly differed from the modern ones.

The Mediterranean, especially the East Mediterranean, is potentially capable of harbouring tropical and tropical-to-warm temperate seaweeds. The paucity there of tropical-to-subtropical seaweeds is probably not the result of glacial extinctions, but rather of the fact that the Mediterranean was refilled, after the Messinian salinity crisis (Miocene, ca 5 Ma ago) with a warm temperate NE Atlantic flora poor in tropical-subtropical species. The cooler conditions in the W Mediterranean allow the growth and survival there of warm temperate species with cooler ranges; this is probably why its flora is richer than that of the E Mediterranean.

BIOGEOGRAPHIC RELATIONSHIPS AMONG NORTH ATLANTIC SEAWEED FLORAS

Since van den Hoek's (1975) analysis of the biogeographic seaweed regions along the coasts of the Northern Atlantic Ocean, a number of studies have appeared which have enlarged and refined our knowledge on the subject (Lawson 1978; Lawson & John, 1987; van den Hoek 1984; Searles 1984; Lüning 1985; Joosten & van den Hoek 1986; Lindstrom 1987; South & Tittley 1986; South 1987; Alvarez *et al.* 1988; Prud'homme van Reine 1988; Prud'homme van Reine & van den Hoek 1988, 1989). Figure 1 summarizes an adjusted version of the biogeographic subdivision of the N Atlantic seaweed floras, in which the following floras and regions are distinguished:

(a) The Arctic cold-temperate N Atlantic flora (comprising the narrowly related Cold-temperate NW Atlantic, Arctic, and Cold-temperate NE Atlantic regions);

(b) the poor Warm-temperate NW Atlantic region;

(c) the rich Warm-temperate NE Atlantic region, in which the Azorean, Canarian and Mediterranean subregions occupy separate and individual positions by their tropical-to-warm temperate "imprint";

(d) the rich Tropical W Atlantic region ;

(e) The poor Tropical E Atlantic region whose subregions of the Cape Verde Islands and mainland tropical Africa are only distantly related.

The biogeographic seaweed regions are not separated by clear-cut boundaries as suggested by Fig. 1, and they share greater or smaller numbers of species and genera. The function of the "regional" approach is to depict the main trends and to generate hypotheses explaining these.

We have tried to find explanations for these generalized distribution patterns, by experimental investigations of the factors causing the distributions of individual species representing each of the distribution types. These investigations include experiments on the species' thermal tolerance and requirements, and on photoperiod responses (Breeman & ten Hoopen 1981; Breeman *et al.* 1984,1988; Breeman 1988, 1989; Cambridge *et al.* 1984, 1987, 1989a, b; van den Hoek 1982a, b, 1987; van den Hoek *et al.* 1989; Klein 1987; Pot *et al.* 1988; Rietema 1982; Rietema & Breeman 1982; Rietema & van den Hoek 1981,1984; Yarish *et al.* 1984, 1986), but also research on the genetic distances between populations, on the basis of DNA-DNA hybridization (Bot *et al.* 1989a, b, c; Olsen *et al.* 1987; Stam *et al.* 1988). A few examples of these representative species will be treated here.

SEAWEED SPECIES:THEIR GEOGRAPHIC DISTRIBUTION IN RELATION TO CLIMATIC FACTORS

Over 60 species have now been tested for their temperature and photoperiod responses (see Breeman 1988, for review). Almost all distribution boundaries could be explained on the basis of the species' temperature responses.

We have previously invoked (e.g. van den Hoek, 1975, 1982a; Breeman 1989) the severity of the Pleistocene climate, to explain present-day aspects of the N Atlantic seaweed biogeography. The temperature responses of the species that will be treated here, will be used to reconstruct their distributions during the last glacial maximum (18,000 b.p.). This will be done on the basis of recent reconstructions of the world's sea surface temperatures during the last glacial maximum (CLIMAP project members, 1981).

There are two fundamentally different types of temperature boundaries, namely: 1) lethal boundaries and 2) growth and/or reproduction boundaries (Hutchins 1947). Lethal boundaries are set by the species' capacity to survive during the adverse season (i.e. the warm season for a cold-water alga, and the cold season for a warm-water alga). Growth and/or reproduction boundaries are set by the species' ability to grow and reproduce during the favorable season (i.e. the warm season for a warm water alga, and the cold season for a cold water alga).

In the following account, a species' geographic boundary will be positioned on either a winter isotherm of the sea surface(corresponding to the mean temperature of the the coldest month), or on a summer isotherm (corresponding to the mean temperature of the warmest month). These isotherms are only indicative of the real temperature requirements of the species, which will deviate to a greater or lesser degree from the isotherm values, depending on the nature of shore. The experimentally determined temperature requirements will not be treated here; the reader is referred, for these, to the original publications.

REPRESENTATIVE SPECIES OF THE ARCTIC COLD-TEMPERATE N ATLANTIC FLORA

Chorda tomentosa Lyngb. (Figs 2, 3)

The brown alga *Chorda tomentosa* (Laminariales) is a typical representative of the "Arctic cold-temperate North Atlantic flora". In its strongly heteromorphic life history, a macroscopic sporophyte alternates with a microscopic oogamous gametophyte.

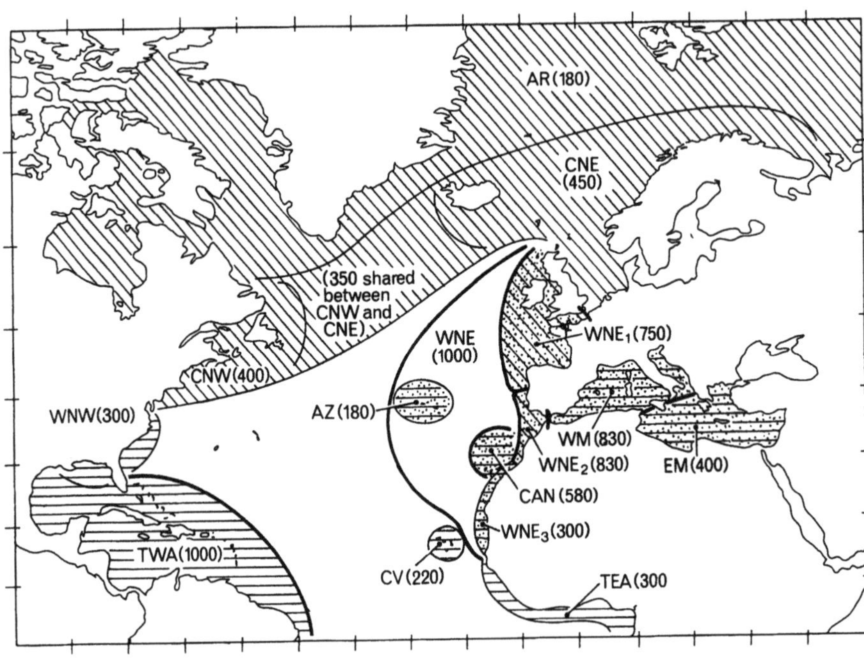

Fig. 1. Biogeographic subdivision of seaweed floras in the N Atlantic Ocean. Oblique hatching: species predominantly with Arctic-temperate distribution. Horizontal hatching: species predominantly with tropical and tropical-to-warm temperate distribution. Stippled: species with warm-temperate distribution. AR: Arctic region. AZ: Azorean subregion. CAN: Canarian-Madeiran subregion. CNE: Cold-temperate NE Atlantic region. CNW: Cold-temperate NW Atlantic region. CV: Cape Verdian Islands subregion. EM: East Mediterranean subregion. TEA: Tropical E Atlantic region. TWA: Tropical W Atlantic region. WNE: Warm-temperate NE Atlantic region. WNE_1: Warm-temperate NE Atlantic subregion 1 (W coasts of UK, Ireland and France). WNE_2: Warm-temperate NE Atlantic subregion 2 (W coasts of Iberia and Morocco). WNE_3: Warm-temperate NE Atlantic subregion 3 (subtropical NW African coasts). WNW: Warm-temperate NW Atlantic region. AR, CNW and CNE are closely allied and form together the Arctic-N Atlantic cold temperate flora. Between brackets: approximate numbers of species. Thick lines delimit relatively rich, thin lines relatively poor floras

In the N Atlantic Ocean, *Chorda tomentosa* has a composite southern boundary. On the east coast of the Atlantic Ocean, the southern boundary is a winter reproductive boundary, of the microscopic gametophyte, near the 10° C winter isotherm; south of this boundary, temperatures are always too high for sexual reproduction. On the west coast of the Atlantic Ocean, the southern boundary is a summer lethal boundary close to the 24°C summer isotherm (Lüning 1984; Maier 1984; Novaczek *et al.* 1986; Breeman 1988). The predominance of lethal boundaries on the NW Atlantic coast, reflects the pronounced seasonality, with annual temperature fluctuations of the sea surface of more than 17°C in inshore waters (on the NE Atlantic coast the seasonality is 6-7°C).

In the glacial N Atlantic, the distribution area of *Chorda tomentosa* was probably extremely reduced in comparison with the present day, as it was compressed between the southern limits of the coasts covered by continental glaciers, and the southern reproduction boundary along the 10° C winter isotherm (compare Fig. 3 with Fig. 2). Along the NW Atlantic coast, the distribution area became extremely reduced. It is likely that the whole Arctic cold-temperate N Atlantic flora became imprisoned in this extremely reduced distribution area, and this may have resulted in species extinctions, especially in the NW Atlantic.

Note that *Chorda tomentosa* did not have a composite southern boundary in the glacial N Atlantic, but a southern reproduction boundary at both sides of the ocean. This reflects the absence of a pronounced seasonality of the sea surface in the glacial NW Atlantic (with annual temperature fluctuations of 6-7°C) (CLIMAP-project members, 1981).

Composite southern boundaries are exhibited by various other seaweed species in the N Atlantic Ocean (Figs 4, 6) (van den Hoek 1982a, b; Breeman 1988).

Cladophora sericea (Huds.) Kütz. (Figs 4, 5)

The green alga *Cladophora sericea* (Cladophorales) is also a member of the "Arctic cold-temperate North Atlantic flora". It has a cold-temperate distribution, but extends into the northern portion of the Warm-temperate NE Atlantic region.

Like *Chorda tomentosa*, *Cladophora sericea* has a composite southern boundary: a winter growth boundary close to the 14°C winter isotherm on the east coast, and a summer lethal boundary near the 25°C summer isotherm on the west coast. The winter growth boundary is set by *Cladophora sericea*'s propensity for forming sporangia, instead of vegetative axes, at higher temperatures (Breeman 1988; Cambridge *et al.* 1989b).

On both the east and west coasts of the North Atlantic, the northern boundary of *Cladophora sericea* is a summer growth boundary near the 9°C summer isotherm.

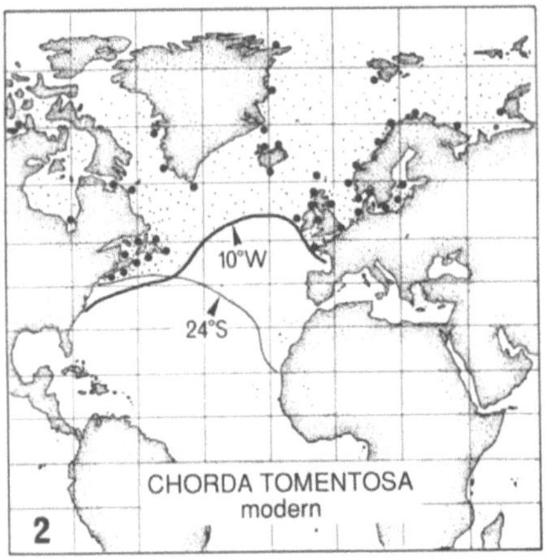

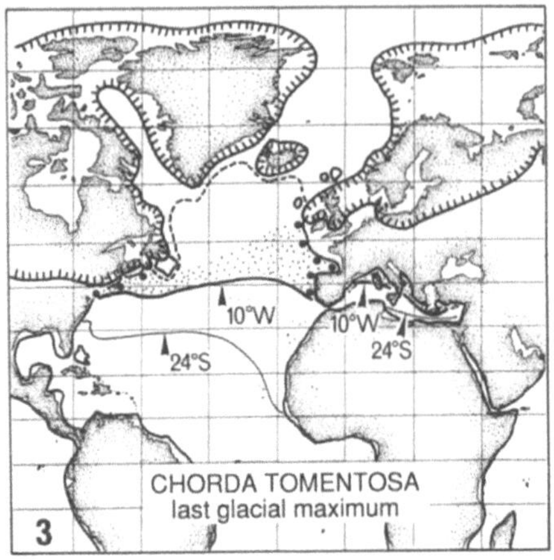

Fig. 2. Distribution of *Chorda tomentosa* (dots) in the N Atlantic Ocean in relation to limiting temperatures. Thick line: limiting winter isotherm. Thin line: limiting summer isotherm. Isotherms according to Gorshkov (1978). Stippled area: potential thermal distribution area. Distribution after Lüning (1985)
Fig. 3. Reconstructed distribution of *Chorda tomentosa* (dots) in the N Atlantic Ocean during the last glacial maximum. Thick contour line around continents: glacial coasts at lower sea level. Hatched contour lines: boundaries of continental glaciers. Broken line: boundary of pack ice in summer. Thick isotherm: limiting winter isotherm. Thin isotherm: limiting summer isotherm. Stippled area: potential thermal distribution area (sea surface temperatures and geography of Ice-age earth based on CLIMAP-project members, 1981)

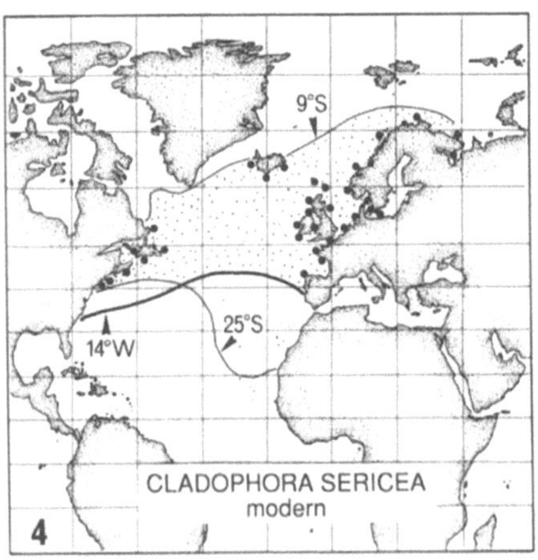

Fig. 4. Distribution of *Cladophora sericea* in the N Atlantic Ocean in relation to limiting temperatures. For explanation of symbols see text to Fig. 2. Distribution after Cambridge *et al.* (1989b) and van den Hoek (1963, 1982c)
Fig. 5. Reconstructed distribution of *Cladophora sericea* in the N Atlantic Ocean during the last glacial maximum. For explanation of symbols, see text to Fig. 3

In the glacial N Atlantic (Fig. 5), the distribution area of *Cladophora sericea* was as reduced as that of *Chorda tomentosa* (Fig. 3) on the west side of the ocean, whereas it had about the same extent, as at present, on the east side of the ocean; there, it could shift its position to the coasts of SW Iberia and NW Africa, and to those of the Mediterranean. Survival of *C.sericea,* and other cold-to-warm temperate species with comparable distributions, could have been precarious in the NW Atlantic during glaciations, but would probably not have been so precaious in the NE Atlantic.

The NW Mediterranean became hospitable to cold-to-warm temperate species, such as *Cladophora sericea,* and even to Arctic-to-cold temperate species like *Chorda tomentosa*, which might have been able to extend into the Mediterranean during the glaciations. This idea is supported by the occurrence of fossil Pleistocene faunas in the W Mediterranean of Arctic cold-temperate molluscs (Pérès 1985).

Isolated populations of some temperate NE Atlantic species with upper lethal temperatures near 20°C, survive in some cool deep water areas (ca 50-100 m) of the W Mediterranean (e.g. *Laminaria ochroleuca* and *Saccorhiza polyschides* (Phaeophyceae)), in the Straits of Messina region (van den Hoek 1982a).

Dumontia contorta Ruprecht

Dumontia contorta has a similar distribution to *Cladophora sericea* (van den Hoek *et al.* 1989); both species also occur in the N Pacific Ocean. Their migration from the Pacific Ocean to the Atlantic Ocean, after the Pliocene opening of the Bering Seaway (3.5 Ma ago), would have necessitated 5-10°C higher summer temperatures than the present ones in the Arctic Ocean, as these species need these summer temperatures for sufficient growth (van den Hoek *et al.* 1990).

Rhodochorton purpureum (Lightf.) Rosenv. (Figs 6, 7)

In the North Atlantic, this species is a common member in the "Arctic cold-temperate North Atlantic flora". Along the NE Atlantic shores, it extends into the northern portions of the Warm-temperate NE Atlantic region in W Iberia; in the N Pacific Ocean, and in the southern hemisphere, *Rhodochorton purpureum* occupies comparable climatic zones (Fig. 6); in the North Atlantic Ocean, *Rhodochorton purpureum* has, again, a composite southern boundary. On the east coast of the North Atlantic Ocean, the southern boundary is a winter reproductive boundary of the tetrasporophyte close to the 15°C winter isotherm; on the west coast, the southern boundary is a summer lethal boundary close to the 23°C summer isotherm (Stegenga 1978; van den Hoek, 1982a, b; Breeman *et al.* 1984; Breeman 1988; West 1969, 1970, 1972).

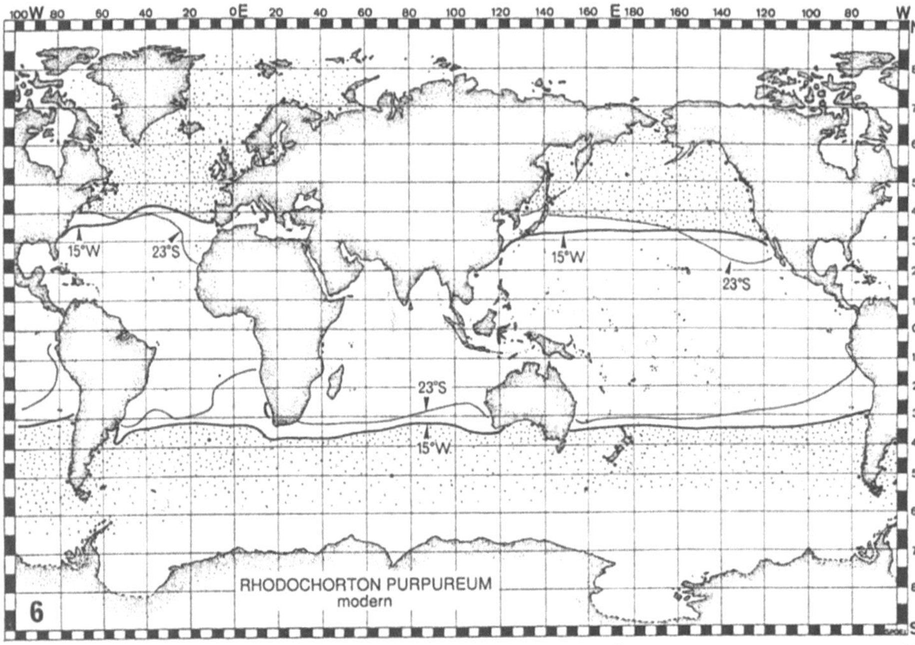

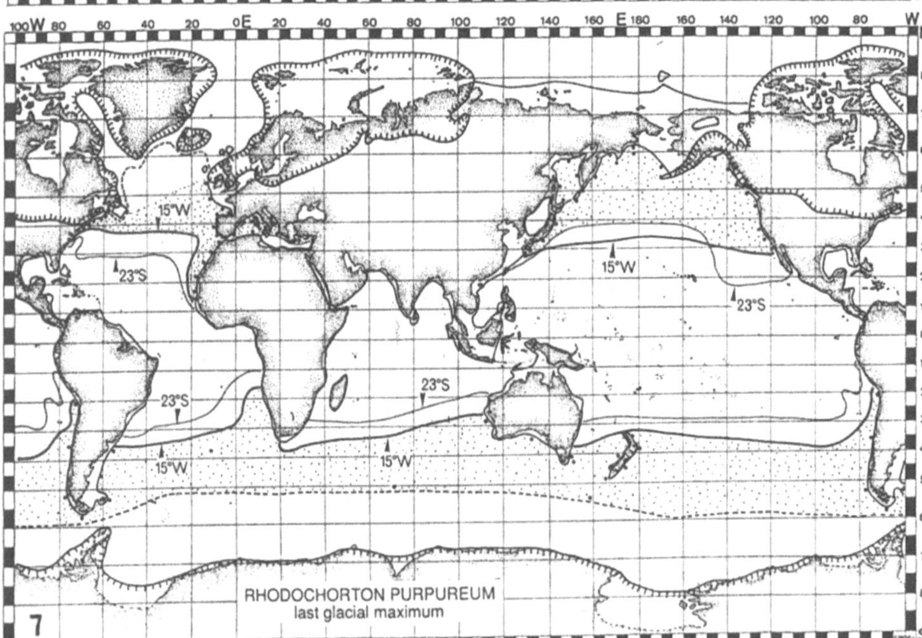

Fig. 6. Global distribution of *Rhodochorton purpureum* in relation to limiting temperatures. For explanation of symbols see text to Fig. 2. Isotherms according to Gorshkov (1976, 1978, 1980). Distribution as in van den Hoek, 1982 b, with additional data from Ricker (1987)
Fig. 7. Reconstructed global distribution of *Rhodochorton purpureum* during the last glacial maximum. For explanation of symbols, see text to Fig. 3

In the glacial N Atlantic, the distribution of *Rhodochorton purpureum* (Fig. 7) was probably quite similar to that of *Cladophora sericea*. Both species had a strongly reduced distribution area on the west coast, and a wide, southerly shifted distribution area on the east coast.

Figure 7 also depicts the strikingly different consequences of the glaciations for the geographic distribution of cold-to-warm temperate species, such as *Rhodochorton purpureum*, in the N Atlantic Ocean and the N Pacific Ocean. It is likely that on the NW Pacific coast, the distribution area was not compressed and reduced during the glacial maximum (as it was in the NW Atlantic), but expanded in a southerly direction. On the NE Pacific coast, the distribution area probably did not shift into a southerly direction, and retained its approximate interglacial latitudinal extent.

Glacial conditions in the N Pacific Ocean were apparently not adverse to temperate seaweeds, a point that has been stressed in several previous publications (van den Hoek 1975; Briggs 1974). This difference from the N Atlantic Ocean was the result of the glacial re-emergence of the Bering Land Bridge (which separated the N Pacific from the severe conditions in the frigid Arctic), and of the virtual absence of glaciers on even the most northerly coasts in the N Pacific.

An interesting question to ask would be: how did the amphipolar, amphi-oceanic distribution of *Rhodochorton purpureum* come about? The species may have been exchanged between the N Pacific and the N Atlantic since the late Cenozoic opening (3.5 Ma ago, McKenna 1983) of the Bering Seaway. The distribution of *Rhodochorton purpureum* in the temperate zones of both hemispheres cannot be explained on the basis of the present day temperature conditions of the oceans, as the species would be unable to survive the high temperatures during a passage through the tropical belt. Only a limited number of cold water species have succeeded in crossing the tropical belt, which suggests that conditions permitting this were infrequent, and of short duration. It is possible that passage across the tropical belt was brought about by rare dispersal events, in periods with temperatures in the tropics just low enough for survival, but not for growth and reproduction.

We have invoked Pleistocene temperature lowering to explain the passage of *Rhodochorton purpureum* and other amphipolar temperate species across the equator, especially along the Pacific and Atlantic east coasts, where the tropical belts are relatively narrow (van den Hoek 1982a, b). Survival of *R. purpureum* would have required an equatorial temperature lowering of at least 5°C from present day temperatures (from ca 28 to 23°C), and reproduction, a temperature decrease of at least 13°C (from ca 28 to 15°C). This is also valid for the amphipolar species (or species complex) *Desmarestia viridis* (O.F. Müll.) Lamour. (van den Hoek *et al.* 1990). Figure 7 shows that glacial summer temperatures remained too high for survival of *Rhodochorton purpureum* in a broad latitudinal belt on both sides of the equator.

Winter temperatures would have been (just) low enough for survival, but transport across the equatorial belt is likely to have taken more than a few months. Therefore, the species probably crossed the equator in an earlier geological period, when temperatures were still lower. This topic will be further pursued in the section on *Dictyosphaeria cavernosa* (see below).

Figures 6 and 7 also indicate that, in the southern hemisphere, the latitudinal ranges of cold-to-warm temperate species, such as *Rhodochorton purpureum*, were only slightly shifted towards the equator during the glaciation (as compared to their recent ranges), and that this shift expanded their distribution

REPRESENTATIVE SPECIES OF THE WARM-TEMPERATE NE ATLANTIC REGION

Polyneura hilliae (Grev.) Kylin (Figs. 8, 9)

This foliose red alga is a representative endemic of the Warm-temperate NE Atlantic region (Fig. 1).

The northern boundary in the NE Atlantic is a composite boundary. It is a summer growth boundary near the 12°C summer isotherm in N Scotland, and a winter lethal boundary near the 6°C winter isotherm in the eastern English Channel. Low lethal winter temperatures exclude the species from the North Sea. The southern boundary is a summer lethal boundary near the 19°C summer isotherm (Yarish *et al.* 1986, 1987; Breeman 1988). In Fig. 8, the imaginary southern growth/reproduction boundary along the 18°C winter isotherm, is given. As this boundary is situated far beyond the lethal boundary it is irrelevant to the actual geographic distribution.

The stippled area in Fig. 8 gives the potential distribution of *Polyneura hilliae* within its lethal and growth/reproduction boundaries. The figure illustrates that the species is excluded from the inner Bay of Biscay by intermittently high summer temperatures; and from the North Sea by excessively low winter temperatures. The figure also shows that *Polyneura hilliae* is incapable of inhabiting the coast of the NW Atlantic Ocean. The species is apparently too stenothermous to survive the enormous annual temperature fluctuations that are so characteristic of this coast.

In the glacial N Atlantic *Polyneura hilliae* (Fig. 9) would have been able to occupy, on the NE coast, a more southerly, and somewhat wider, distribution area than it does in the recent North Atlantic. It probably extended into the Mediterranean, and might even have occurred on the warm temperate NW Atlantic coast as a result of the absence of strong seasonality there. Present day temperature conditions of the N Atlantic sea surface apparently restrict the occurrence of *Polyneura hilliae* to the Warm-temperate NE Atlantic region.

Cladophora hutchinsiae (Dillw.) Kütz. (Figs 10, 11)

This is another representative endemic of the Warm-temperate NE Atlantic region.

The northern boundary on the East coast of the N Atlantic is a composite boundary: it is a summer growth boundary close to the 13°C summer isotherm in northern Scotland, and a winter lethal boundary near the 6°C winter isotherm in the English Channel. Low winter temperatures exclude the species from the North Sea.

Cladophora hutchinsiae extends into the Mediterranean, where its easternmost boundary is a summer lethal boundary near the 26°C summer isotherm (Cambridge *et al.* 1984, 1989a).

In the Atlantic Ocean, the 26°C summer isotherm can be considered as the potential southern lethal boundary (Fig. 10). A potential growth boundary is situated somewhat more to the south, along the 23°C winter isotherm. The lack of records in the southernmost portion of its potential distribution area in the NE Atlantic ocean is possibly due to undersampling of the NW African coasts.

The potential distribution area of *Cladophora hutchinsiae* in the North Atlantic (the stippled area in Fig.10) does not include the NW Atlantic coast, with its too pronounced seasonality.

In the glacial N Atlantic *Cladophora hutchinsiae* (Fig. 11) probably had a more southerly distribution, and occurred throughout the Mediterranean. The species could have occurred along NW Atlantic coasts, as a result of the modest seasonality there during the glacial maximum. Apparently, the distribution of *Cladophora hutchinsiae* can be explained as the result of an interplay between its temperature responses and the recent temperature conditions of the sea surface in the N Atlantic Ocean.

REPRESENTATIVE SPECIES OF THE WARM-TEMPERATE NW ATLANTIC REGION

This depauperate seaweed region harbours a small number of species endemic to the NW Atlantic, which spill over onto the tropical margin in the south, and extend into the Cold-

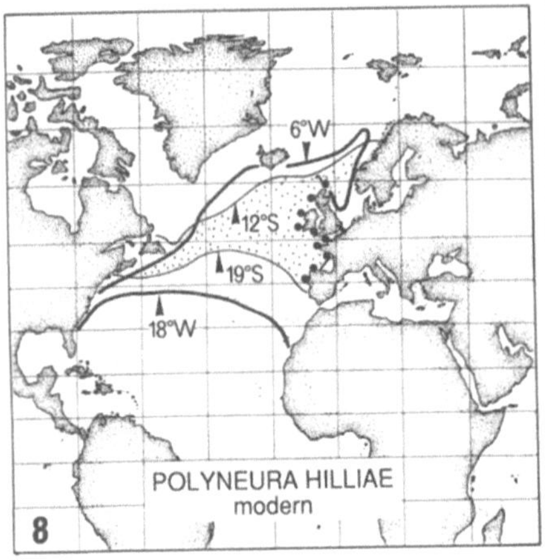

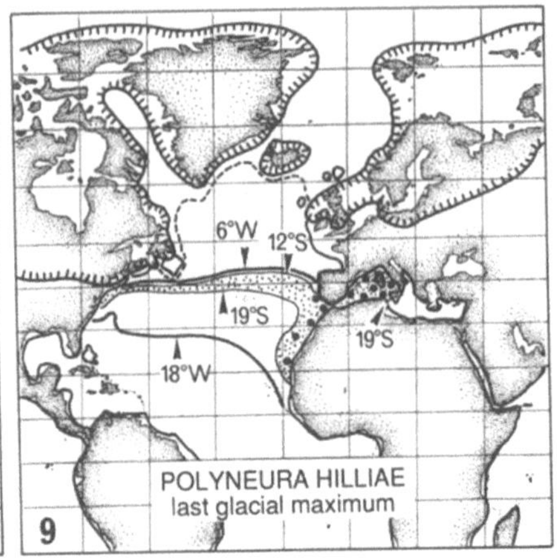

Fig. 8. Distribution of *Polyneura hilliae* in relation to limiting temperatures. For explanation of symbols, see text to Fig. 2. Distribution after Yarish *et al.* (1986)
Fig. 9. Reconstructed distribution of *Polyneura hilliae* during the last glacial maximum. For explanation of symbols, see text to Fig.3

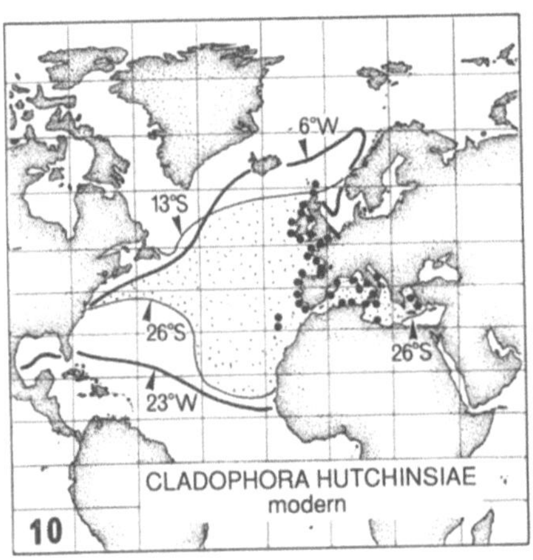

Fig. 10. Distribution of *Cladophora hutchinsiae* in relation to limiting temperatures. For explanation of symbols, see text to Fig. 2. Distribution after Cambridge *et al.* (1989a) and van den Hoek (1963)
Fig. 11. Reconstructed distribution of *Cladophora hutchinsiae* during the last glacial maximum. For explanation of symbols, see text to Fig. 3

temperate NW Atlantic region in the north (van den Hoek 1982a; Yarish *et al.* 1984; Breeman 1988). These species have wide temperature tolerances which contrast with the narrow temperature tolerances of the NE Atlantic endemics, and which enable them to survive the pronounced seasonality along the NW Atlantic shores.

Grinnellia americana (C. Ag.) Harvey (Figs 12, 13)

The foliose red alga *Grinnellia americana* (Ceramiales) is an endemic of NW Atlantic shores, with a disjunct occurrence along the coast of Venezuela.

The northern boundary is a summer growth/reproduction boundary close to the 18°C summer isotherm. As the lower lethal temperature is below 0°C, the species is capable of surviving the low winter temperatures (near 0°C) at its northern boundary (Yarish *et al.* 1984). The southern boundary is a summer lethal boundary near the 28°C summer isotherm (Yarish *et al.* 1987). In Fig. 12 the imaginary southern growth boundary, approximating the 26°C winter isotherm, is given.

The isolated occurrence of the species along the coast of Venezuela is facilitated by coastal upwelling, with temperatures low enough for winter growth and summer survival (Griffiths & Simpson 1972).

The stippled area in the N Atlantic Ocean, in Fig. 12, gives the thermal distribution area in this ocean. The figure suggests that the species would be potentially capable of occupying a wide distribution area on the NE Atlantic coasts, including the Mediterranean.

For comparison, the thermal distribution area of *Campylaephora hypneoides* J. Ag. (Rhodophyta, Ceramiales) which has a similar distribution in the NW Pacific Ocean, is given in Fig. 12 (van den Hoek 1982a). This endemic species of the NW Pacific has temperature responses comparable to those of *Grinnellia americana*. It is, apparently, also potentially capable of occupying a wide distribution area along that ocean's NE coast. Despite the wide ecological windows, both species are lacking on the opposite, eastern sides of the two oceans, and this suggests that long distance transoceanic dispersal is precarious if not impossible.

In the glacial N Atlantic, *Grinnellia americana* probably had a much wider NW Atlantic distribution than at present, because it could have been distributed throughout the Caribbean (Fig. 13); even then, however, it was excluded from equatorial latitudes by adverse high temperatures. It is therefore unlikely that the paucity of warm-temperate NW Atlantic endemics is the result of adverse glacial conditions; rather, present day and previous interglacial (and earlier) adverse conditions (sediment shores with scarce offshore reefs), seem to be responsible. This idea is strengthened by the presence of many more warm temperate NW Pacific endemics in the rich seaweed flora of Japan. On NE Atlantic coasts the potential glacial

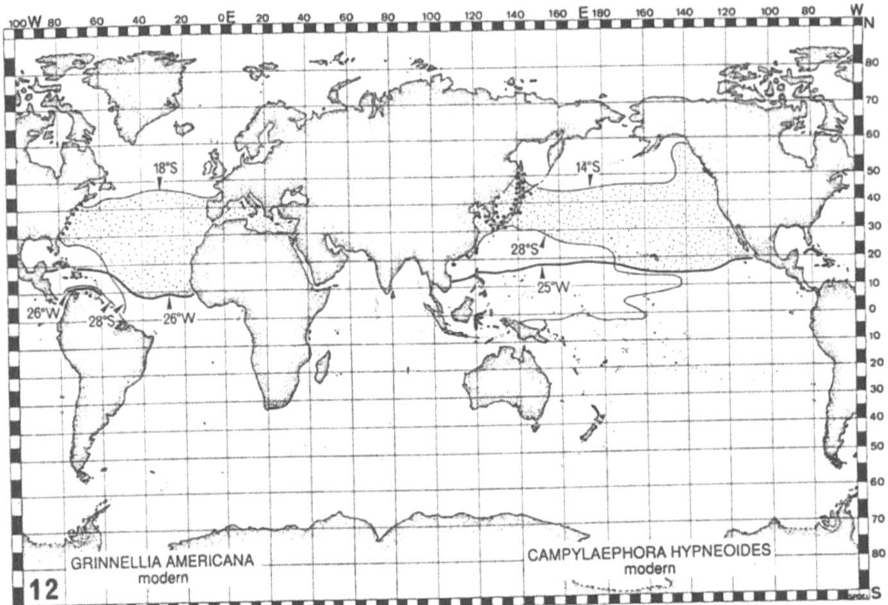

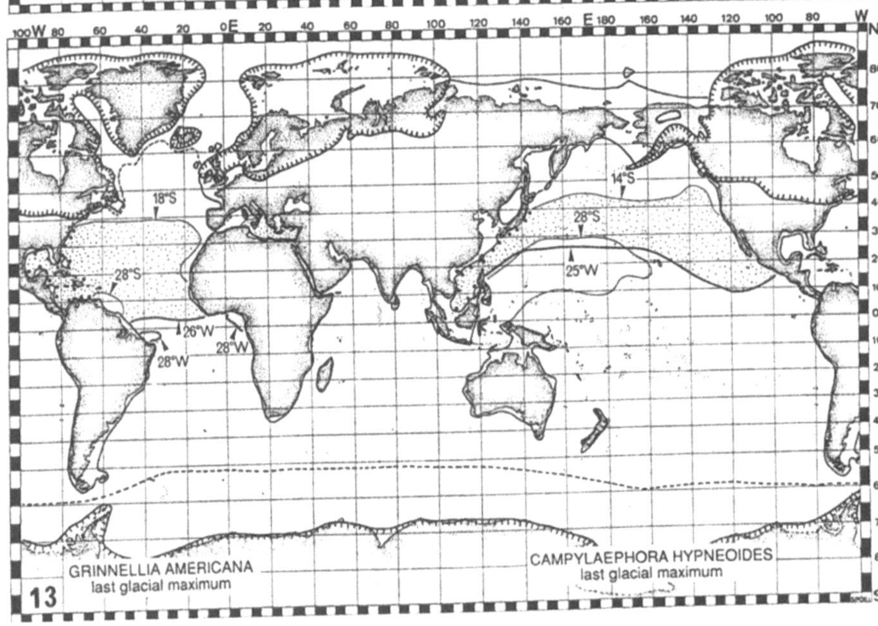

Fig. 12. Distribution of *Grinnellia americana* in the N Atlantic Ocean, and of *Campylaephora hypneoides* in the N Pacific Ocean, both in relation to limiting temperatures. For explanation of symbols, see text to Fig. 2. Distribution of the first species according to Phillips & Springer (1960), and Yarish *et al.* (1984); distribution of the second species according to van den Hoek, 1982a, with additional data from Kang (1966); Kawai & Kurogi (1982); Kim & Lee (1985); Lee & Lee (1982); Noda (1987); Tseng (1983)

Fig. 13. Reconstructed distributions of *Grinnellia americana* in the N Atlantic Ocean and of *Campylaephora hypneoides* in the N Pacific Ocean during the last glacial maximum. For explanation of symbols, see text to Fig. 3

distribution area of *Grinnellia americana* based on temperature was considerably reduced (Fig. 13).

REPRESENTATIVE SPECIES OF THE TROPICAL W ATLANTIC REGION

Cladophora submarina Crouan *et* Crouan *ex* Schramm *et* Mazé (Figs 14,15)

Cladophora submarina is restricted to the tropical W Atlantic region, where it inhabits wave-battered limestone rocks (van den Hoek, 1982c).

The northern boundary is a winter lethal boundary, near the 15°C winter isotherm in offshore water, but near the 23°C winter isotherm of the inshore water of southern Florida (Cambridge *et al.* 1987). The enormous difference between a mean winter temperature of 23°C and an extreme low temperature of 10-15°C, in southern Florida, is due to infrequent cold spells caused by polar continental air masses that intrude into this area (Walker *et al.* 1982). These cold spells can cause spectacular coral kills, and massive fish mortality; they are responsible for an abrupt northward termination of the tropical marine biota in southern inshore Florida. On deep offshore reefs which are not subjected to these cold spells, tropical seaweeds, and even some corals, penetrate into the Warm-temperate NW Atlantic region, as far north as North Carolina, close to where winter temperatures are as low as 10°C (MacIntyre & Pilkey 1969; Searles 1984).

The species' upper lethal temperature is in the 30-35°C interval, and this is probably why it does not occur in protected bays, where temperatures may rise above 30°C. Two other tropical species from the Caribbean appeare to have similar upper lethal temperatures (McLachlan & Bird 1984; Breeman 1988), and this suggests that many tropical seaweeds are living at only a few degrees centigrade below their upper lethal temperature. Thus, their geographic distributions may be restricted not only by temperatures that are too low, but also by those that are too high. The stippled area in Fig. 14 gives the potential thermal distribution of *Cladophora submarina* across the Atlantic Ocean. Apparently the species is capable of inhabiting wide stretches of the warm-temperate to tropical E Atlantic coast, including the eastern Mediterranean. The absence of *C. submarina* and *G. americana* from the E Atlantic coasts, is apparently not caused by modern climatic factors, but because the central Atlantic functions as a barrier to dispersal.

In the glacial N Atlantic, the tropical W Atlantic distribution area probably did not differ much from the present distribution (Fig. 15). The potential thermal distribution area in the NE

Atlantic was much displaced to the south, during the glacial maximum, under the influence of the intense upwelling along the NW African coast. Survival in the E Mediterranean would have been precarious for a tropical species such as *Cladophora submarina,* with its winter lethal boundary near the 15°C winter isotherm.

Dictyosphaeria cavernosa (Forssk.) Børg. (Figs 16, 17)

The species is widely distributed along the tropical west coasts of the Atlantic and Pacific Oceans, as well as along the coasts of the Indian Ocean; it is absent from the tropical east coasts of the Atlantic and Pacific Oceans.

Based on experimental work with Japanese material (Enomoto & Okuda 1981), the northern boundary of the species can be interpreted as a combination of a reproduction boundary, along the 23°C summer isotherm, and a lethal boundary, along the 10°C winter isotherm (Sano *et al.* 1981). In Florida, the northern lethal boundary is situated close to the 20°C winter isotherm of the open ocean, which corresponds with intermittent cold spells of lower than 10°C in the inshore waters (see above, *C. submarina*).

The potential distribution area of *Dictyosphaeria cavernosa* (stippled area in Fig. 16) includes large portions of the Pacific and Atlantic east coasts, including the warm temperate

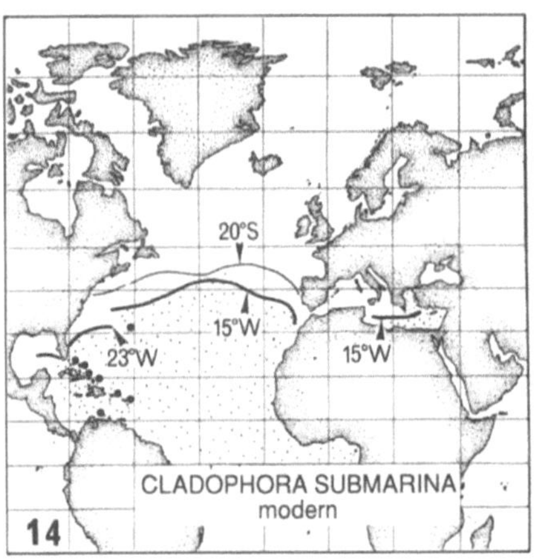

Fig. 14. Distribution of *Cladophora submarina* in the N Atlantic Ocean in relation to limiting temperatures. For explanation of symbols see text to Fig. 2. Distribution after van den Hoek (1982c)

Fig. 15. Reconstructed distribution of *Cladophora submarina* in the N Atlantic Ocean during the last glacial maximum. For explanation of symbols, see text to Fig. 3

coasts of the Mediterranean; clearly, the species is not absent from these coasts because of present-day adverse temperature conditions.

A plausible explanation for the present day distribution of *Dictyosphaeria cavernosa* would be that, after the break up of the Tethys Ocean (ca 14-20 MA ago), populations of *D. cavernosa* initially persisted on both the east and west coasts of the Atlantic and Pacific Oceans, and east coast populations later became extinct because of adverse temperature conditions. It is attractive to invoke the same equatorial temperature lowering, used to explain the passage across the equator by amphipolar cold water species such as *Rhodochorton purpureum,* as the cause of extinction of *D. cavernosa* along the east coasts of the Atlantic and Pacific Oceans. Extinction of *D. cavernosa* would have required a temperature lowering of about 5°C (from 28-23°C), down to a temperature where reproduction is no longer possible; this is about the same lowering of temperature required for a precarious passage through the tropics, by cold water species such as *R. purpureum.*

We have suggested previously (van den Hoek 1975, 1982a), that lowered temperatures along the equatorial Atlantic and Pacific east coasts during the Pleistocene glaciations, could have caused the extinction of tropical seaweeds. The reconstructed glacial distribution of *Dictyosphaeria cavernosa* (Fig.17) is not directly in support of this hypothesis, no more than the reconstructed glacial distribution of the cold-water species *Rhodochorton purpureum* could explain its passage across the equator. Glacial winter temperatures remained considerably above the lower limit for survival of *D. cavernosa.* They were, however, too low for reproduction in winter along most of the tropical eastern Atlantic (and Pacific) coasts. As high grazing pressure in the tropics causes high turnover rates in populations (e.g. a life span of individual plants of less than one month, in the foliose brown alga *Lobophora variegata*; Ruyter van Steveninck & Breeman 1987), year-round active growth and reproduction may well be a prerequisite for survival of tropical populations; thus, persistence might have been precarious for *D. cavernosa* on the eastern sides of the oceans during the glaciation. Alternatively, extinction in a more distant geologic past is a possibility. Fossil evidence, in support of this idea, show the definitive extinction of the last Tethyan corals in the E Atlantic by the late Miocene (ca 5 Ma) (Boekschoten & Best 1981, 1988).

A separation, in the Miocene or earlier, of the Atlantic and Pacific populations of *Dictyosphaeria cavernosa* is supported by the high degree of genetic divergence between an Hawaiian and a Caribbean population, as measured by DNA-DNA hybridization and protein immunological distance (Olsen-Stojkovich *et al.* 1986; Olsen *et al.* 1987).

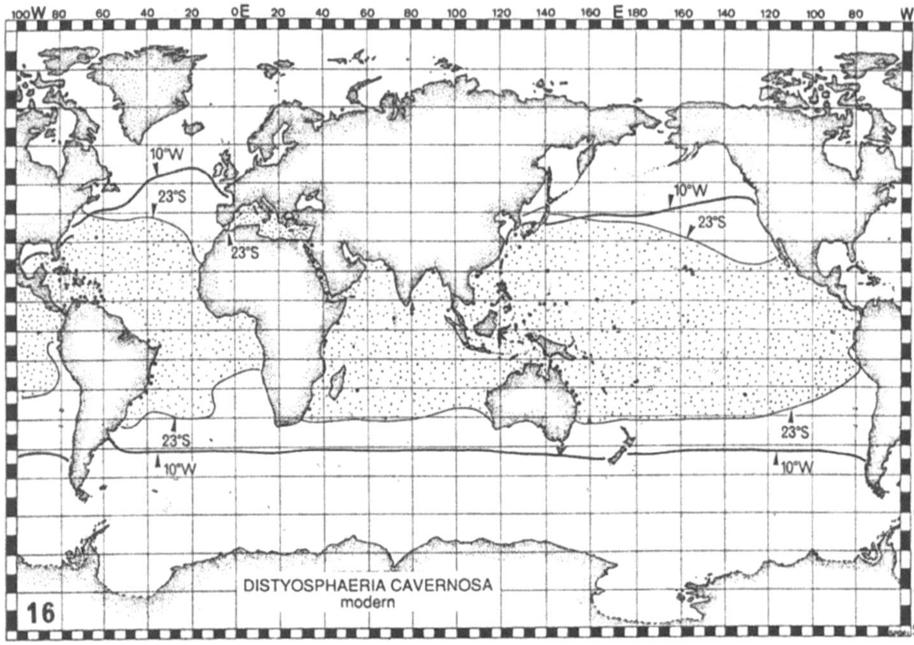

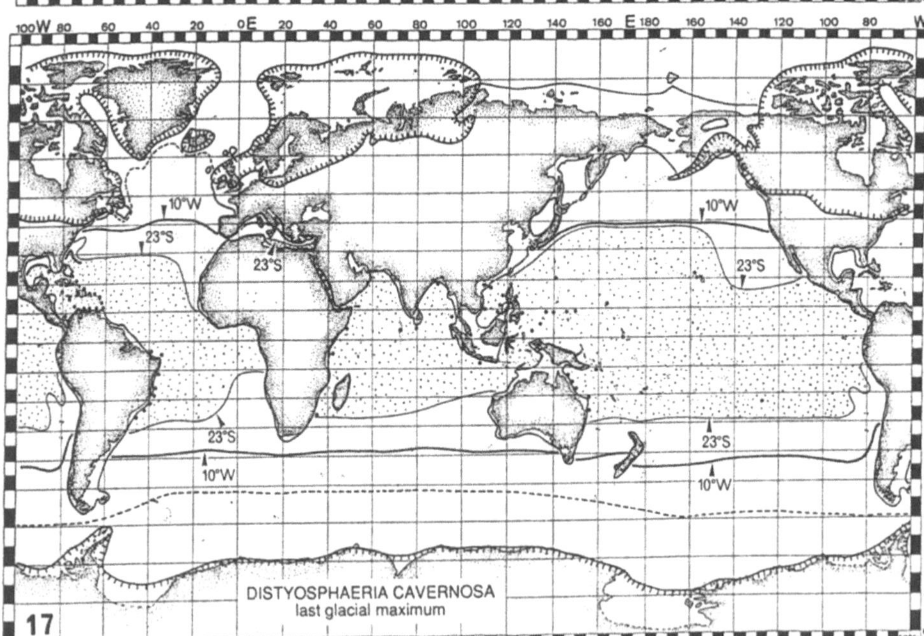

Fig. 16. Global distribution of *Dictyosphaeria cavernosa*. For explanation of symbols, see text to Fig. 2. Distribution according to van den Hoek (1989)

Fig. 17. Reconstructed global distribution of *Dictyosphaeria cavernosa* during the last glacial maximum. For explanation of symbols, see text to Fig. 3

SPECIES WITH AN AMPHIATLANTIC TROPICAL-TO-WARM TEMPERATE DISTRIBUTION

A considerable number of seaweed species is restricted (or almost restricted) to the tropics in the W Atlantic, but has a wide tropical-to-warm temperate distribution in the NE Atlantic (van den Hoek 1982a, b; Cambridge *et al.* 1987). Species of this distribution type are primarily responsible for the tropical-to-warm temperate "imprint", that is so characteristic of the southern and Mediterranean portions of the Warm-temperate NE Atlantic region. Many species in this category have world-wide distributions in the tropics and adjacent areas. A small number is widely distributed in the Caribbean but is restricted, in the NE Atlantic, to the Canaries; some of these also occur in the W Mediterranean (Prud'homme van Reine & van den Hoek 1989; Prud'homme van Reine 1988).

Centroceras clavulatum (C. Ag.) Mont. (Figs 18, 19)

Centroceras clavulatum is an abundant, turf forming, filamentous red alga that is widely distributed throughout the tropics, and in adjacent warm temperate regions (van den Hoek *et al.* 1990). It differs from *Dictyosphaeria cavernosa* in also being widely distributed along the Atlantic and Pacific east coasts.

The potential thermally determined modern and glacial distribution areas of *D. cavernosa* and *C. clavulatum* (stippled areas in Figs 16-19) are quite similar, and largely overlap. Both species could have survived glacial conditions on the equatorial Atlantic and Pacific east coasts. This similarity poses the question: why did *D. cavernosa* become extinct on these coasts and not *C. clavulatum*? A plausible reason might be that the sea surface temperatures along the equatorial east coasts of both oceans were lowered, in the late Miocene, to values below the reproduction limit (ca 23°C) of *D. cavernosa*, but remained above the growth limit (20-15°C) of *C. clavulatum.*

It is likely that *D. cavernosa* is representative of the group of tropical species with a W Atlantic-Indo-W Pacific distribution, while *C. clavulatum* is representative of the large group of species with a pantropical to warm-temperate cosmopolitan distribution.

Cladophora coelothrix Kütz. (Figs 20, 21)

Cladophora coelothrix is a second example of the amphi-Atlantic tropical-to-warm temperate distribution group. It is restricted, in the NW Atlantic, to the tropics, while it ranges

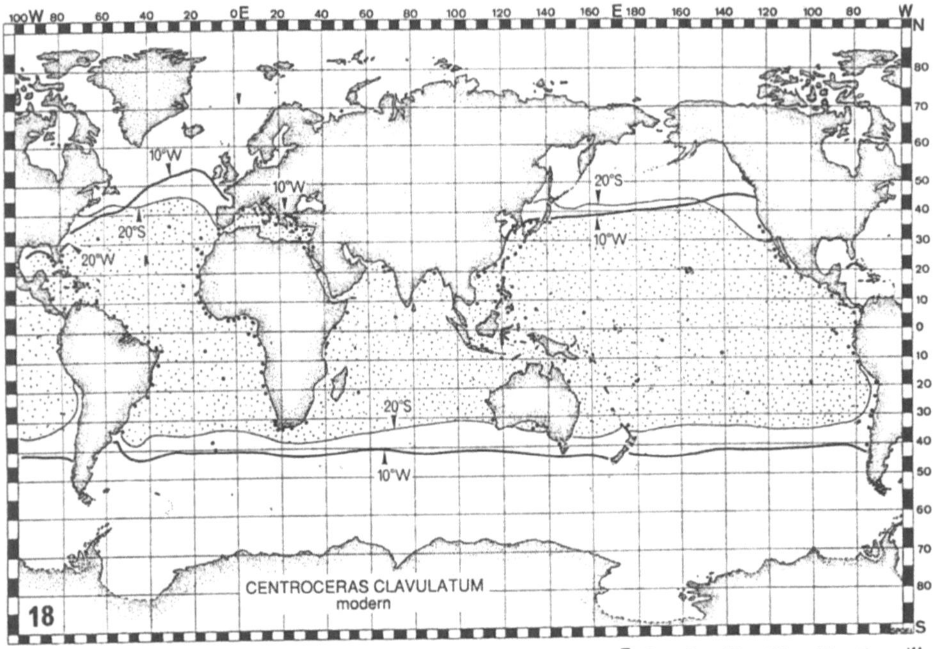

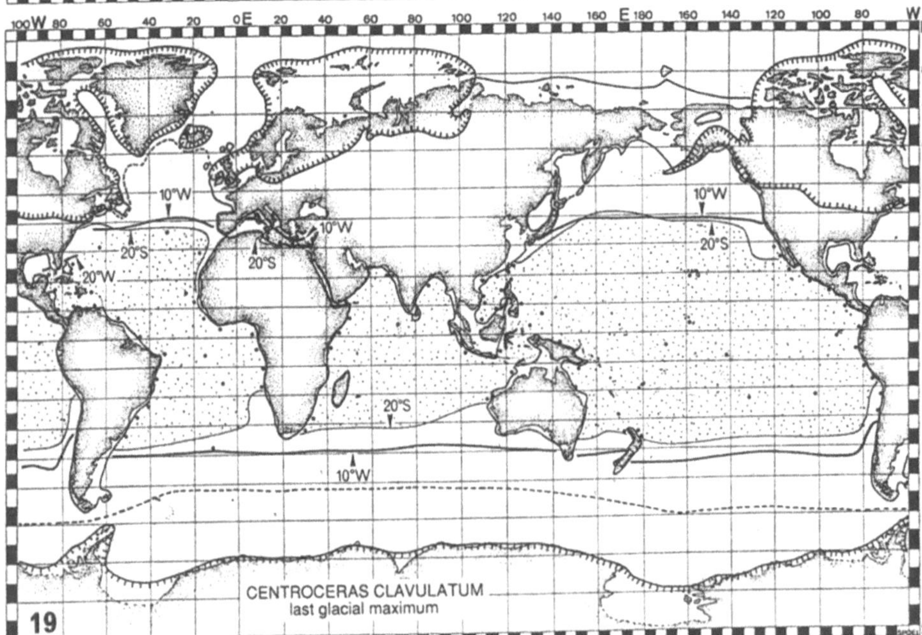

Fig. 18. Global distribution of *Centroceras clavulatum*. For explanation of symbols, see text to Fig. 2. Distribution after van den Hoek (1989)
Fig. 19. Reconstructed global distribution of *Centroceras clavulatum* during the last glacial maximum. For explanation of symbols, see text to Fig. 3

in the NE Atlantic, from the tropics to as far north as W Ireland, near the northern boundary of the Warm-temperate NE Atlantic region.

Experimental data, relevant to the geographic distribution, are available for a Caribbean strain and a W European strain, which differ in their temperature responses (Cambridge *et al.* 1987).

The temperature responses of the Caribbean strain are similar to those of *Centroceras clavulatum*. The northern boundary, near the 10°C winter isotherm (in Florida, near the 20°C winter isotherm), can be interpreted as a winter lethal boundary. A potential summer growth boundary is situated near the 20°C summer isotherm. The W European population of *Cladophora coelothrix* has a composite northern boundary, comparable to those of NE European endemic species like *C. hutchinsiae*. A winter lethal boundary, near the 6°C winter isotherm, excludes *C. coelothrix* from the North Sea, whilst a summer growth boundary lies near the 15°C summer isotherm on the Irish W Coast. Both strains have upper growth and survival limits in the 30-35°C interval, and this correlates with their distribution in the tropical E and W Atlantic Ocean.

Similar differences in temperature ranges between populations on both sides of the ocean have now been discovered in at least three other amphi-Atlantic tropical-to-warm temperate species (unpublished results); even at the high temperature side, the maximum temperatures for growth and survival are slightly different. This suggests that selection

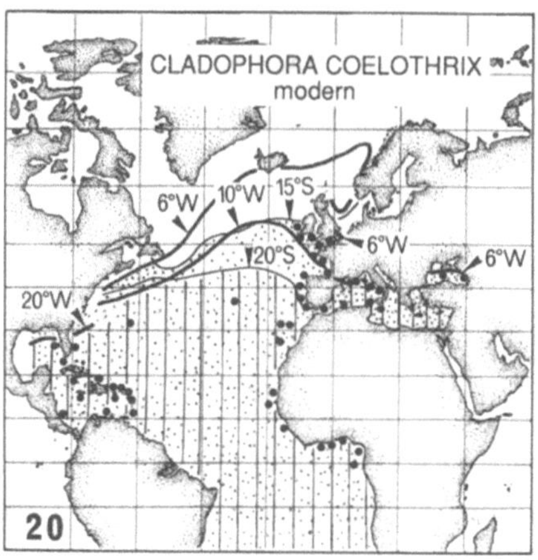

Fig. 20. Distribution of *Cladophora coelothrix* in the N Atlantic Ocean. Stippled area: potential thermal distribution area of W European strain. Hatched area: potential thermal distribution area of Caribbean strain; also see text to Fig. 2. Distribution after Cambridge *et al.* (1987)

Fig. 21. Reconstructed distribution of *Cladophora coelothrix* in the N Atlantic Ocean during the last glacial maximum. For explanation of symbols, see text to Figs 3, 20

pressure favours a shift towards tolerance for lower temperatures along NE Atlantic shores, thus enabling what were originally tropical to subtropical species, to occupy suitable rocky shores in the warm temperate latitudes of the NE Atlantic.

Selection pressure may not have operated similarly in the NW Atlantic for two reasons: first, a shift towards tolerances for lower temperatures would necessitate the colonization of shores that are (at least at present) and were (in previous interglacials) hostile to benthic seaweeds, because they consist of sandy substratum, with rare, rocky, sublittoral reefs. In addition, the shift would have to have been considerable, in order to enable the alga to overcome the steep temperature gradient in southern Florida (mentioned under *Cladophora submarina*).

In other words, the genetic isolation of populations by the central Atlantic Ocean, and the difference in selection pressure on both sides of the ocean, would facilitate the origin of pairs of warm-temperate and tropical sister species.

In the glacial Atlantic (Fig. 21), the distribution area of the W Atlantic population of *Cladophora coelothrix* probably differed little from the modern area. The distribution area of the E Atlantic population, however, was divided into East Mediterranean and NW African portions, which were thermally isolated from one another. The species might also have survived on the Canary Islands.

This type of intermittent thermal isolation of tropical-to-warm temperate species in the eastern Mediterranean may have promoted speciation, and the origin of endemic species. Even at present, at least 25 species out of 120 tropical-to-subtropical species in the NE Atlantic, exhibit this type of distribution, and are imprisoned in the Mediterranean by summer temperatures in the Atlantic (bordering the entrance to the Mediterranean) too low for growth or reproduction .

Cladophora pellucida (Huds.) Kütz. and *C. pellucidoidea* van den Hoek (Figs 22, 23)

Another example of a pair of narrowly related species which exhibit an amphi-Atlantic tropical-to-warm temperate distribution, is formed by *Cladophora pellucida* and *C. pellucidoidea* (Breeman 1988; Cambridge *et al.* 1989a). After their genetic isolation by the central Atlantic Ocean, these sister species not only diverged with regard to their thermal responses, but also developed minute morphological differences (van den Hoek 1982c).

The questions now arise: in which geological epoch, and by which causes, did the ancestral species of *C. pellucida* and *C. pellucidoidea* (and other comparable species pairs) arise, and start their amphi-Atlantic divergence?

We suggest that these hypothetical ancestral species inhabited the broad warm-temperate belt spanning the N Atlantic, during the early Miocene, when this belt was much more northerly than at present. After the subsidence of the Scotland-Greenland Ridge in the Middle Miocene (18-14 Ma ago), the general cooling trend in the N Atlantic shifted the warm temperate latitudes on the NW and NE Atlantic coasts to more southerly positions, so that their seaweed floras became isolated by the central Atlantic Ocean. At the same time, the steep latitudinal temperature gradient, with its pronounced seasonality, developed along the NW Atlantic coast (Kennet 1982; Berggren & Schnitker 1983; Thiede & Eldholm 1983). The considerable genetic divergence (as measured by DNA-DNA hybridization) between a strain from W France of *C. pellucida*, and a strain from Georgia of *C. pellucidoidea*, supports this early separation (Bot *et al.* 1989c).

The *Cladophora pellucida* complex includes several other species, and exhibits a highly disjunct tropical-to-warm temperate distribution in both hemispheres (Fig. 22). *Cladophora feredayi* Harvey, for instance, is a warm temperate species from southern Australia and New Zealand. The thermal tolerances of this species were investigated in our laboratory (Cambridge, unpublished results). It apparently has an even narrower thermal range than *C. pellucida*, and this may be typical of southern Australian endemics.

The reconstructed glacial distribution of *C. feredayi* (Fig. 23) differs only slightly from the modern distribution (though it was somewhat extended in NW Australia). This illustrates that the impact of the glaciations on the seaweed biogeography of temperate seaweed floras in the southern hemisphere, was probably minimal.

CONCLUSIONS AND DISCUSSION

The reconstructions of the glacial distribution areas of individual species belonging to diverse distribution types, lead to the conclusion that species belonging to the Arctic-cold temperate N Atlantic flora with upper reproductive and/or growth limits in the 10-15°C range were faced with such an extreme reduction in their distribution area that many of them could have become extinct during glaciations. This was particularly acute along the Atlantic NW coast where distributions became excessively reduced (Figs 3, 5, 7). Thus, the much more severe conditions in the glacial N Atlantic than in the N Pacific, may provide a plausible explanation for the much greater richness of the cold-temperate N Pacific flora.

The much greater compression of the distribution areas on the Atlantic NW coasts, suggests that some species may have become extinct there, and not on the Atlantic NE coasts.

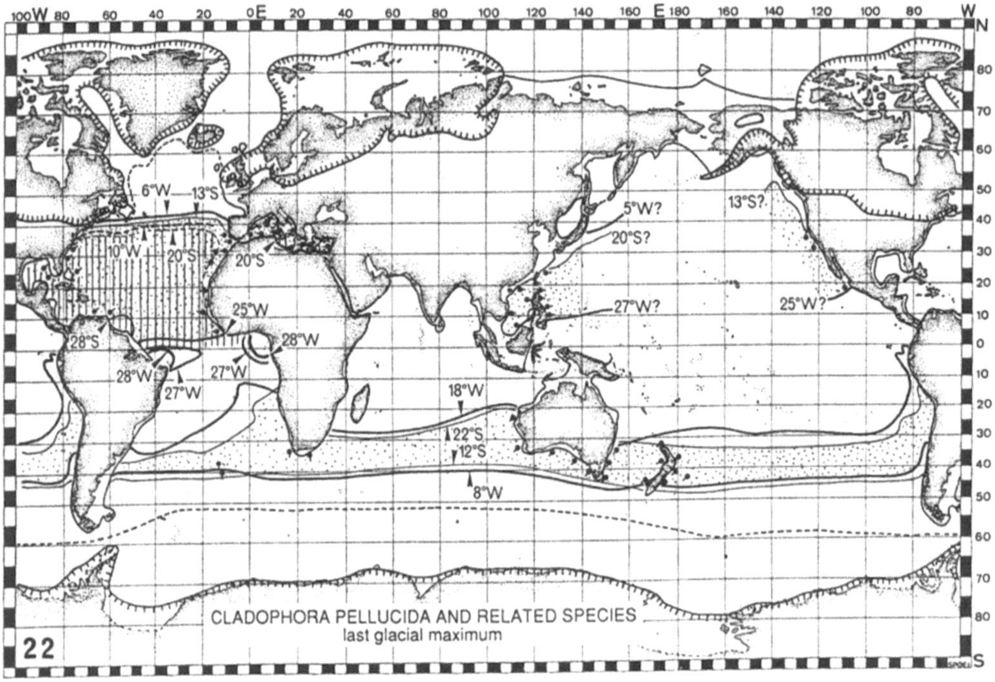

6°W
13°S
10°W
20°S
20°S
25°W
28°S
28°W
28°W
27°W
27°W
18°W
22°S
12°S
8°W
5°W?
20°S?
27°W?
13°S?
25°W?
CLADOPHORA PELLUCIDA AND RELATED SPECIES
last glacial maximum
22

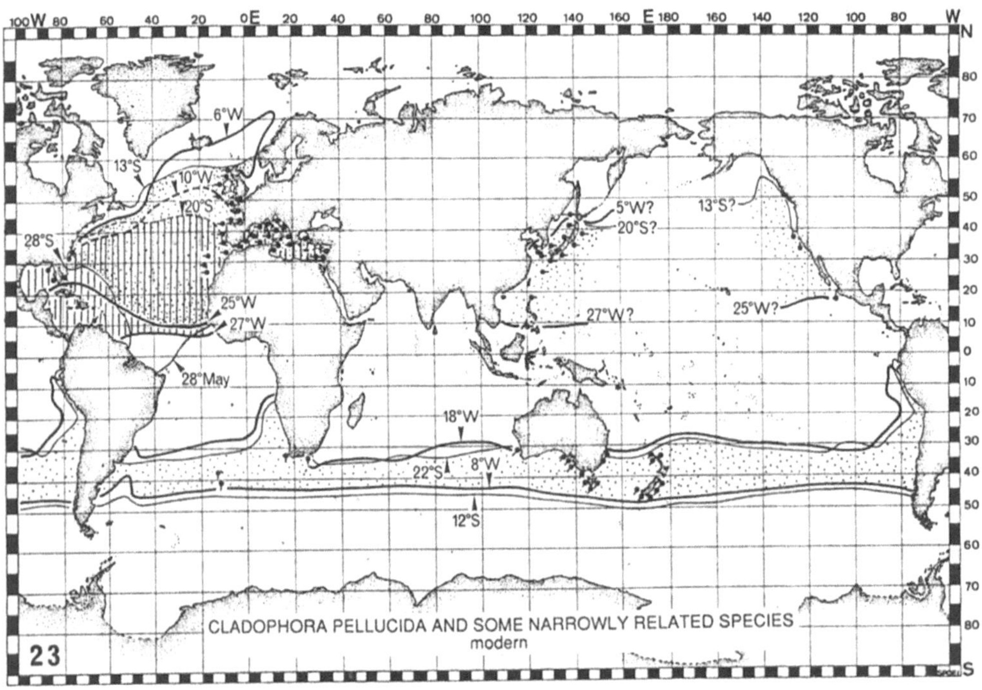

6°W
13°S
10°W
20°S
28°S
25°W
27°W
28°May
18°W
22°S
8°W
12°S
5°W?
20°S?
27°W?
13°S?
25°W?
CLADOPHORA PELLUCIDA AND SOME NARROWLY RELATED SPECIES
modern
23

There is a marked, though not dramatic, difference in species richness on Cold-temperate E and W coasts (Fig. 1). Examples may be provided by the genus *Laminaria*. Using results from DNA-DNA hybridization, Stam *et al.* (1988) suggested that the N Atlantic species *L. digitata* (Huds.) Lamour., *L. saccharina* (L.) Lamour., *L. hyperborea* (Gunn.) Foslie, *L. ochroleuca* La Pylaie and the Mediterranean *L. rodriguezii* Bornet are genotypically closely related, and diverged from a common ancestor in the Mid-Miocene (15-19 Ma ago). This predates the definitive submergence of the Bering Land Bridge in the Pliocene (3.5 Ma ago). As evolutionary radiation of *Laminaria,* and related laminarialean genera, probably took place in the N Pacific, the above five species are thought to have descended from N Pacific predecessors, and to have migrated separately to the N Atlantic, after the opening of the Bering Seaway. *Laminaria digitata* and *L. saccharina* occur on both the NW and NE Arctic-to-cold temperate N Atlantic shores, and this is consistent with their experimentally determined temperature responses. *Laminaria hyperborea* and *L. ochroleuca* are restricted to temperate NE Atlantic shores, however, where the latter species has a somewhat "warmer" range. These temperature responses would allow the occurrence of *L. hyperborea* on the Atlantic NW coast (Breeman 1988,1989); its absence there is possibly the result of glacial extinction (for *L. ochroleuca* no experimental data are available). *Laminaria rodriguezii* (probably a relative of the NE Pacific *L. sinclairii*) is thought to have escaped glacial extinction only in the Mediterranean (Stam *et al.* 1988).

It is unlikely that adverse glacial conditions, as such, are responsible for the relative poverty of the Warm-temperate NW Atlantic region, as its present flora would not have been forced to extreme latitudinal displacement or area compression (Figs 12, 13). In this region the main difference between glacial and interglacial periods is the strongly enhanced seasonality during interglacials, and it is an attractive idea that this caused extinction, of all but the most eurythermous species. In the NW Pacific, strong seasonality (which also persisted to some extent during glacial times) apparently did not prevent the development of a rich flora of eurythermous endemic species. This suggests another reason for the floristic poverty of the

---

Fig. 22. Distribution of *Cladophora pellucida* and some narrowly related species. -● = *C. pellucida*; ✓ = *C. pellucidoidea*; ● = *C. radiosa*; ● = *C. feredayi*; ● = *C. "pellucida"* from Sydney; ● = *C. "pellucida"* from the Solomon Islands (Womersley & Bailey, 1970); ●- = *C. japonica*; ● = *C. graminea*. Stippled area in the N Atlantic: potential thermal distribution area of *C. pellucida*. Hatched area in the N Atlantic: potential thermal area of *C. pellucidoidea*. Stippled area in the Southern Hemisphere: potential thermal area of *C. feredayi*. Thick lines; limiting winter isotherms. Thin lines: limiting isotherms of the warmest months of the year. Limiting isotherms in the N Pacific are tentative (query marks) as experimental evidence is lacking. See also text to Fig.2. Distribution according to Abbott & Hollenberg (1976); Baardseth (1941); Cambridge *et al.* (1989a); Chamberlain (1965); Cordero (1977); Dawson (1951); Hanisak & Blair (1988); van den Hoek (1963, 1982c); van den Hoek & Womersley (1984); Kang (1966); Konno (1973); Modelo & Umezaki (1987); Nakaniwa (1975); Sakai (1964); Seagrief (1984); Silva *et al.* (1987)

Fig. 23. Reconstructed distribution of *Cladophora pellucida* and some narrowly related species during the last glacial maximum. For an explanation of symbols, see text to Figs 3, 22

Warm-temperate NW Atlantic region: namely, the adverse nature of the long sediment shores, with few scattered offshore reefs.

Species endemic to the NW Atlantic coast, such as *Grinnellia americana* (Fig. 13), could potentially inhabit a broad area along NW African coasts and in the Mediterranean (Fig. 12), an area which would, however, be considerably reduced during glaciations (Fig. 13). Its absence there indicates that the central Atlantic Ocean can function as a barrier to long distance dispersal.

The rich Warm-temperate NE Atlantic region is inhabited by amphi-Atlantic, cold-to-warm temperate species in the northern portions (Figs 4, 6), and by generally amphi-Atlantic tropical-to-warm temperate species in the southern portions (Figs 18, 20). The region's primary feature is, however, the occurrence of stenothermous endemic species, which are excluded from the Warm-temperate NW Atlantic region (with its pronounced seasonality) (Figs 8, 10, 22) because of their narrow thermal survival range. Some species in this category can probably be interpreted as belonging to pairs of sister species, each pair consisting of a stenothermous NE Atlantic species with a "cooler" warm-temperate range, and a stenothermous NW Atlantic species with a "warmer" tropical range. The evolutionary advantage of the "cooler" species would be its capacity to occupy the vast rocky coasts of the Warm-temperate NE Atlantic region, and that of the "warmer" species to occupy the vast Caribbean coasts. An example of such a species pair is that of *Cladophora pellucida* (NE Atlantic) and *C. pellucidoidea* (NW Atlantic) (Figs 22, 23). The steepening of the latitudinal temperature gradient, since the Middle Miocene, could have created the central Atlantic barrier, after which the two sister species could have diverged both genetically (Bot *et al.* 1989c), and with regard to their temperature responses (Cambridge *et al.* 1989a). In other examples, the populations on both sides of the ocean diverged only in their temperature responses, but not in their morphologies (e.g. *Cladophora coelothrix*, Fig. 20) (Cambridge *et al.* 1987).

An additional source of warm-temperate NE Atlantic endemics could be sought in the close relatives inhabiting NE Pacific shores, that possibly succeeded in reaching the NE Atlantic after the opening of the Bering Seaway in the Pliocene (3.5 Ma ago). Examples are *Polyneura hilliae* (with the NE Pacific *P. latissima* as a relative) and *Cladophora hutchinsiae* (which species is also recorded from the NE Pacific) (Scagel *et al.* 1986; Scagel 1966). We suspect that many of the numerous endemic species of the NE Pacific coast have temperature responses comparable to those of the NE Atlantic endemics; these responses exclude them from NW Pacific coasts which have a pronounced seasonality. Experimental evidence in support of this hypothesis is lacking.

Still other endemic species of the Warm-temperate NE Atlantic region are entirely restricted to it, and lack close relatives elsewhere; examples are *Saccorhiza polyschides* and species of the related genus *Phyllariopsis* (Breeman 1988).

The reconstructed glacial distributions of various representatives of the Warm-temperate NE Atlantic region, strongly suggest that the glaciations did not adversely affect its flora (Figs 7, 9, 11, 19, 21), although they caused latitudinal shifts in distribution in the Atlantic Ocean, and longitudinal distributional shifts in the Mediterranean. It is probable that even the Canarian and Azorean seaweed floras remained unaffected by glaciation; however, the thermal isolation of warm-water, tropical-to-subtropical species in the Mediterranean, by cold upwelling in the Atlantic entrance of this sea, was much enhanced, and affected more species during the glaciations (Figs 19, 21). After the glaciations, populations of cool water species could also have formed disjunct populations in cooler parts of the Mediterranean (Figs 3, 5, 7): for instance, *Laminaria ochroleuca* and *Saccorhiza polyschides* in scattered, cool, deep (60-100 m) stations of the W Mediterranean (van den Hoek 1982a). Possibly this thermal isolation may have promoted speciation in the Mediterranean, and the origin of endemic species.

The reconstructed glacial distributions of tropical and tropical-to-warm temperate species, do not support the hypothesis of massive extinctions, by lethal low winter temperatures during the glaciations, of tropical species along Atlantic and Pacific east coasts (Figs 15, 17, 19). This hypothesis was invoked in earlier discussions to explain the low diversity of seaweed floras on these coasts (van den Hoek 1975). An alternative explanation for the absence of tropical seaweeds with a W Atlantic-Indo-W Pacific distribution (e.g. *Dictyosphaeria cavernosa*, Fig. 16), could be the progressive extinction of tropical seaweeds, resulting from the reduction in complexity and length of the E Atlantic (and E Pacific) tropical coasts during the Miocene, and the concomitant steep cooling trend (Thunnell & Belyea 1982; Kennett 1982; Prud'homme van Reine & van den Hoek 1988). This temperature lowering could also explain the passage of amphipolar species, such as *Rhodochorton purpureum*, across the equator. The reconstructed glacial distribution of the latter species indicates that it could not have survived equatorial summer temperatures of glacial periods (Fig. 7).

The Mediterranean, especially the Eastern Mediterranean, is apparently capable of harbouring tropical and tropical-to-warm temperate seaweeds (Figs 14-17); the success of recent tropical immigrants through the Suez Canal, provides concrete examples of this (Lipkin 1972; Mayhoub 1976). The paucity of tropical and subtropical seaweeds presently in the Mediterranean is probably not, or only partially, the result of glacial extinctions; rather it is indicative of the fact that the Mediterranean was refilled, after the Messinian salinity crisis (Miocene, ca 5 Ma ago) with a warm temperate NE Atlantic flora, poor in species with a tropical extension. The cooler conditions in the W Mediterranean allow the growth and survival, there, of additional warm temperate species with cooler ranges, and this is probably the cause of its flora being richer than that of the E Mediterranean.

REFERENCES

Abbott IA, Hollenberg GJ (1976) Marine algae of California. Stanford Univ Press

Alvarez M, Gallardo T, Ribera MA, Gomez Garreta A (1988) A reassessment of northern Atlantic biogeography. Phycologia 27: 221-233

Baardseth E (1941) The marine algae of Tristan da Cunha. Results Norw Scient Exped Tristan da Cunha 1937-1938, no 9: 1-173

Berggren WA, Schnitker D (1983) Cenozoic marine environments in the North Atlantic and the Norwegian-Greenland Sea. In: Bott MPH, Saxov S, Talwani M, Thiede J (eds) Structure and development of the Greenland-Scotland Ridge. Plenum Press, New York, p 495-548

Boekschoten GJ, Best M Wijsman (1981) Pocillopora in the Miocene reef at Baixo, Porto Santo (Eastern Atlantic). Proc Kon Ned Akad Wetensch Ser B 84: 13-20

Boekschoten GJ, Best M Borel (1988) Fossil and recent shallow water corals from the Atlantic islands off western Africa. Zoöl Med 62: 99-111

Bot PVM, Stam WT, Boele-Bos SA, Hoek C van den, Delden W van (1989a) Biogeographic and phylogenetic studies in three North Atlantic species of *Cladophora* (Cladophorales, Chlorophyta) using DNA-DNA hybridization. Phycologia 28: 159-168

Bot PVM, Holton RW, Stam WT, Hoek C van den (1989b) Molecular divergence between North-Atlantic and Indo-Pacific *Cladophora albida* (Cladophorales, Chlorophyta) isolates as indicated by DNA-DNA hybridization. Mar Biol 102: 307-313

Bot PVM, Brussaard CPD, Stam WT, Hoek C. van den (1989c) Evolutionary relationships between four species of *Cladophora* (Cladophorales, Chlorophyta) based on DNA-DNA hybridization. J Phycol (submitted)

Breeman AM (1988) Relative importance of temperature and other factors in determining geographic boundaries of seaweeds: experimental and phenological evidence. Helgol Meeresunters 42: 199-241

Breeman AM (1990) Expected effects of changes in seawater temperatures on the geographic distribution of seaweed species. In: Wolff WJ, Brouns JJWM, Beukema JJ (eds). Expected effects of climatic changes on marine coastal ecosystems. Balkema Rotterdam (in press)

Breeman AM, Bos S, Essen S van, Mulekom LL van (1984) Light-dark regimes in the intertidal zone and tetrasporangial periodicity in the red alga *Rhodochorton purpureum*. Helgol Meeresunters 38: 365-387

Breeman AM, Hoopen A ten (1981) Ecology and distribution of the subtidal red alga *Acrosymphyton purpuriferum* (J. Ag.) Sjøst (Rhodophyceae, Cryptonemiales). Aquat Bot 11: 143-160

Breeman AM, Meulenhoff EJS, Guiry MD (1988) Life history regulation and phenology of the red alga *Bonnemaisonia hamifera*. Helgol Meeresunters 42: 535-551

Briggs JC (1974) Marine zoogeography. McGraw-Hill, New York.

Cambridge ML, Breeman AM, Hoek C van den (1989a) Temperature responses of European and American warm temperate *Cladophora* species in relation to their distribution in the North Atlantic Ocean: *C. pellucida*, *C. hutchinsiae*, and *C. pellucidoidea*. Aquat Bot (submitted)

Cambridge ML, Breeman AM, Hoek C van den (1990) Temperature responses and distribution limits of two temperate *Cladophora* species in the North Atlantic. Phycologia 29: 74-85

Cambridge ML, Breeman AM, Kraak S, Hoek C van den (1987) Temperature responses of tropical to warm-temperate *Cladophora* species in relation to their distribution in the North Atlantic Ocean. Helgol Meeresunters 41: 329-354

Cambridge ML, Breeman AM, Oosterwijk R, Hoek C van den (1984) Temperature responses of some North Atlantic *Cladophora* species (Chlorophyceae) in relation to their geographic distribution. Helgol Meeresunters 38: 349-363

Chamberlain YM (1965) Marine algae of Gough Island. Bull Brit Mus Nat Hist 3: 175-232, pl. 16-19

CLIMAP-project members (1981) Seasonal reconstructions of the earth's surface at the last glacial maximum. The Geological Society of America Map and Chart Series, MC-36

Cordero PA (1977) The marine algae of Batan Island, Northern Philippines. II Chlorophyceae. Fish Res J Philipp 2: 19-55
Dawson EY (1951) A further study of upwelling and associated vegetation along Pacific Baja California, Mexico. Sears Found J Mar Res 10: 39-58
Enomoto S, Okuda K (1981) Culture studies of *Dictyosphaeria* (Chlorophyceae, Siphonocladales) 1 Life history and morphogenesis of *Dictyosphaeria cavernosa.* Jap J Phycol 29: 225-236
Gorshkov SG (ed) (1976) World Ocean Atlas 1: Pacific Ocean. Pergamon, Oxford
Gorshkov SG (ed) (1978) World Ocean Atlas 2: Atlantic and Indian Oceans. Pergamon, Oxford
Gorshkov SG (ed) (1980) World Ocean Atlas 3: Arctic Ocean. Pergamon, Oxford
Griffiths RC, Simpson JG (1972) Upwelling and other oceanographic features of the coastal waters of Northeastern Venezuela. Serie recursos y explotacion pesqueros (Venezuela, Ministerio de Agricultura y Cria Oficina Nacional de Pesca) 2(4): 3-72
Hanisak MD, Blair SM (1988) The deep-water macroalgal community of the East Florida continental shelf (USA). Helgol Meeresunters 42:133-163
Hoek C van den (1963) Revision of the European species of *Cladophora*. Brill, Leiden
Hoek C van den (1975) Phytogeographic provinces along the coasts of the northern Atlantic Ocean. Phycologia 14: 317-330
Hoek C van den (1982a) The distribution of benthic marine algae in relation to the temperature regulation of their life histories. Biol J Linn Soc 18: 81-144
Hoek C van den (1982b) Phytogeographic distribution groups of benthic marine algae in the North Atlantic Ocean. A review of experimental evidence from life history studies. Helgol Meeresunters 35: 153-214
Hoek C van den (1982c) A taxonomic revision of the American species of *Cladophora* (Chlorophyceae) in the North Atlantic Ocean and their geographic distribution. Verh Kon Ned Akad Wetensch. 2de reeks 78. North-Holland Publishing Cy, Amsterdam
Hoek C van den (1984) World-wide latitudinal and longitudinal seaweed distribution patterns and their possible causes, as illustrated by the distribution of Rhodophytan genera. Helgol Meeresunters 38: 227-257
Hoek C van den (1987) The possible significance of long-range dispersal for the biogeography of seaweeds. Helgol Meeresunters 41: 261-272
Hoek C van den, Breeman AM, Stam WT (1990) The geographic distribution of seaweed species in relation to temperature: present and past. In: Wolff WJ, Brouns JJWM, Beukema JJ (eds), Expected effects of climatic changes on marine coastal ecosystems. Balkema, Rotterdam (in press)
Hoek C van den, Womersley HBA (1984) The genus *Cladophora*. In Womersley HBA, The marine benthic flora of southern Australia. Part I. Woolman Government Printer, South Australia, p 185-213
Hutchins LW (1947) The bases for temperature zonation in geographical distribution. Ecol Monogr 17: 325-335
Joosten AMT, Hoek C van den (1986) World-wide relationships between red seaweed floras: a multivariate approach. Bot Mar 29: 195-214
Kang JW (1966) On the geographical distribution of marine algae in Korea. Bull Pusan Fish Coll 7(1): 1-25, pl, I-XII
Kawai H, Kurogi M (1982) Benthic marine algal flora on the Okhotsk coast of Hokkaido. Environmental Science Hokkaido 5: 79-90 (Japanese, with English summary)
Kennett JP (1982) Marine geology. Prentice Hall, Englewood Cliffs, NJ
Kim YH, Lee IK (1985) The structure analysis of intertidal algal community in Muchangpo, Western coast of Korea. Kor J Bot 28: 149-164 (Korean, with English summary)
Klein B (1987) The phenology of *Dumontia contorta* (Rhodophyta) studied by following individual plants *in situ* at Roscoff, northern Brittany. Bot Mar 30: 187-194
Konno K (1973) On the marine algae from the coast of Oga Peninsula facing the Japan Sea. Bull Jap Soc Phycol 21: 1-11
Lawson GW (1978) The distribution of seaweed floras in the tropical and subtropical Atlantic Ocean: a quantitative approach. Bot J Linn Soc 76: 177-193

Lawson GW, John DM (1987) The marine algae and coastal environment of Tropical West Africa (2nd ed). Beih Nova Hedw 93: 1-415
Lee YP, Lee IK (1982) Vegetation analysis of marine algae in Jeju Island. Proc Coll Natur Sci SNU 7: 73-91 (Korean, with English abstract)
Lindstrom SC (1987) Possible sister groups and phylogenetic relationships among selected North Pacific and North Atlantic Rhodophyta. Helgol Meeresunters 41: 245-260
Lipkin Y (1972) Marine algal and seagrass flora of the Suez Canal (The significance of this flora to the understanding of the recent migration through the canal). Isr J Zool 21: 405-446
Lüning K (1984) Temperature tolerance and biogeography of seaweeds: the marine algal flora of Helgoland (North Sea) as an example. Helgol Meeresunters 38: 305-317
Lüning K (1985) Meeresbotanik. Thieme, Stuttgart
Macintyre IG, Pilkey OH (1969) Tropical reef corals: tolerance of low temperatures on the North Carolina continental shelf. Science 17: 374-375
Maier I (1984) Culture studies of *Chorda tomentosa* (Phaeophyta, Laminariales). Br Phycol J 19: 95-106
Mayhoub H (1976) Recherches sur la végétation marine de la côte syrienne. Etude expérimentale sur la morphogénèse et le développement de quelques espèces peu connues. Thèse Université de Caen
McKenna MC (1983) Cenozoic paleogeography of North Atlantic land bridges. In: Bott MHP, Saxov S, Talwani M, Thiede J (eds) Structure and development of the Greenland Scotland Ridge. Plenum Press, New York, p 351-399
McLachlan J, Bird CJ (1984) Geographical and experimental assessment of the distribution of *Gracilaria* species (Rhodophyta, Gigartinales) in relation to temperature. Helgol Meeresunters 38: 319-334
Modelo RR, Umezaki I (1987) Green algae (Chlorophyta) in the Visayas, Philippines. In: Umezaki I (ed) Scientific survey of marine algae and their resources in the Philippine Islands. p 32-41
Nakaniwa M (1975) Marine algae along the coast of Ibaraki Prefecture. Bull Jap Soc Phycol 23: 99-110
Noda M (1987) Marine algae of the Japan Sea. Kazama Shobo, Tokyo
Novaczek I, Bird CJ, McLachlan J (1986) The effect of temperature on development and reproduction in *Chorda filum* and *C. tomentosa* (Phaeophyceae, Laminariales) from Nova Scotia. Can J Bot 64: 2414-2420
Olsen-Stojkovich J, West JA, Lowenstein JM (1986) Phylogenetics and biogeography in the Cladophorales complex (Chlorophyta): some insights from immunological distance data. Bot Mar 29: 239-249
Olsen JL, Stam WT, Bot PVM, Hoek C van den (1987) scDNA-DNA hybridization studies in Pacific and Caribbean isolates of *Dictyosphaeria cavernosa* (Chlorophyta) indicate a long divergence. Helgol Meeresunt 41: 377-383
Pérès JM (1985) History of the Mediterranean biota and the colonization of the depths. In Margalef R (ed) Key Environments, Western Mediterranean. Pergamon, Oxford, p 198-232
Phillips RC, Springer VG (1960) Observations on the offshore benthic flora in the Gulf of Mexico off Pinellas County, Florida. Amer Mid Nat 64: 362-381
Pot R, Klein AWO, Rietema H, Hoek C van den (1989) A field study on the growth and development of *Dumontia contorta*. Helgol Meeresunters 42: 553-562
Prud'homme van Reine WF (1988) Phytogeography of seaweeds of the Azores. Helgol Meeresunters 42: 165-185
Prud'homme van Reine WF, Hoek C van den (1988). Biogeography of Cape Verdian seaweeds. Cour Forsch-Inst Senckenberg 105: 35-49
Prud'homme van Reine WF, Hoek C van den (1989) Biogeography of Macaronesian seaweeds. Cour Forsch-Inst Senckenberg 106 (in press)
Ricker RW (1987) Taxonomy and biogeography of Macqarie Island seaweeds. British Museum (Natural History), London.
Rietema H (1982) Effects of temperature and photoperiod on *Dumontia contorta* (Rhodophyta). Mar Ecol Prog Ser 8: 187-196

Rietema H, Breeman AM (1982) The regulation of the life history of *Dumontia contorta* in comparison to that of several other Dumontiaceae (Rhodophyta). Bot Mar 25: 569-576

Rietema H, Hoek C van den (1981) The life history of *Desmotrichum undulatum* (Phaeophyceae) and its regulation by temperature and light conditions. Mar Ecol Prog Ser 4: 321-335

Rietema H, Hoek C van den (1984) Search for possible latitudinal ecotypes in *Dumontia contorta* (Rhodophyta). Helgol Meeresunters 38: 389-399

Ruyter van Steveninck ED de, Breeman AM (1987) Deepwater vegetations of *Lobophora variegata* (Phaeophyceae) in the coral reef of Curaçao: population dynamics in relation to mass mortality of the sea urchin *Diadema antillarum*. Mar Ecol Prog Ser 36: 81-90

Sakai Y (1964) The species of *Cladophora* from Japan and its vicinity. Scient Pap Inst Algol Res Hokkaido Univ 5(1): 1-104, pl I-XVII

Sano O, Ikemori M, Arasaki S (1981) Distribution and ecology of *Acetabularia calyculus* along the coast of Noto Peninsula. Jap J Phycol 29: 31-38

Scagel RF (1966) Marine algae of British Columbia and Northern Washington, Part I. Chlorophyceae (green algae). National Museum of Canada Bull no 207, Biol Series No 74: 1-257

Scagel RF, Garbary DJ, Golden L, Hawkes MW (1986) A synopsis of the benthic marine algae of British Columbia, northern Washington and Southeast Alaska. Phycological Contribution No. 1, Dept of Botany, Univ of British Columbia Vancouver

Seagrief SC (1984) A catalogue of South African green, brown and red marine algae. Mem Bot Surv S Afr 47: 1-72

Searles RB (1984) Seaweed biogeography of the Mid-Atlantic coast of the United States. Helgol Meeresunters 38: 259-271

Silva PC, Meñez EG, Moe RL (1987) Catalog of the benthic marine algae of the Philippines. Smithsonian Contrib Mar Sci 27: 1-179

South GR (1987) Biogeography of the benthic marine algae of the North Atlantic Ocean. Helgol Meeresunters 41: 273-282

South GR, Tittley I (1986) A checklist and distributional index of the benthic marine algae of the North Atlantic Ocean. Huntsman Marine Laboratory and British Museum (Natural History), St. Andrews and London

Stam WT, Bot PVM, Boele-Bos SA, Rooy JM van, Hoek C van den (1988) Single copy DNA-DNA hybridizations among five species of *Laminaria* (Phaeophyceae): phylogenetic implications. Helgol Meeresunters 42: 252-267

Stegenga H (1978) The life histories of *Rhodochorton purpureum* and *Rhodochorton floridulum* (Rhodophyta, Nemaliales) in culture. Br Phycol J 13: 279-289

Thiede J, Eldholm O (1983) Speculations about the paleodepth of the Greenland-Scotland Ridge during late Mesozoic and Cenozoic times. In: Bott MPH, Saxov S, Talwani M, Thiede J (eds), Structure and development of the Scotland-Greenland ridge, new methods and concepts. Plenum Press, New York, p 445-456

Thunnell R, Belyea P (1982) Neogene planktonic foraminiferal biogeography of the Atlantic Ocean. Micropaleontology 28: 381-398

Tseng CK (1983) Common seaweeds of China. Science Press, Bejing

Walker BD, Roberts HH, Rouse LJ, Huh OK (1982) Thermal history of reef-associated environments during a record cold-air outbreak event. Coral Reefs 1: 83-87

West JA (1969) The life histories of *Rhodochorton purpureum* and *Rhodochorton tenue* in culture. J Phycol 5:12-21

West JA (1970) A monoecious isolate of *Rhodochorton purpureum*. J Phycol 6: 368-370

West JA (1972) Environmental regulation of the reproduction in *Rhodochorton purpureum*. In: Abbott IA, Kurogi M (eds) Contribution to the systematics of the benthic marine algae of the North Pacific. Jap Soc Phycol , Kobe, p 213-230

Womersley HBA, Bailey A (1970) Marine algae of the Solomon Islands. Phil Trans Roy Soc London B 259: 257-352

Yarish C, Breeman AM, Hoek C van den (1984) Temperature, light, and photoperiod responses in some Northeast American and West European endemic rhodophytes in relation to their geographic distribution. Helgol Meeresunters 38: 273-304

Yarish C, Breeman AM, Hoek C van den (1986) Survival strategies and temperature responses of seaweeds belonging to different distribution groups. Bot Mar 29: 215-230
Yarish C, Kirkman H, Lüning K (1987) Lethal exposure times and preconditioning to upper temperature limits of some temperate North Atlantic red algae. Helgol Meeresunters 41: 323-327

BEHAVIOURAL, MORPHOLOGICAL AND GENETIC CHANGES IN SOME NORTH ATLANTIC POPULATIONS OF THE BARNACLE *SEMIBALANUS BALANOIDES*

E. Bourget, N. Martel, L. Lapointe and D. Bussières
GIROQ, Département de biologie
Université Laval, Ste-Foy, Québec G1K 7P4
CANADA

Abstract - We use the intertidal barnacle *Semibalanus balanoides* as a model to examine geographical patterns of enzyme polymorphism in a cosmopolitan organism in the North Atlantic. On a smaller geographical scale, we focus on North American populations and show that behavioural, morphological and genetic differences occur between the Gulf of St. Lawrence and the North Atlantic populations of *Semibalanus*. We discuss the possible role of selective environmental pressures (e.g. ice scouring), and relative isolation of the Gulf population, particularly in relation to differences in larval settling behaviour between the Gulf of St. Lawrence (where the animals are cryptic), and the open Atlantic coast (where settlement is greatest on exposed substrata). The factors considered are latitude, water circulation and possible past and present dispersal routes.

INTRODUCTION

Very few marine algal species are preserved as fossils except those with hard skeletal parts (e.g. the frustule of diatoms, corallines), and a few sparse individuals. These species are amenable to evolutionary biogeographical modelling, but they do not constitute the bulk of the marine algal species. Thus, algal biogeography models can gain from those of co-occurring animals, which are better preserved in the fossil record, or which have been more thoroughly studied. It is solely in this context that we approach the recent post-glacial history of *Semibalanus* (= *Balanus*) *balanoides* in the North Atlantic, and particularly on the North American shores.

Semibalanus balanoides is one of the most eurythermal marine shore invertebrates. It is common on most rocky shores of the North Atlantic, north of Cape Hatteras (Wells *et al.* 1960), and also of northwestern Spain (see Fig. 1). On the east coast of the Pacific, it occurş north of Sitka (Pilsbry 1916; Shelford 1930), on the Alaskan Peninsula, and on the southern coast of Alaska. It has been reported from the shores of Japan (Hiro 1935), but is absent from the high Arctic. Its northern records were extensively presented by Barnes (1957), and critically reviewed by Petersen (1966). Its northern limit was correlated with the maximum summer mean monthly temperature of 4°C (which still allows for successful

NATO ASI Series, Vol. G 22
Evolutionary Biogeography of the
Marine Algae of the North Atlantic
Edited by D. J. Garbary and G. R. South

reproduction (Hutchins 1947)), and also with the southern limit of the pack ice (Barnes 1957). According to Feyling-Hanssen (1953), *Semibalanus* probably had a truly circumpolar distribution during a warmer interglacial period. The first fossil *Semibalanus* was observed in middle Miocene.

Crisp (1977), Scheltema (1977) and others (see Battaglia & Beardmore 1977) have discussed the likelihood that species with large-scale epicontinental distributions (thousands of km), and with limited dispersal capabilities (e.g. 5 or 6 weeks of meroplanktonic life compared to years, for teleplanic larvae), will develop clinal patterns, or genetically different populations. This may be caused by limited gene flow between distant populations along the latitudinal gradient, or by biogeographical barriers (e.g. large distances, strong directional flow). The results of Flowerdew (1983a, b), on the effective number of alleles

Fig. 1. Worldwide distribution of *Semibalanus* (= *Balanus*) *balanoides* (modified from Feyling-Hanssen 1953)

in *Semibalanus* populations of USA, Newfoundland and Iceland, are in accordance with this model. The relative frequencies of Gpi^{182} and Mpi^{100} alleles in the population are also in accordance with the above model (Dando 1987). Flowerdew (1983a, b) presents evidence for three or four races of *S. balanoides* throughout its entire range of distribution (Europe, Pacific, USA/Newfoundland/Ireland), according to which, the Newfoundland and Iceland populations are considered a subpopulation of the American Atlantic population (Flowerdew 1983b). The enzyme data on latitudinal variations along the North American seaboard remains very fragmentary, as barnacles from the shores of the Gulf of St. Lawrence, Labrador, Baffin Is., as well as from western and eastern Greenland (thousands of km of arctic and subarctic shores, covering the entire northern half of the distribution of *Semibalanus* along the eastern seaboard of North America) remain unstudied.

Recent studies on the barnacle *Semibalanus balanoides* in eastern Canada have shown that morphological, behavioural and genetic changes have occurred in the Gulf of St. Lawrence, compared to the adjacent Atlantic population. This paper summarizes the evidence that has accumulated to date, supporting the hypothesis of a separate population in the Gulf. This leads us to speculate about the time scale, and possible mechanisms by which a distinct population could become established and/or maintain itself in the Gulf, and possibly in subarctic and arctic regions (sensu Dunbar 1968). In the particular context of this workshop, in which the time scales are in millions of years, and the spatial scales are those of oceans, or indeed, of the world ocean, this paper illustrates that (1) changes in populations can occur very quickly given the right ecological conditions, (2) the spatial scales involved may be relatively small, not necessarily whole oceans, (3) ecological factors are important factors of change.

MATERIALS AND METHODS

Further field experiments on settlement cues

Following the results of Letourneux & Bourget (1988), experiments were planned to test the larval response to potential settlement cues (other than barnacle extracts, see Crisp & Meadows 1962, 1963) used at large scale ($\approx$1 m) in nature. Since settling larvae use their antennae for tactile exploration of the substratum surface (Nott & Foster 1969; Walker & Lee 1975) to detect the appropriate characteristics for a settling site, all substances assayed were adsorbed onto the surface of the test panels. These surfaces, consisting of 20 cm x 20 cm Arborite panels (with a central experimental quadrat of 10 X 10 cm), were coated with the assayed substances and allowed to dry, prior to being offered, with control panels (see below), to settling larvae during one tidal cycle. The settled larvae were counted to assess the intensity of the settling response.

Two types of experiments were carried out between 20-31 May 1989, at St. Andrews, New Brunswick (Fig. 2). In the first set of experiments, periphyton was collected above and below the upper limit of barnacles. This material was crudely filtered to remove the larger particulate material, and applied onto test panels in concentrations (C) (smaller than natural concentrations) equivalent to C, C/2, C/4. To test the hypothesis that cellular extracts might be responsible for promoting larval settlement, the material was disrupted using an ultrasonificator, and coated onto Arborite test panels in concentrations (C). The control panels were coated with artificial seawater, filtered using GF/C (1.2 μm) glass fibre filters. This experiment was repeated 6 times using 10 replicates per treatment, and per experiment. In a second set of experiments, the hypothesis, that *Achnanthes brevipes parvula* cells or extracts are involved in promoting settlement of barnacle larvae, was tested in the following way: a pure strain of *Achnanthes brevipes parvula*, obtained from the Culture Collection of Algae, University of Texas at Austin (USA), was cultured to densities approximating 200,000 cells ml^{-1}. The pure cultures were applied directly, or after ultrasonification (as above), onto test panels. Three types of controls were used: (1) the

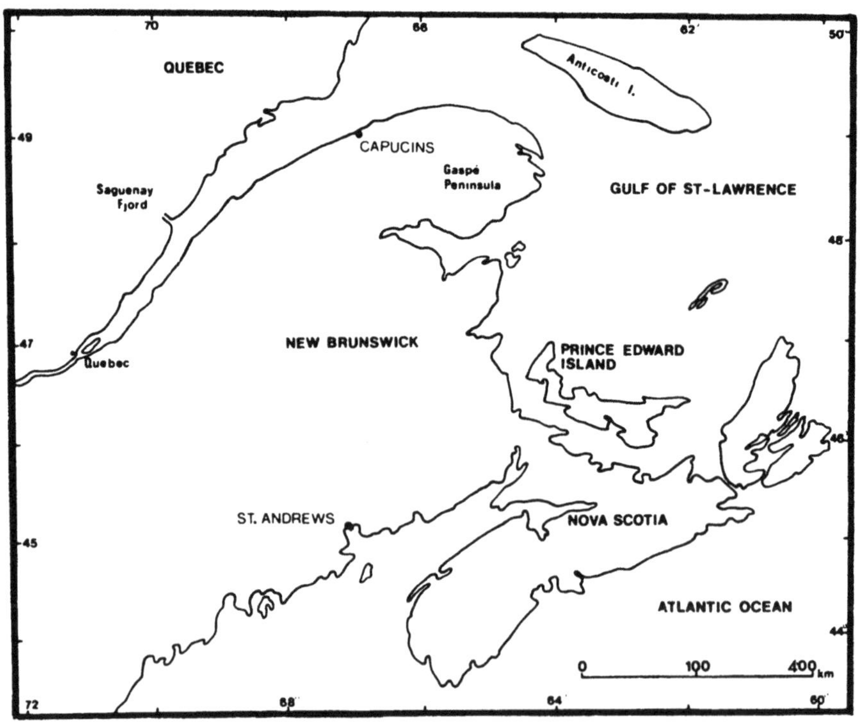

Fig. 2. A. Study sites, at Capucins, on the shores of the Gulf of St. Lawrence, and at St. Andrews, N. B., on the Atlantic coast

culture medium of *Achnanthes* (CMA), (2) cultures of *Phaeodactylum tricata*, a pelagic diatom species (density approx. 200,000 cells ml^{-1}), and (3) seawater filtered through 1.2 µm glass filters. For comparison, all experiments were carried out in the presence of panels similarly treated with barnacle extracts (see Chabot & Bourget 1988). The experiments were repeated 6 times.

Genetic studies on Gulf and Atlantic coast populations

During the spring of 1988, on exposed and cryptic microhabitats, barnacles were collected from the shores of the Gulf of St. Lawrence, at Capucins, and from St. Andrews, N.B. (Fig. 2). These animals were maintained at the Huntsman Marine Science Centre (St. Andrews, N.B.) in flowing seawater, for a period of two weeks (May 88). This acclimation period was followed by a 24 h unfed period, after which they were killed, by freezing on dry ice. The samples were kept in the laboratory at -70°C for a maximum of 6 months. Enzymatic polymorphism was studied by electrofocusing on polyacrylamide ultrathin gels (300 µm). The pH gradients for allozyme preparation were obtained using a mixture of Pharmalyte of pH 2.5-5/5-6 (1:1), for the study of PGM and MPI, and of 5-6 for CAE-3 and GPI. A power supply (LKB 2197) was used with a Hoefer Scientific Isobox. After a 5 min prefocussing at 5W, the protein extracts were applied onto the surface of the gel and focalized during 50 min by increasing the voltage by 5 to 20W, at 10 min intervals, up to the maximum of 2000V. The specific staining was carried out immediately after focalisation was completed.

Morphological studies on Gulf and Atlantic coast populations

Juvenile (<1 yr old) and adult (> 1 yr old) barnacles, from crevices and exposed surfaces at Capucins and at St. Andrews, were collected at random during autumn 1987. The barnacle shell parietes are very plastic. The useful taxonomical criteria are the internal structure of the shell walls (Darwin 1854, Bourget 1977, 1987), the shape of the opercular plates, as well as the mouth parts (see Dando & Southward, 1980). Fig. 3 illustrates the various measurements taken on the scutum (Fig. 3 A), tergum (Fig. 3 B), mandible (Fig. 3 C) and maxilla (Fig. 3 D), in order to determine the extent of morphological differences between the Gulf and Atlantic coast barnacle populations.

RESULTS AND DISCUSSION

The distribution of adult barnacles on the eastern seaboard of Canada

In the Gulf of St. Lawrence, *Semibalanus* occupy the walls of crevices and the basal portion on the sides of boulders; they are seldom observed elsewhere. When they are, however, it is usually in association with very small cracks and depressions in the substratum (see Bergeron & Bourget 1986). Similar distribution patterns have been reported for *Semibalanus* on the east coast of Greenland (Feyling-Hanssen 1953; Petersen 1962,1966), suggesting similar causal relationships. In contrast, on the Canadian Atlantic and New England coasts, *Semibalanus* is observed on exposed substrata (Bourget 1988). This difference in pattern was, at first, thought to result from the direct action of strong ice scouring in the Gulf (Stephenson & Stephenson 1972; Bergeron & Bourget 1984),

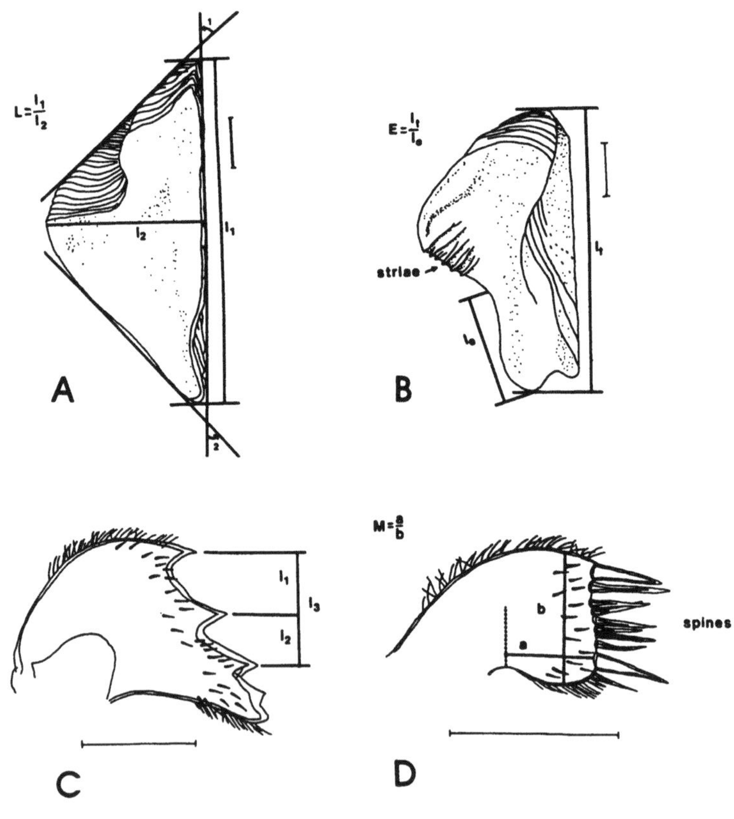

Fig. 3. Morphometric measurements taken on the scutum (A), tergum (B), mandible (C), and maxilla (D) of *Semibalanus balanoides*

compared to infrequent and less damaging ice scouring effects on the open Atlantic coast (see Wethey 1983). However, further studies (Bergeron & Bourget 1986; Chabot & Bourget 1988) showed that these distribution differences were largely due to differences in larval behaviour at settlement.

Differences in microhabitat selection at settlement

Three types of evidence support the hypothesis that the behaviour of the larvae at settlement is responsible for the observed patterns of distribution: (1) in uncontrolled situations in nature, *Semibalanus* larvae in the Gulf settle preferentially (> 95%, see Bergeron and Bourget 1986; Chabot & Bourget 1988) in crevices. (2) in semi-controlled situations, settlement in natural undenuded crevices, or in crevices denuded and treated (chemically, with NaOCl, and thermally, using blowtorches) to remove all possible chemical inducers of settlement (e.g. conspecifics, barnacle extracts), indicated a strong larval preference for crevices, and avoidance of exposed surfaces in the Gulf, and vice versa on the Atlantic coast (Bourget 1988). (3) in controlled situations, in artificial crevices with and without settlement cues (in the form of barnacle extracts and conspecifics), larval preference for exposed surfaces was maintained on the Atlantic coast (Chabot & Bourget 1988).

The experiments alluded to above indicated that the larvae are capable of selecting precise but different microhabitats in two geographically distinct regions. A key question, therefore, is: in the absence of known chemical inducers (extracts, conspecifics), what guides the larvae in making a precise microhabitat selection? The question is important in distinguishing between environmentally- or a genetically-controlled behaviour. This question was addressed by Letourneux & Bourget (1988), by examining the cues used by the settling larvae at three different (spatial and temporal) scales: at a scale of one meter, corresponding, in nature, to the selection of a precise level on the shore, presumably during the "broad exploration" phase (see Crisp 1974), (2) at a scale of one mm, corresponding to selection of potential sites during the "close exploration" phase, and (3) at a scale of tens of microns, corresponding to the selection of a microsite during the "final inspection" phase.

The experiments were carried out at St. Andrews, N. B. At the larger scales (> 1 mm), they involved comparisons of microcharacteristics of selected and unselected sites (biotic: populations of macroalgae, bacteria; physical: micropits, detritus, etc.). One of the major problems in studying larval selection at large scales (1 m) arises from the fact that it is virtually impossible to know whether a site not selected (i.e. without settled larvae) has been explored, and therefore has been discriminated against. This difficulty was overcome by Letourneux & Bourget (1988), by choosing for detailed analysis, sites immediately above (thus systematically rejected), and others below the upper limit of settlement (sites that were acceptable and, indeed, selected by the settling larvae). Indirectly, this experiment also points to the biotic and physical characteristics responsible for the zonation patterns on the shores, before modifying factors intervene. At the mm scale, it was assumed, on the basis of the exploratory behaviour of the larvae at settlement (see Crisp 1974), that all unoccupied sites within a mm of the precise settling site had been explored, and rejected, by the larvae. Thus, the characteristics of the substratum 1 mm in front of the larva, were compared, using the Scanning electron microscope, to those immediately surrounding the attached antennae of the settled larva.

Statistical analyses (logistic regressions) of the results clearly suggest that, at the large scale (1 m), either the larvae were discriminating against *Urospora wormskjoldii* , which was located above the upper limit of settlement of *Semibalanus*, or for an attached benthic diatom, *Achnanthes brevipes parvula,* which was located within the barnacle zone. A comparison of the microcharacteristics of the substratum inside and outside crevices in the Gulf of St. Lawrence at the time of settlement, suggested microhabitat selection based on detection of *Achnanthes* cells.

The most striking results of this study, however, were observed at smaller (1 mm and µm) scales. At mm scales, comparison of microcharacteristics of unselected and selected sites showed that the larvae were now discriminating against the diatom *Achnanthes,* and apparently selecting a clean site for attachment. An alternate hypothesis, however, is that the larvae were, themselves, ridding the surface of the substratum of organic and inorganic material, in order to increase adhesion, and facilitate the "landing" of the newly metamorphosed spat after rotation (see Letourneux & Bourget 1988).

Finally, the hypothesis that larvae were selecting clean surfaces, in order to maximize attachment by selecting topographically complex surfaces, was confirmed by comparing the microheterogeneity of sites (N = 60), immediately under the attached cyprid antennae, with that of adjacent sites. The results showed clearly that, at the site of attachment, the antennae were attached where microheterogeneity is greatest, presumably to limit the spread of cement, as well as to increase adhesion. Moreover, it was shown that the larvae start to make significant selections of microsites, when the metrics of various lengths, used to measure heterogeneity of the microsites (fractal dimensions: see Letourneux & Bourget 1988), reached approximately 35-45 µm, a size approximating the size of the antennular disc used by the larvae to attach to the substratum.

The settling response in relation to periphyton and particularly to *Achnanthes*

With respect to microhabitat selection, the most interesting scale remains the meter scale, and the potential response of the settling larvae to both periphyton and *Achnanthes*.

A series of experiments was, therefore, carried out to test the response of the larvae to these potential settling cues in nature. Test panels, treated with various concentrations of whole or disrupted periphyton and *Achnanthes* cells, and with various control substances (*Phaeodactylum* cells, filtered seawater, barnacle extracts), were offered to the settling larvae. The response was measured by counting the number of larvae, after exposure of the panels for one tidal cycle (see Materials and Methods).

The results shown in Fig. 4 indicate no significant difference between treatments, nor between treatments and controls, except for the treatment with barnacle extracts (Friedman 2-way analyses of variance, Conover 1980); thus, no significant effect of either periphyton or of *Achnanthes* was found in these experiments. The opposite results, obtained by Letourneux & Bourget (1988), would suggest that the larvae need the attached diatoms, rather than extracts or adsorbed cells on the substratum. However, in light of the hierarchical selection process exhibited by the larvae at settlement (see Bourget 1988 and Fig. 5), it is possible that, in this experiment, the larvae are subjected to two conflicting signals. During the broad exploration phase, the larva could be detecting the suitable cue (e.g. *Achnanthes*, its debris or extracts), but since, during close exploration, the larva is seeking a clean site, it cannot find one because the entire surface of the panel has been coated with cells and their extracts. This hypothetical, but not unlikely scenario, points to the difficulty of testing hypotheses involving different hierarchical scales, particularly when opposite responses are expected at each scale. The results of these experiments suggest that, for a meaningful test of settlement responses at the large scale, the appropriate small scale characteristics must also be provided. The latter results, however, could also be interpreted as meaning that the initial correlative responses, obtained between settlement and microalgae (Letourneux & Bourget, 1988), were biologically non significant, and that the behavioural differences observed were genetically controlled. Only further experiments can clarify the question.

Comparisons of populations using enzyme studies

The local distribution and behavioural differences observed between the populations of the Gulf of St. Lawrence, and the southern Atlantic coast of Canada (and probably also New England), suggest possible genetic differences between populations. A comparison of the *Semibalanus* populations of Capucins (Gulf of St. Lawrence) and St. Andrews, N. B. (Atlantic coast of Canada), was carried out using enzyme electrofocusing. After an initial screening of more than 40 enzymes, using starch gel electrophoresis (data graciously provided by R. Vadas, U. of Maine), 13 were further studied (18 loci); eventually, we concentrated our study on 4 potentially polymorphic loci - MPI, CAE-3, PGM, and GPI. Results of contingency comparisons of the allele frequencies indicate that allele frequencies at the MPI loci differed significantly between the two populations of *Semibalanus*. No intra-population differences (crevices vs open exposed surfaces) were observed at a given station. There were no significant inter- or intra-population differences at the other loci (Martel, Bourget and Chagnon, in prep). The results obtained for the Gulf and St. Andrews populations are compared to those of Flowerdew (1983b) and Dando (1987)

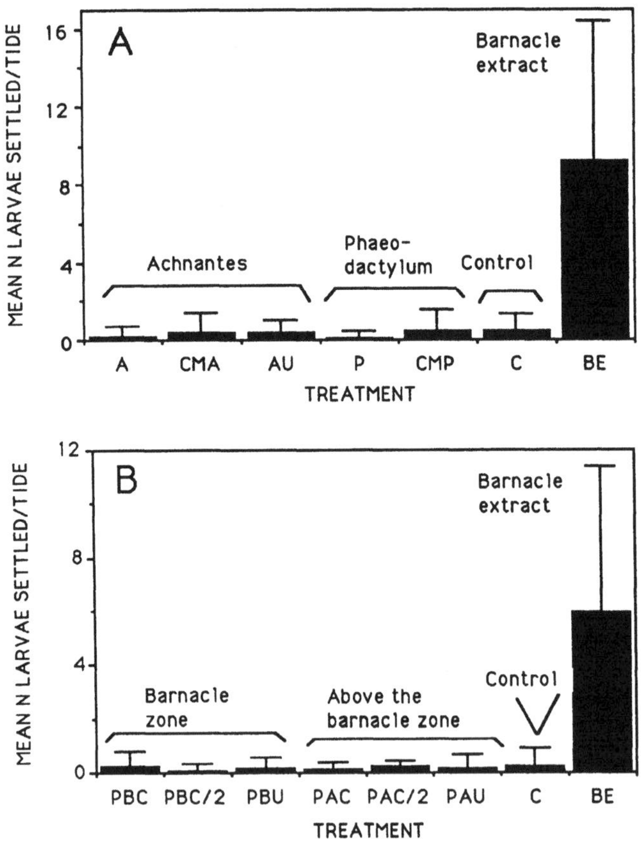

Fig. 4. Results of bioassays carried out to test the effectiveness of periphyton and the diatom *Achnanthes parvula* as settling cues for *Semibalanus balanoides*. A: *Achnanthes parvula*, CMA: *Achnanthes* culture medium, AU: *Achnanthes* treated ultrasonically, CMP: *Phaeodactylum* culture medium; PCB: Periphyton from the barnacle zone; PBU: Periphyton from the barnacle zone treated ultrasonically, PAC: Periphyton from above the barnacle zone, PAU: Periphyton from above the barnacle zone treated ultrasonically

(Fig. 6). Assuming correspondence between the GPI and MPI alleles identified in this study, and those of Flowerdew (1983b), the figure shows that the Gulf population is close to the Newfounland/Iceland group, whereas the St. Andrews population is closer to the New England population.

Table. 1. Comparison of medians and means of the various descriptors of shape of the scutum, tergum, mandible and maxilla of juvenile of adult barnacles between populations of Capucins (Gulf of St. Lawrence) and St. Andrews, N.B. (Atlantic coast). Values obtained from exposed and crevice microhabitats from a given locality were not significantly different and, therefore, were pooled for comparisons. Differences of medians (presented with range of values) were tested using chi square tests while differences between means (presented with s.d.) were tested using Student t tests. * $p < 0.05$

	ST. ANDREWS		CAPUCINS	
	Juveniles	Adults	Juveniles	Adults
Scutum		*	*	
Ratio L	1.97 (1.85-2.10)	2.12 (1.88-2.43)	2.12 (1.93-2.17)	2.05 (1.88-2.29)
	n=10	n=32	n=10	n=30
Angle 1	36.5^0 (32^0-44^0)	43.0^0 (37^0-49^0)	45.0^0 (43^0-47^0)	46.0^0 (40^0-58^0)
	n=10	n=31	n=10	n=30
Angle 2	35.5^0 (33^0-44^0)	39.0 (32^0-55^0)	37.5^0 (35^0-42^0)	46.0^0 (40^0-58^0)
	n=10	n=31	n=10	n=30
Tergum			*	
Ratio E	2.09 (1.93-2.62)	2.15 (1.56-2.59)	2.17 (1.89-2.37)	2.27 (2.01-2.62)
	n=10	n=31	n=10	n=30
N. striae	4 (3-5)	5 (4-6)	5 (3-7)	6 (4-8)
	n=30	n=30	n=30	n=29
Maxilla				
Ratio M		0.71 (0.57-0.81)		0.71 (0.65-0.85)
		n=14		n=18
N. spines		10 (7-14)		10 (9-12)
		n=20		n=20
Mandible				
Ratio L_1/L_3		0.55 ± 0.02		0.55 ± 0.03
		n=19		n=18
Ratio L_2/L_3		0.45 ± 0.03		0.45 ± 0.02
		n=19		n=18
Ratio L_1/L_2		1.23 ± 0.13		1.23 ± 0.12
		n=19		n=19

Comparisons of shell morphology

Superficially, the *Semibalanus* individuals from the Gulf of St. Lawrence differ markedly from those of the neighbouring Atlantic coast, by their larger size (diameter approx. 3 X larger) and much faster spat growth rate in their respective environments (St. Andrews, n=30: 0.018 ± 0.002 mm d^{-1}; Capucins, n=30: 0.10 ± 0.01 mm d^{-1}) . However, comparisons using measurements of the habitual taxonomic characters (relative dimensions of the scutum and tergum, number of striae on the tergum, number of spines on the maxilla, as well as the proportions of the maxilla and mandibular teeth (see Dando & Southward 1980)), after rejecting those characters that varied significantly during development (juvenile vs adult, e.g. ratio L of scutum at St-Andrews), showed that significant differences between the two populations were observed in the proportions of the tergum (Table 1).

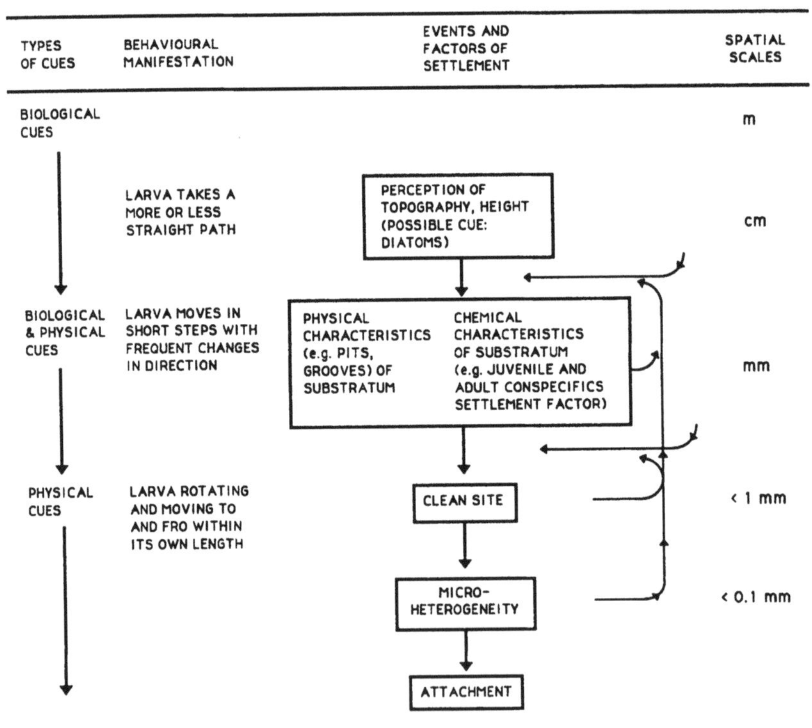

Fig. 5. Hierarchical sequence of events leading to the settlement of *Semibalanus balanoides* cypris larvae (modified after Bourget, 1989)

The distinction between the Gulf and adjacent southern North Atlantic populations of *Semibalanus*

Although only one convenient arthropod morphological character separates the two populations examined, the combination of enzyme characters and behavioural traits described above, suggests some distinctions between the two populations. However, the degree of separation between the Gulf and Atlantic coast populations, based on detailed analysis of Capucins and St. Andrews populations, (one out of four enzymes, one morphological character) is limited, and suggests distinctions below the species level, providing these distinctions were maintained spatially.

HISTORICAL CONSIDERATIONS. We can speculate about the historical and environmental events which might have led to the observed behavioural, morphological

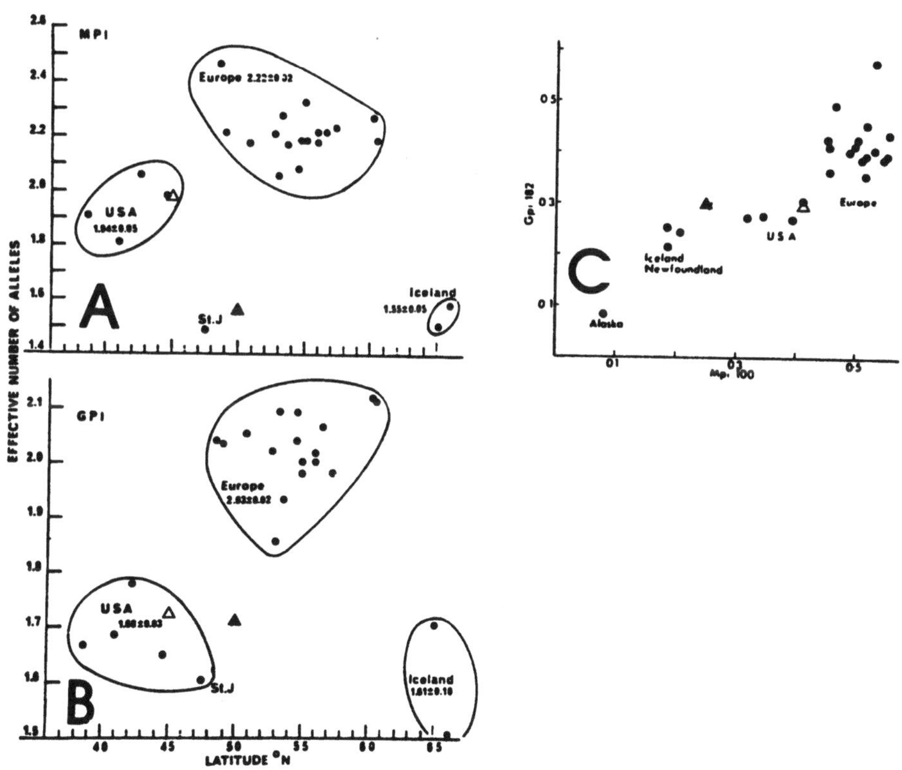

Fig. 6. Effective number of alleles at two polymorphic loci in relation to latitude (°N) in *Semibalanus balanoides* in the North Atlantic (A and C) . B. Relative frequencies of Gpi182 and Mpi100 alleles in the populations of *Semibalanus balanoides* . Modified from Flowerdew 1983a and Dando 1987 to include samples from Capucins (▲) and St. Andrews (△), N. B.

and genetic distinctions observed. First, we can consider the possible fate of *Semibalanus* during the last glaciation; then, since colonization is possible solely by larval dispersion, we can examine possible invasion routes, relative isolation of the Gulf system, as well as

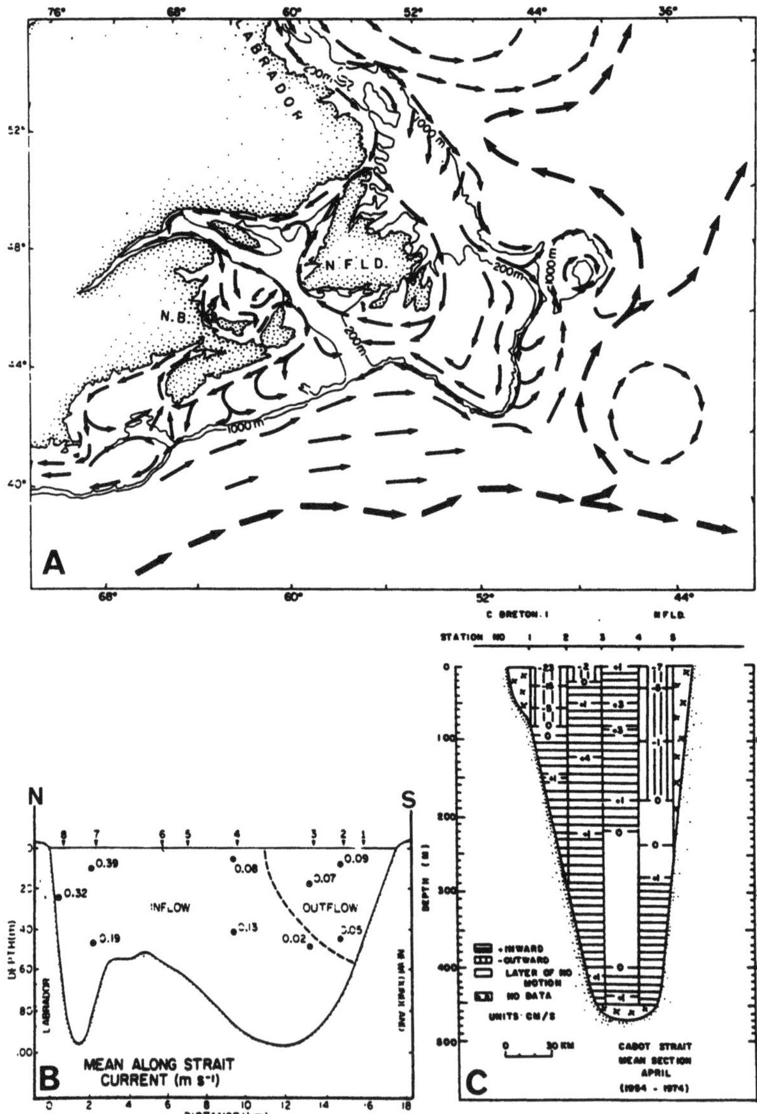

Fig. 7. Water circulation along the Atlantic coast of Canada. A. Model of surface circulation (from O'Boyle *et al.* 1984). B. Profile of current velocity in the Strait of Belle Isle in late summer 1980 (from Petrie *et al.* 1988). C. Profile of geostrophic current velocity in Cabot Strait in April- May (from El-Sabh 1977)

selective forces potentially leading to the patterns observed, particularly the behavioural ones.

Like most North Atlantic shore animals during the last glaciation, *Semibalanus* saw its northern limit of distribution shift considerably southward (most likely out of the Gulf of St. Lawrence), with the progression of the cold climate and water conditions associated with the Inlandis advance. Hutchins' (1947) analysis, showing that the northern limit of *Semibalanus balanoides* was set by the summer mean monthly maximum isotherm of 4°C, below which successful reproduction cannot occur, and studies by Barnes (1957), showing a close correspondence between the southern limit of the summer pack ice in the North Atlantic and the present limit of *Semibalanus*, support the hypothesis of a southern retreat of *Semibalanus* out of the Gulf during the last glaciation. Although fossil shells of *Semibalanus* were not observed in the remains of the Champlain Sea (which was invaded by the Goldthwait Sea some 11 000 yrs B. P.), the presence of *Balanus crenatus*, *Macoma balthica*, *Mytilus edulis*, *Hiatella arctica*, *Mya arenaria* and other normally co-occurring northern intertidal species (Dionne 1977; Wagner 1967, 1970) indicate suitable conditions for *Semibalanus* . They also suggest a rapid reinvasion, after the retreat of the Inlandis, most likely through Cabot Strait (see recent deglaciation maps by Fulton and Andrews 1987, and discussion on post-glacial patterns by Bousfield & Thomas 1975) by the most northern and cold resistant elements of the species, those which remained nearest to the glacial front.

CONDITIONS FOR THE ESTABLISHMENT OF DISTINCT POPULATIONS. At least two conditions are necessary for the establishment and maintenance of genetic differences in a system such as the Estuary-Gulf System: (1) strong selective pressures, and (2) relative genetic isolation from adjacent populations. The following discussion should not be interpreted as suggesting that the present behavioural patterns observed in the Gulf have emerged since the last glaciation (though they might have), but rather as showing (1) that for *Semibalanus*, the above two conditions have been met, at least since Post-hypsithermal times (3,000 yrs B. P.), and possibly for much longer, and (2) that environmental conditions prevailing since the Post-hypsithermal could have led to the present cryptic behaviour of the larvae in the Gulf.

First, it should be stressed that the cryptic distribution of *Semibalanus* appears to be a widespread trait in subarctic and Arctic populations. Secondly, this trait appears to be maintained over widely varying local environmental conditions, and despite large variations in the abundance of competent larvae (Bourget 1988).

SELECTIVE FORCES. It is not too difficult to imagine the possible historical factors which may have led to the differences observed and could still maintain them. The results of Archambault & Bourget (1983) and Bergeron & Bourget (1986), clearly indicate that, in the Estuary-Gulf System, major reductions in rocky shore population abundances occur during winter months, i.e. between December and May. Further analysis by Bourget *et al.* (1985) suggests that this reduction takes place at the time of ice breakup. Population studies on the barnacle population of the Gulf indicate that less than 5% of the larvae settle outside crevices; of these, more than 90% are killed during the first winter, before they are able to reproduce, compared to less than 30% in crevices (see Bergeron & Bourget 1986). These results show that there are strong selective pressures from winter conditions (probably ice scouring), on exposed *Semibalanus* populations in the Estuary-Gulf of St. Lawrence System, and, most likely, on more exposed northern shores as well. It seems a reasonable hypothesis to suggest that the observed behavioural patterns could have evolved from such intense selective pressures, providing that some degree of isolation existed from the non-ice-scoured populations.

RELATIVE ISOLATION OF THE GULF POPULATIONS. Only molecular studies, in regions of potential contact and overlap will be able to determine how distinct the Gulf and Atlantic populations remain in potential regions of overlap (e.g. Cabot and Belle Isle Straits). However, based on water current patterns and velocity during the meroplanktonic phase, it should be possible to estimate the relative importance of larval importation from the North Atlantic into the Gulf of St. Lawrence. As pointed out by Bousfield (1954), the spawning season of *Semibalanus* differs greatly along the east coast of North America. For the populations considered, it is reasonable to estimate the maximum duration of larval life of *Semibalanus* to be approximately six weeks. Thus, if six weeks is taken as the maximum duration of larval life in all regions considered here, then knowing the direction and velocity of residual currents across the Cabot and Belle Isle Straits, where water exchanges occur, it should be possible to estimate the likelihood of larvae entering the Gulf.

In Belle Isle Strait, the inflow of water originates from the Labrador current, which extends along the lower north shore of the Gulf and Estuary, as far as the mouth of the Saguenay Fjord (see Tremblay & Lauzier 1940; El-Sabh 1977; Petrie *et al.* 1988). The current data presented by Petrie *et al.* (1988) indicate that the mean inflow of water in July, August and September in the upper 20 m of the water column (Fig. 7 B), where most of the *Semibalanus* larvae are found (see Bousfield 1954, 1955), is approximately five times greater than the outflow. This ratio is increased if deeper water layers are considered in the calculations.

We can roughly estimate the mean maximum distance potentially travelled by *Semibalanus* larvae entering the Gulf through Belle Isle Strait. Taking 0.4 m s^{-1} (the mean maximum inflowing current speed through the Strait; Petrie *et al.* 1988) and a maximum duration larval life of 42 days (= 6 weeks), the larvae could travel up to 1078 km per breeding season (one per year). Obviously, the exceptionally strong current of Belle Isle Strait is not maintained spatially and the distance travelled by the larvae must be much smaller. This calculation illustrates, however, that the larvae could potentially travel large distances along shores. Similarly, based on a current speed of approximately 4 cm s^{-1} (see El-Sabh 1977, Fig. 7 C) in Cabot Strait, and a 6 week duration of larval life along the Atlantic coast of Nova Scotia, it can be estimated that the larvae entering through Cabot Strait could travel up to 145 km per year. These results are obtained without taking wind effects into consideration; however, the profiles of water flow in Cabot Strait (Fig. 7 C), indicate that, in April, when the larvae are released, inflowing waters are more than 100 km off the nearest Cape Breton coast, and more than 30 km distant from the Newfoundland shore. Thus in April/ May, when dispersal occurs in this region, a larva from the Atlantic coast that successfully colonized the shores of the Gulf would have had to travel a minimum of 30 km in the offshore direction, followed by another 30 km in the inshore direction, to settle. The likelihood of finding competent barnacle larvae 30 km offshore is apparently limited. A maximum offshore colonization distance of 20 km was observed for *Semibalanus* on North Sea oil rigs (see Flowerdew 1983b). *Semibalanus* has also been observed to colonize the buoys moored in Jacques Cartier Passage, along the north shore of the Gulf at a maximum distance of 10 km from shore (P. Ardisson, Univ. Laval; pers. comm.). Thus, the direction and patterns of flow in Cabot Strait suggest little, if any, importation of *Semibalanus* larvae from the Atlantic. Furthermore, the predominant circulation patterns around eastern Newfoundland, shown in Fig. 7 A, suggest that only the larvae released along the east coast could potentially find their way to the mouth of Cabot Strait. In short, the likelihood of *Semibalanus* larvae entering the Gulf through Cabot Strait appears very limited indeed, and should it happen, it is possible that ice scouring would rapidly eliminate all spat in their preferred habitat (exposed surfaces) during their first winter on the shores of the Gulf.

With respect to the cryptic behaviour of the population in the Gulf of St. Lawrence, the hydroclimatic conditions (circulation patterns, strong seasonal climatic variations and ice scouring) suggest that homogeneity of the population could be maintained, as the only significant external input of larvae would come from northern regions, with presumably similar cryptic behavioural patterns at settlement. Thus, these patterns of larval dispersion and very strong selective pressures due to ice scouring are probably sufficient to maintain genetic integrity of the Gulf populations of *Semibalanus* (including possibly the Newfoundland/Iceland populations) in the face of southern non-cryptic populations.

ACKNOWLEDGEMENTS

The authors are grateful to Drs. R.N. O'Boyle, M.I. El Sabh, M.W. Flowerdew, P.R. Dando, B. Petrie for the permission to reproduce published figures, and to Dr. G. Rosenberg for linguistic improvements of the manuscript. This study was supported by grants from NSERC and FCAR to Edwin Bourget.

REFERENCES

Archambault D, Bourget E (1983) Importance du régime de dénudation sur la structure et la succession des communautés intertidales de substrat rocheux en milieu subarctique. Can J Fish Aquat Sci 40: 1278-1292

Barnes H (1957) The northern limits of *Balanus balanoides* (L.). Oikos 8: 1-14

Battaglia B, Beardmore JA (1977) Marine organisms. Genetics, Ecology and Evolution, Plenum Press, New York

Bergeron P, Bourget E (1984) Effet du froid et des glaces sur les peuplements intertidaux des régions nordiques particulièrement dans l'estuaire du Saint-Laurent. Oceanis 10: 279-304

Bergeron P, Bourget E (1986) Shore topography and spatial partitioning of crevice refuges by sessile epibenthos in an ice disturbed environment. Mar Ecol Prog Ser 28: 129-145

Bourget E (1977) Shell structure in sessile barnacles. Naturaliste Can 104: 281-323

Bourget E (1987) Barnacle shells: Composition, structure and growth. In: Southward AJ (ed) Barnacle Biology, AA Balkema Rotterdam, p 267

Bourget E (1988) Barnacle larval settlement: The perception of cues at different spatial scales. In: Chelazzi G, Vannini M (eds) Behavioural adaptation to intertidal life. Plenum Publishing Corporation, New York, p 153

Bourget E, Archambault D, Bergeron P (1985) Effets des propriétés hivernales sur les peuplements épibenthiques intertidaux dans un milieu subarctique, l'estuaire du Saint-Laurent. Naturaliste can 112: 131-142

Bousfield EL (1954) The distribution and spawning seasons of barnacles on the Atlantic coast of Canada. Bull Nat Mus Can 132: 112-154

Bousfield, EL (1955) Ecological control of the occurrence of barnacles in the Miramichi estuary. Bull Nat Mus Can 137: 1-69

Bousfield EL, Thomas MLH (1975) Postglacial changes in distribution of littoral marine invertebrates in the Canadian Atlantic region. Proc N S Inst Sci 27, Supplement 3, 47-60

Chabot R, Bourget E (1988) Influence of substratum heterogeneity and settled barnacle density on the settlement of cypris larvae. Mar Biol 97: 45-56

Conover WJ (1980) Practical nonparametric statistics. Second edition. John Wiley & Sons. New York

Crisp DJ (1974) Factors influencing the settlement of marine invertebrate larvae. In: Grant PT Mackie AN (eds), Chemoreception in marine organisms, Academic Press New York, p 177

Crisp DJ (1977) Genetic consequences of different reproductive strategies in marine invertebrates. In: Battaglia B, Beardmore JA (eds), Marine organisms, Genetics, Ecology and Evolution. Plenum Press, New York, p 257

Crisp DJ, Meadows PS (1962) The chemical basis of gregariousness in cirripedes. Proc Roy Soc B 156: 500-520

Crisp DJ, Meadows PS (1963) Adsorbed layers: the stimulus to settlement in barnacles. Proc Roy Soc B 158: 364-387

Dando PR (1987) Biochemical genetics of barnacles and their taxonomy. In: Southward AJ (ed), Barnacle Biology, Crustacean Issues 5, AA Balkema Rotterdam, p 73

Dando PR, Southward AJ (1980) A new species of *Chthamalus* (Crustacea: Cirripedia) characterized by enzyme electrophoresis and shell morphology: with a revision of other species of *Chthamalus* from the western shores of the Atlantic ocean. J Mar Biol Ass U K 60: 787-831

Darwin C (1854) A monograph of the sub-class Cirripedia, II, The Balanidae, the Verrucidae etc. Ray Society London

Dionne JC (1977) La mer de Goldthwait au Québec. Géogr Phys Quart 31: 61-80

Dunbar M (1968) Ecological development in polar regions: A study in evolution. Prentice Hall Englewood Cliffs, N. J.

El-Sabh MI (1977) Oceanographic features, currents, and transport in Cabot Strait. J Fish Res Board Can 34: 516-528

Farquharson WI, Bailey WB (1966) Oceanographic study of Belle Isle strait 1963. BIO Rep 66-9, Dartmouth, Canada

Feyling-Hanssen RN (1953) The barnacle *Balanus balanoides* (Linné, 1776) in Spitsbergen. Norsk Polarinstitutt Skr No 98: 1-64

Flowerdew MW (1983a) Electrophoretic investigation of populations of the cirripede *Semibalanus balanoides* (L.) around the North Atlantic seaboard. Crustaceana 45: 260-278

Flowerdew MW (1983b) The circumboreal *Balanus balanoides* (L.) and its subpopulations. In: Oxford GS and Rollinson D (eds). Protein polymorphism: Adaptive and taxonomic significance Systematics Association Special Volume no. 24 , Academic Press, London and New York, p 75

Fulton RJ, Andrews JT (1987) Naissance, développement et disparition de la calotte glaciaire laurentidienne. Géogr Phys Quat 41: 180-318

Hiro F (1935) The fauna of Akkeshi Bay II. Cirripedia J Fac Sci Hokkaido Univ, ser 6, Zoology 4: 213-229

Hutchins LW (1947) The basis for temperature zonation in geographical distribution. Ecol Monogr 17: 325-335

Letourneux F, Bourget E (1988) The importance of physical and biological settlement cues used at different spatial scales by the larvae of the barnacle, *Semibalanus balanoides*. Mar Biol 97: 57-66

Nilsson-Cantell C (1978) Marine invertebrates of Scandinavia, No. 5. Cirripedia, Thoracica and Acrothoracica. Universitets fovlaget, Oslo

Nott JA Foster BA (1969) On the structure of the antennular attachment organ of the cypris larva of *Balanus balanoides* (L.). Phil Trans roy Soc B 256: 115-134

O'Boyle RN, Sinclair M, Conover RJ, Mann KH, Kohler AC (1984) Temporal and spatial distribution to biological, hydrological, and physiographic features. Rapp P-v Réun Cons Int Explor Mer 183: 27-40

Petersen GH (1962) The distribution of *Balanus balanoides* (L.) and *Littorina saxatilis*, Olivi, var. *groenlandica*, Mencke in northern west Greenland. Medd Gronland 159: 1-47

Petersen GH (1966) *Balanus balanoides* (L.) (Cirripedia). Life-cycle and growth in Greenland. Medd Gronl 150: 1-116

Petrie B, Toulany B and Garrett C (1988) The transport of water, heat and salt through the strait of Belle Isle. Atmosphere-Ocean 26: 234-251

Pilsbry HA (1916) The sessile barnacles (Cirripedia) contained in the collection of the U.S. National Museum; including a monograph of the American species. Bull US Nat Mus 93: 1-366

Scheltema RS (1977) On the relationship between dispersal of pelagic veliger larvae and the evolution of marine prosobranch gastropods. In: Battaglia B, Beardmore JA (eds) Marine organisms: genetics, ecology, and evolution Plenum Press, New York and London, p 303

Shelford VE (1930) Geographic extent and succession in Pacific North American intertidal (*Balanus*) communities. Publ Puget Sd Mar (Biol.) Sta 7: 217-224

Stephenson TA, Stephenson A. (1972) Life between tidemarks on rocky shores. W. H. Freeman and Company, San Francisco
Tremblay JL, Lauzier LM (1940) L'origine de la nappe d'eau froide dans l'estuaire du Saint-Laurent. Naturaliste Can 67: 5-23
Wagner FJE (1967) Published references to Pleistocene Champlain sea faunas 1837-1966 and list of fossils. Geol Surv Can Paper 67-16
Wagner FJE (1970) Faunas of the Pleistocene Champlain sea. Geol Surv Can Bulletin 181. 104 pp
Walker G, Lee VE (1975) Surface structure and sense organs of *Balanus balanoides* (L.) as seen by scanning and transmission electron microscopy. J Zool (Lond) 178: 161-172
Wells HW, Wells MJ, Grey IE (1960) On the southern limits of *Balanus balanoides* (L.) in the Western Atlantic. Ecology 11: 587-590
Wethey DS (1983) Geographic limits and local zonation: the barnacles *Semibalanus* (*Balanus*) and *Chthamalus* in New England. Biol Bull Mar Biol Lab, Woods Hole 165: 330-341

HERBIVORY AND THE EVOLUTION OF NONGENICULATE CORALLINE ALGAE (RHODOPHYTA, CORALLINALES) IN THE NORTH ATLANTIC AND NORTH PACIFIC

R. S. Steneck
Department of Botany and Plant Pathology
Department of Oceanography
Darling Marine Center
University of Maine
Walpole, Maine 04576
U.S.A.

Abstract - A comparison of the crustose coralline (Corallinales, Rhodophyta) floras of the Boreal eastern North Pacific (Prince William Sound, Alaska to Pt. Conception, California) and Subarctic western North Atlantic (Cape Cod, Massachusetts to the Strait of Belle Isle, Newfoundland) provides an excellent means of comparison for identifying specific adaptations to herbivory. The floras of both regions have about the same number of genera (5) and abundant species (5 - 7) but no species are common to both regions and a few genera are found in only one region. Only chitons, limpets and sea urchins have species known to be capable of grazing corallines in these regions . The diversity of these coralline-grazers is greater in the Pacific Northwest (13 spp. total, 8 important) compared to the North Atlantic (4 spp. total, 2 important). Only the sea urchin, *Strongylocentrotus droebachiensis*, is found in both regions.

Given the apparent taxonomic dissimilarity between the two regions, the anatomical and morphological characteristics found under specific herbivore regimes of grazing frequency and grazing intensity are remarkably similar. In both regions, habitats having no coralline-grazers are dominated by corallines having thin unbranched thalli (i.e., without protuberances or excrescences), and raised conceptacles. Habitats having frequent low-intensity grazing from limpets or chitons are often dominated by thicker forms occasionally having a multi-layered epithallus. In regions where the most intense grazing is from sea urchins, the coralline flora is often dominated by plants having stout protuberances (excrescences), raised conceptacles and a single layered epithallus. Manipulative experiments demonstrate the adaptive nature of these characteristics. Such anatomical and morphological similarities corresponding to the specific herbivore environment suggest that herbivory is a powerful selective force in the evolution of corallines and possibly responsible for the apparent convergent evolution of these cold-water floras.

INTRODUCTION

Nongeniculate, encrusting coralline algae (Rhodophyta, Corallinales) are often the only macroalgae capable of surviving the intense herbivory common throughout shallow rocky regions of the North Atlantic (e.g., Hagen 1983; Steneck 1986). Such a conspicuous and ubiquitous association begs the question, to what extent has herbivory influenced the evolution of this flora? Although the exact origin of current coralline-herbivore species associations in the North Atlantic is unknown, a possible key to understanding factors responsible in shaping their evolution, is to identify some of the adaptations that allow species of these groups to coexist.

NATO ASI Series, Vol. G 22
Evolutionary Biogeography of the
Marine Algae of the North Atlantic
Edited by D. J. Garbary and G. R. South

"Adaptation", however, is a problematic term, because it refers to both the state of being fit (an ecological concept), and the process by which fitness is acquired (an evolutionary concept). Thus, a biologist can explain the adaptations that allow birds to fly, but a paleontologist will have difficulty in explaining precisely the process, timing and conditions under which those characters evolved. This is true for most organisms, and is especially true if the goal is to understand the evolutionary role an ecological process (e.g., herbivory) has had on a regional flora.

My premise is that functionally similar characters, when found among disjunct and distantly related organisms living under similar strong ecological pressures, indicate the adaptive value of those characters. In other words, convergent evolution indicates an evolutionary adaptive response to similar ecological conditions. This concept is not new, and Simpson (1953) stated:

> "The very widespread occurrence of parallelism and convergence is the strongest sort of evidence for the efficacy of selection and for its adaptive orientation of evolution."

Elsewhere, I have examined coralline-herbivore adaptations by considering global patterns of adaptation (Steneck 1986), and long-term evolutionary trends among herbivores (Steneck 1983) and corallines (Steneck 1985), over the past 300 million years. In this paper I examine convergent aspects of morphology and anatomy evident among species in the coralline floras of the Boreal eastern North Pacific (Washington state) and the Subarctic western North

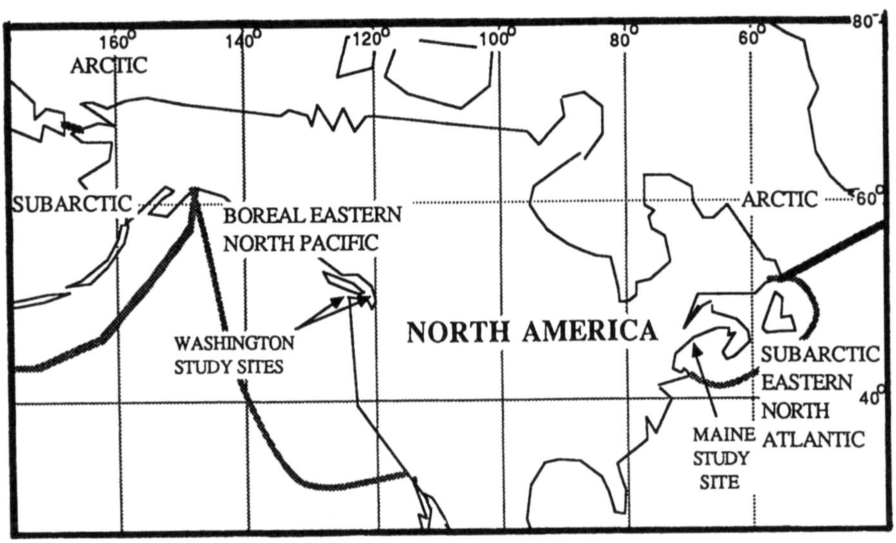

Fig. 1. Diagrammatic map of North America showing study sites and the biogeographic provinces within which they are located

Atlantic (Gulf of Maine) (Fig. 1), relative to the specific grazing pressures they receive. Coralline taxonomy and herbivore ecology have been studied in both regions but, to date, no study has sought to examine quantitatively, at the species level, the relationship between these coralline floras and the specific grazing pressures upon them.

For this examination, I draw from several published studies and provide some new data including results from manipulative experiments conducted in both oceans, that sought to determine the adaptive nature of specific characters related to herbivory in these regions.

TAXONOMIC AFFINITIES, EVOLUTIONARY ORIGINS AND BIOGEOGRAPHIC RANGES

The coralline floras: diverse and distant origins

The focus here is on abundant species of nonparasitic, rock dwelling nongeniculate corallines, also called "crustose corallines", because this group is most commonly associated with herbivores (reviewed in Steneck 1986). Latitudinal and biogeographic patterns will be confined to the northern hemisphere.

The three subfamilies of nongeniculate corallines have distinctive patterns in their global distributions (Adey 1970 a, b; Johansen 1981). Coralline species of the subfamily Melobesioideae dominate the Arctic and the Subarctic regions of the Northern Hemisphere (Table 1, Adey 1970b; Johansen 1981; Steneck 1986; Woelkerling 1988). In the North Atlantic, *Clathromorphum* and *Leptophytum* are thought to be primarily restricted to cold water habitats of the Arctic and Subarctic (Adey 1970 a, b; Lebednik 1977), whereas *Phymatolithon* occurs in Boreal and cold Temperate habitats and *Lithothamnion* is distributed throughout the tropics (Adey 1966, 1970 b, 1971; Adey & Adey 1973, Adey & Macintyre 1973; Steneck 1986). *Lithophyllum*, of the subfamily Lithophylloideae, is widespread as a genus, and is found from the Subarctic to the tropics (Adey 1970 b).

The crustose coralline flora of the Boreal eastern North Pacific is probably more diverse than that of the Subarctic western North Atlantic but, to date, only the intertidal flora has been well documented (Mason 1953; Johansen in Abbott & Hollenberg 1976; Steneck & Paine 1986). Subfamily level diversity is higher in the North Pacific than in the North Atlantic (Table 1) because of the three mastophoroid species of the genus *Pseudolithophyllum* (*sensu* Adey 1970 b), that dominate considerable portions of the rocky intertidal zone (Steneck & Paine 1986).

There has been considerable taxonomic debate over the status of the genus *Pseudolithophyllum*. *Pseudolithophyllum* Lemoine 1913 conforms to the current concept of the lithophylloid genus *Lithophyllum* (see Woelkerling 1988 for review), but *Pseudolithophyllum* (Lemoine 1913) Adey 1970 b conforms to the current concept for the mastophoroid genus *Spongites* Kützing 1841 (see Woelkerling 1985). Other distinctive coralline genera that conform to this genus include *Hydrolithon, Porolithon, Paragoniolithon* and many *Neogoniolithon* species; all were determined to be heterotypic synonyms of *Spongites* (Woelkerling 1985, 1988). However, for the purpose of this study, I will retain the taxonomic concept of *Pseudolithophyllum* (Lemoine 1913) Adey 1970 b used in Steneck and Paine (1986) because it is distinctly different from most of the other genera combined under *Spongites*. There is no question that the three *Pseudolithophyllum* species in the eastern North Pacific are members of the subfamily Mastophoroideae.

Table 1. Predominant nongeniculate, nonparasitic corallines of the Subarctic western North Atlantic and Boreal eastern North Pacific. Abbreviations for species are in parentheses

Subfamily	EASTERN NORTH PACIFIC	WESTERN NORTH ATLANTIC
Lithophylloideae Setchell		
	**Lithophyllum impressum* (Lpm. i)	°*Lithophyllum orbiculatum* (Lpm. o)
Mastophoroideae Setchell		
	*¥ *Pseudolithophyllum neofarlowii* (Pslpm. n) *¥ *Pseudolithophyllum whidbeyense* (Pslpm. w) *¥ *Pseudolithophyllum muricatum* (Pslpm. m)	
Melobesioideae (Areschoug) Bizzozero		
	**Lithothamnion phymatodeum* (Ltm p)	**Lithothamnion glaciale* (Ltm g) **Lithothamnion lemoineae* (Ltm l)
	°*Mesophyllum vancouveriense* (Meso. v)	**Clathromorphum circumscriptum* (Cl. ci) °*Clathromorphum compactum* (Cl. co)
		**Phymatolithon lenormandii* (Phy ln) **Phymatolithon laevigatum* (Phy lv) **Phymatolithon rugulosum* (Phy r)
	Leptophytum adeyii (Lpt. a)	**Leptophytum laeve* (Lpt. l) *Leptophytum foecundum* (Lpt. f)

* Abundant species (Adey 1970 a; Steneck 1978; Steneck & Paine 1986)
° Occasionally abundant (Adey 1970 a; Steneck 1978; Steneck & Paine 1986)
¥ *Pseudolithophyllum* concept of Adey 1970b (see text)

It is significant that mastophoroids are abundant in the Boreal North Pacific, because this subfamily today is best represented in the Tropics (Adey 1970 b; Johansen 1981; Adey *et al* 1982; Steneck 1986). Since the Subarctic regions of of the eastern North Pacific (i.e., Alaska and the Bering Sea, Fig. 1) are dominated by melobesioids (e.g., *Clathromorphum,* and *Mesophyllum*) (Adey 1970 b; Lebednik 1977), it is tempting to conclude that this subfamily evolved in the Arctic. However, melobesioids evolved in, and became abundant throughout, the Tropics during the Cretaceous and early Tertiary (Fig. 2, Johnson 1961; Johnson & Kaska 1965). Their relative unimportance in the Tropics today may be attributable to the remarkable recent evolutionary and ecological success of mastophoroids. The fossil record shows mastophoroids to have had low diversity during much of its geological history (Fig. 2, Johnson 1961; Johnson & Kaska 1965; Adey & Johansen 1972). Today, there are more Mastophoroideae species than the other two subfamilies combined (Steneck 1986). This estimate may be conservative, because taxonomic studies of tropical regions lag behind those of the cold water regions of the northern hemisphere (e.g., there is no monograph covering the Caribbean).

In sum, all three subfamilies of the crustose coralline algae evolved under essentially tropical conditions during the Mesozoic Era. Regardless of where various subfamilies are successful today, the fossil record reveals that their evolutionary divergence occurred between 70 and 200 million years ago (Fig. 2). This provides an excellent opportunity to examine convergent characteristics among coralline species of different subfamilies that have been evolutionarily separated since the Mesozoic Era.

The herbivore faunas

Relatively few groups of herbivores have the ability and propensity to graze encrusting coralline algae (Steneck 1983, 1986). The three major groups in the cold water regions of the world are chitons, limpets and regular echinoids (Table 2). Although all the species listed in Table 2 can graze corallines, relatively few do so, frequently or intensively enough to be considered "significant grazers" (see footnote in Table 2).

The dominant coralline grazers of the western North Atlantic (*Tonicella rubra, Tectura testudinalis, Strongylocentrotus droebachiensis* ; Table 2) have a circumpolar distribution and all range from the Arctic to Massachusetts (occasionally to Long Island Sound) in the western North Atlantic (Gosner 1978). The urchin, *S. droebachiensis* , extends into the Boreal eastern North Pacific to Oregon (Morris *et al.* 1980). The limpet, *T. testudinalis,* has been found in southeast Alaska (Lindberg 1979), but not in our target habitat on the outer coast of Washington (personal observation, Fig. 1).

Other than the urchin *S. droebachiensis* (which extends from the the North Bering Sea to the Oregon coast), the major herbivores of corallines in the Boreal eastern North Pacific (Table 2) are distributed between the South Bering Sea - Aleutian Islands region of Alaska to the north, and Pt. Conception, California USA and Baja California, Mexico to the south (Morris *et al.* 1980, Fig. 1).

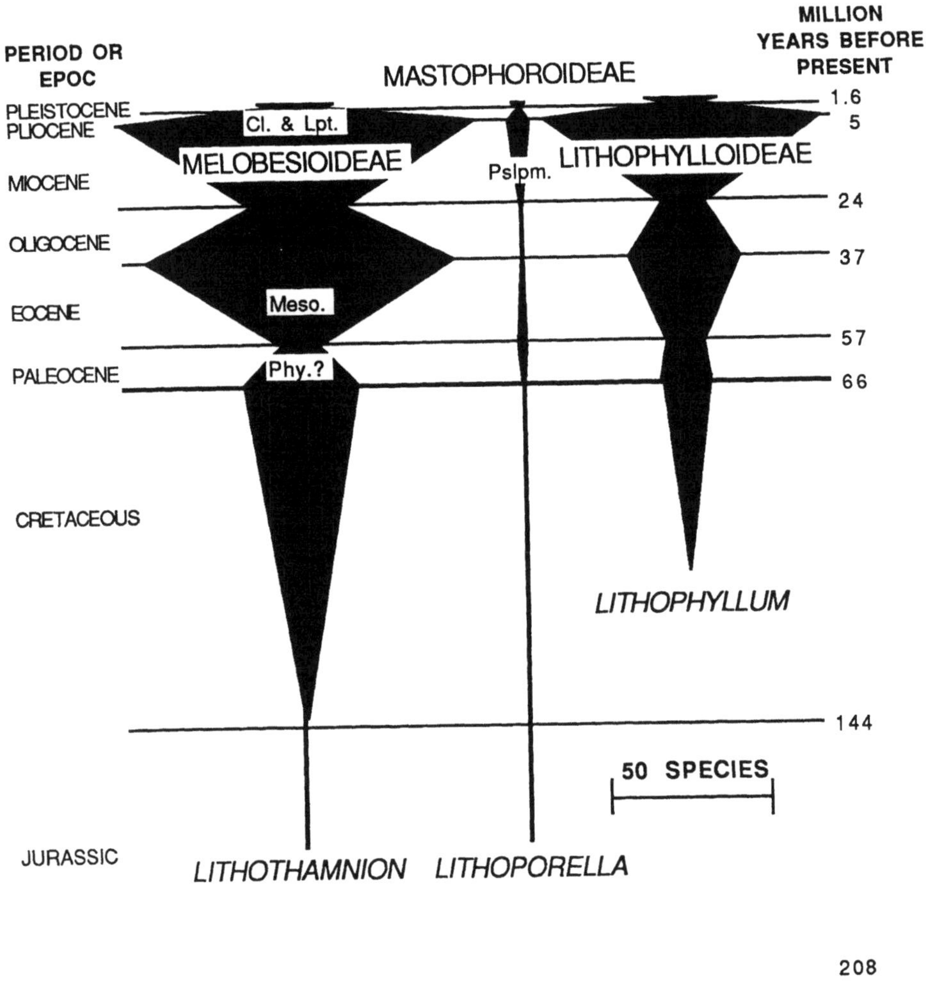

Fig. 2. Geological record of the three nongeniculate subfamilies of coralline red algae. There is some question about the taxonomic affinities of several of the fossil coralline genera (see Johansen 1981). Excellent preservation of the multiporate conceptacles and sporangial sori characteristic of the Melobesioideae subfamily make its record least equivocal. Species records for Melobesioideae (Johnson 1962, 1963; Johnson & Kaska 1965, Adey and Johansen 1972; Wray 1977), Mastophoroideae (Johnson & Ferris 1948, Johnson 1957; Johnson & Kaska 1965; Poignant 1977; Wray 1977) Lithophylloid (Johnson & Adey 1965; Wray 1977). The reported ancestral genus for each group is listed at base of each subfamily plot. Origins of dominant genera for both geographic regions (based on Adey & Johansen 1972) are listed by abbreviations used in Table 1

PATTERNS IN THE DISTRIBUTION, ABUNDANCE AND MORPHOLOGY OF CORALLINES AND ASSOCIATED CORALLINE HERBIVORES

Coralline zonation

SPECIES. Study sites in the central Gulf of Maine are taken as representative of the biota of the western North Atlantic. Depth zonation patterns in Maine (Table 3 a) have been reported by Adey (1970 a) and Steneck (1978) and conform well to the recent studies of Dethier and Steneck (in prep). Study sites in Washington state are taken as representative of the flora of the Boreal eastern North Pacific. Intertidal zonation patterns in this region (Table 3 b) have been

Table 2. Predominant herbivores of corallines in the Subarctic western North Atlantic and the Boreal eastern North Pacific. Abbreviations for species are in parentheses

Herbivore Group	BOREAL EASTERN NORTH PACIFIC	SUBARCTIC WESTERN NORTH ATLANTIC
Polyplacophora (Chitons)		
	*******Tonicella lineata* (To. l)	**Tonicella rubra* (To. r.)
	**Katharina tunicata*	**Tonicella marmorea*
	**Mopalia muscosa*	
	Ischnochiton mertensii	
Patellacean Gastropods (Limpets)		
	** *Lottia pelta* (Lo. p.)	
	Lottia ochracea	
	Lottia limatula	
	Lottia triangularis	
	***Tectura scutum*	***Tectura testudinalis* (Te. t)
	***Acmaea mitra* (Ac. m)	
Regular Echinoids (Urchins)		
	***Strongylocentrotus droebachiensis* (St. d)	***Strongylocentrotus droebachiensis* (St. d)
	**Strongylocentrotus purpuratus*	
	Strongylocentrotus franciscanus	

* Abundant species (Kozloff 1973, Morris *et al* 1980, Paine 1980, Steneck and Dethier in prep.)

** Significant grazers of corallines each region (Paine 1980 Steneck 1982a, 1983, Steneck and Watling 1982, and based on proportion of diet composed of corallines or evidence of widespread identifiable graze marks on collected plants).

qualitatively reported by Steneck & Paine (1986), and agree with recent quantitative studies of Dethier & Steneck (in prep.). For the purposes of this paper, however, I list only the zones in which a species is most abundant or where it is among the dominant elements of the flora.

The crustose corallines of Maine occur predominantly in the subtidal zone (Table 3 a). The only coralline species (outside of tidepools) in the intertidal zone is *Phymatolithon lenormandii*. In the shallow subtidal zone (≤ 10 m), *Lithothamnion glaciale* is the most

Table 3. Zones of maximum abundance for the most abundant corallines in the two biogeographic regions studied (see text)

A. SUBARCTIC
WESTERN NORTH ATLANTIC
(Central Gulf of Maine)

DEPTH ZONE	DOMINANT CORALLINE
+1m	*Phymatolithon lenormandii*
-0 - 1 m	*Lithothamnion glaciale.*, Cl. ci.
-3 m	*Lithothamnion glaciale.*, Phy r, Cl. Ci,
-10 m	*Phymatolithon rugulosum.*
	Clathromorphum circumscriptum, Ltm g & Ltm l
-20 m	*Phymatolithon laevigatum*, Lpt. l.
-30 m	*Leptophytum laeve*

B. BOREAL
EASTERN NORTH PACIFIC
(Coastal Washington State)

DEPTH ZONE	DOMINANT CORALLINE
High Intertidal	*Pseudolithophyllum neofarlowii*
Mid Intertidal	*Pseudolithophyllum whidbeyense* & *Lithothamnion phymatodeum* , Lpm. i
Low (few molluscs)	*Mesophyllum vancouveriense* , Lpt. a.
Low (molluscs)	*Pseudolithophyllum whidbeyense* , *P. muricatum.* *Lithophyllum impressum.*
Low (urchins)	*Lithothamnion phymatodeum* , Pslpm. w.

Underline = Zone where species is most abundant. Species binomials are written out for the zones in which they dominate. Other species having a significant population are listed by their abbreviations (identified in Table 1). Abundances were determined quantitatively by *in situ* surveys (for Maine: Adey 1970 a & Steneck 1978; for Washington: Steneck & Paine (1986) and Dethier & Steneck (in prep.)). Data for Maine is from the central Gulf of Maine whereas Washington State data represents outer coast intertidal environments. Tidepool and sheltered habitats in Washington were not included.

abundant coralline although *Phymatolithon rugulosum and Clathromorphum circumscriptum* also have significant populations (Steneck 1978). At depth (20 - 30 m), *Phymatolithon laevigatum* and *Leptophytum laeve* reach their maximum abundances. *L. laeve* is a strictly deep water plant in this region.

The coralline flora of the outer coast of Washington extends well up into the intertidal zone (Table 3 b) where *Pseudolithophyllum neofarlowii* is the dominant species. *P. whidbeyense* and *Lithothamnion phymatodeum* dominate the mid-intertidal zone. Coralline abundance in the lower intertidal zone to just below the mean-low-water-level zone varies considerably. In regions having few herbivores, erect fleshy macroalgae (e.g., kelps etc.) are abundant and the thin, leafy (i.e., having nonadherent margins), somewhat branched *Mesophyllum vancouveriense* dominates. Other regions, devoid of grazing echinoids (Table 2), occasionally have an abundance of herbivorous molluscs. In this habitat *Pseudolithophyllum muricatum* and the lithophylloid, *Lithophyllum impressum* dominate (Table 3). Regions heavily grazed by urchins are often dominated by the branched coralline *Lithothamnion phymatodeum.*

MORPHOLOGIES. The basic thallus morphologies for rock-dwelling corallines range between branched (thalli having protuberances or excrescenses), thick and thin forms (Steneck 1986). Since all three morphologies grade into one another, triangle diagrams can be used to represent virtually all coralline species (Steneck 1986, Fig. 3). Plotting the coralline species of Maine and Washington state in this way, reveals some of the morphological similarities that exist between the species in the two regions (Fig. 3).

Although no species is common to both regions, the triangle diagrams (Fig. 3) indicate which species are most morphologically similar at this level of analysis. Except for branching crusts, all of which are of the genus *Lithothamnion*, other morphological states are represented by different genera and often different subfamilies. Thick crusts in Washington state include the lithophylloid, *Lithophyllum impressum,* and the mastophoroid *Pseudolithophyllum*

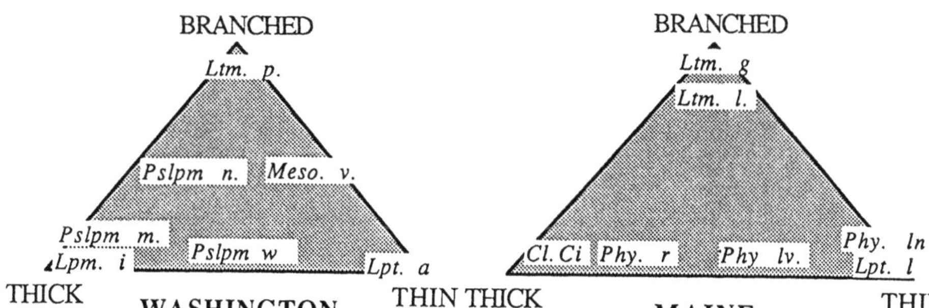

Fig. 3. Triangle diagrams representing the dominant coralline morphologies in the Boreal eastern North Pacific (Washington) and Subarctic western North Atlantic (Maine)

muricatum, whereas that morphology is represented by the melobesioids, *Clathromorphum circumscriptum* and to a lesser extent, *Phymatolithon rugulosum* in Maine. Thin crusts in both regions are of the subfamily Melobesioideae; *Leptophytum* species in both regions, and *Phymatolithon lenormandii* in Maine. Intermediate between thick and thin crusts, are the species of *Pseudolithophyllum whidbeyense* in Washington, and *Phymatolithon laevigatum* in Maine which represent mastophoroid and melobesioid subfamilies respectively.

Herbivores of corallines: zonal patterns and potential impact

Two important components of herbivory are the amount of biomass removed in a single bite (defined as "grazing intensity") and the number of bites per unit time (defined as "grazing frequency"). Herbivores commonly associated with corallines have measurably different impacts on the plants.

Grazing intensity varies as a function of the body size, feeding apparatus, and foraging behavior of the species and individuals involved. Grazing intensity for the important coralline herbivores (Table 2) is relatively easy to measure because bite marks are etched into coralline thalli. Identification of coralline grazers, based on bite marks, is relatively easy. Littorinid snails, even under laboratory conditions where only corallines were available for grazing, do

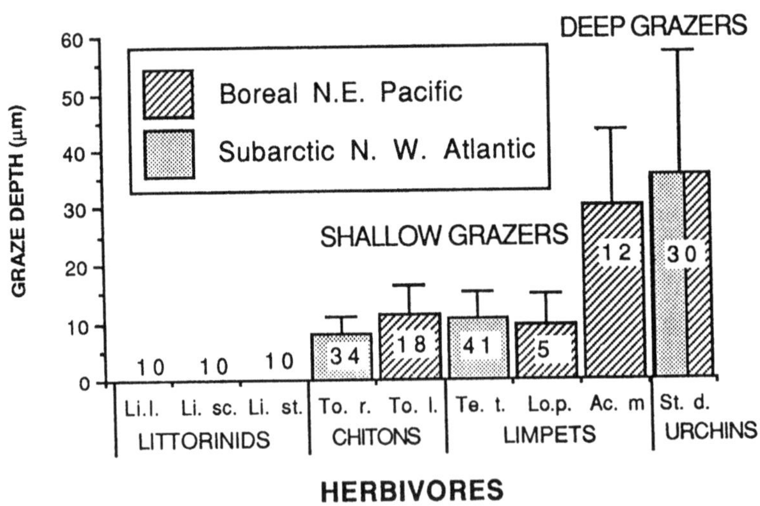

Fig. 4. Grazing intensity as measured by penetration depths into corallines. Data based on cross sections examined using a scanning electron microscope. Sample sizes are represented by numbers within the histograms. Littorinid were included for sake of comparison and because they are abundant in both regions. Abbreviations for the herbivores are in Table 2 except for the littorinids which are: Li l = *Littorina littorea*, Li. sc. = *L. scutulata*, Li. si. = *L. sitkana*

not measurably penetrate coralline thalli (Fig. 4). Chitons and most limpets in both regions have about the same grazing impact on corallines (average penetration depths ranging between 7 and 10 um, Fig. 4). Herbivore species of the two groups are determined by the shape of the bite marks (e.g., chitons are curved, limpets straight) and by the number and size of teeth and spacing between them) . In Washington, *Acmaea mitra* is unique among limpets because, having an additional pair of teeth contacting the substratum, it leaves extraordinarily deep and distinctive graze marks. The grazing intensity of *A. mitra* is not significantly different from that of the green sea urchin *Strongylocentrotus droebachiensis,* although the pentaradial marks made by urchins are very distinctive.

Three distinct grazing intensities on corallines are identified in Fig. 4: 1) non-grazing (i.e., herbivores incapable of excavating coralline thalli; e.g., *Littorinids*), 2) shallow grazers

Table 4 A. Ranking potential grazing pressure on corallines in Maine based on herbivore biomass and grazing intensity (increasing left to right, See Fig. 4). Data based on quadrat sampling at target locations from Steneck & Dethier (in prep.)

Target Habitats	Non-grazing Molluscs	Shallow Grazing Molluscs	Deep Grazing Uurchins[3]	Total Herbivore Biomass	No. Samples	Coralline-Grazing Pressure
+1.6 m Intertidal	1.2 (± 6.1)[4]	0	0	1.2	72[5]	none
0 m Intertidal	56.9 (±17.8)	0	0	56.9	83	none
Coastal 1 m depth	0	0.02 (±.02)	0[6]	0.02	34	low
Coastal 3 m depth	0	0.2 (±.2)	229.2 (±36)	229.2	39	high
Coastal 10 m depth	0	0.02 (±.02)	189.0 (± 13.73)	189.0	76	high
Coastal 20 m depth	0	0.9 (±.4)	9.5 (± 5.3)	9.5	30	mid
Coastal 30 m depth	0	0	0	0	33	none

[1] Biomass (g(dry)/m^2) of herbivores used throughout. Shallow grazing molluscs = *Littorina littorea*, [2]*Tectura testudinalis, Tonicella rubra, T. marmorea.*[3]*Strongylocentrotus droebachiensis*,.
[4] Except for this value (standard deviation) all variances are expressed as standard error because they are averages of a population of means. [5]n = m^2 quadrats. [6]Urchins having a biomass of 56.3 (± 8.75) (g(dry)/m^2) were found in crevices only. This analysis involves plants growing on the upper surfaces of the boulders where there are no urchins.

(most limpets and all chitons) and 3) deep grazers (*Acmaea mitra* and *S. droebachiensis*). These characteristic intensities of grazing transcend phylogenetic relatedness, since chitons and limpets are different classes of molluscs, and *A. mitra* and *S. droebachiensis* are of different phyla. Limpets and chitons in Maine are shallow grazers (Fig. 4), and comparable in intensity to homologous herbivores in Washington (except for *A. mitra*, a relatively rare herbivore (see footnote in Table 4B)).

Grazing frequency will be reflected by patterns of abundance and association. Obviously, grazing must be low, if few or no herbivores are present in a given zone. Table 4 ranks the potential grazing pressures on corallines in Maine (Table 4A), and in Washington (Table 4B), integrating abundance (herbivore biomass) with their grazing abilities (i.e., grazing intensities from Fig 3). In Maine (Table 4A), the highest potential grazing pressure on corallines is found between depths of 3 and 10 m, decreasing toward the upper intertidal zone and toward the lower limit of coralline depth distribution (about 30 m in coastal Maine). In

Table 4B. Ranking potential grazing pressure on corallines in Washington State. Herbivores arranged as in Table 4A but target habitats are arranged from highest to lowest intertidal zone for Tatoosh Island, Washington state (See Fig. 4). Data from Steneck & Dethier (in prep.)

Target Habitats	Non-grazing Moluscs	Shallow Grazing Molluscs	Deep Grazing Urchins[3]	Total Herbivore Biomass[4]	No. Samples	Coralline Grazing Pressure
High	3.9 (± 3.4)	7.1 (± 11.4)	0	11.0	14	low
Mid	0.5 (± .1)	42.1 (± 41.1)	0	42.6	13	mid
Low (few molluscs)[5]						low
Low (with molluscs)	0.2 (± .9)	40.3 (± 56.6)	0	40.5	13	mid
Low (with urchins)	0	3.7[6] (± 7.3)	1952.5[7]	1956.2	8	high

[1] Summed biomass (g(dry)/m^2) of *Littorina* spp., *Onchidella* sp., *Siphonaria* sp.; [2] *Lottia pelta, Tectura scutum, Tonicella lineata, Katharina tunicata*; [3] *Strongylocentrotus purpuratus*; [4] No *A. mitra* were found.; [5] Around holdfast of *L. setchellii*, relatively few herbivores are present and grazing pressure appears low grazing pressure determined by graze marks (Steneck & Paine 1986); [6] *Tonicella lineata*: [7] Urchin biomass was estimated by taking the mean density (275 /m^2) and multiplying it by the dry mass of the mean sized urchin (7.1 g for a test diameter of 3.8 cm (± 1.3) based on a test diameter to dry mass regression for *S. droebachiensis* of: $y = 0.2827 \cdot x^{2.493}$, r = .94, n = 30).

Washington, patchiness in herbivore distribution translates to patchiness in the the potential grazing pressure on corallines (Table 4B). The highest potential grazing pressure is in the low intertidal zone, where urchins have been abundant for a considerable period (Paine 1980, personal observation over 7 years). In other regions in the same zone, where turbulence is high, such as under kelp (*Hedophyllum sessile* or *Laminaria setchellii*), herbivore biomass is

Table 5. Characteristics hypothesized as adaptive for corallines (reviewed in Steneck 1986) and the grazing frequency observed on the thallus of coralline species (% cover of bites). Species are arranged according to the potential grazing pressure determined from Table 4. The alternate character state to sunken conceptacles are raised conceptacles. **A** = MAINE; **B** = WASHINGTON

Potential Grazing Pressure	Thallus Branched	Morphology Thick[1]	Sunken Conceptacle	Thick Epithallus[2]	Number of Adaptations[3]	% Cover Bites[5]
A						
LOW						
Phy. ln.					0	
Lpt. l.					0	0
MID						
Phy. lv.			X		1	
Cl. ci.		X	X	X	3	100
HIGH						
Ltm. g.	X				1[4]	15
Phy. r.		X	X		2	
B						
LOW						
Pslpm n.	-	X			1	59
Meso. v.	-				0+	7
MID						
Pslpm. w		X			1	50
Pslpm. m.		X	X		2	79
Lpm. l.		X	X	X	3	57
HIGH						
Ltm. p.	X				1[4]	9

[1] $>$ 1000 um, [2] $\geq$ 4 cells, [3]Tally across columns (Maine data: Adey 1964, 1966, 1970 a: Washington data Steneck & Paine 1986). [4]Branches effectively reduce the importance of other adaptations to herbivory (e.g., Fig. 6 see text). [5]Proportion of surface covered with graze marks (Maine: Steneck 1982 a, and unpublished data, Washington: Steneck & Paine 1986). In Washington, *Lithothamnion phymatodeum* is most abundant in mid zone but it dominates low intertidal zones where urchins are abundant (Table 4 B). *Mesophyllum vancouveriense* can produce excrescences (branches) but often grows unbranched or sparsely branched.

low, and somewhat dependent on the local mollusc abundance. Higher in the intertidal zone, herbivores of corallines decrease in abundance.

SPECIFIC HERBIVORE ASSOCIATIONS AND CORALLINE ADAPTATIONS TO HERBIVORY

A more direct means of determining grazing frequency consists of examining field-collected corallines for the proportion of the thallus surface covered by graze marks (Table 5,

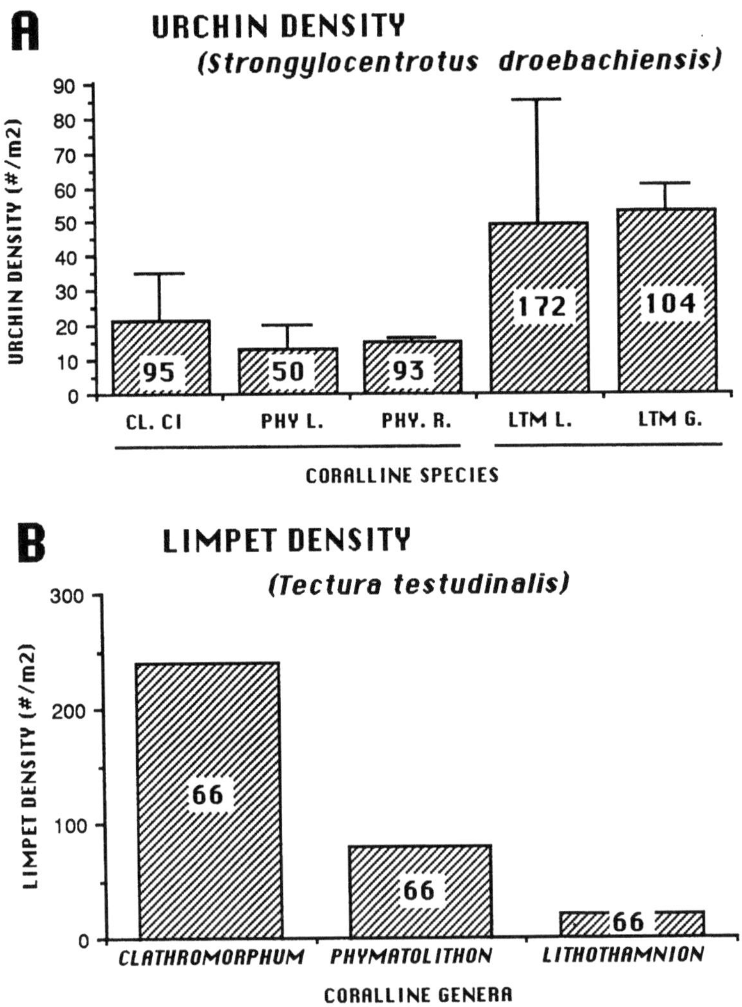

Fig. 5. Herbivore densities on individual coralline species in Maine. Data are from quadrat samples taken at 10 m, Pemaquid Pt., Maine. Sample sizes (.25 m^2 quadrats) are indicated by numbers within the histogram. Lines indicate significant differences for urchin data ($p < .05$ SNK). Coralline abbreviations are listed in Table 1. Limpet data taken from Steneck 1982a

right column). This, however, does not correspond directly to the ranking of potential grazing pressures determined in Table 4. This may be due to some of the morphological and anatomical characteristics of the corallines and behavioral characteristics of the herbivores.

Branched thalli

The rugosity or microrelief of a coralline thallus affects an herbivore's grazing ability. Urchin densities in Maine are significantly higher on the branched *Lithothamnion* species, than they are on the unbranched species of *Clathromorphum* and *Phymatolithon* (Fig. 5 A). The abundance of urchins does not translate into a higher percent cover of bites (Table 5), because bite marks are confined to branch tips, leaving most of the thallus ungrazed. The probable reason for the association is the elevated biomass of filamentous and foliose algal forms, that survive in the refugia created between the branches of *Lithothamnion*. For example, samples taken from the branched thallus of *Lithothamnion* plants revealed 12 species of filamentous and microalgae, having an average biomass of 54 g/m^2, compared to 3 species and 0.6 g/m^2 for *Phymatolithon* and 1 species and 0.002 g/m^2 for *Clathromorphum*.

A similar pattern of reduced bite marks on the branched thallus of *L. phymatodeum*, in zones having abundant urchins (Table 5, right side), was observed in Washington. The effectiveness of branches in reducing the damaging impact of herbivores was tested in an experiment conducted in collaboration with Dr. Robert T. Paine (University of Washington). For this experiment, pieces of the branched coralline *L. phymatodeum,* and the unbranched species *Pseudolithophyllum whidbeyense,* were transplanted into epoxy putty in experimental regions in the intertidal zone, in which herbivores were either removed or allowed to remain. Small plastic nubs were embedded into the epoxy around half of the transplanted crusts of both species; the other half had no such nubs. These nubs were about the same size, and placed in the same density, as naturally occurring branches on the *Lithothamnion* species, thereby creating artificial branches for both species. After one year, both species having artificial branches showed significantly higher rates of growth under grazed conditions (Fig. 6). No differences in growth were observed for either species in the ungrazed controls.

Substrate-specific distributions of limpets (*T. testudinalis*) within a habitat in Maine, reveals a pattern opposite to that observed for urchins (Fig. 5, Steneck 1982). Limpet abundance is lowest on the branched thallus of *Lithothamnion*. The smooth surface of this species offers an excellent substratum for limpet grazing and homesite establishment (Steneck 1982). Conversely, the branched thallus of *Lithothamnion* is a poor surface for both of those activities. To examine the impact of branched morphologies on the survivorship of limpets under *Asterias* spp. (seastar) predation, limpets were placed on corallines that were buried in sand except for the morphologies (species) being tested. Survivorship of limpets on bare,

Clathromorphum and *Phymatolithon* substrata was not significantly different, and ranged from 12 to 16 days (16.0 d ± 19.3, 12.5 d ±16.6, 16.0 d ±16.6, respectively). In contrast, survivorship of limpets on the branched thallus of *Lithothamnion* was 25% less (4.4 days ± 5.4) than the survival times on the other substrata.

It appears that the branched morphology of *Lithothamnion* protects this plant from injury (Table 5) in zones having potentially high grazing pressure from deep-grazing herbivores (Table 4). The effectiveness of branched morphologies in reducing injury may be a contributing factor to the abundance of this form (Table 3) in urchin grazed habitats in both regions.

Unbranched thalli

Limpet abundance on *Clathromorphum* (Fig 5b) corresponds with the pattern of limpet graze marks reported for this species in Maine (Table 5) and throughout its range (Steneck 1982, Table 5). Several adaptations were hypothesized to be adaptive to this sort of high frequency and low intensity grazing (Steneck 1986): these include thick thallus, sunken conceptacles and a thick protective epithallus. Both floras (Table 5) show a positive correspondence between the number of adaptations and the proportion of the thallus that is grazed (Fig. 7) among the unbranched corallines. Thus it appears that the *Clathromorphum* in Maine, and the *Lithophyllum* of Washington state, have convergent morphologies, and similar adaptive strategies (*sensu* Grime 1979).

Herbivore-mediated competition among ecologically equivalent species pairs

The paired interactions of *Clathromorphum* and *Phymatolithon* in Maine, and *Lithophyllum impressum and Pseudolithophyllum whidbeyense* in Washington, are worth additional consideration. These two pairs of coralline species are the dominant, unbranched corallines to exist under conditions of mid to high levels of grazing (Table 5). Besides the similarities already discussed, these pairs are morphologically similar and are represented in the bottom middle-to-left portions of both triangle diagrams in Fig. 3.

Zones of intense herbivory (Table 4), often have a high or complete cover of encrusting coralline algae. Under such conditions, competition for space is intense. Competitive interactions involving corallines have shown that competitive superiority (one species overgrowing another) often correlates with the relative thickness of the competing crusts (Buss 1980, Steneck 1983). In both cases, the naturally thicker species (*Clathromorphum* and *Lithophyllum* Fig. 3) have multilayered epithalli, which overlie and protect the meristem responsible for vertical growth. In Washington, studies of coralline competition, in which shallow-grazing herbivorous mollusc densities (*L. pelta, T. scutum* and *T. lineata*) were manipulated, resulted in *L. impressum* overgrowing *Pseudolithophyllum whidbeyense* under grazed conditions, but this dominance was reversed when herbivores were removed (Paine 1984).

To examine the competitive relationships between those two species in Washington state, under a natural gradient of grazing frequency and grazing intensity, a study was conducted in tidepools along an intertidal gradient on San Juan Island (Fig. 8, Steneck, Hacker and Dethier in prep). *Lithophyllum impressum* and *Pseudolithophyllum whidbeyense* dominate those pools, with competitive superiority between the two reversing along the gradient. *L. impressum* is competitively superior in the upper intertidal pools, but loses out to

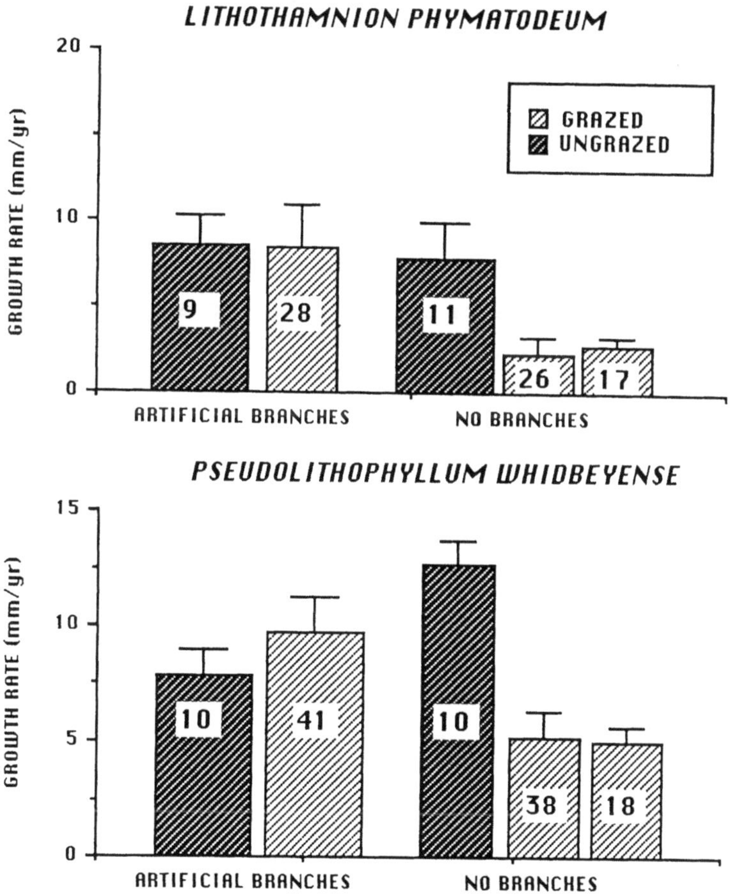

Fig. 6. The effect of artificial branches on the growth rates of *Lithothamnion phymatodeum* and *Pseudolithophyllum whidbeyense* under conditions with and without grazers (Steneck & Paine in prep.). Sample sizes are represented by numbers within the histograms

P. whidbeyense in pools of the lower intertidal zone Fig. 8a). This reversal corresponds to a decrease in grazing frequency (Fig. 8b) and an increase in grazing intensity (Fig. 8c). "Deep graze marks" refers to those penetrating through the thick (25 um) multilayered epithallus of *L. impressum* (Steneck & Paine 1986) with a single bite (Fig. 4). This case, and the example of the study by Paine (1984), show that the dominance of *L. impressum* depends on frequent, low intensity grazing and that competitive dominance declined when frequency declined (herbivores are removed) or when intensity increased (toward lower intertidal zone, Fig. 8c).

An analogous situation of herbivore-mediated competition was observed in Maine between *Clathromorphum circumscriptum* and *Phymatolithon spp.* (Steneck 1982b) two anatomically similar but distantly related coralline species (Table 1, Fig. 2). Fig. 9 shows that competitive dominance of *Clathromorphum* over *Phymatolithon* corresponds with the frequency (Fig. 5, Table 5) of low intensity grazing (Fig. 4) created by the limpet *T. testudinalis*. This was experimentally simulated under grazer free conditions in which cobbles, having competing plants of the two coralline species, were treated with a graded series of increasing grazing intensity. The treatments, applied weekly, ranged from no handling, to gentle epiphyte removal, to scrubbing with nylon brushes, and to the scrubbing of each coralline with steel brushes for one minute. The most intense artificial grazing simulated the intensity created by limpets (as determined, following inspection, by the use of scanning electron micrographs). With increased artificial grazing intensity, the competitive dominance of *C. circumscriptum* over *Phymatolithon* increased.

Evidence from both regions indicates that corallines, possessing thick multilayered epithalli, have an adaptive advantage over other corallines under a narrow range of conditions of

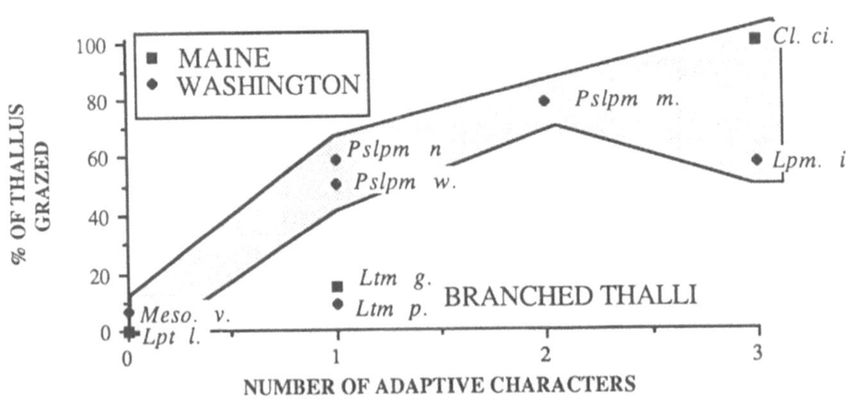

Fig. 7. Number of adaptive characters hypothesized to be effective against grazers (Table 5) relative to the percent of the coralline thallus covered with graze marks (based on scanning electron micrographs). Data and methods in Steneck & Paine 1986)

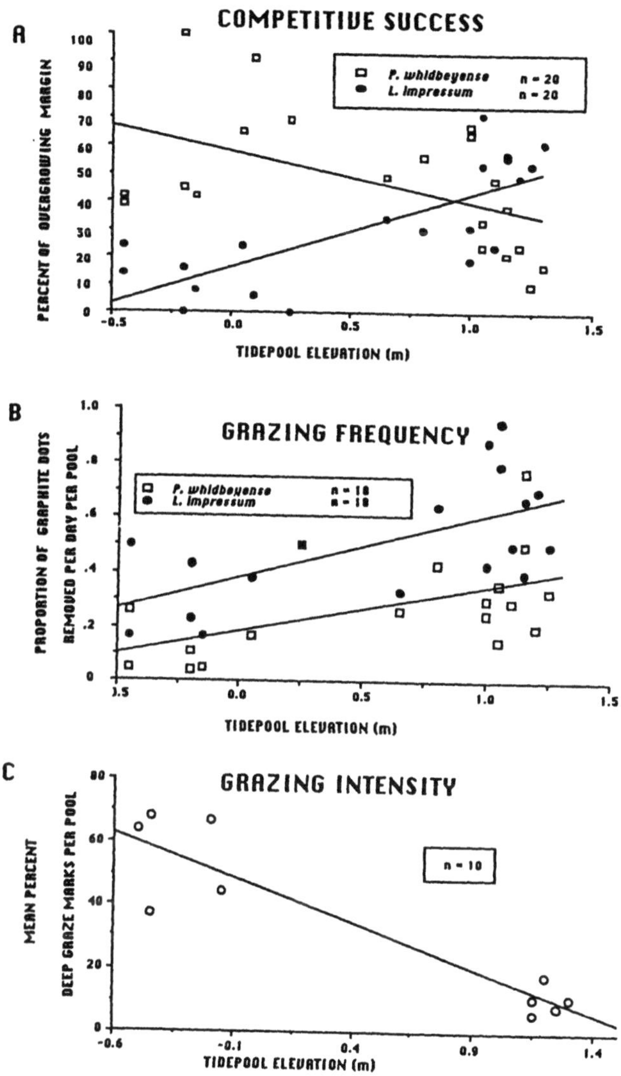

Fig. 8. Overgrowth competition between *Lithophyllum impressum* and *Pseudolithophyllum whidbeyense* in tidepools distributed along an intertidal gradient of grazing frequency and grazing intensity on San Juan Island, Wa (Steneck, Hacker and Dethier in prep). A) Competitive ability measured as percent of overgrowing margin. Regressions are significant ($P < 0.03$ for arcsin transformed data). B) Frequency of grazing by limpet species (*L. pelta* and *T. scutum*) on both corallines species. Grazing frequency determined by placing equal number of graphite pencil dots on corallines and monitoring and replacing them daily. Rate of dot disappearance is taken to reflect grazing frequency. Regressions are significant ($P < 0.008$ for arcsin transformed data). The two slopes are not significantly different (ANCOVA: $F = 0.497$, $P = 0.486$ for percentage data). A test of equal slope elevations showed significant differences between species over tide height (ANCOVA: $F = 16.8$, $P < 0.001$). C) Mean proportion of deep graze marks compared to all graze marks made by large (> 2 cm) *L. pelta* and *T. scutum* on field-collected corallines. The regression ($y = 45.7 - 28.6x$, $r = 0.92$ for percentage data) is significant ($P < 0.001$). Deep graze marks are those > 25 um (thickness of the epithallus of *Lithophyllum impressum* Steneck & Paine 1986)

frequent, low intensity grazing such as that provided by small limpets. Experiments in Maine, where limpets were elevated to unnaturally high population densities, resulted in the loss of the protective epithallus (Steneck 1982). In Washington, there is similar evidence that, as grazing intensity increases, the protective epithallus may be penetrated and its advantage lost (Fig. 8).

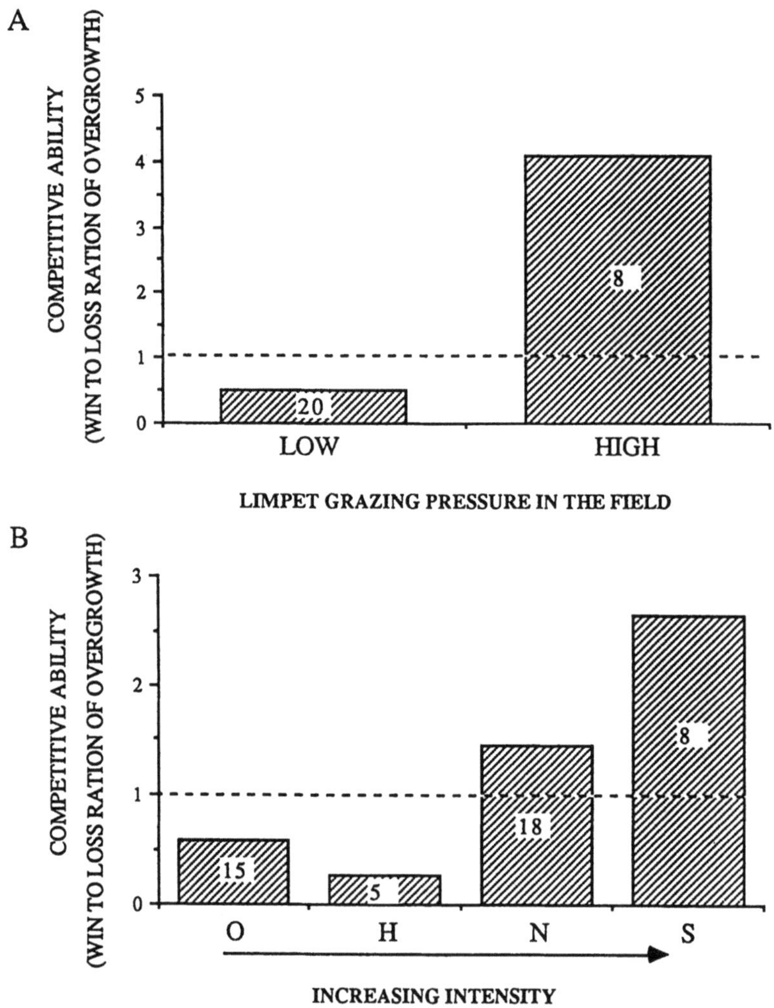

Fig. 9. Competitive ability of *Clathromorphum circumscriptum* over *Phymatolithon spp.* as a function of grazing intensity (data from Steneck 1982b). Competitive ability is assessed by measuring overgrowth. A) The results of competition under field conditions of low (10.7 limpets/$m^2 \pm$ 13 SD) and high (70.4 limpets/$m^2 \pm$ 88) limpet grazing pressures. B) Competition results determined under experimental conditions for a graded series of artificial grazing intensities administered weekly. All corallines were grown under the same physical conditions in a raft without macroherbivores. "O" indicates no handling at all, "H" indicates epiphytes were gently rubbed off the corallines by hand, "N" indicates the corallines were scrubbed with a nylon brush for one minute, "S" indicates the corallines were scrubbed with a steel brush for one minute. The dotted line indicates win to loss ratios of one; or the point where wins equal losses

In both regions, corallines lacking the thick, multilayered epithallus are capable of more rapid lateral growth (Steneck 1985) than their counterparts Therefore, under conditions of very low grazing pressure, these crusts can overgrow the thick crusts having multilayered epithalli (e.g., see Paine 1984). Under very intense grazing (i.e., from deep grazers, Fig. 4) crusts capable of rapid lateral growth can recover quickly, and remain the most abundant crust under those conditions (see Table 3).

CONCLUSIONS

In comparing the coralline floras of the western North Atlantic and eastern North Pacific, several cases of distantly related corallines (Fig. 2) having convergent morphological (Fig. 3) and/or anatomical features (Table 5) are evident. The adaptive nature (ecologically speaking) of certain characters, such as the possession of branched morphologies or multi-layered epithalli, is demonstrable from both field observations and manipulative experiments. The evolutionary question of exactly how or when these characters evolved in these floras remains to be answered. While it is impossible at this point to demonstrate the evolutionary origin of these characters, it is possible that the mechanistic explanation provided here suggests how they persist, and often dominate, under the current and relatively recent past herbivore environment.

I offer the following general model which summarized the case presented here: Under conditions of potentially high primary production, the frequency and intensity of herbivory become prime determinants of the suites of adaptive characters likely to be successful among nongeniculate coralline algae. When grazing frequency and intensity are very low, relatively fast growing, thin-thallused forms with raised conceptacles are likely to be the most abundant; under conditions of high grazing frequency but low intensity, thick, smooth crusts with multilayered epithalli and sunken conceptacles are likely to dominate. With higher intensity, but relatively low frequency, moderately thick crusts, capable of rapid lateral growth, are likely to succeed; under more frequent and more intense herbivory, corallines having branched thalli are more likely to dominate. This pattern is not likely to hold under tropical conditions, where parrotfish grazing creates a quantum jump in grazing intensity, in which the coralline's branches become "bite-sized morsels" for these megaherbivores (Steneck 1986).

ACKNOWLEDGEMENTS

I received help from many people on all aspects of the work presented here. Principal among these were: Mike Brittsan, Grant Dawson, Megan Dethier, Sally Hacker, Bruce

Milliken, Bob Paine, Cathy Pfister, Susan Forward Stilwell, Roberta Townsend, and Jeff Williams. To all I am grateful. This research was primarily funded by grants from the National Science Foundation under Grant Nos. OCE-8315136 and OCE-8600262 (with Megan Dethier) with additional funding from NOAA's UM/UNH Sea Grant (R/FM - 169) and grants from the NOAA's National Undersea Research Center (Univ. of Connecticut at Avery Point). This is contribution number 213 of the Darling Marine Center of the University of Maine.

REFERENCES

Abbott, IA, Hollenberg, GJ (1976) Marine algae of California. Stanford University Press, Stanford, California

Adey WH (1964) The genus *Phymatolithon* in the Gulf of Maine. Hydrobiologia 24: 277-420

Adey WH (1966) The genera *Lithothamnium, Leptophytum* (nov. gen.) and *Phymatolithon* in the Gulf of Maine. Hydrobiologia 28: 321-370

Adey WH (1970 a) The crustose corallines of the northwestern North Atlantic including *Lithothamnium lemoineae* n. sp. J Phycol 6: 225-229

Adey WH (1970 b) A revision of the Foslie crustose coralline herbarium. K Norske Vidensk Selsk Skr 1: 1-46

Adey WH (1971) The sublittoral distribution of crustose corallines on the Norwegian coast. Sarsia 46: 41-58

Adey WH, Adey PJ (1973) Studies on the biosystematics and ecology of the epilithic crustose Corallinaceae of the British Isles, Br Phycol J 8: 343-407

Adey WH, Johansen HW (1972) Morphology and taxonomy of Corallinaceae with special reference to *Clathromorphum, Mesophyllum*, and *Neopolyporolithon* gen. nov. (Rhodophyceae, Cryptonemiales). Phycologia 11: 159-180

Adey WH, Macintyre IG (1973) Crustose coralline algae: a re-evaluation in the geological sciences. Geol Soc Am Bull 84: 883-904

Adey WH, Townsend RA, Boykins WT (1982) The crustose coralline algae (Rhodophyta: Corallinaceae) of the Hawaiian Islands. Smithson Contr Mar Sci 15: 1-74

Buss LW (1980) Competitive intensity and size frequency distribution of interacting populations. Proc Nat Acad Sci USA 77: 5355-5359

Gosner KL (1978) A field guide to the Atlantic seashore. Houghton Mifflin, Boston

Grime, J P (1979) Plant strategies and vegetation processes. John Wiley and Sons, New York

Hagen N (1983) Destructive grazing of kelp beds by sea urchins in Vestifjorden, Northern Norway. Sarsia 68: 177-190

Johnson JH (1961) Limestone building algae and algal limestones. Johnson, Boulder, Colorado

Johansen HW (1981) Coralline algae, a first synthesis. CRC Press, Boca Raton, Fl

Johnson JH (1957) Calcareous algae. Prof Pap U S Geol Surv 280: 209-246

Johnson J H (1962) The algal genus *Lithothamnium* and its fossil representatives. Colo Sch Mines 515: 1-111

Johnson JH (1963) The algal genus *Archaeolithothamnion* and its fossil representatives. J Paleont 37: 175-211.

Johnson JH (1965) The genus *Lithophyllum* and its fossil representatives. In Johnson JH, Adey WH Studies of *Lithophyllum* and related algal genera. Colo Sch Mines Q 60: 1-105

Johnson JH, Adey WH (1965) Studies of *Lithophyllum* and related algal genera. Colo Sch Mines Q 60: 1 - 105

Johnson JH, Ferris BJ (1948) Eocene algae from Florida. J Paleontol 22: 764-766

Johnson JH, Kaska H V (1965) Fossil algae from Guatamala. Prof Contr Colo Sch Mines 1: 1-52

Kozloff EF (1973) Seashore life of Puget Sound, the Strait of Georgia, and the San Juan Archipelago. University of Washington Press, Seattle

Kützing FT (1841) Ueber die 'Polypiers calciféres' des Lamouroux. F. Thiele, Nordhausen
Lebednik PA (1977) The Corallinaceae of northwestern North America. 1. *Clathromorphum* Foslie emend Adey. Syesis 9: 59-112
Lemoine MP (1913) Mélobésiées. Revision des Mélobésiées antarctiques. In: Charcot J (ed.) Deuxéme Expédition antarctique française (1908-1910) Sciences Naturelles, Vol. 1 Botanique. Masson, Paris, p 1-67, 2 pl
Lindberg DR (1979) Variations in the limpet, *Collisella ochracea*, and the northeastern Pacific distribution of *Notoacmaea testudinalis* (Acmaeidae). The Nautilus 93: 50-56
Mason LR (1953) The crustaceous coralline algae of the Pacific coast of the United States, Canada and Alaska. Univ Calif Publ Bot 26: 313-390
Morris RH, Abbott DP, Haderlie EC (1980) Intertidal invertebrates of California. Stanford Univ. Press. Stanford, Ca
Paine RT (1980) Food webs: linkage, interaction strength and community infrastructure. J Anim Ecol 49: 667-685
Paine R T (1984) Ecological determinism in the competition for space. Ecology 65: 1339-1348
Poignant AF (1977) The Mesozoic red algae: a general survey. In: Flügel (ed) Fossil Algae Recent Results and Developments. Springer-Verlag, Berlin
Simpson, GG (1953) The major features of evolution. Simon & Schuster, New York
Steneck RS (1978) Factors influencing the distribution of crustose coralline algae (Rhodophyta, Corallinaceae) in the Damariscotta River estuary, Maine. MS thesis. Univ. Maine, Orono
Steneck RS (1982a) A limpet-coralline alga association: adaptations and defenses between a selective herbivore and its prey. Ecology 63: 507-522
Steneck RS (1982b) Adaptive trends in the ecology and evolution of crustose coralline algae (Rhodophyta, Corallinaceae). Ph. D. Dissertation. Johns Hopkins Univ., Baltimore
Steneck, RS (1983) Escalating herbivory and resulting adaptive trends in calcareous algal crusts. Paleobiology 9: 44-61
Steneck RS (1985) Adaptations of crustose coralline algae to herbivory: patterns in space and time. In: Toomey DF, Nitecki MH (eds.) Paleoalgology. Springer-Verlag, p 352-366
Steneck RS (1986) The ecology of coralline algal crusts: convergent patterns and adaptive strategies. Ann Rev Ecol Syst 17: 273-303
Steneck RS, Paine RT (1986) Ecological and taxonomic studies of shallow-water encrusting Corallinaceae (Rhodophyta) of the boreal northeastern Pacific
Steneck RS, Watling L (1982) Feeding capabilities and limitations of herbivorous molluscs: a function group approach. Mar Biol 68: 299-319
Woelkerling WJ (1985) A taxonomic reassessment of *Spongites* (Corallinaceae, Rhodophyta) based on studies of Kützing's original collection. Phycologia 20: 125-153
Woelkerling WJ (1988) The coralline red algae: an analysis of the genera and subfamilies of nongeniculate Corallinaceae. Oxford University Press, London
Wray JL (1977) Calcareous Algae. Elsevier, Amsterdam

THE APPLICATION OF DNA-DNA HYBRIDIZATION IN SEAWEEDS AND DNA DIVERGENCE BETWEEN SOME AMPHI-ATLANTIC *CLADOPHORA* (CLADOPHORALES, CHLOROPHYTA) SPECIES

P.V.M. Bot[1,2] and C. van den Hoek[2]
Department of Genetics[1] and Department of Marine Biology[2]
University of Groningen
P.O. Box 14
9750 AA Haren (Gn)
The Netherlands

Abstract - This paper evaluates the relationship between strategy of genome evolution and the robustness of DNA hybridizations data. Birds and primates differ significantly from *Drosophila*, plants and *Cladophora* in the ratio of NPH to ΔT_{me} values. This difference is discussed from an evolutionary and molecular perspective. We conclude that the intensity and frequency of genome evolutionary processes (such as DNA turnover) not only hamper the ability to infer phylogenetic relationships, but also the ability to estimate time of divergence from the data. In the second section of this paper, single copy DNA of West European isolates of four *Cladophora* species: *C. sericea*, *C. albida*, *C. pellucida* and *C. vagabunda* was hybridized with DNA of conspecific West Atlantic isolates (in the case of *C. pellucida* with a related species, *C. pellucidoidea*). The high levels of DNA similarity found in the cold-temperate North Atlantic region are interpreted as resulting from a disruption in the biogeographic patterns during glacial periods and the re-establishment of the same or comparable patterns during interglacial periods. At lower latitudes, amphi-Atlantic isolates show more DNA divergence. Thermal isolation of the warm-temperate East and West Atlantic regions may date from the middle Miocene onwards. The effectiveness of the central Atlantic region as a dispersal barrier is thought to vary both as function of the adaptive plasticity of the species concerned and environmental changes over time.

INTRODUCTION

Most eukaryotic genomes contain DNA that is many times in excess of the gene-coding requirements of the organism. An important portion of this noncoding DNA is composed of families of various types of repetitive sequences. In addition, the vast majority of unique or single copy sequences may not be transcribed. The study of the evolution of eukaryotic genomes is therefore primarily the study of the evolutionary behavior of noncoding DNA. The rate of evolutionary change in noncoding DNA is much more rapid in plant genomes than in the genomes of higher animals (Flavell 1980, 1982; Murray *et al.* 1981; Walbolt & Cullis 1985; Antonov 1986; Antonov *et al.* 1988). A manifestation of the evolutionary flexibility of plant

NATO ASI Series, Vol. G 22
Evolutionary Biogeography of the
Marine Algae of the North Atlantic
Edited by D. J. Garbary and G. R. South

genomes is, for example, the broad quantitative variation in the amount of repeated and single copy DNA sequences between closely related species (Narayan & Rees 1977; Flavell *et al.* 1974), and the wide range of genome sizes found in angiosperms (Ohri & Khoshoo 1986).

In this paper we explore the extent to which differences in strategy of genomic evolution affect DNA-DNA hybridization data. To this end we compare data of widely distant taxonomic groups and discuss how mode and rate of genomic evolution may influence both the robustness of DNA hybridization interpretation, as well as the ability to use the data in estimating divergence time.

In a second section, we evaluate data of DNA-DNA hybridizations between amphi-Atlantic *Cladophora* species, and relate the data to a hypothetical, peripatric model for trans-Atlantic speciation.

MATERIALS AND METHODS

Cladophora isolates and species distributions

Cladophora species and geographic isolates used in this study are listed in Table 1. Axenic cultures from cloned apical cells were grown according to Bot *et al.* (1989a).

Cladophora sericea is a cold temperate species. On the E Atlantic coast, the southern distribution boundary is close to the 14°C winter isotherm; on the W Atlantic coast, near the 25° C isotherm. The distribution boundary in the N Atlantic corresponds with the 9° C summer isotherm (Cambridge unpubl.).

Cladophora albida is a warm to cold temperate species with rare records from the tropics. The northern distribution boundary is close to the 13°C summer isotherm (Cambridge *et al.* 1984).

Cladophora pellucida is a stenothermous E Atlantic endemic. The species is distributed between the 13°C summer isotherm in the north, and the 28°C summer isotherm in the south (Cambridge unpubl.).

Cladophora pellucida is represented on the W Atlantic coast by a close related species, *Cladophora pellucidoidea*, which is endemic to the Caribbean but extends into the warm temperate western Atlantic region (van den Hoek 1982).

Cladophora vagabunda is a cosmopolitan species with tropical to warm-temperate temperature requirements. In the N Atlantic the species is distributed along the coasts of

America, the Caribbean, W Europe and W Africa. Its northern distribution boundary on both sides of the Atlantic corresponds with the 15°C summer isotherm (Cambridge unpubl.).

DNA-DNA hybridization

Details of DNA isolation and purification, DNA shearing, fragment length determination, preparation of radioactive tracers, and analysis of reassociation kinetics have been described previously (Bot *et al.* 1989a). A brief overview of the DNA-DNA Hybridization procedure is given below.

Single copy, or, more correctly, low copy tracer sequences were isolated from total native DNA, radioactively labelled and allowed to form hybrid double stranded DNA in the presence of an excess of native driver DNA from the same or a different species. After incubation for

Table 1. Origins of North Atlantic *Cladophora* isolates

Species and isolates	Distribution type	Collection site
C. sericea	cold-temperate	
S 84-35		Roscoff (France)
S 84-30		Roscoff
S 83-14		South Iceland
S 82-12		Labrador (Canada)
S 82-6		Newfoundland (Canada)
C. albida	warm-temperate	
A 83-4		Roscoff
A 83-3		Roscoff
A 83-6		Roscoff
A 83-c		Connecticut (USA)
C. pellucida	warm-temperate E Atlantic endemic	
P 84-23		Roscoff
P 83-4		Roscoff
C. pellucidoidea	warm-tropical W Atlantic endemic	
Po 85-100		Georgia (USA)
C. vagabunda	warm-temperate to tropical	
V83-17		Roscoff
V82-2		Curaçao (Caribbean)
V82-10		Curaçao

several days at an appropriate temperature, single stranded DNA was separated from double stranded DNA to determine the amount of radioactivity incorporated into double stranded DNA. NPH (normalized percentage of hybridization), is the percentage of the tracer sequences that forms stable hybrids, normalized to the homologous percentage (tracer and driver DNA from the same species), which, in turn is assumed to be 100%. Subsequently, double stranded (hybridized) DNA is heated from 60 to 95°C in 5°C increments. The amount of tracer DNA released at each temperature was measured. Denaturation or melting temperature (T_{me}) is the temperature at which 50% of the radioactive tracer is released. ΔT_{me} is the difference between the T_{me} of the homologous hybridization (tracer and driver from the same individual) and the T_{me} of the heterologous hybridization (tracer and driver from different individuals or different species).

RESULTS AND DISCUSSION

DNA-DNA hybridization and molecular evolution

It is impossible to make a direct comparison between all the sequences of different eukaryotic genomes. This is because of the enormous complexity. For example, the haploid genome size of *Cladophora* species ranges between about 2.8 x 10^8 and 7.0 x 10^8 base pairs, while 10^{11} base pairs and more are found in the haploid genomes of some salamanders and plants (Bot *et al.* 1989a,b; Comings & Berger 1969; Sparrow *et al.* 1972; Bennet & Smith 1976). In addition, there are problems with the interpretation of direct sequence comparisons. The amount of change accumulated in the course of evolution varies considerably in different nuclear sequences. Moreover, the same sequence can evolve at different rates in different species lineages (Fitch 1986; Li *et al.* 1985). This inconsistence in the internal rates of molecular evolution is an important reason why the choice of genomic regions being sequenced for phylogenetic purposes is focussed on the genes for ribosomal RNA. These genes have the combined advantage of a high abundance and a relatively slow, more or less constant, rate of evolutionary change (Cedergren *et al.* 1988).

DNA-DNA hybridization is an indirect method of comparing genomes. As discussed below, this approach has limitations, but we argue that the technique has the significant advantage that it provides quantitative estimates of the average amount of DNA divergence between large fractions of genomes. Both the data sets (NPH and ΔTme) generated by DNA hybridizations can be considered as indicators of divergent DNA evolution. NPH represents the proportion of tracer sequences that show more than 75-80% homology to its complementary counterparts in other genomes. NPH reduction in heterologous DNA comparisons is mainly caused by three factors: (1) differences in relative and absolute numbers of deletions and insertions between the genomes compared, (2) retardation of the reassociation rate of homologous, but not identical sequences (more divergent sequences reassociate more slowly than less divergent ones), (3) the presence of highly divergent sequences (more than 20-25%).

The average temperature at which hybridized sequences denature or melt (T_{me}), gives insight into the fidelity of base pairing (e.g., the number of binding sites between the two strands of the DNA molecule). T_{me} depression is related to the accumulated number of nucleotide substitutions in homologous sequences (Britten 1986). However, T_{me} is also affected by the length of the homologous regions that sequences have in common. T_{me} depression is, therefore, a function of the distribution of nucleotide substitutions (random or clustered), of the accumulated number of sequence rearrangements and translocations, and of the ability to separate single copy sequences from short, repetitive elements (Tautz *et al.* 1986).

To summarize, DNA hybridization data are the net outcome of a variety of genome evolutionary processes. This suggests that the data may be distorted by the extent to which these processes have characterized the evolution of a genome. To address this point we have made a graph (Fig. 1) of T_{me} depression against NPH reduction based on DNA hybridization data for primates (Sibley & Ahlquist 1984, data published by Sarich *et al.* 1989), birds (Sheldon 1987), *Drosophila* (Schulze & Lee 1986), Umbelliferae and Iridaceae (Antonov *et al.* 1988), and *Cladophora* (Bot *et al.* 1989a, b, 1990, and unpubl.). The data agree on one point: both NPH and T_{me} decrease with phylogenetic distance. However, differences between slopes demonstrate that NPH reduction for a given T_{me} depression is much greater for *Cladophora* and the two angiosperm families than for birds and primates, and even greater than for *Drosophila*, which is thought to be a rapidly evolving group (Britten 1986; Caccone *et al.* 1987a). This variation cannot be explained in terms of a difference in the average rate of random nucleotide substitutions because this would not result in a change of slope. However, the different ratios between NPH and ΔT_{me} can be understood when differences in the internal dynamics or mechanisms of genomic evolution are invoked. The high NPH reduction, relative to T_{me} depression, in *Cladophora* may reveal a pervasive strategy in the evolution of plant genomes, namely a high rate of sequence amplification and DNA turnover (Bedbrook *et al.* 1980; Murray *et al.* 1981; Flavell 1980, 1982). It is beyond the scope of this paper to explain DNA turnover, the mechanisms of which have been extensively reviewed (Dover *et al.* 1982, 1986 a,b). The essence of these studies is that identical sequences are formed because of periodic amplification events. The members of such a family of amplified sequences diverge from each other because of nucleotide substitutions, translocations and rearrangements. Such modifications may occur to the extent that the amplified sequences eventually qualify as nonrepeated sequences, thus increasing the net amount of single copy DNA. To maintain genomic integrity, sequences are partially deleted. Amplifications, translocations, deletion of older DNA, and the retention of evolutionarily younger (i.e. more recently amplified) DNA are the major aspects of DNA turnover.

The overall effect of these processes is to cause a reduction in the amount of single copy DNA that taxa share in common. This reduction is primarily in NPH, but reflected most clearly as a reduction in the ratio between NPH and ΔT_{me}. The low ratios found in *Cladophora* can be explained by assuming that a large fraction of the genomic DNA is evolving by DNA turnover. This may also be a reason for the high levels of intra-and interspecific DNA divergence found, not only in *Cladophora* (Bot *et al.* 1989b), but also in *Drosophila* (Schulze & Lee 1986; Caccone *et al.* 1987a), and in other taxonomic groups with low NPH/ΔT_{me} ratios (Narayan & Rees 1988; Britten *et al.* 1978; Antonov 1986; Antonov *et al.* 1988; Caccone & Powell 1987b). The relatively high NPH/ΔT_{me} ratios in the DNA hybridization data of primates and birds are consistent with the slow rate of DNA turnover observed in the genomes of these organisms (Burr & Schimke 1980; Sherwood & Patton 1982; Romanini, 1985), and may be a reason for the low level of DNA differentiation found within and between species of birds and primates (Bledsoe 1987; Sheldon 1987; Sibley & Ahlquist 1987).

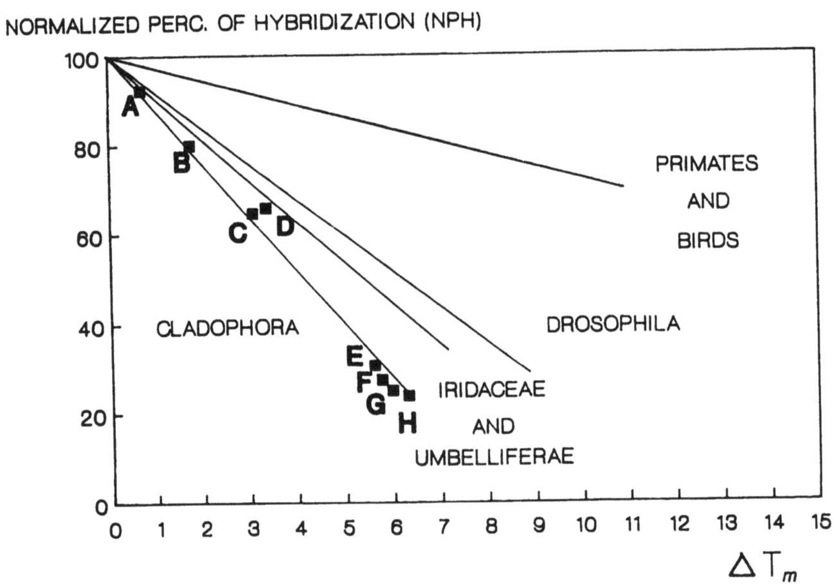

Fig. 1. Relationship between hybridization percentages and reduction of the melting temperature of hybridized DNA sequences. Primates, birds (herons), *Drosophila* (Mulleri complex), Umbelliferae and Iridaceae. *Cladophora*: DNA divergence between A: amphi-Atlantic, B: Australian and Japanese, F: N Atlantic and Pacific *C. albida* isolates. D: *C. sericea* and *C. flexuosa*, E: *C. sericea* and *C. ruchingeri* C: W European and Caribbean *C. vagabunda* isolates. G: W European and Australian species of the *C. pellucida* lineage and H: *C. pellucida* and *C. pellucidoidea*

NPH reduction is not, however, the only effect of DNA turnover. The processes involved change the overall composition and age distribution of sequences, and thereby, even when the genome size remains constant, increase the amount of nonequivalency between the total single copy sequence populations of taxa. Thus, even when genome size remains constant, such processes increase the amount of nonequivalency between the total single copy sequence populations of taxa. This leads to differences in the effective concentration of homologous sequences shared between tracer and driver DNA. As a result, the ratio between the rate of driver self-reassociation and the rate of hybrid formation increases. Consequently, driver sequences are removed from the unreacted pool by driver self-reassociation, and tracer fragments that are potentially capable of forming hybrids are prevented from doing so for lack of driver sequences with which to react. The tracer fragments that remain unhybridized disturb not only NPH, but also the average T_{me} of hybridized DNA (Bledsoe & Sheldon 1989; Springer & Krajewski 1989). The ability to infer phylogenetic relationships from this type of data will therefore depend on the extent to which rates and modes of DNA turnover have characterized the evolution of a genome. High rates of DNA turnover not only affect the topology of reconstructions, but also hamper the possibility of estimating phylogenetic distance.

Another consideration is the relationship between DNA hybridization and the hypothesis of a molecular evolutionary clock. The idea of a molecular clock assumes that long term average rates of nucleotide substitution in single copy sequences equals the rate of neutral substitutions, and that the fixation of neutral substitutions is stochastically constant over time (Britten 1986; Kimura, 1987). As a result, the number of substitutions in single copy sequences of related taxa provides a measure directly proportional to the age of common ancestry (Sibley & Ahlquist 1984, 1987; Templeton 1985; Ruvolo & Smith 1986). It is evident from the reservations discussed above that the extent to which nucleotide substitutions have contributed to reduced NPH and T_{me} values is obscured by the intensity and frequency of other genomic evolutionary processes. Thus, even if the underlying rate of neutral substitutions is stochastically constant per unit time (which is doubtful), rates derived from DNA hybridization data will be overestimated by unknown factors.

The relationship between DNA hybridization and the constancy of the observed rate of DNA change is also problematic. The genome of each organism contains a mixture of sequences that are newly evolved, and those that are retained from the ancient lineage. The latter sequences have maintained their identity for a long time, which implies on an a priori basis a low average rate of evolutionary change. Low NPH values will increasingly reflect this type of ancient, conserved sequence. Thus, the lower the NPH value, the slower the rate of evolution has been in the sequences that hybridize. The observed degree of DNA divergence therefore is not linear over time (i.e., rates measured over short intervals seem to be faster than rates measured over long intervals).

In general, the different evolutionary mechanisms that surround genomic change have an impact on the data of DNA hybridizations. Implicit in this is that particular care must be taken to understand as completely as possible the evolutionary forces in a genome before using the data to imply phylogenetic relatedness. Another aspect is that a universal clock-like basis of DNA change, as predicted by the molecular clock hypothesis, is inadequate to explain the complex interplay between different genomic evolutionary processes. Consequently, generalizations based on similarities in rates of DNA evolution between such widely distant organisms, as for example, animals and seaweeds (Olsen *et al.* 1987; Stam *et al.* 1988; van den Hoek *et al.* 1990), are extremely tentative, and likely to give rise to erroneous conclusions about the time of separation. At this point the question arises as to how to interpret DNA hybridization data in seaweeds. We believe that a suitable approach would be to consider the observed amount of DNA divergence as the total of a series of genomic changes which is the consequence of genetic isolation, and has a tendency to increase with time. Thus, the extent of DNA divergence observed between the genomes of different seaweed species, or geographic isolates of these species (and this applies to each individual level of DNA divergence), has no predictive value with respect to the age of common ancestry, without additional information about the time of separation.

DNA-DNA hybridization and the geography of some N Atlantic *Cladophora* species

The amphi-Atlantic distribution of a seaweed species can be explained in two ways: 1. the species became separated during the formation of the Atlantic Ocean; 2. by trans-Atlantic migration. In the first case, the time of isolation between E and W Atlantic populations equals the age of the Atlantic Ocean (165 million years for the central part (Beurlen 1974), and 20 to 30 million years for the northernmost sector (Thiede & Eldholm 1983)). In the case of migration, the separation between E and W Atlantic populations not only postdates the age of the Atlantic Ocean, but migration may also lead to the evolution of complex amphi-Atlantic distribution patterns. We will illustrate this with a simple theoretical model in which two coasts are separated by an ocean across which there is unidirectional migration. One population of a species occurs on one of these coasts. If the chance of migration is not zero, the occurrence of the species on the opposite coast will only be a matter of time. As long as there is the potential for genetic exchange new migrants will disappear in the recipient population. However, when the amount of migration is insufficient to offset genetic changes, the recipient population will diverge from the donor population and a sequence will start which will end with a state of total reproductive isolation. From that time onwards new migrants will found a new population, and a new sequence begins. Given sufficient time a monophyletic lineage may develop with such relationships as those depicted in Fig. 2, in which we ranked transoceanic speciations with increasing age from D to A. Note, that the ranking in Fig. 2 is based on hypothetical levels of genetic incompatibility and not on speciation in the sense of morphological change. Of course, in reality the situation is much more complex, and aberrations of this model can be caused by a great number of biological and non-biological factors. Nonetheless, relationships of the A or B type indicate isolation, and C or D type relationships indicate recent introduction of migrants.

It appears that when the DNA comparisons between amphi-Atlantic *Cladophora* species (Table 2) are compared with this hypothetical model, DNA relationships in the cold-temperate N Atlantic region are found to be of type D (*C. sericea* and *C. albida*). Between the endemic species from the warm-temperate E and W Atlantic coasts the relationship is found to be of type A (between *C. pellucida* and *C. pellucidoidea*), and the *C. vagabunda* isolates from Roscoff and Curaçao showed a relationship is of type C.

From north to south the W Atlantic coast becomes increasingly isolated (at least with respect to the *Cladophora* species in this study) from the E Atlantic coast. This geographical trend is in agreement with the general pattern of seaweed distribution in the N Atlantic, which shows a decline in floristic similarity between the E and W coast towards the south (Joosten &

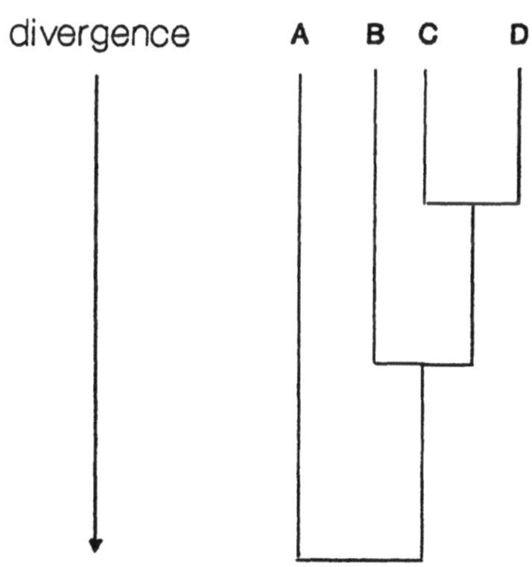

Fig. 2. Hypothetical DNA phylogeny of the possible effects of trans-oceanic migration and long-range genetic relationships (see discussion)

van den Hoek 1986; South 1987; Alvarez *et al.* 1988). The same is true for the genus *Cladophora* as a whole. All species occurring in the cold-temperate region are amphi-Atlantic, and endemic species occur on both sides only at the lower latitudes (van den Hoek 1982).

As discussed above, a calibration of divergence time based on DNA hybridization data derived from other organisms will likely produce inaccurate dating. We therefore chose to analyze the data in Table 2 in relation to the geological history of the present distribution area of the species, and their temperature responses. In this treatment, we make no *a priori* assumptions regarding the relationship between the observed degree of DNA differentiation and the age of divergence.

In seaweed species from the cold-temperate N Atlantic region distributions have been seriously affected by Pleistocene glacial and interglacial oscillations. van den Hoek & Breeman (1990) have reconstructed the distribution areas of some seaweeds during glacial periods. They have suggested that, during the last glaciation, the survival of *C. sericea* on the NW Atlantic coast would have been precarious, whereas the species could have survived on the NE Atlantic coast. This view is corroborated by the absence of significant DNA divergence between isolates of *C. sericea* from Roscoff, Iceland and Canada, which suggests that the species was introduced (or reintroduced) on the NW Atlantic coast after the last glaciation.

Cladophora albida is excluded from the far northern waters of the Atlantic by the cool

Table 2. DNA-DNA hybridizations between amphi-Atlantic *Cladophora* species. Hybridizations were performed according to procedures developed for seaweeds (Bot *et al.* 1989a)

Hybridization	Hybridization percentage	Tm°C	Number of hybridizations
S 84-35*	100.0	0	14
S 84-30	100.7	-0.2	6
S 83-14	98.9	-0.2	6
S 82-6	101.1	+0.1	6
S 82-12	99.6	-0.1	6
A 83-4*	100.0	0	23
A 83-6	98.4	-0.1	3
A 83-3	101.2	+ 0.3	6
A 83-C	92.4	-0.8	14
P 84-23*	100.0	0	6
P 83-4	100.0	+0.2	3
Po 85-100	21.7	-5.9	6
V 83-17*	100.0	0	18
V 82-2	64.7	-2.8	6
V 82-10	68.2	-3.1	6

*Isolates used for the isolation of single copy sequences.

summer temperatures (Cambridge *et al.* 1984). The rarity of the species in tropical waters cannot be explained from its upper survival temperature (Breeman 1988). This temperature response profile enables the species to shift to the south during glacial periods and back to the north during inter-glacials. There is, therefore, no reason to assume that the present amphi-Atlantic occurrence of *C. albida* is the result of recent or subrecent trans-Atlantic migration. DNA comparisons between amphi-Atlantic isolates of this species showed a small but significant reduction, both in the amount of DNA hybridization and in the melting temperature of sequences that hybridized (Table 2). This suggests a degree of isolation. At the observed high level of hybridization, however, modifications in subsets of unstable, probably rapidly evolving sequences, can significantly affect the data (Bot *et al.* 1989b). We have no information about the rate of change in these sequences, and thus cannot estimate the time required for the evolution of the observed degree of divergence. This is not the only reason; the type D or C relationship between the E and W Atlantic *C. albida* isolates suggests a recent subdivision of the distribution range of the species or trans-Atlantic genetic interactions. Although we cannot reject the first possibility, the second alternative is the more likely. In that case, however, the rate of divergent DNA evolution is correlated with unknown factors, such as the intensity of genetic interactions and population size. For this reason we limit our conclusion with respect to the data, with the assumption that the DNA divergence between the amphi-Atlantic *C. albida* isolates is the result of attenuated gene flow across open ocean.

In previous studies (Bot *et al* 1989 a,b) we determined the level of DNA divergence between European and Australian isolates of *C. albida*, and between *C. pellucida* and two of its Australian sister species (*C. feredayi* and *C. pellucida* sp). We were surprised to find that, in these two distantly related *Cladophora* lineages, we could trace common ancestors within the same relative time span. On the basis of these similarities in DNA data and distribution pattern we estimated that the final closure of the eastern section of the Tethys sea during the middle Miocene (12 million years ago; Rögl & Steininger 1984) initiated speciation in these *Cladophora* lineages. The level of DNA divergence between *C. pellucida* and *C. pellucidoidea* is of the same order of magnitude as that observed between C. pellucida and the Australian representatives of the *C. pellucida* lineage. Assuming equal rates of DNA evolution, this suggests that the separation of these species also dates back to the middle Miocene. In the Atlantic Ocean, the Middle Miocene was marked by cooling and a steepening of the latitudinal temperature gradients (Kennett 1982; Valentine 1983; Berggren & Schnitker 1983; Thiede & Eldholm 1983). Climatic changes of this type may have resulted in a southward shift of the distributional area, and, in the case of stenothermous species, a significant reduction in the range (the *C. pellucida* lineage is predominantly comprised of stenothermous species; Cambridge, unpubl. data). It is possible that during the Miocene an ancestral, probably also stenothermous, *C. pellucida*-like species was subdivided into a warm-temperate species in the E Atlantic (the present *C. pellucida*) and a temperate to warm-tropical species (the present *C.*

pellucidoidea) in the Caribbean and the warm temperate region of the W Atlantic coasts (Bot *et al.* 1990).

The biogeographical history of *Cladophora vagabunda* approaches the hypothetical model for transoceanic speciation indicated in Fig. 2. The central sector of the Atlantic Ocean has, in contrast to the northern sector, a more constant and longer evolutionary history. In addition, *C. vagabunda* has a wide temperature response capacity, and has evolved thermal ecotypes (Breeman 1988). In view of the wide distribution of *C. vagabunda* (as *C. fascicularis*) in the tropical sector of the pacific, it is not unrealistic to consider the Caribbean population as an Atlantic donor population. In that case, and according to the model, we might expect to find diverged founder populations of *C. vagabunda* on the W European coast. DNA hybridizations between seaweeds are extremely laborious, and do not allow for the comparison of large numbers of isolates. However, the above hypothesis is supported by preliminary results of allozyme electrophoresis which showed that on the E Atlantic coast *C. vagabunda* is comprised of an assembly of different population types. This is perhaps also the reason why van den Hoek (1963, 1982) observed such a wide morphological range in this species. The implication of this is that the observed level of DNA differentiation may depend on the isolate being used. This is a serious problem, particularly in a seaweed genus like *Cladophora* in which there appears to be no causal connection between the degree of DNA change and the degree of change of important morphotaxonomic traits (Bot. unpubl.). Clearly, in a situation like this, DNA hybridization data cannot be interpreted without additional information on the population structure on both sides of the Atlantic.

The evolution of thermal ecotypes in *C. vagabunda* demonstrates that selection pressure can cause adaptation to physical factors in the environment. Furthermore, it is an indication of low gene flow because selection cannot produce specialized ecological units when significant gene flow occurs (Breeman 1988). On the other hand, this factor reduces the ability to determine time of isolation from DNA hybridization data, since the data may reflect both genetic isolation and the effects that adaptation to a changing or new environment may have on genomic characters (Walbot & Cullis 1983, 1985). Nevertheless, it can be stated that the E and W Atlantic isolates of the eurythermous *C. vagabunda* lineage are separated in a more recent period than the E and W Atlantic species of the stenothermous *C. pellucida* lineage. This implies that there is a relationship between the degree of isolation of the E and W Atlantic coasts and the thermal response capacity of the species on these coasts.

In conclusion, the biogeographical history of the amphi-Atlantic *Cladophora* species in this study can be summarized as follows: all estimates with respect to the time of isolation between E and W Atlantic isolates indicate genetic relations postdating the age of the N Atlantic Ocean. The high degree of DNA similarity in the cold temperate region suggests a strong role for the Pleistocene glacial and interglacial periods. The possible effects of these climatic alterations can be summarized as follows: a disruption of existing biogeographic patterns during glaciations, resulting in extinctions on one or both sides, and a restoration of the same or comparable biogeographic patterns by trans Atlantic migration during interglacial periods. The present interglacial may have been important for the introduction (or reintroduction) of species with a temperature profile comparable with that of *C. sericea* on the NW Atlantic coasts. The longer and more constant history of the central Atlantic region, may have allowed the evolution of differentiated biogeographical patterns. In the development of these patterns two factors may have played an important role: first, the environmental differentiation between the central E and W Atlantic coasts, which probably gave rise to the high number of endemic stenothermous seaweed species on the warm-temperature sector of the NE Atlantic; second, the ecological window and plasticity of different species enabling them to cope with the various environments that occur within the central Atlantic region.

ACKNOWLEDGEMENTS

We would like to thank Dr. G.R. South for inviting us to the NATO workshop in St. Andrews, at which the above paper was presented. M.L. Cambridge kindly provided data on the temperature responses of some *Cladophora* species. This study was supported by the Foundation for Fundamental Biological Research (BION), which is subsidized by the Netherlands Organization for the Advancement of Pure Research (ZWO).

REFERENCES

Alvarez M, Gallardo T, Ribera MA, Gomez Garetta A (1988) A reassessment of Northern Atlantic seaweed biogeography. Phycologia 27: 221-223

Antonov AS (1986) Molecular analysis of plant genomes; Conserved and diverged DNA sequences. In: Dutta SK (ed) DNA systematics. Vol. II. CRC Press, Florida, p 171-183

Antonov AS, Valiejo-Roman KM, Pimenov MG, Beridze NA (1988) Non-equivalency of genera in Angiospermae: evidence from DNA hybridization studies. Pl Syst Evol 161: 155-168

Bedbrook JR, O'Dell M, Flavell RB (1980) Amplification of rearranged repeated DNA sequences in cereal plants. Nature 288: 133-137

Bennett MD, Smith JB (1976) Nuclear DNA amounts in angiosperms. Proc R Soc London, Ser B 274: 227-274

Berggren WA, Schnitker D (1983) Cenozoic marine environments in the North Atlantic and the Norwegian Greenland sea. In: Bott MPH, Saxov S, Thiede J (ed) Structure and development of the Greenland-Scotland ridge, new methods and concepts. Plenum Press, N.Y., p 495-548

Beurlen K (1974) Die geologische Entwickelung des Atlantischen Ozeans. Geotekt Forsch 46: 1-69

Bledsoe AH (1987) DNA evolutionary rates in nine pre-married passerine birds. Mol Biol Evol 4: 559-571

Bledsoe AH, Sheldon FH (1989) The metric proportions of DNA-DNA hybridization dissimilarity measures. Syst Zool 38: 93-105

Bot PVM, Stam WT, Boele-Bos SA, Hoeck van den C, Delden van W (1989a) Biogeographic and phylogenetic studies of *Cladophora* (Cladophorales, Chlorophyta) using DNA-DNA hybridization. Phycologia 28: 159-168

Bot PVM, Holton R, Stam WT, Hoeck van den C (1989b) Molecular divergence between North-Atlantic and Indo-West Pacific *Cladophora albida* (Cladophorales, Chlorophyta) isolates as indicated by DNA-DNA, hybridization. Mar Biol 102: 307-313

Bot PVM, Burssaard CPD, Stam WT, Hoeck van den C (1990) Evolutionary relationships between four species of *Cladophora* (Cladophorales, Chlorophyta) based on DNA-DNA hybridization. J Phycol (submitted)

Breeman AM (1988) Relative importance of temperature and other factors in determining geographic boundaries in seaweeds: experimental and phenological evidence. Helgol Meeresunt 42: 199-241

Britten RJ, Ceta A, Davidson EH (1978) The single-copy sequence polymorphism of the sea urchin *Strongilocentrotus purpuratus*. Cell 15: 1175-1186

Britten, RJ (1986) Rates of DNA sequence evolution differ between taxonomic groups. Science 231: 1393-1398

Burr HE, Schimke RT (1980) Intragenomic DNA sequence homologies in the chicken and other members of the class Aves: DNA reassociation under reduced stringency conditions. J Mol Evol 15: 291-308

Caccone LA, Amato GD, Powell RJ (1987a) Intraspecific DNA divergence in *Drosophila*: a study on parthenogenetic *D. mercatorum*. Mol Biol Evol 4: 343-350

Caccone A, Powell RJ (1987b) Molecular evolutionary divergence among North American cave crickets. II. DNA-DNA hybridization. Evolution 41: 1215-1238

Cambridge ML, Breeman AM, Oosterwijk R, van den Hoek C (1984) Temperature responses of some North Atlantic *Cladophora* species (Chlorophyceae) in relation to their geographic distribution. Helgol Meeresunt 38: 249-363

Cedergren R, Gray WG, Abel Y, Sankoff D (1988) The evolutionary relationships among known lifeforms. J Mol Evol 28: 98-112

Comings DE, Berger RO (1969) Gene products in *Amphiuma*: an amphibian with an excessive amount of DNA. Biochem Genet 2: 319-333

Dover GA, Brown SDM, Coen ES, Dallas J, Strachan T, Trick M (1982) The dynamics of genome evolution and species differentiation. In: Dover GA, Flavell RB (eds) Genome Evolution. Academic Press, London, p 343-366

Dover GA (1986a) The spread and success of non-Darwinian novelties. In: Karlin S, Nevo E (eds) Evolutionary processes and theory. Academic press, New York, p 199-237

Dover GA (1986b) Molecular drive in multigene families: how biological novelties arise, spread and are assimilated. Trends Genet 2: 159-165

Fitch WM (1986) The estimate of total nucleotide substitutions from pairwise differences is biased. Phil Trans R Soc Lond [Biol] 312: 317-324

Flavell RB, Bennett MD, Smith JB, Smith DB (1974) Genome size and the proportion of repeated nucleotide sequences in plants. Biochem Genet 12: 257-269

Flavell RB (1980) The molecular characterization and organization of plant chromosomal DNA sequences. Ann Rev Pl Physiol 31: 569-596

Flavell RB (1982) Sequence amplification, deletion and rearrangement; major sources of variation during species divergence. In: Dover GA, Flavell RB (eds) Genome evolution. Academic Press, London, p 301-324

Hoeck C van den (1963) Revision of the European species of *Cladophora*. Brill, Leiden.

Hoeck C van den (1982) A taxonomic revision of the American species of *Cladophora* (Chlorophyceae) in the North Atlantic Ocean and their geographic distribution. North-Holland Publishing Co., Amsterdam

Hoek C van den, Breeman AM (1990) Seaweed biogeography of the North Atlantic: where are we now? In: Garbary DJ, South, GR (eds), Evolutionary biogeography of the marine algae of the North Atlantic. Springer-Verlag, Berlin, p 55-86

Hoeck C van den, Breeman AM, Stam WT (1990) The geographic distribution of seaweed species: present and past. In: Wolff WJ, Brouns JJWM, Beukema JJ (eds) Expected effects of climatic changes on marine coastal ecosystems. Balkema, Rotterdam (in press)

Joosten AMT, Hoeck van den C (1986) World-wide relationships between red seaweed floras: a multivariate approach. Bot Mar 29: 195-214

Kennett JP (1982) Marine Geology. Prentice-Hall, New York

Kimura M (1987) Molecular evolutionary clock and the neutral theory. J Mol Evol 26: 24-33

Li W-H, Luo CC, Wu CI (1985) Evolution of DNA sequences. In: McIntyre RJ (ed) Molecular evolutionary genetics. Plenum, New York, p 1-84

Murray MG, Peters DL, Thompson WF (1981) Ancient repeated sequences in the pea and mung bean genomes and implications for genome evolution. J Mol Evol 17: 31-42

Narayan RKJ, Rees H (1977) Nuclear DNA divergence among *Lathyrus* species. Chromosoma 63: 101-107

Olsen JL, Stam WT, Bot PVM, Hoeck C van den (1987) scDNA-DNA hybridization studies in Pacific and Caribbean isolates of *Dictyosphaeria cavernosa* (Chlorophyta) indicate a long divergence. Helgol Meeresunt 41: 377-383

Ohri D, Khoshoo TN (1986) Plant DNA: contents and systematics. In: Dutta SK (ed) DNA systematics. Vol. II. CRC Press, Florida, p 1-19

Romanini MGM (1985) On the meaning of genome size in mammals. Acta Zool Fennica 170: 37-39

Rögl F, Steininger FF (1984) Neogene Paratethys Mediterranean and Indo-pacific seaways. Implications for the palaeobiogeography of marine and terrestrial biotas. In: Brenchley PJ (ed) Fossils and climate. John Wiley & Sons, Chichester, p 171-200

Ruvolo M, Smith TF (1986) Phylogeny and DNA-DNA hybridization. Mol Biol Evol 3: 285-289

Sarich VM, Schmid CW, Marks J (1989) DNA hybridization as a guide to phylogenies: a critical analysis. Cladistics 5: 3-32

Schulze LDH, Lee CS (1986) DNA sequence comparison among closely related *Drosophila* species in the Mulleri complex. Genetics 113: 287-303

Sheldon FH (1987) Rates of single copy DNA evolution in herons. Mol Biol Evol 4: 56-69

Sherwood SW, Patton JL (1982) Genome evolution in pocket gophers (genus *Thomomys*). II. Variation in cellular DNA content. Chromosoma 85: 163-179

Sibley LCG, Ahlquist JE (1984) The phylogeny of the hominoid primates as indicated by DNA-DNA hybridization. J Mol Evol 20: 2-15

Sibley CG, Ahlquist JE (1987) DNA hybridization evidence of hominoid phylogeny: results from an expended data set. J Mol Evol 26: 99-121

South GR (1987) Biogeography of the benthic marine algae of the North Atlantic Ocean - an overview. Helgol Meeresunt 41: 273-282

Sparrow AH, Price HJ, Underbrink AG (1972) A survey of DNA content per cell and per chromosome of prokaryote and eukaryote organisms: some evolutionary considerations. Brookhaven Symp Biol 23: 451-493

Springer M, Krajewski C (1989) DNA hybridization in animal taxonomy: A critique from first principles. Quart Rev Biol 64: 291-318

Stam WT, Bot PVM, Poele-Bos SA, Rooij van JM, Hoeck van den C (1988) Single copy DNA-DNA hybridizations among five species of *Laminaria* (Phaeophyceae): phylogenetic and biogeographic implications. Helgol Meeresunt 42: 251-267

Tautz D, Trick M, Dover GA (1986) Cryptic simplicity in DNA is a major source of genetic variation. Nature 322: 652-656

Templeton AR (1985) The phylogeny of the hominoid primates: a statistical analysis of the DNA-DNA hybridization data. Mol Biol Evol 2: 420-433

Thiede J & Eldholm O (1983) Speculations about the paleodepth of the Greenland-Scotland ridge during late Mesozoic and Cenozoic times. In: Bott MPH, Saxov S, Thiede J. (ed) Structure and development of the Greenland-Scotland ridge, new methods and concepts. Plenum Press, N.Y., p 445-456

Valentine JW (1984) Climate and evolution in the shallow sea. In: Brenchley PJ (ed) Fossils and climate. John Wiley & Sons, Chichester, 265-277

Walbot V, Cullis CA (1983) The plasticity of the plant genome - is it a requirement for success? Pl Mol Biol Rep 1: 3-11

Walbot V, Cullis CA (1985) Rapid genomic change in higher plants. Ann Rev Pl Physiol 36: 367-396

GENETICS OF MORPHOLOGY AND GROWTH IN *LAMINARIA* FROM THE NORTH ATLANTIC OCEAN - IMPLICATIONS FOR BIOGEOGRAPHY

B. Egan[1*], Z. Garcia-Ezquivel[2], B. H. Brinkhuis[2*] and C. Yarish[1]

[1]Dept. of Ecology & Evolutionary Biology
University of Connecticut,
641 Scofieldtown Rd,
Stamford, CT 06903
U.S.A.

[2]Marine Sciences Research Center
State University of New York
Stony Brook
NY 11794-5000
U.S.A.

ABSTRACT - F_1 sporophytes produced by inter- and intraspecific crosses of North Atlantic *Laminaria* were reared in greenhouse culture conditions from February through June, 1989. Growth rates and morphometric variation among different families were assessed throughout this period in order to evaluate phenetic and distributional relationships. Significant differences in growth rates were observed among families, with slower rates associated with offspring from the Faeroe Islands and Helgoland, and maximum rates observed in Long Island Sound sporophytes. Between-site and interspecifically-derived offspring showed intermediate growth rates. Morphometric measurements also varied with the type of cross and source of isolates, but stipe length and blade width characteristics seem to be sex-linked. Although no sterility barriers were observed among isolates, and genetic contribution to the measured characters was demonstrated in F_1 sporophytes, this study substantiates the presence of temperature races of *L. saccharina* and *L. longicruris* along the North Atlantic coast.

INTRODUCTION

The genus *Laminaria* inhabits temperate and circum-arctic waters in the northern, and temperate waters in the southern, hemispheres. It is presently believed that the genus originated in the northern Pacific (Lüning 1985). Wilce (1965) concluded there were only three species in the Section Simplices from the North Atlantic. These are *L. solidungula* J. Ag. (with a distinct discoid holdfast whereas all other species have haptera), *L. saccharina* (L.) Lamour. and *L. longicruris* de la Pyl. Wilce (1965) proposed that all hollow-stiped

* Deceased

NATO ASI Series, Vol. G 22
Evolutionary Biogeography of the
Marine Algae of the North Atlantic
Edited by D. J. Garbary and G. R. South

plants (including *L. faeroensis* Børg. from the Faeroe Islands) be identified as *L. longicruris* and all solid-stiped plants as *L. saccharina*.

The use of the hollow stipe as a taxonomic character was first questioned by Mann (1971) who had difficulty separating *L. longicruris* from *L. agardhii* Kjellm. (= *L. saccharina*). Chapman (1973) shared similar doubts when he found a clinal relationship in stipe length and hollowness with environmental conditions. According to Chapman (1974), the two taxa are conspecific.

Geographic variation among seaweed populations has resulted in considerable taxonomic confusion where morphological or physiological characters have been used to delineate species (Mathieson *et al.* 1981; Rice & Chapman 1985). The definition of what constitutes a species lies at the heart of taxonomic disputes. The classically accepted definition states that true biological species are reproductively isolated. Ecotypic status is frequently assigned to a taxon when "species" cannot be determined due to a lack of sterility barriers (Yarish *et al.* 1990a,b). By definition, an ecotype is the product of a genotypic response of a species to a particular habitat (environmental) condition and it implies interfertility with other members of the species (Turesson 1922). Both physiological and morphological ecotypes have been described for various algae (e.g., Gerard & Mann 1979; Yarish *et al.* 1979; Mathieson *et al.* 1981; Russell & Fielding 1981; Espinoza & Chapman 1983; Bolton 1983; Innes 1984; Gerard 1988; Gerard & Du Bois 1988). Ecotypic variation in seaweeds has been considered as evidence of genetic differentiation among populations, whose members are perhaps in the process of speciation (Innes 1984).

Crossing experiments offer a means of assessing differentiation when correlated to specific environmental differences. However, interpretation of such experiments is not straightforward. Genetic differentiation may evolve without affecting compatibility, and intersterility does not necessarily indicate accumulation of a large number of genetic differences (Innes 1984). Increased incompatibility is often associated with greater geographical separation (Rueness 1973; Müller 1979; West *et al.* 1978; Guiry & West 1983), but there are numerous examples where genetic differentiation occurs among populations separated by only short distances (e.g., Russell & Morris 1970; DeSilva & Burrows 1973; Espinoza & Chapman 1983; Innes 1987).

Interfertility was reported in crossing experiments between *L. saccharina* and *L. longicruris* (Lüning *et al.* 1978; Bolton *et al.* 1983). No crosses between *L. saccharina* and *L. longicruris* from the western Atlantic Ocean have been reported. Furthermore, all of the crosses reported thus far have been made with materials collected from near the middle of the latitudinal distribution ranges of the genus. Preliminary data on crosses between *L. saccharina* and *L. longicruris* from Long Island Sound, U.S.A. were presented recently (Yarish *et al.* 1990).

Bolton & Lüning (1982) concluded that there was a lack of within-species differences in the genus to different temperature regimes in isolates from the North Atlantic Ocean. However, physiological differentiation, e.g. seasonal nutrient/growth interactions (Gagne *et al.* 1982), light/photosynthesis interactions (Gerard 1988), photosynthesis/temperature interactions (Gerard & Du Bois 1988), has been reported in *Laminaria* populations along the western Atlantic coastline.

In this paper, we report on crossing experiments between *L. saccharina* and *L. longicruris* from the North Atlantic, and evaluate the importance of selection of isolates from the widest possible latitudinal ranges in biogeographical studies.

MATERIALS AND METHODS

Biogeography of Laminaria

Observations on stipe morphology have been synthesized and used for the plotting of the biogeographic distribution of *L. saccharina* (Fig.1) and *L. longicruris* (Fig. 2) in the North Atlantic and Arctic Oceans. The figures also include mean sea surface temperature (^{o}C) in

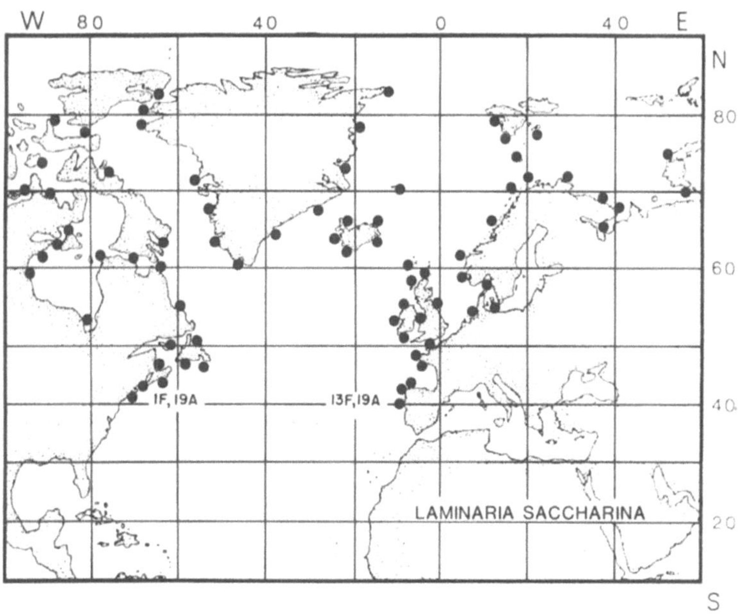

Fig. 1. The distribution of *Laminaria saccharina* in the North Atlantic. Data sources are listed in the text. Temperatures presented on map are February (F) and August (A) mean surface water isotherms from Sverdrup *et al.* (1942)

February (F) and August (A) from Sverdrup *et al.* (1942). The distribution data are based on the following sources: Batters (1889); Børgesen (1903-08); Christensen & Thomsen (1974); Egan & Yarish (1988); Guiry (1978); Hoek (1982a, b); Hylander (1928); Irvine (1974, 1982); Jónsson (1912); Kain (1976); Kylin (1947); Lee (1980); Lund (1959); Lüning (1975, 1985); Lüning *et al.* (1978); Parke (1948); Parke & Dixon (1976); Taylor (1957); Turner (1802); Wilce (1963, 1965). Wilce (1965) indicated that the southern limit of *Laminaria* distribution in the western Atlantic is Long Island, New York. However, Egan & Yarish (1988) have recently reported discovery of a deep-water population of *L. saccharina* off the coast of New Jersey.

Laminaria longicruris-type plants have been observed as far west as the Thimble Islands, Branford, CT, U.S.A. (Fig. 3) in water 5 m deep. Mean stipe lengths of 140 cm have been reported (Yarish & Egan 1989). Significant stands of *L. saccharina*-type plants with stipes < 10 cm long occur as far west as Stamford, CT in water depths rarely exceeding 4 m.

Two species of *Laminaria* are found on Long Island, *L. saccharina* and *L. digitata*. *Laminaria digitata* is found only at Montauk Point (Fig. 3). *Laminaria saccharina* is found primarily along the north shore of the Island, wherever cobble and rock deposited during the last glaciation occurs; it also occurs at Montauk Point. Plants were found in 1969 by Yarish as far west on the south shore as the jetty at Shinnecock Inlet (unpublished data); Liddle (pers. comm.) reported its continued presence there in the winter of 1985. Populations are rarely

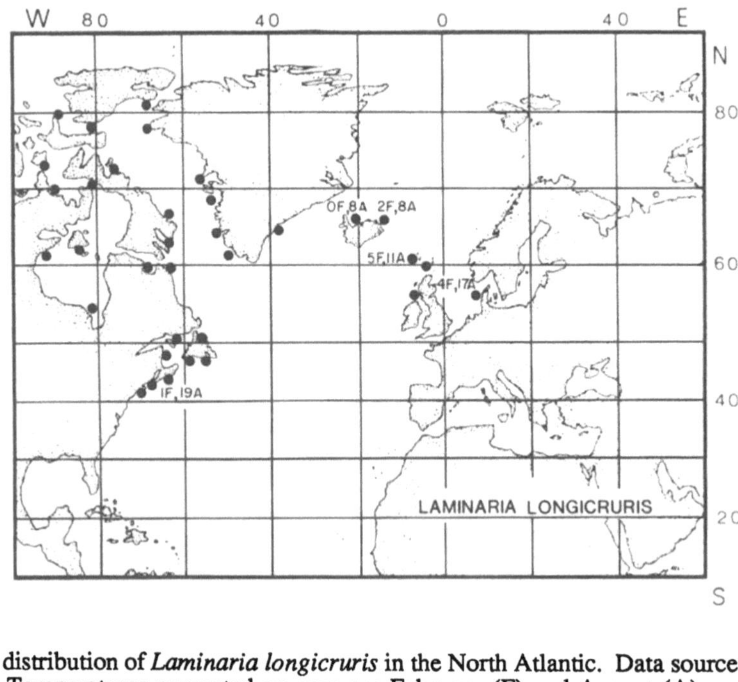

Fig. 2. The distribution of *Laminaria longicruris* in the North Atlantic. Data sources are listed in the text. Temperatures presented on map are February (F) and August (A) mean surface water isotherms (Sverdrup *et al.* 1942)

found at depths greater than 5 m along the north shore. All of the plants have stipe lengths and morphology typical of *L. saccharina*. Stipes reach a mean length of 4.5 cm (Lee & Brinkhuis 1986). Occasionally, longer-stiped plants can be found at Montauk Point, a more exposed environment than other north shore locations. Some plants with stipes up to 60 cm long have been collected, but these have not been observed to be hollow.

Crossing Experiments

Crossing experiments were conducted both within and between culture isolates of *Laminaria saccharina* and *L. longicruris*. The crosses and sources of the isolates are presented in Table 1. Female and male gametophytes of each plant were maintained for at least 6 months in red light (5 μmol m^{-2} s^{-1} photon fluence rate) at 10°C. For experimentation, each gametophyte culture was ground in an Osterizer blender. The resultant slurry was passed through a 0.74 um Nitex mesh filter. Filtrates were allowed a recovery period of 1 week in crystallizing dishes at 10°C under white fluorescent light (40 μmol m^{-2} s^{-1} photon fluence rate) before crosses were conducted. When sporophytes became evident, cultures were trans-

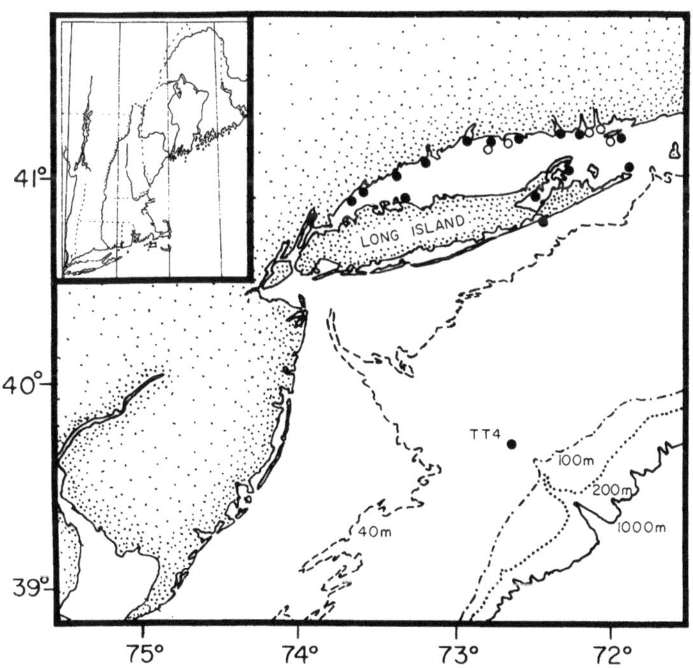

Fig. 3. The distribution of *L. longicruris* (open circles) and *L. saccharina* (closed circles) in the Long Island Sound, U.S.A. region. Data sources are listed in the text

ferred to 10 l polycarbonate aquaria and maintained in an incubator at 10°C. Sporophytes approximately 10 cm long were transferred to greenhouse culture.

Growth rates were determined by measuring blade length changes over two week intervals, as well as using the hole-punch technique (Parke 1948). Measurements of stipe length and blade width were also recorded bi-weekly.

Statistical significances in gross growth rates among the twenty-one experimental crosses were determined using the Kruskal-Wallis test (Sokal & Rohlf 1981). The test was applied on three separate dates (27/4, 11/5, 25/5/1989), after heteroscedasticity was confirmed using Bartlett's test for homogeneity of variances (Sokal & Rohlf 1981). Multiple comparisons of the maximum meristematic growth rates were carried out using the Games and Howell test for equality of pairs of means with unequal variances (Sokal & Rohlf 1981).

Discriminant analysis (Sokal & Rohlf 1981) was applied to the crosses in order to obtain discriminant functions that maximized the separation of groups based on data from stipe lengths, widest blade dimensions, blade widths at 10 cm, and meristematic growth rates. The derived canonical function coefficients were standardized by conditional (within groups) standard deviations. Clustering was performed on the same data using untransformed measurements. A dissimilarity matrix of average "taxonomic" distance coefficients was used for constructing clusters based on an unweighted pair-group arithmetic average clustering method (UPGMA). The distance coefficients and UPGMA clusters (data from only one date, 27/4/1989, when plants were eight weeks in greenhouse culture, and before blade erosion became significant) were performed on NTSYS-pc (Rohlf 1988).

Gametophyte and young sporophyte cultures were maintained at The University of Connecticut at Stamford, CT (Yarish & Egan 1989). The flow-through marine greenhouse (see Brinkhuis & Hanisak 1982) at The State University of New York at Stony Brook, NY (SUSB) was used for maintaining large sporophytes.

RESULTS

All gametophyte crosses produced successful progenies sporophytes. The latter grew to approximately 10.0 cm in laboratory culture. Growth was slow during the first four weeks after plants were transferred to greenhouse culture. Most blades grew exponentially after reaching approximately 30.0 cm with maximum growth occurring in the eighth or tenth week (May). All morphometric measurements varied with type of cross (self-crosses and reciprocal crosses) and sources of the isolates.

ferred to 10 l polycarbonate aquaria and maintained in an incubator at 10°C. Sporophytes approximately 10 cm long were transferred to greenhouse culture.

Growth rates were determined by measuring blade length changes over two week intervals, as well as using the hole-punch technique (Parke 1948). Measurements of stipe length and blade width were also recorded bi-weekly.

Statistical significances in gross growth rates among the twenty-one experimental crosses were determined using the Kruskal-Wallis test (Sokal & Rohlf 1981). The test was applied on three separate dates (27/4, 11/5, 25/5/1989), after heteroscedasticity was confirmed using Bartlett's test for homogeneity of variances (Sokal & Rohlf 1981). Multiple comparisons of the maximum meristematic growth rates were carried out using the Games and Howell test for equality of pairs of means with unequal variances (Sokal & Rohlf 1981).

Discriminant analysis (Sokal & Rohlf 1981) was applied to the crosses in order to obtain discriminant functions that maximized the separation of groups based on data from stipe lengths, widest blade dimensions, blade widths at 10 cm, and meristematic growth rates. The derived canonical function coefficients were standardized by conditional (within groups) standard deviations. Clustering was performed on the same data using untransformed measurements. A dissimilarity matrix of average "taxonomic" distance coefficients was used for constructing clusters based on an unweighted pair-group arithmetic average clustering method (UPGMA). The distance coefficients and UPGMA clusters (data from only one date, 27/4/1989, when plants were eight weeks in greenhouse culture, and before blade erosion became significant) were performed on NTSYS-pc (Rohlf 1988).

Gametophyte and young sporophyte cultures were maintained at The University of Connecticut at Stamford, CT (Yarish & Egan 1989). The flow-through marine greenhouse

Table 1. Self-crosses and reciprocal crosses of *Laminaria saccharina* (Ls) and *L. longicruris* (Ll). Sources of isolates are Connecticut, USA (CT); Nova Scotia, Canada (NS); Igloolik Island, Canadian Arctic (IGL); Faroe Islands (FAR); Helgoland, FRG (H). Codes of isolates are those listed in the culture collections of C. Yarish, Connecticut, USA and K. Lüning, Helgoland, FRG

Culture collections code	Self crosses	Reciprocal crosses
BL32, Yarish	Ll X Ll CT	Ll CT X Ls CT
LlNS, Yarish	Ll X Ll NS	Ll CT X Ls H
1101/1102, Lüning	Ll X Ll IGL	Ll CT X Ls NS
1145/1146, Lüning	Ll X Ll FAR	Ll CT X Ll NS
LsCT, Yarish	Ls X Ls CT	Ll NS X Ls CT
1011/1012, Lüning	Ls X Ls NS	Ls NS X Ls CT
1005/1006, Lüning	Ls X Ls H	Ls H X Ls CT

(see Brinkhuis & Hanisak 1982) at The State University of New York at Stony Brook, NY (SUSB) was used for maintaining large sporophytes.

RESULTS

All gametophyte crosses produced successful progenies sporophytes. The latter grew to approximately 10.0 cm in laboratory culture. Growth was slow during the first four weeks after plants were transferred to greenhouse culture. Most blades grew exponentially after reaching approximately 30.0 cm with maximum growth occurring in the eighth or tenth week (May). All morphometric measurements varied with type of cross (self-crosses and reciprocal crosses) and sources of the isolates.

Meristematic Growth

In most cases the net growth rate was the same as the gross growth rate during the first six weeks. However, erosion rate became a significant factor in the following weeks. The critical growth period (when gross growth = erosion, or net growth = 0) was reached earlier in crosses involving plants with distributions occurring further north from LIS. Erosion rate was usually greater than net growth rate after the eighth week.

Connecticut (CT) plants always showed a sustained positive net growth, while crosses from Nova Scotia (NS), Igloolik (IGL), Helgoland (H), and Faeroe Islands (FAR) showed different degrees of adaptation to Long Island Sound (LIS) conditions.

During the peak growing period, *L. saccharina* (CT) showed a mean net growth rate of 1.6 cm d-1, which was at least 2X greater than that of either NS or H plants (Fig. 4 a-c). Reciprocal crosses involving CT and NS, and CT and H plants had net growth values between those observed in self-crosses. Growth was more vigorous and was sustained for a longer period when the female parent came from CT (e.g. Fig. 4 d-g).

The same behavior was observed in *L. longicruris*. *Laminaria longicruris* (CT) grew (1.5 cm d-1) more rapidly than the self-crosses of IGL (1.15X), NS (1.25X), and FAR (1.5X) (Fig. 4 h-k). After the tenth week, growth of NS and IGL plants slowed considerably (Fig. 4 i, k) as water temperatures began to exceed 12°C. Blades began to disintegrate rapidly resulting in negative growth rates throughout the following weeks (e.g. Fig. 4 k). *Laminaria longicruris* (FAR) was never able to adapt to greenhouse culture conditions. Blade erosion was always greater than growth which ultimately resulted in the loss of the plants. A first batch of plants completely disintegrated within six weeks; a second batch, used in the tenth week, behaved in a similar manner (Fig. 4 j). Reciprocal crosses of CT X NS had higher sustained growth rates than either of the self-crosses (Fig. 4 l, m).

Different patterns were observed in the reciprocal crosses of both species from CT, NS, and H. *Laminaria longicruris* (CT) X *L. saccharina* (NS) resulted in heterosis for growth rate (Fig. 4 r, s). In contrast, offspring of *L. longicruris* (NS) X *L. saccharina* (CT) were

Table 2. Grouping of intra-, and interspecific *Laminaria* F1 sporophytes, based on multiple comparisons (Games and Howell test; Sokal & Rohlf 1981) of their maximum growth rates in greenhouse culture conditions. See Table 1 for codes

Date	Cross	Max growth rate (cm d^{-1})	Sample size	Variance	Group observations
25/5	LlCT X LsCT	1.48	16	0.6500	Highest variances,
11/5	LlCT X LsH	1.59	16	0.6053	undetectable differences
27/4	LsH X LlCT	1.25	14	0.1584	with the rest of the
11/5	LlCT X LlNS	1.68	7	0.2304	crosses
27/4	LsCT X LlCT	2.57	32	0.4010	Fast growers
27/4	LsCT X LsNS	2.56	30	0.3528	
27/4	LsNS X LlCT	1.86	32	0.0502	Intermediate growers
27/4	LlCT X LsNS	2.05	31	0.0912	
27/4	LsCT X LsCT	1.64	32	0.1730	
11/5	LsCT X LsNS	1.44	32	0.3466	
27/4	LlIGL X LlIGL	1.43	31	0.0595	
25/5	LlNS X LlCT	1.93	29	0.5640	
27/4	LlNS X LsCT	1.87	23	0.3376	
25/5	LlCT X LlCT	1.83	20	0.0292	
27/4	LlNS X LlNS	1.17	28	0.1096	
27/4	LsH X LsCT	1.07	31	0.1289	
27/4	LsNS X LsCT	0.95	27	0.0246	
25/5	LlFAR X LlFAR	0.62	10	0.0112	Slow growers
11/5	LsH X LsH	0.65	6	0.0086	
11/5	LsNS X LsNS	0.80	4	0.0086	
22/6	LsCT X LlNS	0.55	3	0.0110	

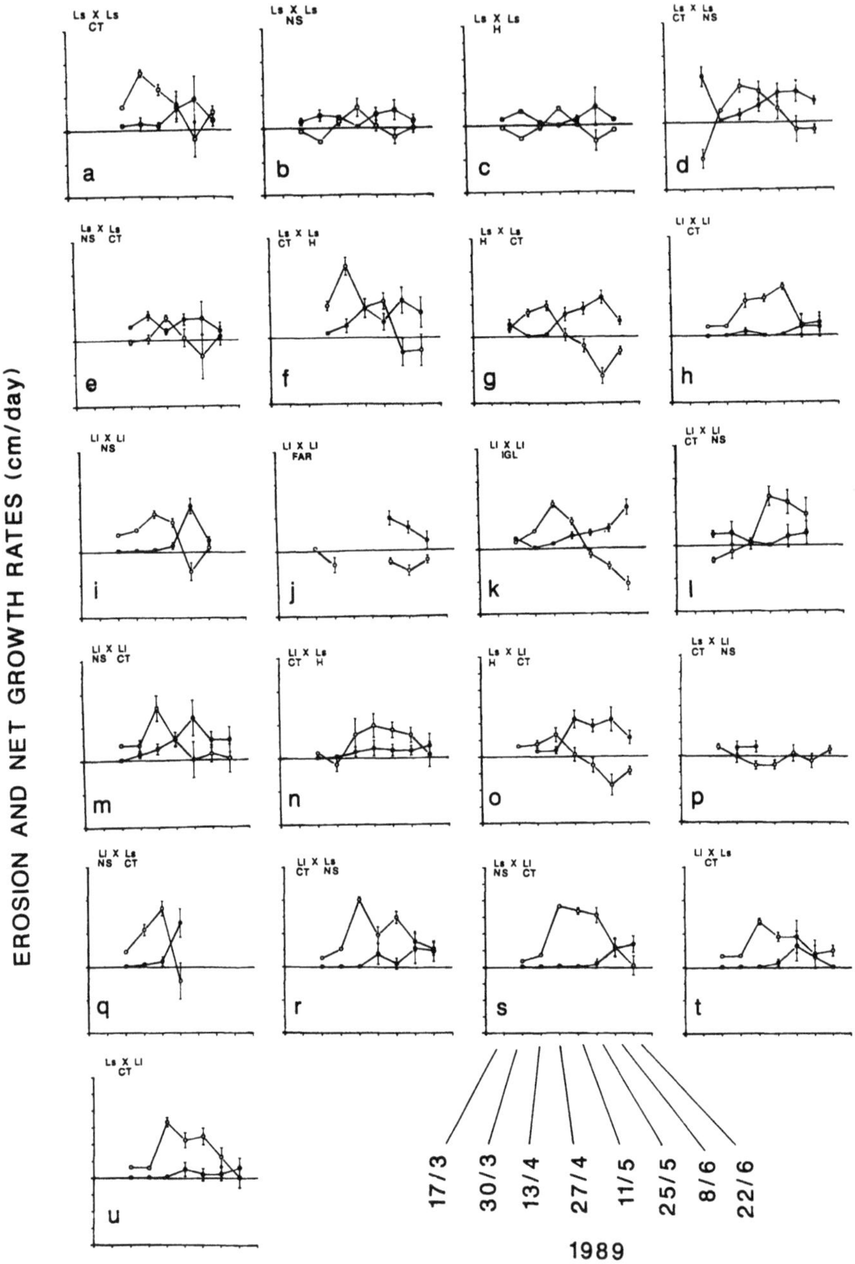

Fig. 4. Erosion (closed circles) and net growth (open circles) rates of F1 progenies from crosses of *L. longicruris* and *L. saccharina* (Table 1) in greenhouse culture. Bars represent 95% confidence limits. Initial sample sizes were n = 32. However, sample sizes decreased over time due to loss of plants, or culture tank thinning to compensate for increasing sizes of plants. Sample sizes for 27/4/89, for example, are presented in Table 2. First species code listed for each cross represents the female parent in that cross

weaker and net growth rates decreased to below 0 cm d-1 after the sixth or eighth week of greenhouse culture (Fig. 4 p, q). Connecticut X Helgoland offspring produced a hybrid growth pattern, with higher growth rates and longer periods of positive increments associated with the female *L. longicruris* (CT) (Fig. 4 n, o). Connecticut *L. longicruris* X *L. saccharina* also exhibited hybrid growth patterns (Fig. 4 t, u).

Significant differences (P<0.01) were observed in comparisons between gross growth rates. Six crosses (30%), including *L. longicruris* and *L. saccharina* from Connecticut, as well as *L. longicruris* from Igloolik, formed a main group with significantly different (P<0.05) maximum growth rates from the others (Table 2). The group was characterized by individuals with intermediate growth rates (1.6-1.9 cm d-1). Fastest growing (~2.6 cm d-1) plants were

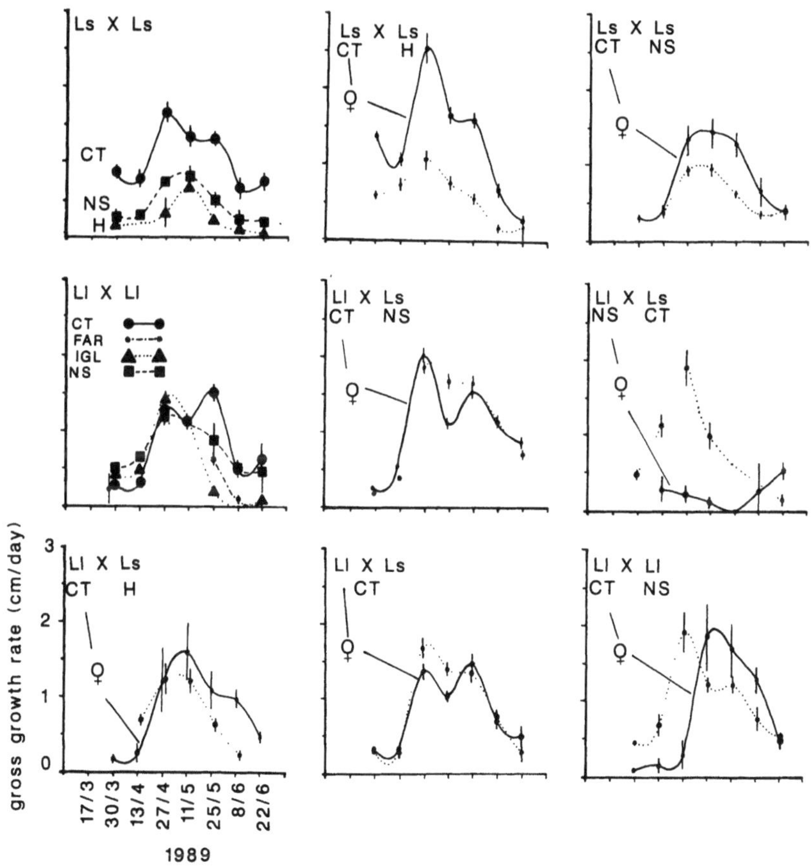

Fig. 5. Gross growth rates of F1 progenies from crosses of *L. longicruris* and *L. saccharina* in greenhouse culture. First species code listed for each cross represents the female. Solid continuous lines represent growth of progenies from crosses in which the initially listed parent in each cross is the female. The dotted lines represents growth of progenies from the reciprocal crosses. Sample sizes are as indicated in the legend for Fig. 1

mainly from crosses with *L. saccharina* female parents. Offspring of intraspecific Nova Scotia and Helgoland crosses were among the group with lowest growth rates (Table 2). A fourth major group, comprising interspecific crosses, did not exhibit any detectable differences with any of the other groups due to high variances (Table 2). Patterns in gross growth rates are summarized in Fig. 5.

Blade width inheritance

Frequency distribution curves of *Laminaria* blades, measured at their widest dimension, showed different shapes and size classes. In general, self-crosses of *L. longicruris* produced offspring with slightly narrower blades than *L. saccharina* (e.g. Fig. 6 a-d). Plants from Nova Scotia produced wider blades (Fig. 6 a-d). *Laminaria saccharina* CT X NS offspring had a strong pattern of site-dependent X-linked (female-linked) inheritance of blade width character (Fig. 6 e, f). Similarly, between-sites crosses of *L. longicruris* (CT X NS) resulted in site-dependent X-linked inheritance (Fig. 6 g, h) with female *L. longicruris* (NS) giving wider blades (Fig. 6 h). Reciprocal crosses of *L. longicruris* X *L. saccharina* (CT) also showed X-linked blade width inheritance. Reciprocal crosses of both species from Connecticut and Nova Scotia were more difficult to interpret since no specific pattern was obvious. The X-linkage was not clearly observed in Connecticut X Helgoland crosses. The shapes of the bases of blades of all crosses and reciprocal crosses were cuneate with one exception; the bases of the blades of *L. longicruris* IGL were always cucullate.

Stipe character inheritance

All self-crosses of *L. saccharina* produced short stipes (0-4 cm; Fig. 7 a-c). Approximately 70% of Connecticut plants had stipes in the 0-2 cm class (Fig. 7 a). *Laminaria longicruris* generally had longer stipes (CT, 2-20 cm; NS, 0-4 cm; IGL, 2-6 cm; Fig. 7 d-f). *Laminaria saccharina* produced > 90% 0-2 cm stipes in Connecticut X Nova Scotia reciprocal crosses, and also in the female Helgoland X Connecticut cross. However, in the latter reciprocal cross with the female *L. saccharina* (CT) 86% of the stipes ranged in length to 8 cm. The remaining 14% was distributed between the classes from 8-14 cm. In the reciprocal crosses between *L. longicruris* and *L. saccharina*, stipe length was linked to maternal inheritance. For example, longest stipes were influenced by the female *L. longicruris* parent in *L. longicruris* (NS) X *L. saccharina* (CT), *L. longicruris* (CT) X *L. saccharina* (NS), and *L. longicruris* (CT) X *L. saccharina* (H) (e.g. Fig. 7 g, h). The reciprocal crosses of *L.*

longicruris (CT X NS) produced offspring with stipe lengths influenced by the Nova Scotia site (Fig. 7 i, j).

Assessing ecotypic boundaries through quantitative characters

Discriminant analysis provided evidence for morphological and functional differences among F1 offspring for all North Atlantic *L. longicruris* and *L. saccharina*. The first three

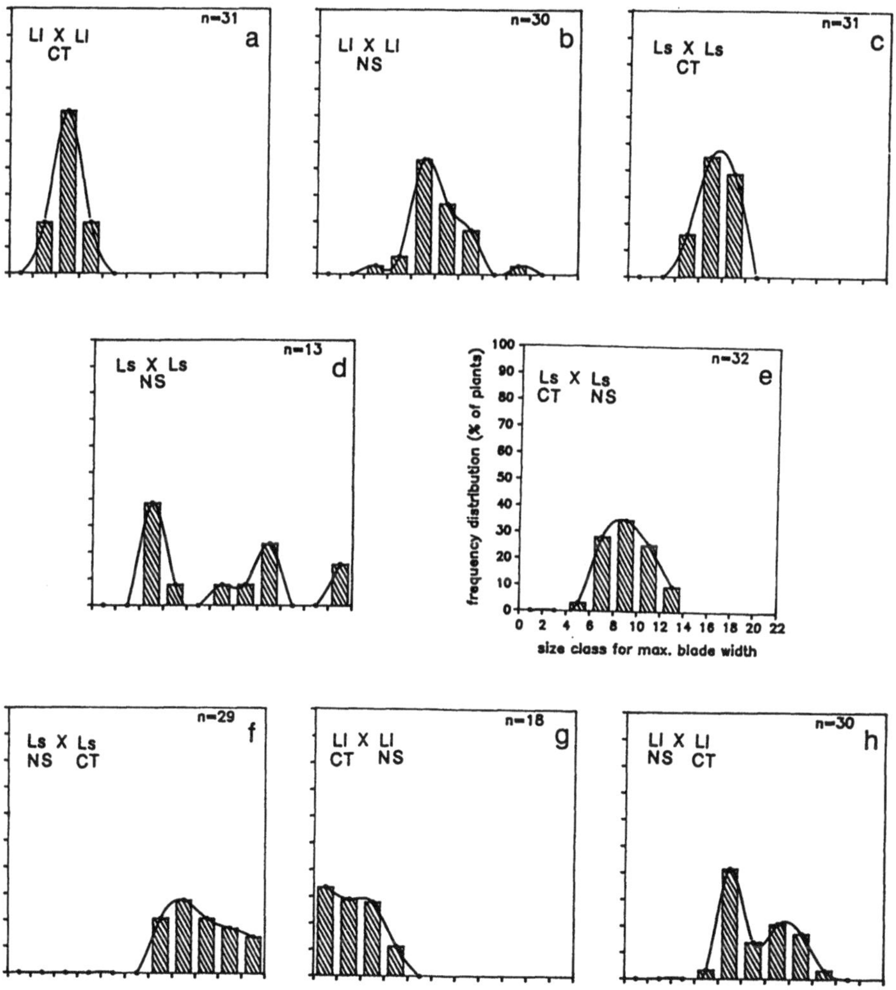

Fig. 6. Frequency distributions for maximum blade width (cm) in some representative F1 progenies from crosses of *L. longicruris* and *L. saccharina*. Sample size for each cross is indicated in the top right corner in each graph. Data are from 27/4/89

functions had considerable power in separating the groups, as demonstrated by the degree of association (canonical correlations, Table 3) between each discriminant function and the set of group variables. The stipe was the most important contributor to function one, followed by growth rate (Table 3). Function two was almost equally weighted by blade and growth variables. Correlations between these characters were not as high as in function one. Plotted group centroids (mean discriminant scores for each cross on their respective functions), defined by the first two functions, illustrate group location in discriminant space (Fig. 8). The first function distinguished crosses with long stipes and fast growth (Fig. 8 f, h, p, o) from those with short stipes and slow growth (Fig. 8 b, c, k, r, s). Function two distinguished offspring with wider blades and medium growth rates (Fig. 8 d, e, g, l, m, q, s) from the rest. The group was comprised of offspring involving the *L. saccharina* female parent from both Connecticut and Helgoland. Despite some overlapping among hybrids, the two discriminant functions separated or associated offspring of selfed parents from Helgoland (c), Nova Scotia (b, i), Igloolik (j) and Connecticut (a, h) (Fig. 8). The cluster in the middle (Fig. 8 a, i, j) was comprised of plants with wide blades, medium length stipes, and moderate growth rates. The analysis clearly showed separation of plants with short stipes (*L. saccharina* NS, H) from

Table 3. Discriminant power of the functions derived from the Discriminant Analysis for morphological and physiological characters of offspring from intra-, and interspecific crosses of *Laminaria* in greenhouse culture. A indicates Standardized discriminant function coefficients

Discriminant Functions	Canonical Correlations	A Stipe	A Widest Bld.	A Bld./10 cm	A Growth
1	0.892	0.790	-0.095	0.273	0.386
2	0.860	-0.666	0.470	0.234	0.555
3	0.718	0.325	-0.179	0.907	-0.706
4	0.486	0.006	-2.020	1.736	0.566

Correlations between canonical functions and conditional (within groups) variables

Functions	Stipe	Widest Bld.	Bld./10 cm	Growth
1	0.911	0.402	0.255	0.644
2	-0.379	0.694	0.650	0.486
3	0.088	0.451	0.711	-0.575
4	-0.139	-0.392	0.076	0.134

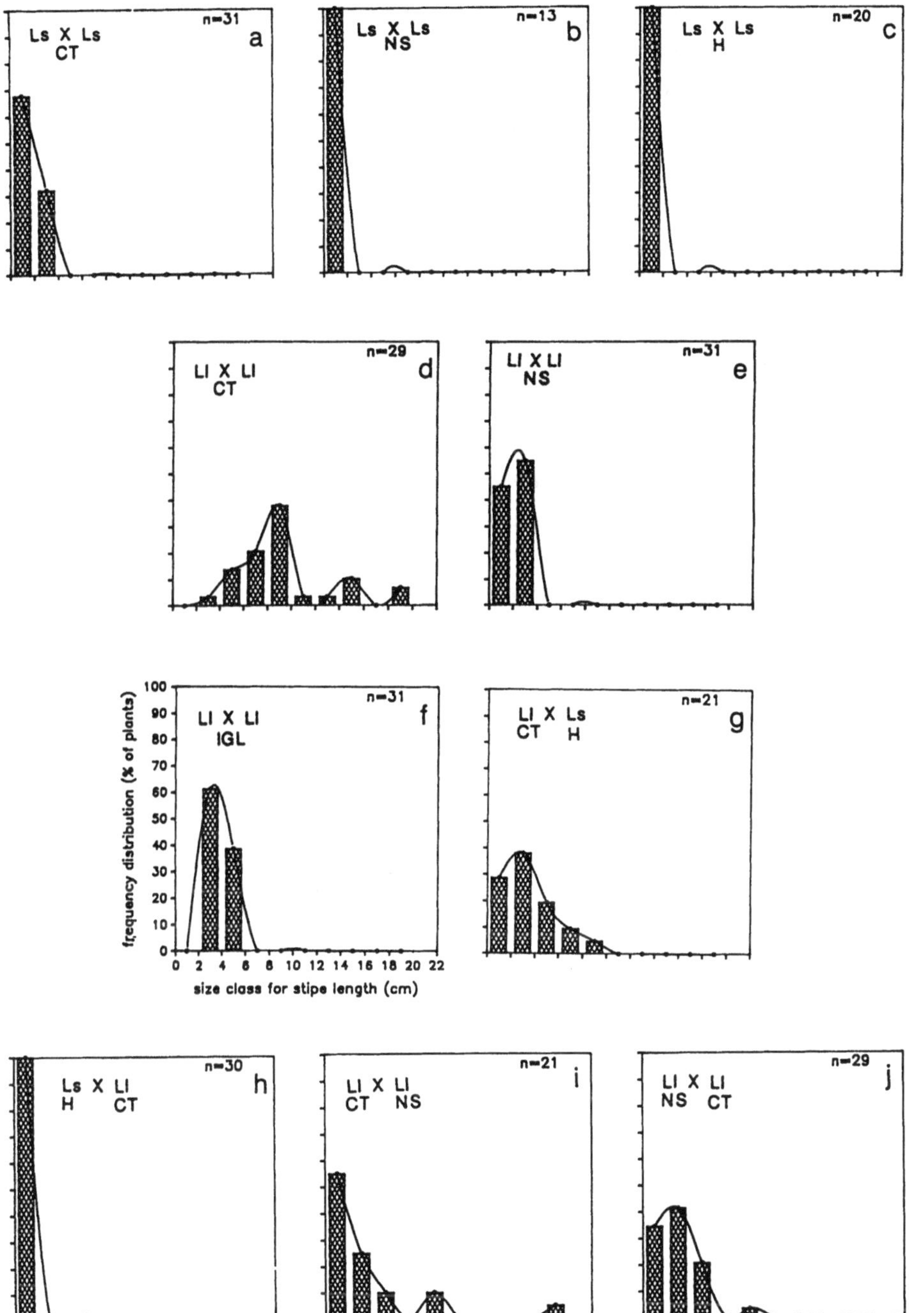

Fig. 7. Frequency distributions for stipe length (cm) in representative F1 progenies from crosses of *L. saccharina* and *L. longicruris*. Sample sizes are presented in the top right coner of each graph. Data are from 27/4/89

those with long stipes (*L. longicruris* CT). In addition, the interbreeding of plants possessing extremes of very short stipes (e.g. *L. saccharina* H), on one hand, and long stipes (e.g. *L. longicruris* CT), on the other, resulted in a maternally driven phenotype (Fig. 8 s, t).

The phenogram (Fig. 9), constructed from a matrix of average taxonomic distances using the same variables (above), shows four clusters comprising 80% of the crosses. The first cluster (Fig. 9 a, l, q, j, i) was characterized by plants with maximum blade widths between 4 and 18 cm, medium-short stipes and medium-high growth rates (1.2-1.8 cm d-1). The cluster includes the self-crosses of *L. saccharina* from Connecticut (A), and *L. longicruris* from both Nova Scotia (I) and Igloolik (J), as well as hybrids from the maternal Nova Scotia phenotype. The second cluster was comprised of hybrids from Connecticut and Nova Scotia interspecific crosses. This group had the highest growth rates of all groups (1.4-2.5 cm d-1) with blade widths resembling those of the first group, and medium-long stipes. Cluster G, D, S, B (Fig. 9) represented hybrids of *L. saccharina* parents only and also included selfed *L. saccharina* from Nova Scotia (B). Stipes in this group were consistently shorter than those of the first group and growth rates were 0.7-1.3 cm d-1. The fourth cluster included offspring of selfed *L. saccharina* from Helgoland (C), and hybrids (R) from *L. saccharina* (CT) X *L. longicruris* (NS). The morphotypes were characterized by having narrow blades, short stipes, and slow growth (0.2-0.3 cm d-1). The erosion rate continuously exceeded the net growth rate in these crosses (see growth above). Selfed *L. longicruris* (CT) was isolated between the second and third clusters. The morphological characteristics of these plants were most similar to those of

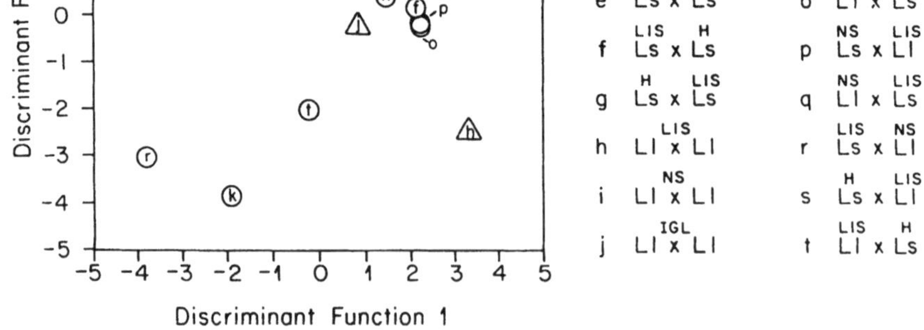

Fig. 8. Discriminant function analysis utilizing data from stipe lengths, widest blade dimensions, blade width at 10 cm above the blade-stipe juncture, and meristematic growth rates of F1 progenies from crosses of *L. longicruris* and *L. saccharina*

the second cluster with growth rates more similar to those of the third cluster. Similarly, growth rates were important in determining other single-grouped hybrids (Fig. 9 T, E, K).

DISCUSSION

The successful production of hybrids from the interspecific crosses between *Laminaria saccharina* and *L. longicruris* support data from previous reports (Chapman 1974,1975; Lüning *et al.* 1978; Bolton *et al.* 1983; Yarish *et al.* 1990). The experiments of Lüning *et al.* (1978) and Bolton *et al.* (1983) have shown that isolates of *Laminaria* from different geographical regions and different oceans hybridize both within and between species, whereas a sterility barrier has been demonstrated between some trans-Atlantic members of the Simplices. Lüning *et al.* (1978) produced hybrids from *L. saccharina* (Helgoland, Isle of Man) X *L. longicruris* (Nova Scotia) crosses, with the exception of a female *L. longicruris* X male *L. saccharina* (Isle of Man) cross. Yarish *et al.* (1990) concluded that the fact that members of the Laminariaceae share a common sex pheromone (Müller *et al.* 1985) is good enough reason to expect interfertility. Lewis *et al.* (1986) reported interfertility between three species of *Macrocystis* in

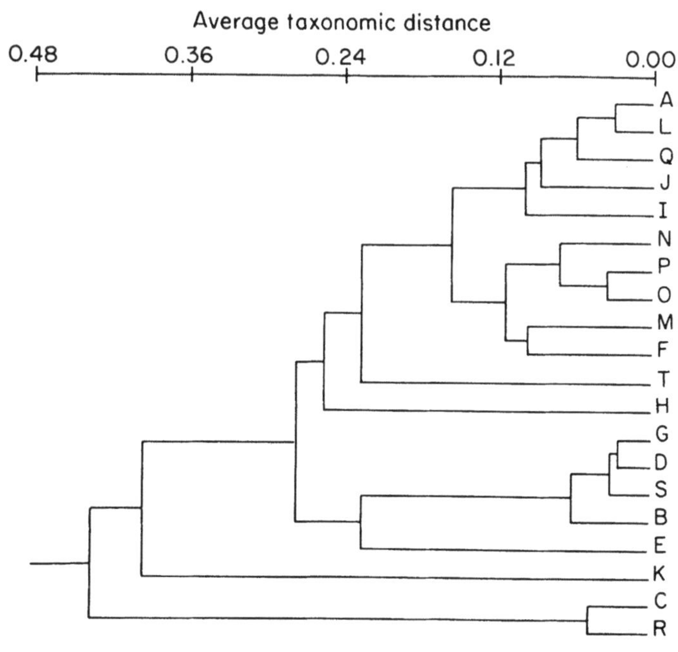

Fig. 9. Phenogram illustrating clusters of progenies produced in crosses of *L. saccharina* and *L. longicruris*. Data sources, and codes representing different crosses are the same as those presented for Fig. 8

California. Sanbonsuga & Neushul (1978) demonstrated interfertility in float-bearing kelps between genera, resulting in intermediate morphological forms. Coyer & Zaugg-Haglun (1982) confirmed that a close phyletic relationship exists between float-bearing kelps of different genera. Examples of morphologically distinct species capable of hybridizing with others are common in many groups of plants (Stace 1980).

Crossing experiments provided further insight into differences in growth rates (Figs 4, 5; Table 2), blade width (Fig. 6), and stipe length (Fig. 7).

Highest growth rates were generally observed in crosses involving Connecticut plants. Yarish *et al.* (1990) presented data suggesting that growth rate was linked to *L. longicruris* in hybrid crosses involving Connecticut *L. longicruris* and *L. saccharina*. Self-crosses of *L. longicruris* demonstrated superior growth. In contrast, there were no significant differences in maximum growth between the species in this study (Table 2) or in a previous study (Egan & Yarish 1988). Interspecific crosses between Connecticut plants, and between Connecticut and Helgoland plants resulted in hybrid growth patterns (Fig. 4). Highest growth was associated with the female parent involving intra-, and interspecific crosses between Connecticut and Helgoland plants. Differences in male and female genetic contributions to growth have also been reported in other studies (Sanbonsuga & Neushul 1978; Neushul 1982; Lewis *et al.* 1986).

Yarish *et al.* (1990) reviewed and compared the growth characteristics of juvenile stages of both *L. saccharina* and *L. longicruris* from Long Island Sound. In summary, they found that in March growth of *L. saccharina* was the poorest of all months, and, at the same time, growth of *L. longicruris* was quite good. *Laminaria longicruris* sporophytes grew at a temperature of 20°C; this was never observed in *L. saccharina*. Development and growth of *L. longicruris* in May was comparable to that of *L. saccharina*, except that the former's sporophytes grew to a larger size, i.e. more quickly. Responses to light and temperature were similar. However, some growth at 5°C was seen in *L. longicruris* sporophytes and growth was not observed in *L. saccharina* at 4°C. In July, *L. longicruris* sporophytes grew best at irradiances above 15 μ mol m-2 s-1 and the best growth was observed at 15°C (except at 70 μ mol m-2 s-1). Although Yarish & Egan (1989) showed superior growth of *L. longicruris* at 17°C, this temperature was not tested in the study of Yarish *et al.* (1990). On the other hand, *L. saccharina* sporophytes grew equally well at all irradiances and best at 11°C.

The growth patterns (Figs 4, 5) of *L. longicruris* from both Igloolik and Nova Scotia are similar to those presented by Bolton & Lüning (1982). However, *L. saccharina* from Helgoland, which was among the fastest growing strains in the experiments of Lüning *et al.* (1978), Bolton & Lüning (1982), and Bolton *et al.* (1983), was among the slowest growing strains in the present study (Figs 4, 5; Table 2). Poor growth observed in some of the progenies of their genetic crosses led Lewis *et al.* (1986) to conclude that differences observed in growth behavior of various strains from different localities were due to differences between the environments from which parental sporophytes were collected and the environment in which F1 sporophytes were grown. Information on the failure of specific isolates from one geographical locality to grow/survive in culture conditions at another probably goes unreported in most cases because such information represents negative data. *Laminaria longicruris* from the Faroes never adapted to Long Island Sound conditions in greenhouse culture (Figs 4, 5).

Bolton & Lüning (1982) found no significant population variation within species of *Laminaria* "despite wide variations in temperature between the original collecting sites". However, they did not include representatives from southern boundary populations. Recent studies have shown differences between southern boundary populations and populations occurring further north. Isolates from southern populations exhibited higher temperature optima for growth of gametophytes and young sporophytes (Lee & Brinkhuis 1988; Egan *et al.* 1989), higher upper temperature limits for reproduction in gametophytes (Egan *et al.*1989), and higher survival temperatures in large sporophytes (Gerard & Du Bois 1988). Lack of population variation in Bolton & Lüning's (1982) study may have been compounded also by their use of small juvenile (<5 cm) sporophytes. Connecticut plants (Figs 4, 5) had longer periods of sustained growth than plants from more northern populations. Plants from Nova Scotia, and in particular from Igloolik, ceased growth as soon as water temperatures reached 12°C in greenhouse culture. Yarish & Egan (1987) showed highly significant differences in growth at 15°C between *L. longicruris* from Connecticut on one hand, and *L. longicruris* from both Igloolik and the Faroe Islands on the other. Bolton & Lüning (1982) concluded that *L. saccharina* had a broad temperature optimum for growth between 10-15°C, whereas that for *L. longicruris* from Nova Scotia occurred at 10°C. Although *L. longicruris* from Connecticut grew best at 17°C, some strains isolated in the vicinity of a thermal plume from a nuclear power plant in eastern Connecticut had a higher optimum temperature (19°C) for growth (Yarish & Egan 1989).

Although phenotypic plasticity, in response to environmental conditions, has been documented for species of Laminariales (e.g. Parke 1948; Mann 1971; Svendsen & Kain 1971; Gerard & Mann 1979; Gerard 1987) and Fucales (Sideman & Mathieson 1985), Guiry *et al.* (1987) cautioned that variability may not be due solely to ecological factors, but in fact may have a genetic origin. Svendsen & Kain (1971) documented the production of an undivided cucullate blade (blade with a bi-lobed base) in *L. hyperborea* (Gunn.) Fosl. in response to sheltered environmental conditions. *Laminaria longicruris* from Igloolik produced only cucullate blades, while the progenies of all other crosses produced cuneate blades (blade with a V shaped base) in the common garden conditions in this study, suggesting a genetic base for the difference.

Lüning *et al.* (1978) found that blade width inheritance varied according to the male parent in hybrid crosses involving *L. saccharina* from Helgoland and *L. longicruris* from Nova Scotia. Sex-linkage to the male gametophyte, for the same character, was also reported by Yarish *et al.* (1990) for crosses between the two species from Connecticut. However, the results of this study (Fig. 6) suggest sex-linkage to the female gametophyte.

Data also suggest that stipe length is linked to the female gametophyte (Fig. 7; Yarish *et al.* 1990). While phenetic variation in stipe morphology has been documented in *L. longicruris* by Chapman (1973), Bolton *et al.* (1983) found no evidence that *L. saccharina* from the Isle of Man, Brittany, Helgoland, or the northeast Pacific can be environmentally induced to produce long, hollow stipes. Thus, they concluded that the ability to produce these characters under particular conditions appears to be genetically determined. Similarly, long or hollow stipes were not produced by *L. saccharina* from Connecticut in previous studies (e.g. Egan & Yarish 1988; Yarish *et al.* 1990). These examples are in contrast to data in Gerard & Mann (1979). The latter suggest that stipe length is inversely related to exposure and water movement. However, the relationship of stipe length to exposure is not discussed and far from clear.

There is some apparent confusion concerning short solid, short hollow, and long hollow stipes in Chapman (1973, 1974). No short, hollow-stiped, mature forms occur in *L. saccharina* or *L. longicruris*. Børgesen's materials from the Faeroes did include short (10-15 cm), hollow stipes in *L. faeroensis*. Kain (1976) stated "inflation of the stipe is not solely the result of continued growth in length and thus diameter, but a particular event which occurs after the stipe is about 0.4 m long". Egan & Yarish (1988) demonstrated that while some stipes could become hollow when 20 cm long, the highest frequency occurred after stipes had attained a length of 40 cm. Parke (1948) encountered plants with hollow stipes at Argyll, Scotland. Kain (1976) noted Parke's observation, but Parke (1948) attributed the finding to the species *L. saccharina*. Kain did not compare any of the features between populations at Devon (short, solid stipe) and Argyll (long, hollow stipe). It appears that Parke's record was really *L. longicruris* (Egan & Yarish 1988). Parke (1948) concluded that the hollow stipe is produced in summer or autumn developed plants because of rapid growth in length and thickness during the first growth period and, as a result, the medullary tissue does not keep pace with that of the cortical tissue so that air pockets develop in the central portion of the stipe. If this is the case, why do *L. saccharina* plants not produce long, hollow stipes? What other environmental conditions could induce this phenomenon? The distribution data presented for *L. longicruris* in Fig. 2 indicates that the form is found where winter water temperatures are less than 5°C. Egan & Yarish (1988) demonstrated that hollow stipes were produced more rapidly in winter developed, than in summer developed plants. This is in contrast to Parke's (1948) conclusions (above). An interesting avenue of future research would be to determine if low temperature is involved, and if it is, how it is involved in hollow stipe production. Egan & Yarish (submitted) suggested recently that the +/- production of bullate blades in western Atlantic populations of Simplices *Laminaria* is temperature related.

Maximum stipe lengths of all crosses conducted in the experiments of Lüning *et al.* (1978) and Bolton *et al.* (1983) were <10 cm. Bolton *et al.* (1983) produced successful hybrids from *L. longicruris* (Nova Scotia) X *L. saccharina* (Brittany and Helgoland) crosses with stipe lengths similar to those reported by Lüning *et al.* (1978). *Laminaria longicruris* X *L. longicruris* (Nova Scotia) also produced short solid stipes. All of these experiments were conducted in holding tanks or in the sea at Helgoland, Germany. While it could be argued that conditions near Helgoland may not have been conducive to stipe elongation, *L. longicruris* from Nova Scotia similarly did not produce long stipes in Long Island Sound water, whereas the Connecticut strain did. Sporophytes used by Lüning *et al.* (1978), Bolton *et al.* (1983), Yarish *et al.* (1990) and also in this study were between 4 and 10 months old. Maximum stipe elongation occurs in spring, while the hollow condition develops at the end of the first year of growth (Egan & Yarish 1988).

Taxonomic delineation in brown seaweeds has been conducted mainly on the basis of phenotypic characters observed in the field and in culture (Russell 1978). Hierarchical analysis in numerical taxonomy has been attempted as a means of simplifying taxonomic criteria and identification of species (e.g., Widdowson 1971; Russell & Fletcher 1975; Prud'homme van Reine 1978; Pankhurst & Tittley 1978; Marsden *et al.* 1983; Rice & Chapman 1985). These methods have had limited success in natural groups, but have failed where phenotypic variability is poorly understood. Many of the criteria used are based on 'gut feeling' or intuitive weighting. Russell (1978) cited numerous instances where the role of genotype has been inadequately separated from phenotype in dictating morphological form. To date, there has been only one attempt at using numerical taxonomic methods in the study of laminarian (*Alaria*) populations (Widdowson 1971). Stipe character was the primary attribute that separated the crosses during the assessment of ecotypic boundaries in this study (Figs 8, 9). Since an ecotype is the product arising as the result of genotypic response of a species to its particular habitat (Turesson 1922), this status implies interfertility. Although the short, solid stipe on one hand, and the long, hollow stipe on the other, appear to be genetically fixed characters, cold winter temperatures may play a role in hollow stipe production, as stated previously.

Crossability is not an all-or-nothing criterion among species of Laminariales (Chapman 1978; Lewis *et al.* 1986) and *Fucus* (Rice & Chapman 1985). The primary conclusion from these studies is that, where species are easily distinguished morphologically, there is usually reproductive isolation, while morphologically confused species are interfertile (Mathieson *et al.* 1981). Bolton *et al.* (1983) suggested that great care should be exercised in assessing the value of hybridization experiments in the taxonomy of the Phaeophyta. Guiry *et al.* (1987)

commented that "there is not much point in having well-founded species that only a few can recognize".

Although one would clearly have difficulty distinguishing between *L. saccharina*, and less than one-year-old plants of *L. longicruris* (Egan & Yarish 1988), it is our opinion that the two entities should probably remain separate, as presented by Wilce (1965). Their dissimilarities appear to outweigh their similarities appreciably. In addition, this study further substantiates the fact that temperature races of *L. saccharina* and *L. longicruris* do exist in the North Atlantic, and that physiological differentiation has occurred within the respective species. Despite this, the phytogeographic boundaries of Simplices *Laminaria* presented by van den Hoek (1982 a, b) remain unchanged. It is also apparent that any species, used in biogeographical studies using culture conditions, should be represented by isolates from its entire distributional range before any definitive conclusions or generalizations are made.

ACKNOWLEDGEMENTS

This research was supported by a grant to C. Yarish and B.H. Brinkhuis from the NOAA Office of Sea Grant, Department of Commerce, under Grant No. NA85AA-D-SG101, the Connecticut Sea Grant College Program. The U.S. Government is authorized to produce and distribute reprints for governmental purposes notwithstanding any copyright notation that may appear herein. Z. Garcia-Eszquivel is indebted to Consejo Nacional de Ciencia y Tecnologia (CONACyT) for their scholarship. The authors are grateful to: M.J. Spring for help with the illustrations; S. Tobin-Charnon, R.G. Rowland and special thanks to K.J. Yarish, J. Brinkhuis, the Egans, the O'Callahans, and D. and M. Barvenik for support and encouragement.

REFERENCES

Batters EAL (1889) A List of the Marine Algae of Berwick-on-Tweed. Henry H. Blair, Alnwick

Bolton JJ (1983) Ecoclinal variation in *Ectocarpus siliculosus* (Phaeophyceae) with respect to temperature growth optima and survival limits. Mar Biol 73: 131-138

Bolton JJ, Lüning K (1982) Optimal growth and maximal survival temperatures of Atlantic *Laminaria* species (Phaeophyta) in culture. Mar Biol 66: 89-94

Bolton JJ, Germann I, Lüning K (1983) Hybridization between Atlantic and Pacific representatives of the section Simplices of *Laminaria* (Phaeophyta). Phycologia 22: 133-140

Børgeson F (1903-1908) II. Marine algae of the Faeroes (1903). III. The algae vegetation of the Faeroese coasts (1908). In: Botany of the Faeroes II, III. Det Nordiske Forlag, p 339-532, 683-834

Brinkhuis BH, Hanisak MD (1982) Development of a marine biomass program in New York. In: Proc Int Gas Research Conf 1981, Govt Institutes Inc, Rockville, MD, p 682-692

Chapman ARO (1973) Phenetic variability of stipe morphology in relation to season, exposure, and depth in the non-digitate complex of *Laminaria* Lamour. (Phaeophyta, Laminariales) in Nova Scotia. Phycologia 12: 53-57

Chapman ARO (1974) The genetic basis of morphological differentiation in some *Laminaria* populations. Mar Biol 24: 85-91

Chapman ARO (1975) Inheritance of mucilage canals in *Laminaria* (Section Simplices) in eastern Canada. Br Phycol J 10: 219-223

Chapman ARO (1978) Experimental and numerical taxonomy of the Laminariales. In: Irvine DEG, Price JH (eds) Modern approaches to the taxonomy of red and brown algae. Academic Press, London, p 423-432

Christensen T, Thomsen HA (1974) Algefortegnelse. Universitets Bogladen Naturfagsbogladen der Kobenhavn

Coyer JA, Zaugg-Haglund AC (1982) A demographic study of the kelp *Pelagophycus porra* (Laminariales, Lessoniaceae), with notes on *Pelagophycus* X *Macrocystis* hybrids. Phycologia 21: 399-407

DeSilva MWRN, Burrows EM (1973) An experimental assessment of the status of the species *Enteromorpha intestinalis* (L.) Link and *Enteromorpha compressa* (L.) Grev. J Mar Biol Ass U K 53: 895-904

Egan B, Yarish C (1988) The distribution of the genus *Laminaria* (Phaeophyta) at its southern limit in the western Atlantic Ocean. Bot Mar 31: 155-161

Egan B, Vlasto A, Yarish C (1989) Seasonal acclimation to temperature and light in *Laminaria longicruris* de la Pyl. (Phaeophyta). J Exp Mar Biol Ecol 129: 1-16

Egan B, Yarish C Productivity and life-history strategies of *Laminaria longicruris* de la Pyl. at its southern limit in the Western Atlantic Ocean. Mar Ecol Prog Ser (in press).

Espinoza J, Chapman ARO (1983) Ecotypic differentiation of *Laminaria longicruris* in relation to seawater nitrate concentration. Mar Biol 74: 213-218

Gagne JA, Mann KH, Chapman ARO (1982) Seasonal patterns of growth and storage in *Laminaria longicruris* in relation to differing patterns of availability of nitrogen in the water. Mar Biol 21: 91-101

Gerard VA (1987) Hydrodynamic streamlining of *Laminaria saccharina* Lamour. in response to mechanical stress. J Exp Mar Biol Ecol 107: 237-244

Gerard VA (1988) Ecotypic differentiation in light-related traits of the kelp *Laminaria saccharina*. Mar Biol 97: 25-36

Gerard VA, Mann KH (1979) Growth and production of *Laminaria longicruris* (Phaeophyta) populations exposed to different intensities of water movement. J Phycol 15: 33-41

Gerard VA, Du Bois KR (1988) Temperature ecotypes near the southern boundary of the kelp *Laminaria saccharina*. Mar Biol 97: 575-580

Guiry MD (1978) A consensus and bibliography of Irish seaweeds. (Bibliotheca Phycologia vol. 44) Cramer, Vaduz

Guiry MD, West JA (1983) Life history and hybridization studies on *Gigartina stellata* and *Petrocelis cruenta* (Rhodophyta) in the North Atlantic. J Phycol 19: 474-494

Guiry MD, Tripodi G, Lüning K (1987) Biosystematics, genetics and upper temperature tolerance of *Gigartina teedii* (Rhodophyta) from the Atlantic and Mediterranean. Helgol Meeresunters 41: 283-295

Hoek C van den (1982a) Phytogeographic distribution groups of benthic marine algae in the North Atlantic Ocean. A review of experimental evidence from life history studies. Helgol Meeresunters 35: 153-214

Hoek C van den (1982b) The distribution of benthic marine algae in relation to the temperature regulation of their life histories. Biol J Linn Soc 18: 81-144

Hylander CJ (1928) The algae of Connecticut. Bull Geol Nat Hist Surv Conn 42: 1-245

Innes DJ (1984) Genetic differentiation among populations of marine algae. Helgol Meeresunters 38: 401-417
Innes DJ (1987) Genetic structure of asexually reproducing *Enteromorpha linza* (Ulvales: Chlorophyta) in Long Island Sound. Mar Biol 94: 459-467
Irvine DEG (1974) The marine vegetation of the Shetland Isles. In: Goodier R (ed) The natural environment of the Shetland Isles. Edinburgh, p 107-113
Irvine DEG (1982) Seaweeds of the Faeroes. 1. The flora. Bull Br Mus Nat Hist (Bot.) 10: 109-131
Jónsson H (1912) The botany of Iceland I. Marine algal vegetation. Frimodt, Copenhagen
Kain JM (1976) New and interesting algae from the Shetland Isles. II. Hollow and solid stiped *Laminaria* (Simplices). Br Phycol J 11: 1-11
Kylin H (1947) Die Phaeophyceen der schwedischen Westküste. Lunds Univ Arssk NF Adv 2, 43(4): 1-99
Lee RKS (1980) A Catalogue of the Marine Algae of the Canadian Arctic. Publ Bot 9, Nat Mus Nat Sci Can, Ottawa
Lee JA, Brinkhuis BH (1986) Reproductive phenology of *Laminaria saccharina* (L.) Lamour. (Phaeophyta) at the southern limit of its distribution in the Northwestern Atlantic Ocean. J Phycol 22: 276-285
Lee JA, Brinkhuis BH (1988) Seasonal irradiance and temperature interaction effects on development of *Laminaria saccharina* gametophytes and juvenile sporophytes. J Phycol 24: 181-191
Lewis RJ, Neushul M, Harger BWW (1986) Interspecific hybridization of the species of *Macrocystis* in California. Aquaculture 57: 203-210
Lund S (1959) The marine algae of east Greenland I. Taxonomical part. Medd Gronl 156: 1-247
Lüning K (1975) Kreuzungsexperimente an *Laminaria saccharina* von Helgoland und von der Isle of Man. Helgol Meeresunters 27: 108-114
Lüning K (1985) Meeresbotanik: Verbreitung, Ökophysiologie und Nutzung der marinen Makroalgen. Thieme Verlag, Stuttgart
Lüning K, Chapman ARO, Mann KH (1978) Crossing experiments in the non-digitate complex of *Laminaria* from both sides of the Atlantic. Phycologia 17: 293-298
Mann KH (1971) Relation between stipe length, environment and the taxonomic characters of *Laminaria*. J Fish Res Bd Can 28: 778-780
Marsden WJN, Evans LV, Callow JA (1983) Analysis of character variation in *Fucus* L. I. Discriminant analysis of morphometric estimates. Bot Mar 26: 383-392
Mathieson AC, Norton TA, Neushul M (1981) The taxonomic implications of genetic and environmentally induced variations in seaweed morphology. Bot Rev 47: 313-347
Müller DG (1979) Genetic affinity of *Ectocarpus siliculosus* (Dillw.) Lyngb. from the Mediterranean, North Atlantic, and Australia. Phycologia 18: 312-318
Müller DG, Maier I, Gassmann G (1985) Survey on sexual pheromone specificity in Laminariales (Phaeophyceae). Phycologia 24: 475-477
Neushul M (1982) Morphology, structure, systematics and evolution of the giant kelp *Macrocystis*. In: Proceedings of the U.S.-China Phycological Symposium. Science Press, Beijing, People's republic of China, p 1-27
Pankhurst RJ, Tittley I (1978) The application of computers to the identification of the Phaeophyta. In: Irvine DEG, Price JH (eds) Modern approaches to the taxonomy of red and brown algae. Academic Press, London, p 325-337
Parke M (1948) Studies on British Laminariaceae. I. Growth in *Laminaria saccharina* (L.) Lamour. J Mar Biol Ass U K 27: 651-709
Parke M, Dixon PS (1976) Checklist of British marine algae, 3rd revision. J Mar Biol Ass U K 56: 527-594
Prud'homme van Reine WF (1978) Criteria used in systematic studies of Sphacelariales. In: Irvine DEG, Price JH (eds) Modern approaches to the taxonomy of red and brown algae. Academic Press, London, p 301-323
Rice EL, Chapman ARO (1985) A numerical taxonomic study of *Fucus distichus* (Phaeophyta). J Mar Biol Ass U K 65: 433-459

Rohlf FJ (1988) Numerical Taxonomy and Multivariate Analysis System. (Handbook) Exeter Publ. Ltd
Rueness J (1973) Speciation in *Polysiphonia* (Rhodophyceae, Ceramiales) in view of hybridization experiments: *P. hemisphaerica* and *P. boldii*. Phycologia 12: 107-109
Russell G (1978) Environment and form in the discrimination of taxa in brown algae. In: Irvine DEG, Price JH (eds) Modern approaches to the taxonomy of red and brown algae. Academic Press, London, p 339-369
Russell G, Morris OP (1970) Copper tolerance in the marine fouling alga *Ectocarpus siliculosus*. Nature (London) 228: 288-289
Russell G, Fletcher RL (1975) A numerical taxonomic study of the British Phaeophyta. J Mar Biol Ass U K 55: 763-783
Russell G, Fielding AH (1981) Individuals, populations and communities. In: Lobban CS, Wynne MJ (eds) The Biology of Seaweeds. Blackwell, Oxford, p 394-411
Sanbonsuga Y, Neushul M (1978) Hybridization of *Macrocystis* (Phaeophyta) with other float-bearing kelps. J Phycol 14: 214-224
Sideman EJ, Mathieson AC (1985) Morphological variation within and between natural populations of non-tidepool *Fucus distichus* (Phaeophyta) in New England. J Phycol 21: 250-257
Sokal RR, Rohlf FJ (1981) Biometry, 2nd edn. WH Freeman & Co, San Francisco
Stace CA (1980) Plant taxonomy and biosystematics. Edward Arnold, London
Sverdrup HU, Johnson MW, Fleming RH (1942) The oceans. Prentice-Hall, Englewood Cliffs
Svendsen P, Kain JM (1971) The taxonomic status, distribution and morphology of *Laminaria cucullata* sense Jorde and Klavestad. Sarsia 46: 1-22
Taylor WR (1957) Marine algae of the northeastern coast of North America. University of Michigan Press, Ann Arbor
Turesson G (1922) The species and the variety as ecological units. Hereditas 3: 100-113
Turner D (1802) A synopsis of the British Fuci, Vol. 1. F. Bush, Yarmouth
West JA, Polanshek AR, Shevlin DE (1978) Field and culture studies on *Gigartina agardhii* (Rhodophyta). J Phycol 14: 416-426
Widdowson TB (1971) A statistical analysis of variation in the brown algae Alaria. Syesis 4: 125-143
Wilce RT (1963) Studies on benthic marine algae in Northwest Greenland. Proc Int Seaweed Symp 4: 280-287
Wilce RT (1965) Studies in the genus *Laminaria*, III. A revision of the North Atlantic species of the Simplices section of *Laminaria*. Bot Gothoburg 3: 247-256
Yarish C, Brinkhuis BH, Egan B, Garcia-Ezquivel Z (1990) Morphological and physiological bases for *Laminaria* selection protocols in Long Island Sound aquaculture. In: Yarish C, Penniman CA, van Patten M (eds) Economically important marine plants of the Atlantic: their biology and cultivation. Connecticut Sea Grant Symposium, p 53-94
Yarish C, Edwards P, Casey P (1979a) A culture study of salinity responses in ecotypes of two estuarine red algae. J Phycol 15: 341-346
Yarish C, Egan PF (1987) Biological studies on *Laminaria longicruris* and its aquaculture potential in Long Island Sound. In: Halasi-Kun GJ (ed) Columbia University seminars on pollution and water resources No. 20. Columbia University Press, New York, p 189-215
Yarish C, Egan B (1989) The biology of *Laminaria longicruris* and its potential for mariculture in Long Island Sound. In: Kanungo K, Brousseau D, Singletary R, Liscek L (eds) The second conference on the aquatic environment: Problems and perspectives. Western Connecticut State University, Danbury, Connecticut, p 55-89
Yarish C, Lee KW, Edwards P (1979b) An improved apparatus for the culture of algae under varying regimes of temperature and light intensity. Bot Mar 22: 395-397

SEXUAL COMPATABILITY AND SPECIES CONCEPT IN *ECTOCARPUS SILICULOSUS* (ECTOCARPALES, PHAEOPHYCEAE) FROM ITALY, NORTH CAROLINA, CHILE, AND NEW ZEALAND

Birgit Stache
Fakultät für Biologie
Universität Konstanz
D-7750 Konstanz 1
West Germany

Abstract- Several sexual strains of *Ectocarpus* with differing morphology originating from the Mediterranean, NW Atlantic, SE and SW Pacific were studied under culture conditions. The sporophytic and gametophytic phases are anisomorphic to a varying degree and both show morphological variation primarily dependent upon temperature. In crossing experiments zygotes formed in all combinations of strains, however, the viability of hybrids between northern and southern strains was reduced. In some crosses the hybrids were incapable of forming intact meio-sporangia but showed normal mitotic spore formation. These strains are considered to be members of one species *Ectocarpus siliculosus*, which is divided into geographical races or subspecies. They are sexually isolated, but can form hybrids, which are fertile on the basis of mito-sporangia.

INTRODUCTION

Phytogeographic distribution patterns are elaborated by using the information from floras and local taxonomic work. In a typical case, the description and taxonomic treatment of a species is based on morphological features of field material such as size, branching pattern, number and shape of reproductive organs etc. A morphological variant is described as a forma, subspecies or new species by subjective judgement of the importance of morphological features. In many cases environmental, seasonal or developmental changes of the phenotype are not considered with this approach.

In genera with a wide distribution, this often leads to a confusing variety of morphologically defined taxa. In phytogeographic distribution patterns on the species level this may result in a picture in which there are a number of apparently disjunct and/or endemic species rather than one (or at least fewer) widely spread species.

NATO ASI Series, Vol. G 22
Evolutionary Biogeography of the
Marine Algae of the North Atlantic
Edited by D. J. Garbary and G. R. South

The study of life histories and phenotypic variation as a response to environmental factors under controlled culture conditions can give additional information about the different morphological types encountered in the field.

A biological species is defined as a population, in which all members are able to interbreed and to produce viable, fertile offspring (Grant 1976). Geographically isolated populations, which are totally interfertile in the experiment, belong to the same biological species and can be referred to as members of one species or as geographical races of a species.

For asexual organisms, the biological species concept cannot be used in this sense. Each isolate and its offspring is a single clone and develops independently from others. Thus, extensive phenotypic and genotypic variation may be present, and the individual clones can be referred to as microspecies. *Pilayella littoralis* (L.) Kjellman is an example of a cosmopolitan species which probably has only asexual reproduction. Isolates from different origins show differences in morphology, however, they have the same type of reproduction, and the same phenotypic response to environmental factors (Müller & Stache 1989).

Sexual compatibility is the main prerequisite for a biological species in biparental organisms. It is expressed on various levels ranging from plasmogamy and karyogamy to the fertility of the hybrids and their ability to produce viable offspring. Sterility and reduced viability of hybrids is a measure of advancing genetic isolation in the parental plants. Various crossing experiments with genera of a wide geographic distribution were made, resulting in different taxonomic consequences (Mathiesen *et al.* 1981).

Members of the genera *Ulva* L. and *Enteromorpha* Link (Chlorophyceae) were found to be intersterile on genera and species level, although under culture conditions they showed a strong phenotypic variation with overlap of morphological forms. In this case culture and crossing experiments confirmed the taxonomic concept of these genera (Mathiesen *et al.* 1981).

In the Phaeophyceae, by contrast, interspecific and intergeneric hybrids are known. A hybrid between *Macrocystis angustifolia* Bory and *Pelagophycus porra* (Leman) Setchell was found to grow normally with an intermediate morphology, but formed abnormal sterile sporangia (Sanbonsuga & Neushul 1978, 1979). Plants with probable hybrid character were also found in the field (Neushul 1962, 1971). In the genus *Fucus* L. morphologically distinct species were crossed with different percentages of fertile hybrid offspring, but only under certain culture conditions. Plants with hybrid character were also reported in the field (Mathiesen *et al.* 1981). This shows that interfertility under culture conditions does not necessarily imply, that two parent plants belong to the same biological species.

Intrageneric crossing experiments in the Rhodophyceae have led to taxonomic revisions in several cases. For example, hybrid formations in the *Ceramium diaphanum* (Lightf.) Roth -*C. strictum* (Kütz.) Harv. complex or between *Callithamnion byssoides* Arnott and *C. halliae* Collins led to the conclusion that those species forming hybrids were conspecific (van der Meer 1986). On the other hand, *Laurencia pacifica* Kylin isolates with the same morphology but different contents of secondary metabolites proved to be intersterile, suggesting sibling species (Howard *et al.* 1980).

Few interspecific hybridization studies have been carried out with taxa separated over large distances. Crosses of *Laminaria saccharina* (L.) Lamour. (Phaeophyceae) from both sides of the North Atlantic and the Northeast Pacific with *Laminaria longicruris* de la Pyl. from the North Atlantic gave rise to normal hybrid sporophytes (Bolton *et al.* 1983), showing a close relationship between the Atlantic and Pacific populations of *L. saccharina*. Crosses of *Palmaria palmata f. mollis* (Setchell & Gardner) Guiry (Rhodophyceae) from the N-Pacific and *P. palmata* (L.) O. Kuntze from the N-Atlantic were not interfertile, indicating a more distant relationship. In this case the Pacific form was raised to a separate species (van der Meer & Bird 1985). *Gigartina teedii* (Roth) Lamour. (Rhodophyceae) from Europe and Brazil were interfertile, while they were non-compatible with other *Gigartina* species from California and France (Guiry 1984).

Intraspecific crossing experiments between populations from different geographical origins can also be expected to give information about the origin and age of a species, and its migration and distribution. The brown alga *Ectocarpus siliculosus* (Dillw.) Lyngbye (Ectocarpales, Phaeophyceae) proved to be especially suitable for such broader studies on geographical distribution and species concept.

The genus *Ectocarpus* is found in subantarctic to warm temperate waters in the Southern Hemisphere and in arctic to warm temperate waters in the Northern Hemisphere (Fig. 26). Traditionally, species in this genus are distinguished by morphological criteria, without considering the life history and sexual compatibility. The life history and reproduction of *Ectocarpus siliculosus* have been studied intensively (Müller 1967). The dioecious haploid gametophytes produce isomorphic but physiologically different female or male gametes which develop after zygote formation into diploid heterosporophytes. These sporophytes reproduce by mitotic plurizoids, while meiospores from unilocular sporangia develop into gametophytes. Unfertilized gametes develop into haploid parthenosporophytes, which form mitotic plurilocular sporangia to reproduce themselves or apomeiotic unilocular sporangia to reestablish the gametophytic phase.

The taxonomic treatment of *Ectocarpus* has varied greatly in the past, e.g. several genera like *Giffordia* Batters or *Feldmannia* Hamel are now recognized, the species of which were previously included in *Ectocarpus*. Russell (1966) studied the morphological variability of *Ectocarpus* in Britain and reduced the number of species to two: *Ectocarpus siliculosus* and *E. fasciculatus* Harv. These two species did not form zygotes in crossing experiments (Müller & Gassmann 1980).

Müller (1979) compared several sexual strains of *E. siliculosus* from different locations. Most combinations were successful in zygote formation and consequently the strains were considered as one species. Two local populations in Atlantic North America were incompatible to varying degrees, and this was interpreted as demonstrating an early stage in the speciation process. In this study the viability of the zygotes and fertility of the hybrids were not investigated. More recent crossing experiments between strains from Italy (Mediterranean) and Chile (Southeast Pacific) gave rise to viable hybrids which showed meiotic sterility. These findings suggested that further crossing experiments should be carried out using a few selected isolates of *Ectocarpus* from geographic areas that were separated by maximal distances.

MATERIALS AND METHODS

Four strains of *Ectocarpus siliculosus* (Dillw.) Lyngbye were selected for this study. They were collected and maintained as unialgal cultures by Müller (1967, 1979, 1988):
- Na from Napoli/Italy (40°51'N, 14°17'E; Müller 1967),
- Wil from Wilmington/North Carolina, USA (34°10'N, 77°56'W; Müller 1979),
- Ch from Puyuguapi/Chile (44°24'S, 72°36'W; Müller 1988),
- NZ collected 1988 in Kaikoura/New Zealand (42°26'S, 173°42'E).

They were kept in stock cultures as gametophytes of both sexes (Na, Wil and NZ) or as parthenosporophytes (Ch) from which gametophytes were derived.

The culture medium utilized natural seawater from the North Sea (German Bight, 28‰) supplemented with a modified Provasoli-enrichment as described by Saga & Gibor (1986). The cultures were kept at 12±1°C under illumination with daylight-type white fluorescent light at a photon irradiance of 10-20 μmol $m^{-2}s^{-1}$ for 16 h per day. The effect of temperature on growth was studied on a temperature gradient plate ranging from 3°C to 26°C with an illumination of 8-12 μmol $m^{-2}s^{-1}$.

For crossing experiments, fertile gametophytes were placed in a small drop of medium with unilateral illumination. The released gametes were phototactic and accumulated at one side of the drop, where they were collected with a pasteur pipette. Small drops of suspensions with male and female gametes were combined on a sterile cover glass in a moist chamber. After several hours the cover glass was placed upside down on a micro slide with a spacer. Under the microscope, zygotes were identified with an ink mark on the cover glass, and the cover glass was incubated in a petri dish with culture medium. If necessary, a mixture of antibiotics (Hoshaw & Rosowski 1973) was added. Subsequently, germlings of heterozygous sporophytes obtained in this way were isolated and subcultured. The hybrid sporophytes were designated in reference to their parent plants, for example NaCh as a hybrid from female Na and male Ch.

RESULTS

Vegetative characters

The four strains of *Ectocarpus siliculosus* compared in this study differ strongly in their morphology (Figs 1-8), but show similar responses to temperature (Figs 13-15).

The gametophytes of all strains are alternately branched and form plurilocular gametangia. Growing at their optimum temperature of 12-15°C, gametophytes of Wil (Fig. 2) and Ch (Fig. 3) form very soft loose tufts, while those of Na (Fig. 1) are coarser and those of NZ (Fig. 4) very compact. The gametangia show significant differences. Those of Wil (Fig. 10) and Ch (Fig. 11) are long and slim, whereas those of Na (Fig. 9) are relatively short, and those of NZ (Fig. 12) are very compact . The filament cells are longer than wide and contain several ribbon-shaped chloroplasts with pyrenoids. At low temperatures (3-5°C, Fig. 13) the gametophytes grow very slowly. The filaments have many very short branches, the filament cells are nearly twice as wide as long and gametangia are compact and small. At high temperatures (20°C, Fig. 15) growth is also reduced, and the branches are inserted at nearly right angles. Wil and Na did not grow or reproduce at temperatures higher than 22°C and Ch and NZ not higher than 20°C.

Partheno- and heterosporophytes in each strain have the same morphology. In the North American strain Wil gametophyte and sporophyte are isomorphic and profusely branched (Figs 2+6). They can only be distinguished by the presence of unilocular sporangia or by confirmation of the gamete character of their plurizoids. In the other strains, gametophytes and sporophytes can be clearly distinguished by their morphology (Figs 1+5, 3+7, 4+8). The sporophytes form long, scarcely branched filaments bearing both unilocular and plurilocular sporangia. The sporophytes show the same response to temperature as the gametophytes. Their optimum growth temperature is 17-20°C. At low temperatures growth is slow and predominantly unilocular sporangia are formed. At high temperatures only a few unilocular sporangia are present, while many plurilocular sporangia are formed. In these experiments the sporophytes survived temperatures 3-6°C higher than the corresponding gametophytes.

Crossing experiments (Table 1)

In all combinations of female and male gametes plasmogamy occurred; however, the viability of the zygotes and the growth of the hybrid sporophytes differed strongly. WilCh,

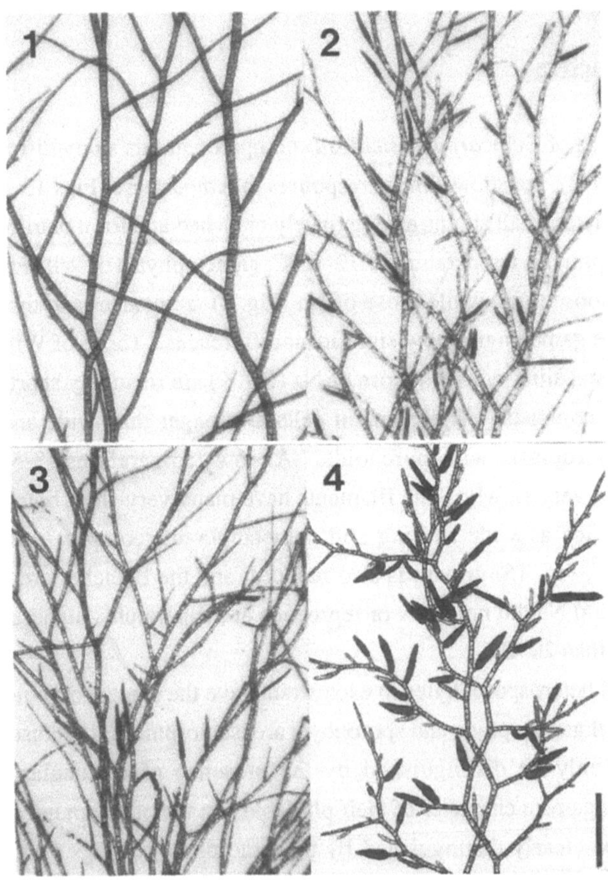

Figs 1-4. Morphology of the four *Ectocarpus* strains. Female gametophytes of Na (Fig. 1), Wil (Fig. 2), Ch (Fig. 3), and NZ (Fig. 4). Scale bar = 250 μm

WilNZ and the reciprocal crosses had a strongly reduced viability of the zygotes. After 2 or 3 cell divisions the development stopped and most of the germlings died after 4-8 weeks. Some plants continued to grow very slowly but did not become reproductive. The crosses NaCh and ChNa showed similar viability as WilCh and ChWil. The slowly growing germlings resemble the hybrids NaCh and ChNa grown by Müller (1988). The hybrids NaNZ and the reciprocal showed better viability. About one half of the zygotes grew up and produced normal plurilocular sporangia. Some plants of NaNZ produced unilocular sporangium initials (Fig. 25) similar to those of NaCh (figs 23, 24) and ChNa described by Müller (1988). They did not develop further and lost most of their pigmentation. Figs 20-22 show normal unilocular sporangia and initials in parthenosporophytes of Na and Ch.

The crosses NaWil and WilNa were growing as well as their internal crosses, and produced pluri- and unilocular sporangia. The hybrids NZCh and ChNZ were less viable

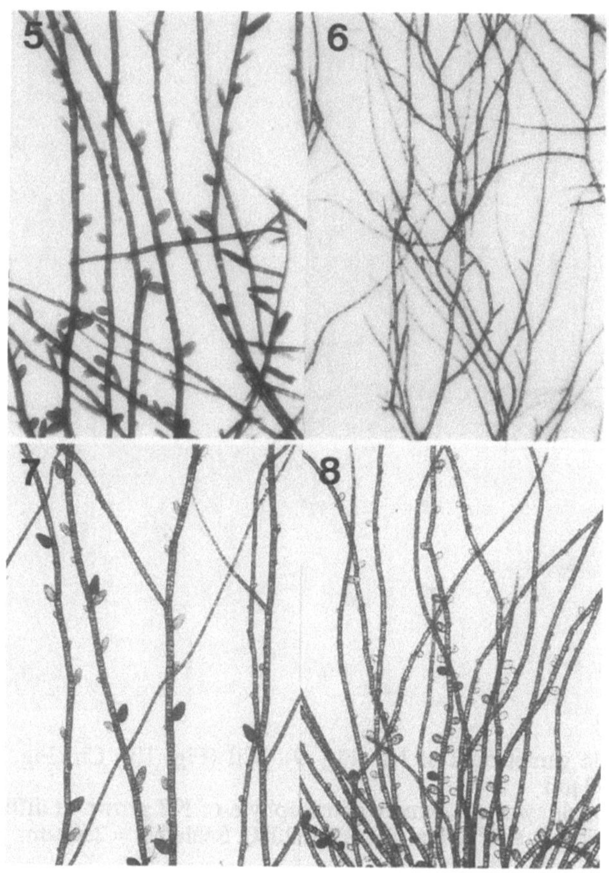

Figs 5-8. Morphology of the four *Ectocarpus* strains. Parthenosporophytes of Na (Fig. 5), Wil (Fig. 6), Ch (Fig. 7), and NZ (Fig. 8). Same Scale as Fig. 4

(≈70% survival of zygotes). The sporophytes produced plurilocular sporangia but unilocular sporangia have not been observed to date.

The morphology of the hybrids shows dominance of morphological characters contributed by the male parent. The sporangia of Ch (Figs 16, 20) and NZCh (Fig. 17) sporophytes, for example, were formed primarily in clusters on adjacent cells, whereas the sporangia on NZ (Fig. 19) and ChNZ (Fig. 18) sporophytes were more or less evenly distributed. A similar dominance of a character contributed by the male parent was found in diploid heterozygous gametophytes (Müller 1970). Their gametes were heterozygous with regard to their sex-determining genes, but reacted as male gametes.

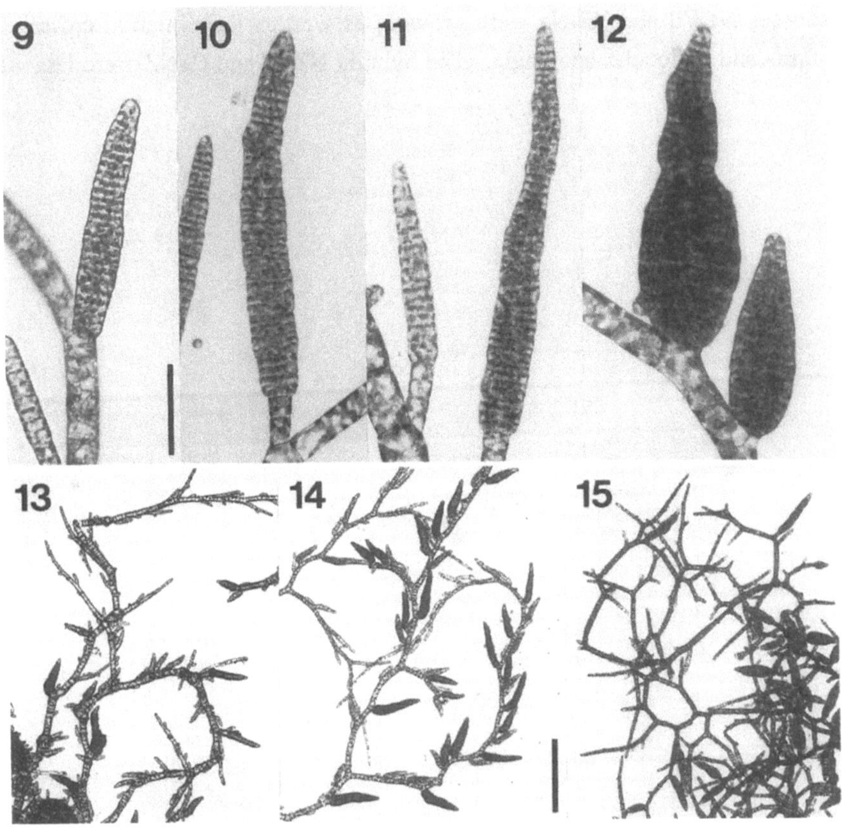

Figs 9-12. Female gametangia of Na (Fig. 9), Wil (Fig. 10), Ch (Fig. 11), and NZ (Fig. 12). Scale bar = 50 µm
Figs 13-15. Morphology of the female gametophyte of NZ grown at different temperatures, Fig. 13 at 3-4°C, Fig. 14 at 7°C, and Fig. 15 at 20°C. Scale bar = 250 µm

DISCUSSION

The distribution of the genus *Ectocarpus* (Fig. 26) extends from arctic to northern warm temperate regions and from subantarctic to southern warm temperate regions. The distribution limit towards the tropics corresponds well with the 20°C isotherm of the coldest month (February) in the Northern Hemisphere and in the Southern Hemisphere roughly with the 15°C - 20°C August isotherm. These varying limits in the Southern Hemisphere and distribution gaps in the Arctic region may reflect the paucity of coasts with suitable substrata for macroalgae. The lethal temperatures of 22°C for the sporophytes and 20°C for the gametophytes corresponds to this distribution with a safety margin of about 2°C. In the warm temperate regions, the gametophytes can grow during the cold season; their offspring of partheno- and heterosporophytes can then survive the warm season. When temperatures drop again, the sporophytes ideally produce unilocular sporangia, and give rise to a generation of gametophytes. Bolton (1983) studied the temperature tolerance of several *Ectocarpus siliculosus* sporophytes. He found an ecoclinal variation and determined lethal temperatures between 20°C and 33°C. He concluded that the warm season in warm temperate regions is

Table 1. Crossing experiments. n, total number of observed zygotes; V, survival of zygotes as percent of total number, in internal crosses the zygotes showed full viability; Pl, plurilocular sporangia; U, unilocular sporangia: def U, abnormal unilocular sporangium initials; - hybrid sporophytes not mature

	Na	Wil	Ch	NZ
Na	n 9 V: full Pl, U	n 44 V 100% Pl	n 14 V 6,8% Pl, def U	 V 50% Pl, def U
Wil	n 12 V 91,6% Pl, U	 V: full Pl, U	n 48 V 4,2% -	n 60 V 10% -
Ch	n 45 V 2,2% Pl, def U	n 36 V 5,5% -	 V: full Pl, U	n 8 V 62,5% Pl
NZ	n 14 V 57,1% Pl	n 51 V 5,9% -	n 20 V 65% Pl	 V: full Pl, U

survived by a microscopic sporophyte. In these regions, macroscopic phases are only recorded for the cold season. Bolton did not examine gametophytes. His Arctic strain, which was the only one with an optimal temperature at 15°C, was of uncertain genetic status, but considering my results, this strain was most likely gametophytic.

The phenotypic variation in the four *Ectocarpus* isolates studied here, illustrate the difficulties of taxonomy. Without prior knowledge of the life history, the gametophytes and sporophytes of the heteromorphic strains Na, Ch and NZ could be considered as different species. The gametophytes of Wil and Ch are very similar, but the sporophytes Wil and Ch could be considered as different taxa.

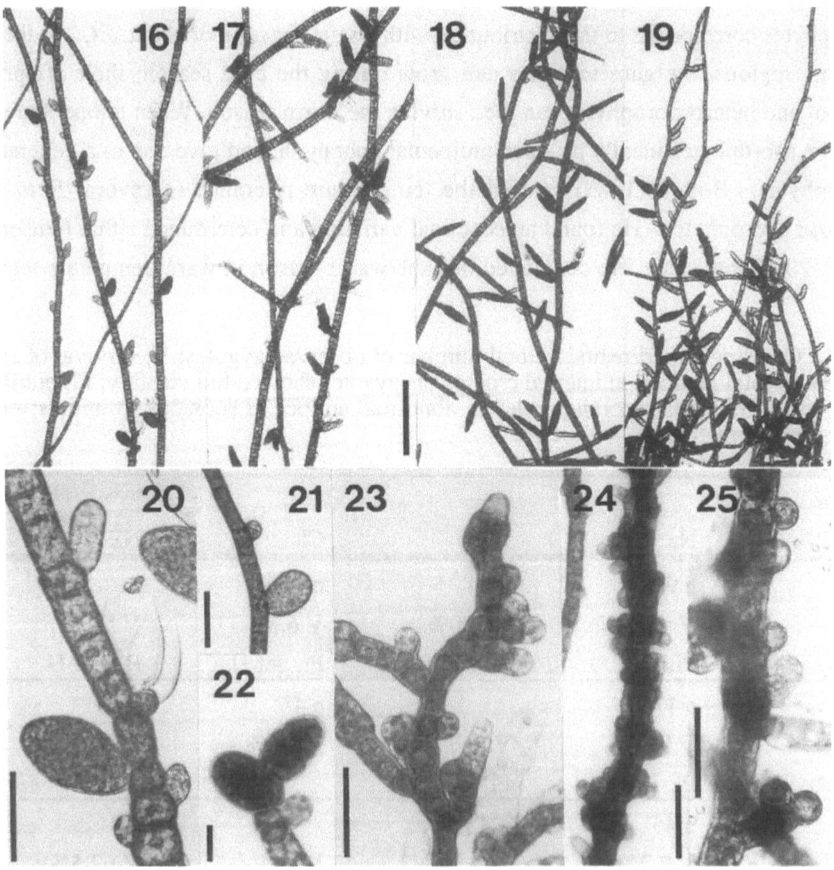

Figs 16-19. Morphology of the parthenosporophytes of Ch (Fig. 16) and NZ (Fig. 19) and the reciprocal hybrids NZCh (Fig. 17) and ChNZ (Fig. 18). Scale bar = 250 µm.
Figs 20-25. Unilocular sporangia and initials on parthenosporophytes of Ch (Fig. 20, scale bar = 50 µm) and Na (Fig. 21, scale bar = 50 µm; Fig. 22, scale bar = 20 µm). Abortive unilocular sporangium initials on the hybrids NaCh (Fig. 23, scale bar = 50 µm; Fig. 24, scale bar = 250 µm) and NaNZ (Fig. 25, scale bar = 50 µm).

The crossing experiments give more information on the relationship between the strains. Interfertility and good growth of the hybrids between Wil and Na show a close relationship. The southern strains NZ and Ch form viable hybrids, but to date it is not clear if they are capable of meiosis, since no unilocular sporangia have been observed. The slightly reduced viability of zygotes suggests that these strains are not as closely related as the northern strains, and that they may be considered as geographical races. Crosses of northern and southern strains show strongly reduced zygote viability, in addition to hybrid sterility. Meiosis in these hybrids is possibly disturbed by partial failure of chromosome pairing during prophase. These hybrids are capable of reproduction based on mitosis in plurilocular sporangia: the hybrid sporophytes can perpetuate themselves as clonal populations. The strains NZ and Ch could be regarded as subspecies, but it does not seem useful to regard them as a separate taxonomic entity because of the great variability of their morphological characters.

The close relationship between Na and Wil suggests a common ancestor, which could have invaded the opening Atlantic through the Bering Strait and the Arctic Ocean. The original

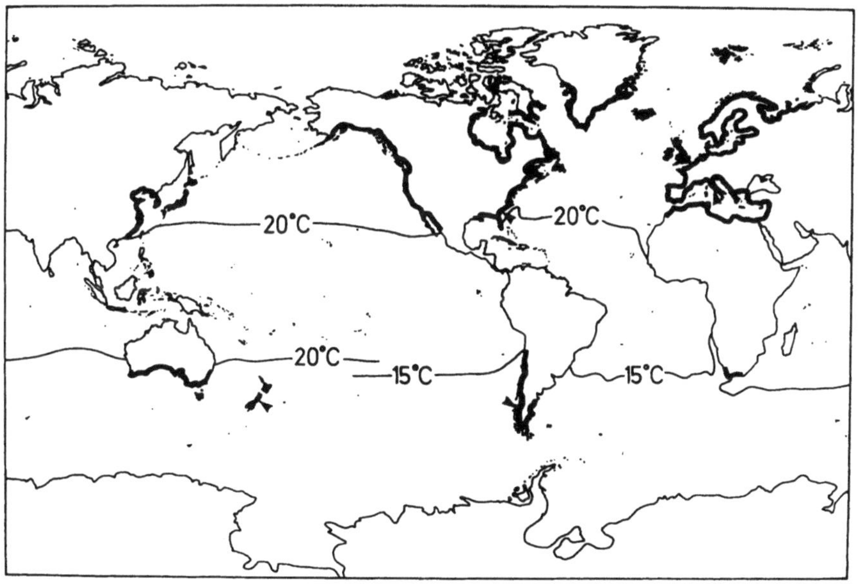

Fig. 26. Distribution of the genus *Ectocarpus* (thick lined coasts). The isotherms are those for the coldest months (February in the Northern Hemisphere, August in the Southern Hemisphere). Arrows indicate the collection sites of the 4 strains in this study: Na - Mediterranean, Wil - Northwest Atlantic, Ch - Southeast Pacific and NZ - Southwest Pacific. (References for distribution: Abbott and Hollenberg, 1976; Asensi, 1966; Caram and Jónsson, 1972; Cardinal, 1964; Gayral, 1966; Jaasund, 1965; Jónsson, 1903; Kornmann and Sahling, 1977; Lee, 1980; Levring, 1937, 1938, 1945, 1960; Lindauer *et al.*, 1961; Lund, 1933; Parke and Dixon, 1976; Pedersen, 1976; Ricker, 1987; Svendsen, 1959; Taylor, 1957; Widdowson, 1973; Womersley, 1976, 1987; Yoshida *et al.*, 1985; Zinova, 1953, 1959)

distribution of this ancestor could have been the North Pacific during the glacial periods of the Pleistocene, when temperate and cold water species may have developed from warm water species. Such a history is suggested for the genus *Laminaria* (Lüning 1985; Bolton *et al.* 1983). During glacial periods, the temperature at the equator dropped and cold water species such as *Laminaria*, *Macrocystis* and possibly *Ectocarpus* could have crossed into the Southern Hemisphere. Once in the Southern Hemisphere, *Ectocarpus* could have spread with the West Wind Drift to South Africa, Australia, and New Zealand. The dissimilarity of NZ and Ch and the better viability of NaNZ and NZNa in comparison to NaCh and ChNa, however, do not support this distributional history. An alternative explanation is that *Ectocarpus* is a very ancient genus that inhabited the coasts of Pangaea. According to this scenario, the cold water forms of the two hemispheres would have developed separately. For the phylogenetically young order Laminariales, Stam *et al.* (1988) found, by single copy DNA-DNA hybridization studies, a divergence time before the opening of the Bering Strait. The genus *Ectocarpus* may well be much older.

The closer relationship between Na and NZ can only be explained by a parallel development. The dispersal of *Ectocarpus* through the Tethys Sea is unlikely, since this area had a tropical character and, at present, the genus *Ectocarpus* is only poorly represented in the tropics.

Expansion of these crossing experiments to include representatives from the North Pacific could give more information about the origin of *Ectocarpus* and the relationships of its vastly disjunct populations.

REFERENCES

Abbott IA, Hollenberg GJ (1976) Marine algae of California. Stanford University Press, Stanford, California

Asensi AO (1966) Guia para reconocer los generos de algas pardas de la Argentina. Contrib del Inst Antar Argen 103: 1-51

Bolton JJ (1983) Ecoclinal variation in *Ectocarpus siliculosus* (Phaeophyceae) with respect to temperature growth optima and survival limits. Mar Biol 73: 131-138

Bolton JJ, Germann I, Lüning K (1983) Hybridization between Atlantic and Pacific representatives of the Simplices section of *Laminaria* (Phaeophyta). Phycologia 22: 133-140

Caram B, Jónsson S (1972) Nouvel inventaire des algues marines de l'Islande. Acta Bot Isl 1: 5-31

Cardinal A (1964) Etudes sur les Ectocarpacées de la Manche. Beih Nova Hedwigia 15 : 1-86

Gayral P (1966) Les algues des côtes françaises (Manche et Atlantique). Edition Doin, Paris

Grant V (1976) Artbildung bei Pflanzen. Verlag Paul Parey, Berlin, Hamburg

Guiry MD (1984) Structure, life history and hybridization studies on *Gigartina teedii* (Rhodophyta) in culture. Br Phycol J 19: 37-55

Hoshaw RW, Rosowski JR (1973) Methods for microscopic algae. In: Stein JR (ed) Handbook of phycological methods, culture methods and growth measurement. University Press, Cambridge, p 53-68
Howard BM, Nonomura AM, Fenical W (1980) Chemotaxonomy in marine algae: Secondary metabolite synthesis by *Laurencia* in unialgal culture. Biochem Syst Ecol 8: 329-336
Jaasund E (1965) Aspects of marine algal vegetation of North Norway. Bot Gothob 4: 1-174
Jónsson H (1903) The marine algae of Iceland. (II. Phaeophyceae). Bot Tidsskr 25: 141-195
Kornmann P, Sahling P-H (1977) Meeresalgen von Helgoland, Benthische Grün-, Braun- und Rotalgen. Helgol Wiss Meeresunters 29: 1-289
Lee RKS (1980) A catalogue of the marine algae of the Canadian Arctic. Nat Mus Canada, Publ Bot 9: 1-82
Levring T (1937) Zur Kenntnis der Algenflora der Norwegischen Westküste. Lunds Univ Årsskr NF Avd 2, 33(8): 1-147
Levring T (1938) Verzeichnis einiger Chlorophyceen und Phaeophyceen von Südafrika. Lunds Univ Årsskr NF Avd 2, 34(9): 1-25 + 4 pl
Levring T (1945) Marine algae from some antarctic and subantarctic islands. Lunds Univ Årsskr NF Avd 2, 41(7): 1-36
Levring T (1960) Contributions to the algal flora of Chile. Lunds Univ Årsskr NF Avd 2, 56(10): 1-85
Lindauer VW, Chapman VJ, Aiken M (1961) The marine algae of New Zealand, II Phaeophyceae. Nova Hedwigia 3: 129-350
Lund S (1933) The marine algae, The Godthaab expedition 1928. Medd Grønland 82(4): 1-18
Lüning K (1985) Meeresbotanik, Verbreitung, Ökophysiologie und Nutzung der marinen Makroalgen. Georg Thieme Verlag, Stuttgart, New York
Mathiesen AC, Norton TA, Neushul M (1981) The taxonomic implications of genetic and environmentally induced variations in seaweed morphology. Bot Rev 47: 313-347
Müller DG (1967) Generationswechsel, Kernphasenwechsel und Sexualität der Braunalge *Ectocarpus siliculosus* im Kulturversuch. Planta (Berlin) 75: 39-54
Müller DG (1970) Diploide, heterozygote Gametophyten bei der Braunalge *Ectocarpus siliculosus*. Naturwissenschaften 57: 357-358
Müller DG (1979) Genetic affinity of *Ectocarpus siliculosus* (Dillw.) Lyngb. from the Mediterranean, North Atlantic and Australia. Phycologia 18: 312-318
Müller DG (1988) Studies on sexual compatibility between *Ectocarpus siliculosus* (Phaeophyceae) from Chile and the Mediterranean Sea. Helgol Meeresunters 42: 469-476
Müller DG, Gassmann G (1980) Sexual hormone specificity in *Ectocarpus* and *Laminaria* (Phaeophyceae). Naturwissenschaften 67: 462
Müller DG, Stache B (1989) Life history studies on *Pilayella littoralis* (L.) Kjellman (Phaeophyceae, Ectocarpales) of different geographical origin. Bot Mar 32: 71-78
Neushul M (1962) Possible intergeneric hybrids in giant kelps (Lessoniaceae). Amer J Bot 49(suppl): 672
Neushul M (1971) *Macrocystis* and related kelps. Nova Hedwigia 32: 223-228
Papenfuss GF (1964) Catalogue and bibliography of antarctic and subantarctic benthic marine algae. In: Milton OL (ed) Biology of the Antarctic Seas. Antarctic Research Series 1: 1-76, American Geographical Union, Washington
Parke H, Dixon PS (1976) Check-list of British marine algae, third revision. J Mar Biol Ass U K 56: 527-594
Pedersen PM (1976) Marine, benthic algae from southernmost Greenland. Medd Grønland 199(3): 1-79
Ricker RW (1987) Taxonomy and Biogeography of Macquarie Island Seaweeds. British Muscum (Natural History), London
Russell G (1966) The genus *Ectocarpus* in Britain, I. The attached forms. J Mar Biol Ass U K 46: 267-294
Saga N, Gibor A (1986) Pure culture of algae (in Japanese). In: Yamada Y, Okada Y (eds) Gendai-Kagaku, Plant Biotechnology. Tokyo Kagaku Dojin, Tokyo, suppl 5: 55-71

Sanbonsuga Y, Neushul M (1978) Hybridization of *Macrocystis* (Phaeophyta) with other float-bearing kelps. J Phycol 14: 214-224
Sanbonsuga Y, Neushul M (1979) Cultivation and hybridization of giant kelps (Phaeophyceae). Proc Int Seaweed Symp 9: 91-96
Stam WT, Bot PVM, Boele-Bos SA, van den Rooji JM, van den Hoek C (1988) .Single-copy DNA-DNA hybridization among five species of *Laminaria* (Phaeophyceae): phylogenetic and biogeographic implications. Helgol Meeresunters 42: 251-267
Svendsen P (1959) The algal vegetation of Spitsbergen. Skr Norsk Polarinst 116: 1-49
Taylor WR (1957) Marine algae of the northeastern coast of North America. University of Michigan Press, Ann Arbor
Tseng CK (1983) Common Seaweeds of China. Science Press, Bejing, China
van der Meer JP (1986) Genetic contributions to research on seaweeds. Prog Phycol Res 4: 1-38
van der Meer JP, Bird CJ (1985) *Palmaria mollis* stat. nov.: a newly recognized species of *Palmaria* (Rhodophyceae) from the northeast Pacific Ocean. Can J Bot 63: 398-403
Widdowson TB (1973) The marine algae of British Columbia and northern Washington: revised list and keys, Part I Phaeophyceae (brown algae). Syesis 6: 81-96
Womersley HBS (1967) A critical survey of the marine algae of southern Australia, II Phaeophyta. Aust J Bot 15: 189-270
Womersley HBS (1987) The marine benthic flora of Southern Australia Part II. Southern Australian Government Printing Division, Adelaide
Yoshida T, Nakajima Y, Nakata Y (1985) Preliminary check-list of marine benthic algae of Japan, I Chlorophyceae and Phaeophyceae. Jap J Phycol 33: 57-74
Zinova AD (1953) Manual of the brown algae of the northern seas of the USSR (in Russian). Isdatestvo Akademia Nauk, SSSR, Moskau, Leningrad
Zinova AD (1959) List of marine algae of southern Sakkalin and the southern Kurile Islands. Issled Dal' Nevost Morei SSSR 6: 146-161 (in Russian)

THE DISTRIBUTION AND EVOLUTION OF THE LAMINARIALES: NORTH PACIFIC - ATLANTIC RELATIONSHIPS

Klaus Lüning and Inka tom Dieck
Biologische Anstalt Helgoland
Zentrale Hamburg
Notkestrasse 31
D-2000 Hamburg 52
Federal Republic of Germany

Abstract - The present paper outlines within order relationships of the Laminariales on the basis of paleoceanographic constraints and recent experimental results, especially those dealing with temperature tolerance and hybridization studies. Within the digitate section of *Laminaria*, it is suggested that at least two Pacific ancestors invaded the North Atlantic via the Bering Strait and the Arctic coasts when the the Bering Land Bridge became submerged during the Pliocene. A single common ancestor from the Pacific is hypothesized for species with a rigid stipe as currently represented in the North Pacific by *L. setchellii*, in the northeastern Atlantic by *L. hyperborea* and *L. ochroleuca*, and in southern Africa by *L. pallida* and *L. schinzii*. A second immigrant from the North Pacific with a flexible stipe related to the North Pacific *L. bongardiana* (= *L. groenlandica* sensu Druehl 1968), is hypothesized to be ancestral to the North Atlantic *L. digitata*, and the two southwestern Atlantic deep-water species *L. brasiliensis* and *L. abyssalis*.

INTRODUCTION

The Laminariales occur at the northernmost location where seaweeds have been investigated. At this site in the Jörgen Brönlunds Fjord (east coast of Greenland, 82°N), the Arctic cold-temperate species *Laminaria saccharina* grows. The southern limit of the Laminariales in the Northern Hemisphere is reached off southern Morocco and the Western Sahara, where the distribution of the warm-temperate species *Laminaria ochroleuca*, *Saccorhiza polyschides* and *Phyllariopsis spp.* ends (Lüning 1985). In the North Pacific, the southern limit of the Laminariales occurs at a latitude of 25-27°N: in South China, *Undaria pinnatifida* extends to the province of Zhejiang; in Baja California, *Macrocystis pyrifera* and *Eisenia arborea* occur in upwelling water southward to Magdalena Bay; in the eastern Philippine Sea (27°N, 145°E), a digitate species of *Laminaria*, newly described as *L. philippinensis*, was discovered in cooler water at 85 m depth (Petrov *et al.* 1973). This parallels the deep-water find of the digitate *L. brasiliensis* at 23°S, near the northern limit of the warm-temperate Eastern

NATO ASI Series, Vol. G 22
Evolutionary Biogeography of the
Marine Algae of the North Atlantic
Edited by D. J. Garbary and G. R. South

South American region at Rio de Janeiro, 100 km off the coast, at 40-100 m depth (Joly & Oliveira Filho 1967).

It is well-known that the Laminariales are absent in tropical regions, reach their highest diversity in the North Pacific, and have few genera in the southern hemisphere. The question remains: to what extent do the present-day distribution patterns of the species of Laminariales reflect a site of origin for the order and its subsequent dispersal and speciation over the world's oceans?

If the Laminariales date back to the Paleozoic, there is no direct geological evidence in the form of a fossil record. This differs from the Dasycladales (Chlorophyta) and Corallinales (Rhodophyta), both with well-established fossil records from the Paleozoic (Wray 1977; Johansen 1971; Woelkerling 1988). Despite a poor fossil record, there are several lines of evidence to strengthen the hypothesis that the order Laminariales is young in geological terms, that it evolved in the Tertiary in the North Pacific, and spread to the Atlantic towards the end of the Tertiary, and then to the Southern Hemisphere.

DISTRIBUTION OF THE LAMINARIALES: OVERVIEW AND BIOGEOGRAPHICAL PROBLEMS

It is a general phenomenon that cold temperate faunas and floras are much more diverse in the North Pacific than in the North Atlantic (Ekman 1953; Briggs 1974; van den Hoek 1982a, b; Lüning 1985; Estes & Steinberg 1988); this is discussed below in the context of the different families of Laminariales. The ancestral family Phyllariaceae provides an exception to this generalization: its representatives occur only in the North Atlantic and Mediterranean. This biogeographical enigma deserves special attention.

Alariaceae

In the North Pacific, there are four genera along North American coasts (*Alaria, Pterygophora, Egregia, Eisenia*), and five genera on Asiatic coasts (*Alaria, Ecklonia, Eckloniopsis, Eisenia, Undaria*). *Alaria* is represented by eleven species in the North Pacific. The North Atlantic flora has only two species of Alariaceae, *Alaria esculenta* (Widdowson

1971), and *Ecklonia muratii*. The latter species occurs on the coasts of Mauritania and Senegal (Feldmann 1937), and is part of a widely distributed genus with disjunct species in South Africa, in the upwelling area of the Gulf of Oman (Hiscock *et al.* 1984), in Australia, New Zealand and Japan (Womersley 1987).

Laminariaceae

In the North Pacific, this family comprises the genera *Agarum, Arthrothamnus, Costaria, Cymathere, Hedophyllum, Kjellmaniella, Laminaria, Pleurophycus,* and *Thalassiophyllum*. *Costularia kurilensis, Feditia simuschurensis* (Gussarova & Petrov 1970, 1972) and *Phyllariella ochotensis* (Petrov & Vozzhinskaja 1966), have been described as new members of the Laminariaceae on Asiatic coasts.

Streptophyllopsis kuroshioensis is a newly-discovered deep-water representative of the Laminariaceae. It grows at 20-50 m depth off the Izu Peninsula on the Pacific coast of Honshu (Kajimura 1981, 1987). *Streptophyllopsis* produces two short-stiped annual blades up to 20 cm long which arise from a dichotomously branched perennial system up to 10 cm high. Sori develop on the blade, and on the annual stipe, which attains a maximum length of 4 cm. The prostrate holdfast is reminiscent of the rhizomatous *Laminaria* spp. on the Pacific coast of North America and in the Mediterranean, and also of *Ecklonia stolonifera* (family Alariaceae).

In the Atlantic, the family Laminariaceae is represented by a single species of *Agarum* limited to the western North Atlantic, and nine widely distributed species of *Laminaria*: *L. saccharina, L. longicruris* in the Simplices section; *L. digitata, L. hyperborea, L. ochroleuca* (also in the Mediterranean), *L. pallida, L. schinzii, L. brasiliensis* in the Digitatae section, and *L. abyssalis*, possibly also in the Digitatae section (see below). In the Mediterranean, at greater depths, there is *L. rodriguezii*, a species with a rhizomatous holdfast and an undivided blade. Several questions may be asked regarding the evolutionary origin of these species: if the North Atlantic obtained its *Laminaria* spp. from the North Pacific, can one still detect Pacific-Atlantic sister groups or sister species relationships? A second question relates to the recently discovered deep-water *Laminaria* spp. off the Brazilian coast: are they more related to the northern Atlantic *L. digitata*, or to *L. pallida* and *L. schinzii* in southern Africa ?

Lessoniaceae

This family does not occur either in the North Atlantic, or on the Asiatic coasts of the North Pacific. The genera of this family endemic to the American North Pacific are

Dictyoneurum, *Dictyoneuropsis*, *Lessoniopsis*, *Macrocystis*, *Nereocystis*, *Pelagophycus* and *Postelsia*. *Lessonia* is the only genus of Laminariales in the southern hemisphere that does not also occur in the northern hemisphere; it is related to *Lessoniopsis*; an ancestor of both may have crossed the equator in cold phases during the Miocene or Pleistocene. The same has probably occurred with *Macrocystis*, which today occurs in the northern and southern hemisphere (also see Nicholson 1979). The giant kelp genera of the Lessoniaceae, i.e. *Macrocystis*, *Nereocystis* and *Pelagophycus* may be polyphyletic, with *Macrocystis*, *Lessonia* and *Alaria* constituting one lineage, and *Nereocystis* and *Laminaria* a second. This has been indicated by an analysis of kelp chloroplast DNA restriction fragments (Fain *et al.* 1988) (see Druehl, 1990 for review).

Chordaceae

Chorda is the only genus in this family, and with its string-like morphology it does not have a kelp like appearance. It has several primitive traits including zoospores with an eyespot and monoecious gametophytes (Maier 1984). *Chorda filum* occurs in the North Pacific and North Atlantic; *C. tomentosa* only in the North Atlantic (see distribution maps in Lüning 1985, 1990).

Pseudochordaceae

This family has been proposed by Kawai & Kurogi (1985) for *Pseudochorda nagaii*, a genus/species formerly included in Chordariales. This alga has a whip-like thallus up to 60 cm long, and distinctive plurilocular antheridia in the monoecious gametophyte. The alga occurs from the Aleutian Islands to the Japan Sea. The presence of *Pseudochorda* and *Streptophyllopsis* (see above), gives Japanese waters the aspect of a "museum for Laminariales", and provides evidence that this order may have originated there. In any case, it shows that the North Pacific still harbours species that are either sister taxa, or ancestral to other species – a point emphasized by Henry & South (1987).

Phyllariaceae

This European-endemic family consists of the warm-temperate species *Saccorhiza polyschides*, *Phyllariopsis brevipes* (*Phyllaria reniformis*), and *P. purpurascens* (Henry &

South 1987). In addition, there is one Arctic cold-temperate species, *Saccorhiza dermatodea*; that is confined to the North Atlantic and adjacent Arctic areas. The Phyllariaceae have several primitive features, which suggests that they diverged from the other Laminariales early in evolution (Henry & South 1987): e.g. no lamoxirene as a sexual pheromone, as found in all other families of the Laminariales (Müller *et al.* 1985); zoospores with an eyespot; a conducting system of "solenocysts" and "allelocysts", analogous, but not identical, to the sieve elements of the Laminariaceae; presence of hair tufts on young lamina (also in *Costaria* and *Phyllariella* in the Laminariaceae, and in *Alaria* and *Undaria* in the Alariaceae). This morphological evidence has important phylogenetic consequences, and there are important biogeographic implications that require explanation. For instance, if the Laminariales originated in the North Pacific, why are the ancestral or sister group Phyllariaceae only present in the Atlantic and Mediterranean ?

In the following discussion, explanations of the distributions of Laminariales will be carried out based on the paleoceanography and presumed relationships of sister taxa.

THE EVOLUTION OF COLD-WATER BIOTAS

Cold-water faunas and floras must have evolved roughly every 200 million years, when the earth cooled during the large Ice Ages (Fig. 1 A). This occurred three times in the last 600 million years, i.e. in the Phanerozoic (Fairbridge 1973; Frakes 1979). Nothing is known of the cold-water algae of the two large Paleozoic Ice Ages, at around 450 million years before the present (= 450 Ma) and 280 Ma, but it may have been difficult for cool-water groups (including Laminariales) to survive from one large Ice Age to the next. Moreover, there is the possibility that the brown seaweeds may not have existed in the Paleozoic, but have evolved since the Mesozoic; this was concluded from the molecular evolution of 5S ribosomal RNA (Lim *et al.* 1986).

The present-day cold-water algal floras must have evolved as a response to climatic deterioration during the Tertiary. Large temperature drops at the Eocene/Oligocene boundary (40 Ma), and in the Miocene (15-10 Ma), heralded the third great Ice Age of the Phanerozoic, i.e. the Pleistocene (Fig. 1 B). The fact that the cold-water faunas and floras of the Northern and Southern Hemisphere consist of fundamentally different sets of taxa, is a consequence of their almost complete geographical isolation. The evolution of cold-water taxa started at high latitudes, and the tropical warm-water girdle, that persisted throughout geological time, essentially prevented the mixing of Northern and Southern Hemisphere cold-water faunas and floras. To cite the larger brown algae as an example: the kelp-like representatives of the Desmarestiales, Ascoseirales and Durvillaeales are the Southern Hemisphere products of the global cooling of the earth during the Tertiary, and the Laminariales are the corresponding Northern Hemisphere product.

There was a period of unipolar glaciation for about 10 million years between mid-Miocene and early Pliocene, in which Antarctica was glaciated, while the Arctic was essentially ice-free (Flohn 1984). During this period, the cold-water biota of the Southern Hemisphere had better opportunities for cold adaptation than biotas in the Northern Hemisphere. Scarlato (1977) pointed out two essential steps in cold-water adaptation: the first step is to broaden the survival

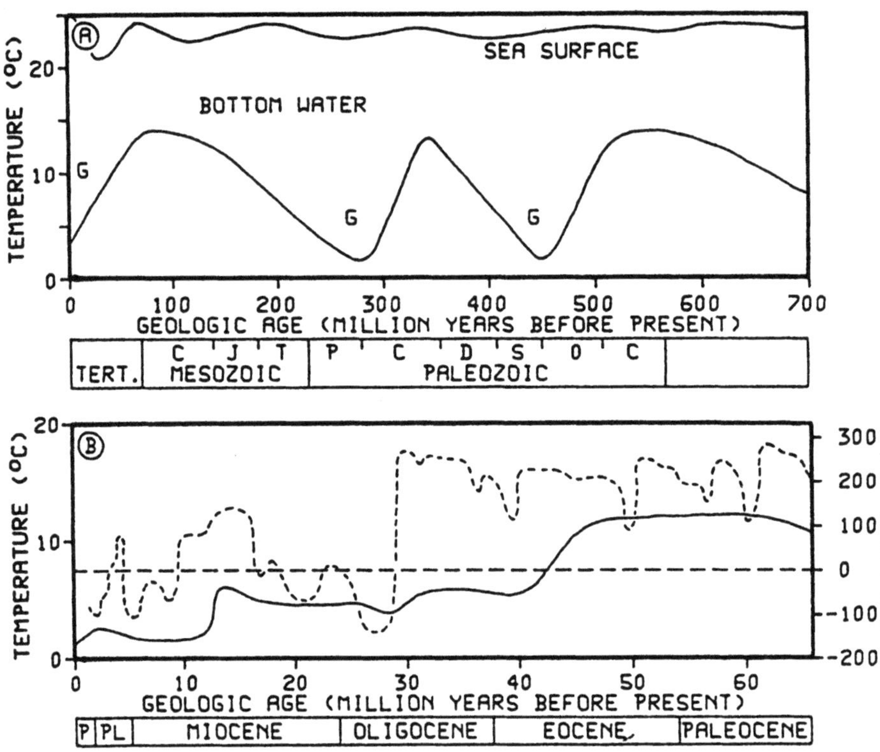

Fig. 1. Paleoclimate, paleotemperatures, and paleo-sea levels. **(A)** Estimated change in mean summer sea surface temperature and in bottom water temperature for tropical latitudes over the past 700 million years. Three periods of extensive glaciation (G) and cold bottom water (Ma = million years before present). (1) Around 450 Ma: late Ordovician-early Silurian; (2) 300 Ma: late Carboniferous-early Permian; (3) around 2 Ma: Pleistocene. Abbreviations, Paleozoic: C Cambrian; O Ordovicium; S Silurian; D Devonian; C Carboniferous; P Permian. Mesozoic: T Triassic; J Jurassic; C Cretaceous; Tert. Tertiary. **(B)** Tertiary and Pleistocene (P). Solid line: High latitude seawater surface temperature. Major cooling phases occurred near the Eocene/Oligocene boundary (40 Ma) and in the Miocene (10 Ma). Broken line: Tertiary eustatic changes of sea levels. Meters above or below present day sea level (=0) are tentative. Pl Pliocene (A after Schopf 1980; B after Savin *et al.* 1975, Vail & Hardenbol 1979; A, B from Lüning 1990)

and reproductive ranges towards the lower end of the temperature scale; this is followed by a loss of ability to thrive at higher temperatures.

Experimental evidence, from studies of algal growth in relation to temperature requirements, show that optimal temperatures for growth are consistently lower in Antarctic than in Arctic algae. The following comparisons are examples of this phenomenon: *Desmarestia anceps* and *Himantothallus grandifolius* from King George Island (South Shetland Islands) have growth optima at 0-5°C (Wiencke & tom Dieck 1989), whereas *Laminaria longicruris* and *L. solidungula*, both fróm Igloolik in the Canadian Arctic have growth optima at 10°C and 5-10°C, respectively (Bolton & Lüning 1982; tom Dieck 1989). Similarly, the upper survival limit of the gametophytes is 15-16°C in the Antarctic-endemic *Himantothallus grandifolius* (Wiencke & tom Dieck 1989), 19-20°C in the Arctic *L. solidungula*, and 23°C in *L. longicruris* (Bolton & Lüning 1982). A higher temperature tolerance is present in the gametophytes of those species whose ancestors probably crossed the equator in deeper, cooler water than during cold phases in the Pleistocene (van den Hoek 1982b) or late Tertiary: 25°C in *L. ochroleuca*; 26°C in *Macrocystis pyrifera*; and 27°C in *Eisenia arborea* (tom Dieck 1989). The even higher temperature tolerance of 27-28°C in *Ecklonia radiata* from Victoria, Australia, 28-29°C in *Undaria pinnatifida*, and 29°C in *Eisenia bicyclis* (both from Japan), and the fact that the gametophytes of all investigated members of the Laminariales survive 20°C or more (tom Dieck 1989) point to a warm-water character for kelp ancestors. The fact that cold adaptation of Laminariales has not progressed as far as that of their Antarctic kelp-like counterparts, is consistent with the observation that the few kelp taxa that crossed the equator belong to the Laminariales, while the Antarctic kelp-like algae remained caught in their cold environment.

Cool-water history of the North Pacific

The paleoceanographic reason for the high diversity of cool-water taxa in the North Pacific is its much longer history as a cool-water habitat as compared to the North Atlantic. In the Mesozoic, the North Atlantic existed only as a tropical near-equatorial sea (Fig. 2). The North Pacific, in contrast, had since the Mesozoic, extensive connections to the early Arctic Ocean, and offered vast coastlines for the development of temperate taxa in the high North (Fig. 2 A).

Phyllariaceae: possible conservation in the Arctic Ocean

The Phyllariaceae, a forerunner or ancestral side group of the Laminariales, may have developed and survived in the early Arctic Ocean, until the North Atlantic could be reached after its break-through to the Arctic Ocean at around 50 Ma (Fig. 2 C, D; Thiede 1979, 1980). A fossil example, pointing to the notion that the early Tertiary Arctic Ocean was a refugium, is found in Paleocene and early Eocene (65-50 Ma) molluscs and ostracods discovered at Ocean Point in north Alaska (Marinkovich *et al.* 1985). This fossil assemblage shows that Mesozoic taxa of the North Pacific had survived at Ocean Point into the Tertiary, and migrated into the North Atlantic, after a seaway had been formed to the Arctic Ocean at around 50 Ma. If an equivalent sequence of events took place in Laminariales, it would explain the occurrence of the primitive family Phyllariaceae in North Atlantic and Mediterranean refuge areas, perhaps with temperatures similar to those along the Arctic Tertiary coasts; while the primary development of the Laminariales took place in the North Pacific.

An alternate explanation is that the Phyllariaceae originated, as part of a vicariant event, at the southwestern part of the range of ancestral Laminariales in Asia, and then dispersed through the Tethys, into the Mediterranean and Atlantic. However, the present temperature tolerances of 25°C of the gametophytes in *Saccorhiza polyschides* (tom Dieck 1989) and less that 25°C in *Phyllariopsis brevipes* (Henry 1987) do not support this idea.

Jolyna, a relict from the early evolution of the Laminariales?

Jolyna laminarioides is a brown alga, up to 90 cm long, in which one to several erect blades arise from a discoid holdfast. The alga was detected east of Rio de Janeiro, in the wave-exposed eulittoral zone near Cape Frio, and bears some features of the Scytosiphonales, e.g. a single chloroplast with a conspicuous pyrenoid in each cell and plurilocular organs (Guimaraes *et al.* 1986). Its highly differentiated medullary filaments are, however, reminiscent of the Phyllariaceae. *Jolyna laminarioides* may represent a relic of the early evolution of the Laminariales. It may, hypothetically, have grown in the early Tertiary Arctic Ocean and adjacent

---

Fig. 2. Continental drift and formation of the Atlantic Ocean. **(A)** 100 Ma (Phase 1, Cretaceous): North Pacific wide open to early Arctic Ocean. **(B)** 60 Ma (Phase 2, early Tertiary): North Pacific closed off from early Arctic Ocean. Note: 1. The paleocontinental maps show, in addition to the coastlines, the present-day 1000 metre submarine contour which is much more enduring than the coastlines. 2. The two endpoints of the 1000 metre submarine contour in the Bering Sea area (circled crosses) approach each other in the time interval from 100 to 60 Ma. The quadrat indicates the approximate present position of the North Pole to emphasize the position of the Pole near the Bering Strait at 60 Ma. **(C)** 55-38 Ma (Phase 2, early Tertiary): The North Pacific closed off from the early Arctic Ocean. Epicontinental (shallow) marine waterways connecting Arctic Ocean to Tethys Sea: Turgai Strait via Eurasia (S) and Cannonball Sea (C - intermittently closed) via North America. Note that Beringia was formerly at a higher paleolatitude. **(D)** 13 Ma (Phase 3, mid-Tertiary): North Atlantic connected to the Arctic Ocean at around 40 Ma; epicontinental seaways dried up; Bering Strait still closed by land-bridge. (A and B from Smith *et al.* 1981; C and D after Hopkins & Marincovich 1984; from Lüning 1990)

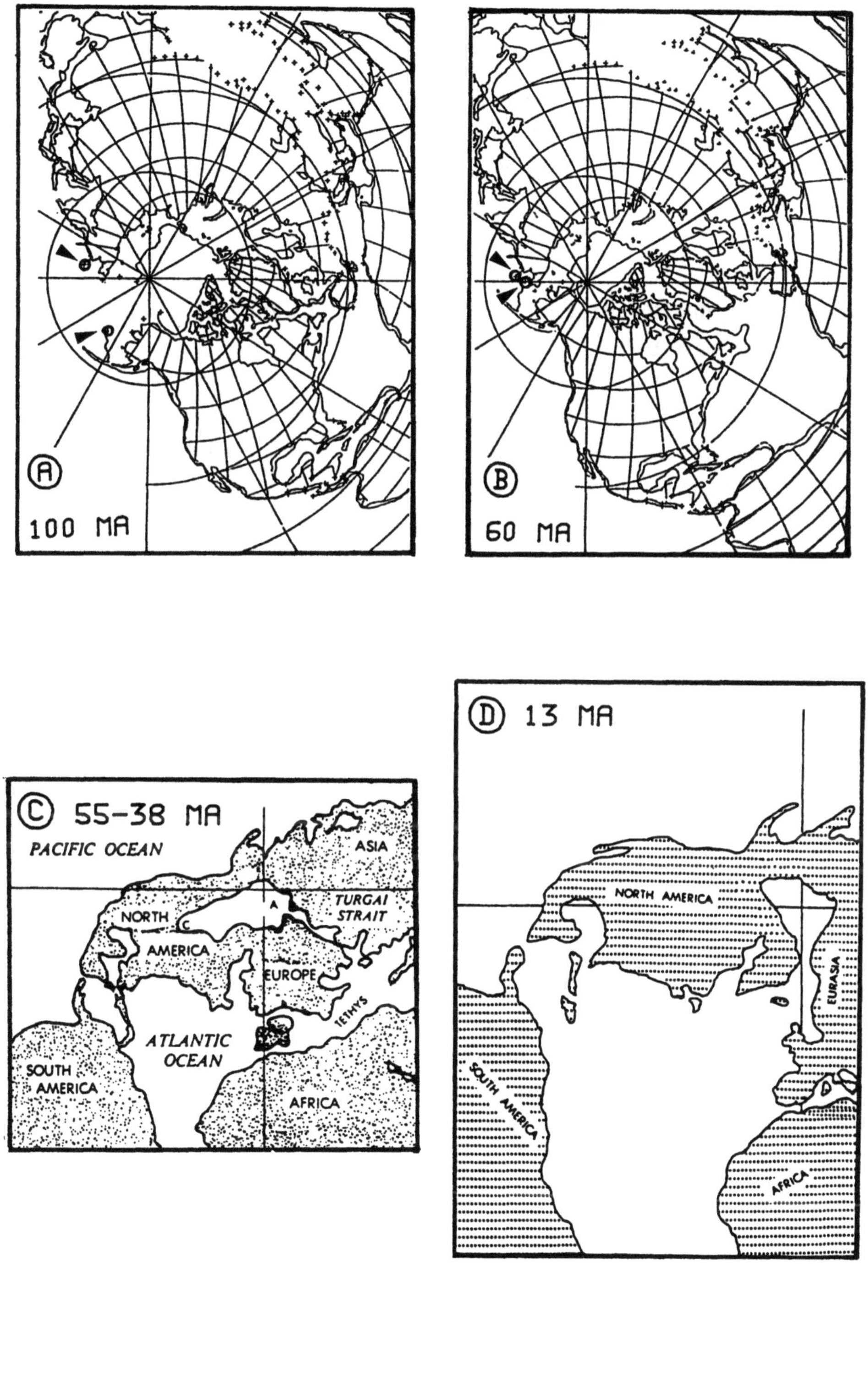
A
100 MA
B
60 MA
C 55–38 MA
PACIFIC OCEAN
ASIA
NORTH
C
A
TURGAI
STRAIT
AMERICA
EUROPE
TETHYS
ATLANTIC
OCEAN
SOUTH
AMERICA
AFRICA
D 13 MA
NORTH AMERICA
EURASIA
SOUTH AMERICA
AFRICA

North Pacific, and spread into the still small North Atlantic, after its opening to the Arctic Ocean at around 50 Ma. Cape Frio, in Brazil, was at that time not far from West Africa, where *Saccorhiza polyschides* and two *Phyllariopsis* spp. are found today. The present day warm-water locations of most Phyllariaceae and *J. laminarioides* seem to provide refugia for these living fossils, where they survive the competition of Laminariaceae, Alariaceae and Lessoniaceae.

North Atlantic invaded by Pacific newcomers after the opening of the Bering Strait at 3 Ma: speciation in *Laminaria*

If kelp dispersal followed the pattern similar to fossil marine invertebrates, it is likely that many cold-temperate and Arctic cold-temperate species invaded the North Atlantic, after the opening of the Bering Strait at 3 Ma (MacNeil 1965; Durham & MacNeil 1967, Hopkins 1967; Strauch 1972; McKenna 1983; Gladenkov 1979, 1986). Did one or several *Laminaria* species come from the North Pacific to the North Atlantic at 3 Ma? Stam *et al.* (1988) performed DNA-DNA hybridization experiments on the following species: *L. digitata, L. hyperborea, L. ochroleuca, L. saccharina*, and *L. rodriguezii*. They concluded that the speciation events resulting in these species occurred at about 15-19 Ma (Miocene) in the North Pacific. The Simplices section of *Laminaria* (undivided blade) is consistent with these conclusions, with *L. saccharina* occurring both in the North Atlantic and North Pacific (Kain 1979). If these conclusions are correct, it should be possible to identify Pacific-Atlantic sister species of *Laminaria*. Relationships among some *Laminaria* species discussed in this paper are shown in Fig. 3.

LAMINARIA DIGITATA AND *LAMINARIA GROENLANDICA*. In the section Digitatae (divided blade), the North Atlantic *L. digitata* finds its counterpart in the North Pacific *L. groenlandica*. Both have a complanate and flexible stipe, and the undivided blade splits after a juvenile phase; both species reach a maximum age of only a few years. The two species differ in some respects, e.g. the juvenile phase, with an undivided blade, ends after a few months in *L. digitata*, but may last longer in *L. groenlandica*; splitting starts from tears at the ends of the blades in *L. groenlandica* (Druehl 1968), and not from internal, longitudinal splits as in *L. digitata*; bullations may occur in *L. groenlandica* reminiscent of *L. saccharina*, but these have never been observed in *L. digitata*. While *L. digitata* from the American and European side of the North Atlantic cross readily, and F_3-sporophytes have been produced, no sporophytes resulted from crossings of *L. digitata* and *L. groenlandica* (tom Dieck 1989).

TAXONOMY OF *LAMINARIA GROENLANDICA*. It is unfortunate that the name, *Laminaria groenlandica*, has been applied to a species that is a clearly-circumscribed endemic species in the North Pacific. This situation was the result of a series of misunderstandings. Setchell & Gardner (1925) called the plant *L. cuneifolia* J. Ag., but Agardh had a North Atlantic representative of the Simplices section in mind, and *L. cuneifolia* J. Ag. is now regarded as synonymous with *L. saccharina* (South & Tittley 1986). Wilce (1959, 1960, 1965), working on *Laminaria* species with undivided blades (Section Simplices) synonymized the Atlantic *L. cuneifolia* with the Atlantic *L. groenlandica* Rosenvinge (1893), and the latter entity is also regarded as synonymous with *L. saccharina*. Druehl (1968) suspected that the Pacific digitate plant should be given its own specific epithet when he stated "In my opinion, we are presently not prepared to answer this question, and for this reason I have elected to retain the name *L. cuneifolia* (= *L. groenlandica*) earlier used (Setchell & Gardner 1925) to describe these plants" (Druehl 1968, p. 543). Today it is clear that the North Pacific *L. groenlandica* should have been related to the Atlantic *L. digitata* and not with *L. saccharina*. This Pacific plant is a distinct entity, and has consistently (and erroneously) been synonymized with the wrong Atlantic

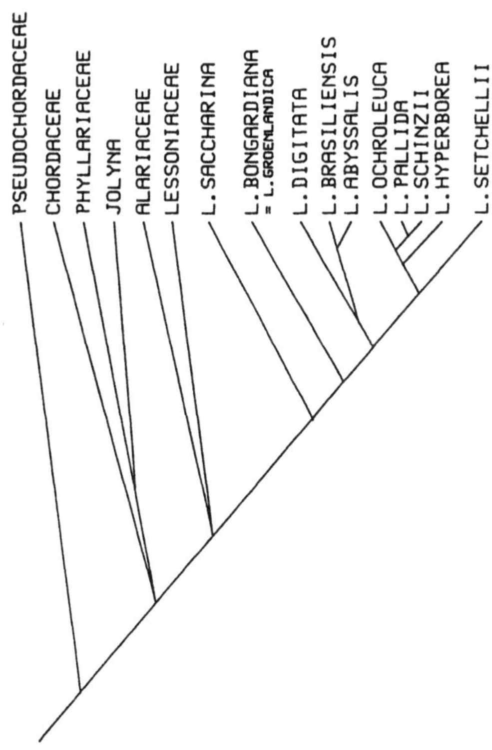

Fig. 3. Evolutionary relationships among some taxa of Laminariales discussed in this paper. See text for discussion of biogeographic implications. Compare with Druehl Fig. 1 (this volume)

partners. The best solution to this problem would be to follow Petrov (1972) who included *L. groenlandica* sensu Druehl (1968) as a synonym of *L. bongardiana* Postels & Ruprecht (1840).

LAMINARIA HYPERBOREA AND *LAMINARIA SETCHELLII*. Because of its rigid stipe, the European-Atlantic *Laminaria hyperborea* forms an upper story kelp forest that overshadows competing algae. Other features different from *L. digitata* include: a) terete stipes; b) a long lifespan (up to 15 years); c) spore formation in the winter, and d) a clear constriction between the old and new blade. These features are also present in the Pacific *L. setchellii*, and are absent in the Pacific *L. groenlandica*. The physiological basis for the constriction between the old and new blade in *L. hyperborea* and *L. setchellii* is a period, during summer, when growth stops. The cessation of growth in *L. hyperborea* was shown to be a genuine long-day reaction (tom Dieck 1989). She also found that *L. hyperborea* and *L. setchellii* could not be hybridized. Small sporophytes were obtained but these were stunted, had an abnormal thickened meristematic zone, and did not grow longer than 1-3 cm after 3-11 months. In contrast, in *L. setchellii*, 20-40 cm long sporophytes with sori were produced, after 33 months, in the same experimental tanks (tom Dieck 1989).

PHYSIOLOGY AND SEASONALITY. The fact that one can still discern Pacific-Atlantic pairs of sister taxa, which can clearly be recognized as separate species, supports the above-mentioned view that more than one species of *Laminaria* dispersed at 3 Ma through the Bering Strait after its opening, and reached the North Atlantic via a then still ice-free Arctic Ocean. A number of photobiological traits among these kelp species reflect the ancestral adaptation to life at high latitudes, with its strong seasonality. These physiological features include a) the occurrence of new blade growth in short days in *L. hyperborea* (Lüning 1986), b) the cessation of growth in long days in *L. hyperborea* and *L. setchellii* (tom Dieck 1989), c) the formation of sori in *L. saccharina* and *L. setchellii* caused by short days in autumn (Lüning 1988; tom Dieck 1989), and d) the survival of vegetative gametophytes of eight species of *Laminaria*, *Saccorhiza dermatodea*, *Saccorhiza polyschides* and *Alaria esculenta* after at least 18 months in the dark (tom Dieck 1989).

The importance of *Laminaria ochroleuca* and its ancestors

ECOLOGY. *Laminaria ochroleuca* occurs as far north as the English Channel, west to the Azores, and south to Morocco. It has a rigid, terete stipe, is long-lived, and forms dense laminarian forests similar to *L. hyperborea*. While the underwater forests of *L. hyperborea* reach northern Norway, in the south they are replaced by forests of *L. ochroleuca* from mid-Portugal to Morocco (John 1971; Braud 1974; Sheppard *et al.* 1978). *L. ochroleuca* also has relationships to *L. digitata*: both species have a smooth stipe, have no constriction between the old and new blade, and sporulate in summer. Since warm-temperate species are probably ancestral to the cold-temperate species, it is possible that *L. ochroleuca* is the sister group of the *L. digitata* (with a distribution from mid-Greenland to Brittany) – *L. hyperborea* complex (Fig. 3).

DISPERSAL ACROSS THE EQUATOR. The presences of *Laminaria* species in the southern hemisphere requires a biogeographic explanation. It is possible that an ancestor of *Laminaria ochroleuca* crossed the equator during one of the glaciation periods (van den Hoek 1982b), or in the late Tertiary. The descendant species are probably *L. pallida* and *L. schinzii* in southern Africa. As to the Brazilian deep-water species *L. brasiliensis* and *L. abyssalis*, one may speculate that the two species (a) may be conspecific and belong to the Digitatae section (*L. abyssalis* had been assigned to the Simplices section by Joly & Oliveira 1967), and (b) may belong to the flexible stiped group along with *L. digitata* (Fig. 3). Crossing of *L. abyssalis* up to a non-reproductive F_1-sporophyte generation was successful with *L. digitata* (tom Dieck, unpublished), and also with the stiff-stiped southern African *L. pallida* and *L. schinzii* (tom Dieck 1989).

Why are the Laminariales lacking in the tropics and why did large algae appear when the earth cooled down?

The tropical regions have persisted throughout geological time, and there has been sufficient time to evolve a diversity of biotas, with their constituent algal species. One example of a tropical biota is provided by the free-floating *Sargassum* community in the Sargasso Sea, with its typical accompanying fauna, well-adapted with regard to colour and morphological traits (Butler *et al.* 1983; Stoner & Greening 1984). Late Eocene fossil remains of this community have been detected in the Carpathian sector of the Tethys Sea, and give it a maximum age of 40 Ma (Jerzmanska & Kotlarczyk 1976). Since the Fucales, probably of Gondwanan origin (Clayton 1988), are present with a few genera in the tropics, they offered better possibilities for mimicry and co-evolution with animals than the Laminariales, which did not invade tropical surface waters.

A second example of algal evolution in a tropical environment relates to the evolution of the tropical fish order Perciformes. The late Cretaceous Perciformes include the four most important families of herbivorous reef fishes which evolved since the Eocene (Steneck 1983). Most of the more prominent marine macroalgae on coral reefs form toxic secondary metabolites, and this is presumed to be a response to herbivory (e.g. Paul & Fenical 1983, 1987). Herbivorous fish are absent in cold-temperate regions, and this may have favoured the evolution and persistence of larger algae in such climates. According to Estes & Steinberg (1988) Laminariales had a special evolutionary advantage in the North Pacific, because, in the late Cenozoic, the intensity of herbivory by kelp-related limpets and herbivorous marine mammals may have been low due to late evolution of these groups. In addition, the activity of sea otters, and other groups of benthic feeding-predatory mammals, may have favoured the evolution of large thalli.

One may also relate the evolutionary increases in size, which occurred in Laminariales, to more global events, and in the end, to plate tectonics. There is a relationship between decreases in global sea levels and decreases in global temperature (Fig. 1 B). This occurs because the emerging shelf areas produced greater cooling by reflecting greater amounts of sunlight (i.e. higher albedo) than the same areas would if covered with water (Seibold & Berger 1982). The relatively low sea levels in the Oligocene (Fig. 1 B) were possibly caused by the rapidly expanding Atlantic Ocean, which had grown deep enough to decrease the average depth of the sea floor, whereas the Mesozoic and early Tertiary transgressions had been caused by the shallower Atlantic (Vail & Hardenbol 1979; Haq *et al.* 1987, Seibold & Berger 1982).

There is an evolutionary trend towards increasing body size within a phylogenetic series (Cope's rule; Cope 1896). Evolution in marine endotherms proceeded, from small forms during periods of marine transgression, to larger forms during the ensuing regression and climatic deterioration (Cooper 1977). Bergmann's size rule (Bergmann 1847; Brown & Gibson 1983; Colinvaux 1986), which states that body size increases towards the Poles during glacial epochs, may thus point to an extreme case of Cope's rule. Accordingly, the advanced families of Laminariales (Laminariaceae, Lessoniaceae and Alariaceae) may have proceeded from small-sized algae originating in the ancient Arctic Ocean (Boreal Gulf), when it was warmer and wide open to the North Pacific, during the Cretaceous. The end result of this evolutionary trend was our modern kelp.

ACKNOWLEDGEMENTS

We thank Dr. David Garbary for valuable suggestions.

REFERENCES

Bergmann C (1847) Über die Verhältnisse der Wärmeökonomie der Thiere zu ihrer Größe. Abt Göttinger Studien 1: 595-708

Bolton JJ, Lüning K (1982) Optimal growth and maximal survival temperatures of Atlantic *Laminaria* species (Phaeophyta) in culture. Mar Biol 66: 89-94

Braud J-P (1974) Etude de quelques paramètres écologiques, biologiques et biochimiques chez un phéophycee des cotes bretonnes, *Laminaria ochroleuca*. Rev Trav Inst Pêches Marit 38: 115-204

Briggs JC (1974) Marine Zoogeography. McGraw-Hill, New York

Brown JH, Gibson AC (1983) Biogeography. Mosby, St. Louis

Butler JN, Morris BF, Cadwallader J, Stoner AW (1983) Studies of *Sargassum* and the Sargassum community. Bermud Biol Stn Res Spec Publ 22: 1-307

Clayton MN (1988) Evolution and life histories in the Phaeophyta. Bot Mar 31: 379-387

Colinvaux PA (1986) Introduction to ecology. Wiley, New York

Cooper MR (1977) Eustacy during the Cretaceous: its implications and importance. Palaeogeo Palaeoclim Palaeoecol 22: 1-60

Cope ED (1896) Primary factors of organic evolution. Chicago, Ill

Druehl LD (1968) Taxonomy and distribution of northeast Pacific species of *Laminaria*. Can J Bot 46: 539-547

Druehl LD (1990) Molecular evolution in the Laminariales: a review. In: Garbary DJ, South GR (eds) Evolutionary biogeography of the marine algae of the North Atlantic. Springer-Verlag, Berlin, p 205-218

Durham JW, MacNeil FS (1967) Cenozoic migrations of marine invertebrates through the Bering Strait region. In: Hopkins DM (ed) The Bering Land Bridge. University Press, Stanford, California, p 312-325

Ekman S (1953) Zoogeography of the sea. Sidwick and Jackson, London

Estes JA, Steinberg PD (1988) Predation, herbivory, and kelp evolution. Paleobiology 14: 19-36

Fain SR, Druehl LD, Baillie DL (1988) Repeat and single copy sequences are differentially conserved in the evolution of kelp chloroplast DNA. J Phycol 24: 292-302

Fairbridge RW (1973) Glaciation and plate migration. In: Implications of continental drift to the earth sciences. Vol. 1. Tarling DH, Runcorn SK (eds). Academic Press, London, p 503-515

Feldmann J (1937) Sur une nouvelle espece de Laminariacee de Mauritanie, *Ecklonia muratii* nov. sp. Bull Soc Hist Nat Afr N 28: 325-327

Flohn H (1984) Climate evolution in the Southern Hemisphere and the equatorial region during the Late Cenozoic. In: Vogel JC (ed) Late Cenozoic palaeoclimates of the Southern Hemisphere. Balkema, Rotterdam, p 5-20

Frakes LA (1979) Climates throughout geologic time. Elsevier, Amsterdam

Gladenkov YB (1979) Comparison of late Cenozoic molluscan assemblages in northern regions of the Atlantic and Pacific Oceans. Int Geol Rev 21: 880-890

Gladenkov YB (1986) Some aspects of the Late Cenozoic history of the Bering Strait in light of stratigraphy data on Iceland. In: Kontrimavichus VL (ed) Beringia in the Cenozoic era. Russian Translation Series 28. Balkema, Rotterdam, p 36-45

Guimaraes SMPB, Braga MRA, Cordeiro-Marino M, Pedrini AG (1986) Morphology and taxonomy of *Jolyna laminarioides*, a new member of the Scytosiphonales (Phaeophyceae) from Brazil. Phycologia 25: 99-108

Gussarova IS, Petrov JE (1970) Genus novum et species nova Laminariacearum ex insulis Kurilensibus. Novit System Plant Non Vasc 7: 87-90

Gussarova IS, Petrov JE (1972) Genus novum ac species nova Laminarialium ex insula Simuschir (Insulae Kurilenses). Novit System Plant Non Vasc 9: 39-44

Haq BU, Hardenbol J, Vail PR (1987): Chronology of fluctuating sea level since the Triassic. Science 235: 1156-1167

Henry EC (1987) The life history of *Phyllaria brevipes* (=*Phyllaria reniformis*) (Phyllariaceae, Laminariales, Phaeophyceae), a kelp with dioecious but sexually monomorphic gametophytes. Phycologia 26: 17-22

Henry EC, South GR (1987) *Phyllariopsis* gen. nov. and a reappraisal of the Phyllariaceae Tilden 1935 (Phaeophyceae, Laminariales). Phycologia 26: 9-16

Hiscock S, Barratt L, Ormond R (1984) The marine algae of Dhofar, Oman - an upwelling system in the Arabian Sea. Br Phycol J 19: 194

Hoek C van den (1982a) Phytogeographic distribution groups of benthic marine algae in the North Atlantic Ocean. A review of experimental evidence from life history studies. Helgol Meeresunters 35: 153-214

Hoek C van den (1982b) The distribution of benthic marine algae in relation to the temperature regulation of their life histories. Biol J Linn Soc 18: 81-144

Hopkins DM (ed.) (1967) The Bering Land Bridge. Stanford University Press, Stanford, California.

Hopkins DM, Marinkovich L (1984) Whale biogeography and the history of the Arctic Basin. Groningen Rijksuniversitet, Netherlands, Work of the Arctic Centre, no. 8, p 7-24

Jerzmanska A, Kotlarczyk J (1976) The beginnings of the Sargasso assemblage in the Tethys? Palaeogeo. Palaeoclim Palaeoecol 20: 297-306

Johansen HW (1971) Effects of elevation changes on benthic algae in Prince William Sound. In: The Great Alaska Earthquake of 1964: Biology. National Academy of Sciences, Washington, p 35-68

John DM (1971) The distribution and net productivity of sublittoral populations of attached macrophytic algae in an estuary on the Atlantic coast of Spain. Mar Biol 11: 90-97

Joly AB, Oliveira Filho EC de (1967) Two Brazilian *Laminarias*. Publ Inst Pesq Mar 4: 1-13

Kain JM 1979) A view of the genus *Laminaria*. Oceanogr. Mar Biol Ann Rev 17: 101-161

Kajimura M (1981) *Streptophyllopsis* a new genus of Laminariaceae, Phaeophyta, from Japan. Mem Fac Sci, Shimane Univ 15: 75-87

Kajimura M (1987) Typification of *Streptophyllopsis kuroshioensis* (Segawa) Kajimura (Phaeophyta, Laminariaceae). Jap J Phycol 35: 19-21

Kawai H, Kurogi M (1985) On the life history of *Pseudochorda nagaii* (Pseudochordaceae fam. nov.) and its transfer from the Chordariales to the Laminariales. Phycologia 24: 289-296

Lim B-L, Kawai H, Osawa S (1986) Molecular evolution of 5S ribosomal RNA from red and brown algae. Jap J Genet 61: 169-176

Lüning K (1985) Meeresbotanik. Thieme-Verlag, Stuttgart

Lüning K (1986) New frond formation in *Laminaria hyperborea* (Phaeophyta): a photoperiodic response. Br Phycol J 21: 269-273

Lüning K (1988) Photoperiodic control of sorus formation in the brown alga *Laminaria saccharina*. Mar Ecol Prog Ser 45: 137-144

Lüning K (1990) Seaweeds. Their environment, biogeography and ecophysiology. Wiley, New York

Maier I (1984) Culture studies of *Chorda tomentosa* (Phaeophyta, Laminariales). Br Phycol J 19: 95-106

MacNeil S (1965) Evolution and distribution of the genus *Mya*, and tertiary migrations of Mollusca. Prof Pap Geol Surv 483-G, 1-51

Marinkovich L, Brouwers EM, Carter LD (1985) Early Tertiary marine fossils from northern Alaska: implications for Arctic Ocean paleogeography and faunal evolution. Geology 13: 770-773

McKenna MC (1983) Cenozoic paleogeography of North Atlantic land bridges. In: Bott HP, Saxov S, Talwani M, Thiede, J (eds) Structure and development of the Greenland-Scotland ridge. Plenum Press, New York, p 351-399

Müller DG, Maier I, Gassmann G (1985) Survey on sexual pheromone specificity in Laminariales (Phaeophyceae). Phycologia 24: 475-477

Nicholson NL (1979) Evolution within *Macrocystis*: northern and southern hemisphere taxa. In: Proceedings of the International Symposium on Marine Biogeography and Evolution in Southern Hemisphere. Vol. 2. N.Z. DSIR Information Series 137, Wellington, p 433-441

Paul VJ, Fenical W (1983) Isolation of Halimedatrial: chemical defense adaptation in the calcareous reef-building alga *Halimeda*. Science 221: 747-749

Paul VJ, Fenical W (1987) Natural products chemistry and chemical defense in tropical marine algae of the phylum Chlorophyta. In: Scheuer PJ (ed) Bioorganic Marine Chemistry. Vol. 1. Springer-Verlag, Berlin, p 1-29

Petrov JE (1972) De systemate specierum nonnullarum Laminariae Lamour. ex oriente extremo. Novit System Plant Non Vasc. 9: 47-58

Petrov JE, Suchovejeva MV, Avdejev GV (1973) Species generis *Laminaria* Lam. e mari phillipinensi nova. Novit System Plant Non Vasc 10: 59-61

Petrov JE, Vozzhinskaja VB (1966) De genere ac specie novis Laminarialium e mari Ochotensi notula. Novit System Plant Non Vasc 4: 100-102

Postels A, Ruprecht F (1840) Illustrationes algarum . . . Pratz, St. Petersburg

Rosenvinge LK (1893) Gronland Havalger. Medd Gronl 3: 764-1006

Savin SM, Douglas RG, Stehli SG (1975) Tertiary marine paleotemperatures. Bull Geol Soc Am 86: 1499-1510

Scarlato OA (1977) Bivalve molluscs and temperature as an agent determining their geographical distribution. Malacologia 16: 247-250

Schopf TJM (1980) Paleoceanography. Harvard University Press, Cambridge, Mass

Seibold E, Berger WH (1982) The sea floor. An introduction to marine geology. Springer-Verlag, Berlin

Setchell WA, Gardner NL (1920-1925) The marine algae of the Pacific coast of North America. II. Chlorophyceae. III. Melanophyceae. Univ Calif Publs Bot 8: 139-374 (1920); 383-739 (1925)

Sheppard CRC, Jupp BP, Sheppard ALS, Bellamy DJ (1978) Studies on the growth of *Laminaria hyperborea* (Gunn.) Fosl. and *Laminaria ochroleuca* De la Pylaie on the French Channel coast. Bot Mar 21: 109-116

Smith AG, Hurley AM, Briden JC (1981) Phanerozoic paleocontinental world maps. University Press, Cambridge

South GR, Tittley I (1986) A checklist and distributional index of the benthic marine algae of the North Atlantic Ocean. Huntsman Marine Laboratory and British Museum (Natural History), St. Andrews and London

Stam WT, Bot PVM, Boele-Bos SA, Rooij JM van, Hoek C van den (1988) Single-copy DNA-DNA hybridizations among five species of *Laminaria* (Phaeophyceae): phylogenetic and biogeographic implications. Helgol Meeresunters 42: 251-267

Steneck RS (1983) Escalating herbivory and resulting adaptive trends in calcareous algal crusts. Paleobiology 9: 44-61

Stoner AW, Greening HS (1984) Geographic variation in the macrofaunal associates of pelagic *Sargassum* and some biogeographic implications. Mar Ecol Prog Ser 20: 185-192

Strauch F (1972) Phylogenese, Adaptation und Migration einiger nordischer Molluskengenera (*Neptunea, Panomya, Cyrtodaria* und *Mya)*. Abh Senckenb Naturforsch Ges 531: 1-211

Thiede J (1979) History of the North Atlantic Ocean: evolution of an asymmetrical zonal paleo-environment in a latitudinal ocean basin. In Deep Drilling Results in the Atlantic Ocean. Continental Margins and Paleoenvironment. Maurice-Ewing-Series (AGU) 3, p 275-296

Thiede J (1980) Palaeo-oceanography, margin stratigraphy and palaeophysiography of the Tertiary North Atlantic and Norwegian-Greenland Seas. Phil Trans R Soc Lond A 294: 177-185

tom Dieck I (1989) Vergleichende Untersuchungen zur Ökophysiologie und Kreuzbarkeit innerhalb der digitaten Sektion der Gattung *Laminaria* Lamouroux (Phaeophyceae). Dissertation, Universität Hamburg

Vail PR, Hardenbol J (1979) Sea-level changes during the Tertiary. Oceanus 22: 71-79

Widdowson TB (1971) A taxonomic revision of the genus *Alaria* Greville. Syesis 4: 11-49

Wiencke C, tom Dieck I (1989) Temperature requirements for growth and temperature tolerance of macroalgae endemic to the Antarctic region. Mar Ecol Prog Ser 54: 189-197

Wilce RT (1959) Studies in the genus *Laminaria*. I. *Laminaria cuneifolia* J.G. Agardh: a review. Bot Not 112: 158-174

Wilce RT (1960) Studies in the genus *Laminaria* II. *Laminaria groenlandica* L. K. Rosenvinge. Bot Not 113: 203-209

Wilce RT (1965). III. A revision of the North Atlantic species of the Simplices section of *Laminaria*. Bot Gothoborg 3: 247-256

Woelkerling WJ (1988) The coralline red algae: an analysis of the genera and subfamilies of nongeniculate Corallinaceae. Oxford University Press, London and Oxford

Womersley HBS (1987) The marine benthic flora of Southern Australia. Part II. South Australian Government Printing Division, Adelaide

Wray JL (1977) Calcareous algae. Elsevier, Amsterdam

MOLECULAR EVOLUTION IN THE LAMINARIALES: A REVIEW

Louis D. Druehl
Department of Biological Sciences
Simon Fraser University
Burnaby, B.C.
Canada V5A 1S6

Abstract -Analyses of specific regions of kelp chloroplast DNA and nuclear ribosomal DNA offer unambiguous estimates of phylogenetic relationships at all levels of taxonomic organization. For example, sequence analysis of the highly conserved small-subunit ribosomal DNA revealed a close evolutionary relationship among the kelp *Costaria costata*, chrysophytes and oomycetes. At the other taxonomic extreme, analysis of restriction fragment length polymorphisms (RFLP) resulting from restriction enzyme digestion of the small-subunit rDNA and associated intergenic spacer have been used to define populations of *C. costata* and identify interspecific hybrids of *Alaria*. RFLP analysis of chloroplast DNA has suggested an inter-generic phylogeny for the Lessoniaceae which challenges the composition of this family. Sequence analysis of rDNA of selected Alariaceae and Lessoniaceae strengthens the cpDNA challenge to the phylogeny of the Lessoniaceae.

INTRODUCTION

Evolutionary relationships among the Laminariales have, until recently, been interpreted largely on the basis of gross morphological features. Whereas these features appear to define morphological genera and families, their value in delineating species is questionable. Kelp species exhibit considerable morphological variation which, for the most part, has been shown to be a plastic phenotypic response to a variable environment (Sundene 1964; Svendson & Kain 1971; Norton *et al.* 1981; Druehl & Kemp 1982; Gerard 1987). The relationships among putative kelp species are further clouded by the potential for interspecific, and possibly intergeneric hybridizations (Schreiber 1930; Saito 1972; Sanbonsuga & Neushul 1978; Cosson & Oliveri 1982; Bolton *et al.* 1983).

Resolution of kelp taxonomic and phylogenetic relationships requires unambiguous characters which are universally distributed and transmitted in a regular and predictable manner from generation to generation. DNA organization provides characters having these properties.

In our laboratory we have employed chloroplast DNA (cpDNA) and nuclear ribosomal

NATO ASI Series, Vol. G 22
Evolutionary Biogeography of the
Marine Algae of the North Atlantic
Edited by D. J. Garbary and G. R. South

DNA (rDNA) in the unravelling of kelp phylogenies. Chloroplast DNA is a slowly evolving molecule inherited from one parent, usually the female. When both parents contribute cpDNA, their genomes do not fuse. Palmer (1987) reviews uses of cpDNA in evolutionary and biosystematic studies. Ribosomal DNA is transmitted in a Mendelian manner. This molecule consists of highly conserved elements coding for small and large ribosomal subunits and less highly conserved spacer regions - the intergenic spåcers (IGS). Sogin *et al.* (1986) review evolutionary diversity of eukaryotic small-subunit rRNA genes.

In this paper I review recent advances in understanding kelp genetic interrelationships

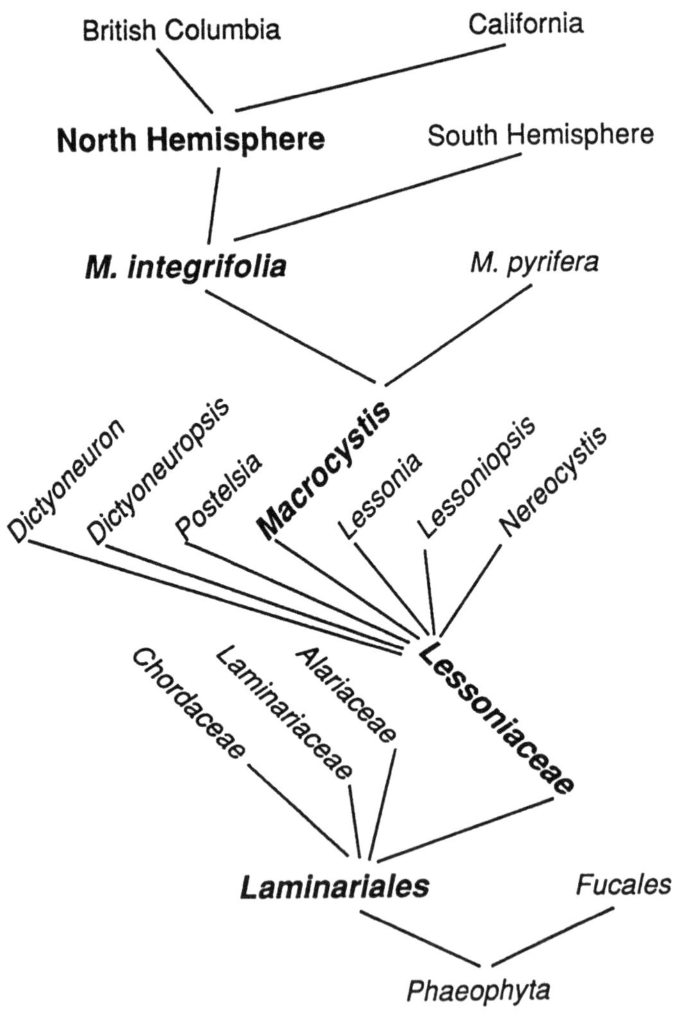

Fig. 1. Taxonomic entities within the Phaeophyta whose cpDNA diversity was assessed through restriction analysis (Fain 1986)

resulting from rDNA and cpDNA studies conducted in our laboratory. Other studies employing molecular DNA techniques on kelp are restricted to three publications Lim *et al.* (1986) and Hori & Osawa (1987) compared 5S ribosomal RNA relationships of *Eisenia bicyclis* (Laminariales) to other brown algal orders and other divisions and phyla. Stam *et al.* (1988) presented a phylogenetic tree for five *Laminaria* species based upon kinetic analysis of single copy DNA. They concluded that these species could not be statistically distinguished and that they arose simultaneously from a common ancestor about 15-19 million years ago.

KELP CHLOROPLAST DNA - FIRST STARTS

The evolution of cpDNA was studied at the population, species, genus, family and order levels (Figs 1, 2). Two (Fain 1986) or three (Fain *et al.* 1988) regions of the kelp chloroplast genome were investigated by restriction analysis.

In the Fain (1986) study, eleven endonucleases (Table 1) were used to restrict two distinct cpDNA fragments. Ninety-one restriction sites were identified, representing approximately 0.32% of the total 170 kb cpDNA molecule.

Percent sequence divergence was calculated according to Nei & Li (1979). *Macrocystis integrifolia* and *M. pyrifera* were indistinguishable, as were the two northern hemispheric populations of *M. integrifolia* (British Columbia and California). Northern (British Columbia) and southern (Chile) hemispheric populations of *M. integrifolia* had 0.46% sequence divergence (Table 2).

Sequence divergence among the seven studied Lessoniaceae genera varied from 0.0% (*Dictyoneuron* vs *Dictyoneuropsis*) to 5.4% (*Postelsia* vs *Dictyoneurum*). At the family level, sequence divergence ranged from a low of 5.4% (Lessoniaceae vs Laminariaceae) to 11.0% (Chordaceae vs Laminariaceae). *Fucus distichus* (Fucales) diverged 10.0 - 12.2% from studied members of the Laminariales.

Fain *et al.* (1988) suggested a kelp phylogeny for five genera, based upon restriction analysis of three cpDNA fragments digested with four enzymes (Table 1). Sixty-nine restriction fragments, representing 0.24% of the 170 kb cpDNA molecule, were observed. Mean sequence divergence varied among the three cpDNA fragments from 2.53 to 7.74%, suggesting a differential conserved evolutionary rate for the three fragments. cpDNA divergence among the five genera and *Fucus* suggests a polyphyletic origin of the Lessoniaceae, with some genera closely related to the Laminariaceae and others to the Alariaceae (Fig. 2).

KELP NUCLEAR DNA - THE SEARCH FOR VARIATION

The conserved nature of cpDNA precluded assessment of genetic interrelationships at the kelp population and species levels (Fain 1986). Bhattacharya & Druehl (1989) sought nuclear DNA (nDNA) variation between two morphologically distinct populations of *Costaria costata* (C.A. Agardh) Saunders through DNA hybridization analysis using randomly cloned fragments of nDNA from wave sheltered plants as probes against nDNA from wave exposed plants. This approach has been employed to estimate inter- and intraspecific sequence divergences (Rose *et al.* 1982; Sapienza & Doolittle 1982; Natvig & Jackson 1987).

nDNA from wave-exposed *Costaria* was restricted with Eco RI, Hind III, and Xba I and then probed with 42 anonymous probes of DNA fragments from sheltered *Costaria*. The mean size of these fragments was 2086 bp and the total nDNA sampled was 87620 bp.

Only one RFLP was resolved between the two studied *Costaria* stands. Screening by the ribosomal DNA (rDNA) cistron of the nematode *Caenorhabditis elegans* demonstrated the presence of this RFLP in the *Costaria* small-subunit rDNA. We speculated that the polymorphism was in the intergenic spacer (IGS) region of the *Costaria* cistron. The successful probe was designated pCc18 and has been successfully cloned and sequenced.

The majority of the anonymous *Costaria* probes hybridized to an unresolvable number of bands creating a smear on the autoradiograph (Table 3). This suggests that these probes contain dispersed sequences and are highly repetitive (Young 1979). Similar proportions of repeated sequences have been revealed in a reannealing kinetics study of fifteen higher plant genomes (Flavell *et al.* 1974).

Bhattacharya & Druehl (1989) also compared morphological variation between the wave exposed and wave sheltered stands of *Costaria* using multivariate statistical analysis. Eleven morphological variables were assessed, and the most discriminating variables are listed in Table 4. The Bartlett's Chi-square statistic placed the probability of association between the two stands at $p < 0.0005$.

This study suggested that *Costaria* responds to environmental heterogeneity by phenotypic plasticity in the absence of resolvable genetic variation.

KELP RIBOSOMAL DNA - NEW OPPORTUNITIES

The discovery of pCc18, a rDNA probe from *Costaria*, provided us with access to a genetic system which has been used to define variation within populations (Black *et al.* 1989) and among divisions and phyla (Sogin *et al.* 1986). Following are resumés of studies on kelp

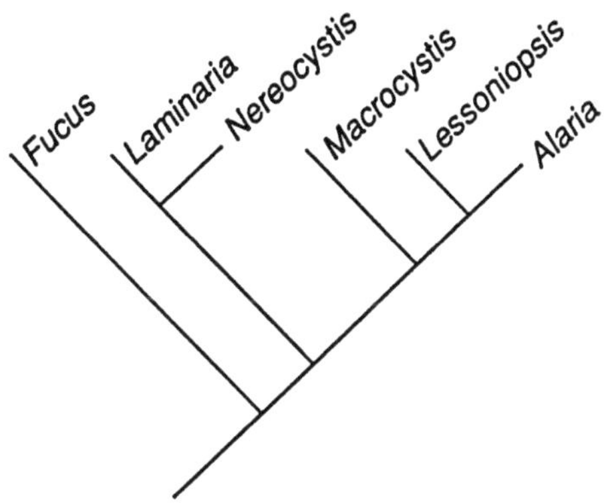

Fig. 2. A phylogeny for five genera of Laminariales and *Fucus* based upon RFLP (after Fain *et al.* 1988)

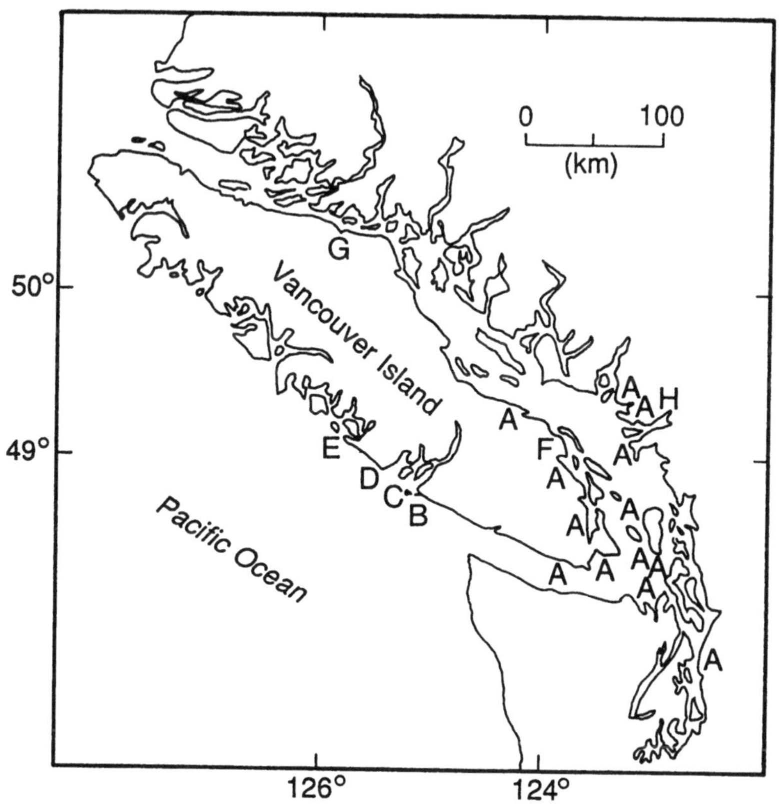

Fig. 3. Distribution of eight rDNA variants (A-H) of *Costaria costata* in the vicinity of Vancouver Island, Canada. (after Bhattacharya *et al.* 1989)

Table 1. Restriction endonucleases used in the analyses of the study of kelp cpDNA (from Fain 1986)

Enzyme	Recognition Site	Interpopulation and Interspecies	Intergenus	Interfamily
Eco RI*	GAATTC	X	X	X
Bam HI	GGATCC	X	X	X
Hind III*	AAGCTT			X
Pst I*	CTGCAG	X	X	X
Sal I	GTCGAC	X	X	X
Xho I	CTCGAG	X		X
Hpa I	GTTAAC		X	X
Pvu 2*	CAGCTG	X	X	X
Xba I	TCTAGA	X	X	X
Kpn I	GGTACC		X	
Sac I	GAGCTC		X	

*Endonucleases employed in Fain *et al.* 1988.

Table 2. Percent sequence divergences as assessed from restricted cpDNA from kelp (after Fain 1986)

Populations of	Percent Sequence Divergences
Macrocystis integrifolia	0.00 - 0.46%
Species of *Macrocystis*	0.00 - 0.00%
Genera of Lessoniaceae	0.00 - 5.40%
Families of Laminariales	5.40 - 10.80%
Laminariales vs. Fucales	10.30 - 12.20%

Table 3. Estimation of relative percentages of dispersed DNA in *Costaria costata* (from Bhattacharya & Druehl 1989)

Hybridization Result	Eco RI	Hind III	Xba I	Mean
Smear	85.7	76.2	80.9	80.9
+ 10 Bands	2.4	9.5	4.8	5.6
5 - 10 Bands	4.8	4.8	7.1	5.6
2-4 Bands	4.8	4.8	7.1	5.6
1 Band	2.4	4.8	0	2.4

rDNA variability, at the population, species and division level, that have been conducted in our laboratory.

Population Analysis of *Costaria*

Bhattacharya *et al.* (1989) revealed eight polymorphisms within the rDNA of twenty stands of *Costaria* (Fig. 3). These polymorphisms are largely due to repeat unit length variation and restriction site heterogeneity within the IGS. Similar polymorphisms have been used as population markers in domesticated lines and in wild plant populations (e.g. Saghai - Maroof *et al.* 1984; Flavell *et al.* 1986; Schaal *et al.* 1987). The distribution of *Costaria* rDNA variants indicated a northward radiation from a dominant southern breeding group (Fig. 3, Variant A).

Alaria Interspecific Genetic Relationships

Mroz (1989) defined distinctive breeding populations of *Alaria marginata*, *A. nana* and *A. tenuifolia* in the Northeast Pacific on the basis of three restriction map variants derived from Pst I, Bgl II, and Dra I restricted rDNA cistron (Fig. 4). Eight other endonucleases failed to differentiate the three studied species.

In addition to the clear relationships between rDNA variant and morphology other

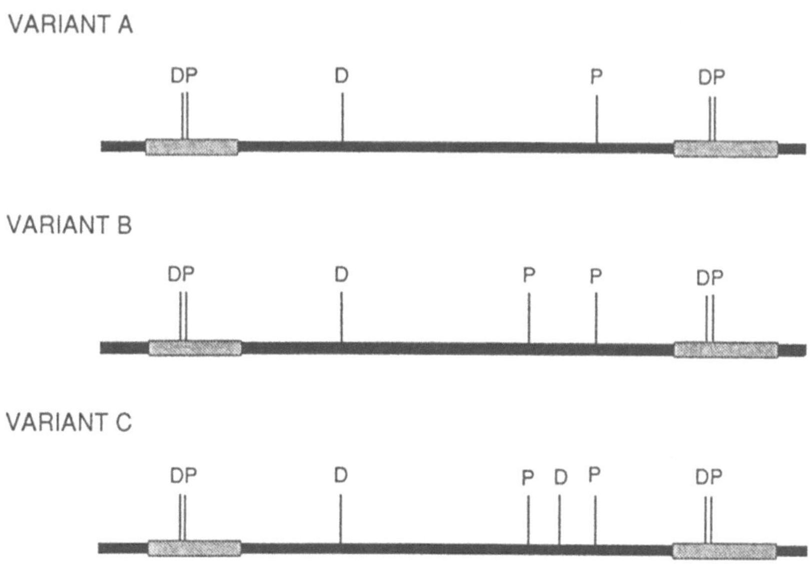

Fig. 4. Restriction maps of three variants of the rDNA repeat units. *Alaria marginata* (Variant A), *A. tenuifolia* (Variant B), *A. nana* (Variant C). P (Pst I), D (Dra I). Solid line indicates intergenic spacer regions (after Mroz 1989)

Table 4. Discriminant function analysis showing the six variables most important in population discrimination of *Costaria* (after Bhattacharya & Druehl 1989)

Step	Variable	F Value
1	Blade Width*	1071.9
2	Basal Blade Thickness	67.4
3	Width Between Middle Ribs*	8.1
4	Dry Weight/Unit Area*	8.8
5	Central Rib Width*	6.4
6	Central Rib Thickness*	5.2

*Measurements 20 cm above blade base.

instructive cases were discovered. Plants having *A. marginata* plus *A. tenuifolia* rDNA variants had either a *A. marginata* or *A. tenuifolia* morphology, depending upon wave exposure (moderate exposure, *A. marginata*; sheltered, *A. tenuifolia*). Plants from adjacent stands near the northern end of Vancouver Island had the *A. tenuifolia* variant and either *A. tenuifolia* (sheltered area) or *A. marginata* (moderately wave exposed area) morphology. On Botany Beach, Vancouver Island, plants with *A. nana* morphology (high intertidal, full wave exposure), *A. marginata* morphology (low intertidal, moderate wave exposure) and an intermediate morphology (located between the two distinct species) all shared the *A. marginata* variant.

These data suggest the dominant role of environment in producing the distinctive species morphotypes. Further, they illustrate the existence of distinct breeding groups with occasional hybridization.

Laminaria Interspecific Genetic Relationships

This enigmatic genus is presently under investigation in our laboratory. In a preliminary study we prepared restriction maps of the rDNA for three *Laminaria* species (Bhattacharya & Druehl 1990). Ten endonucleases were employed and with the exception of one Sac I restriction site in the small-subunit of *L. digitata*, all polymorphisms were restricted to the IGS (Fig. 5). These results suggest that rDNA RFLP markers may be employed to define *Laminaria* breeding groups (species?). Restriction maps of eight more *Laminaria* species are being prepared.

Kelp's Position in the Plant Kingdom

Phylogenetic comparison of the small-subunit rDNA sequence of *Costaria* with those of other algae, vascular plants and oomycetes was conducted by Bhattacharya & Druehl (1988).

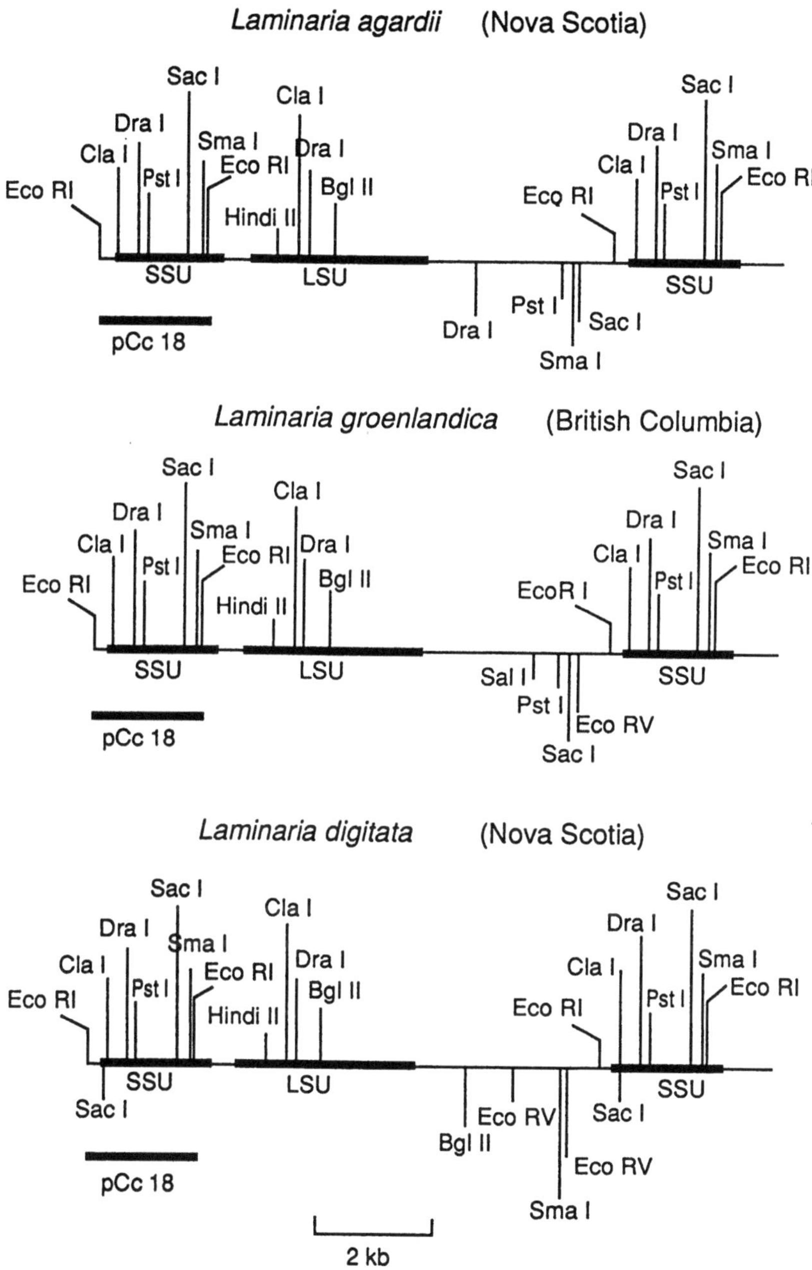

Fig. 5. Restriction maps of the *Laminaria* rDNA cistron. Restriction sites of the intergenic spacer are shown below the line. rDNA coding region restriction sites are shown above the line, SSU (small-subunit), LSU (large subunit). (from Bhattacharya & Druehl 1990)

Table 5. Uncorrected sequence divergences (possible range 1.0-0.0) as assessed from sequenced (648 bp) small-subunit rDNA (G. Saunders unpublished). Laminariales: *Pterygophora californica* Rupr., *Alaria marginata* Post. *et* Rupr. (Alariaceae); *Lessoniopsis littoralis* (Tilden) Reinke, *Macrocystis integrifolia* Bory (Lessoniaceae); Fucales: *Fucus gardneri* Silva

Genus		(1)	(2)	(3)	(4)	(5)
Fucus	(1)	0				
Pterygophora	(2)	0.62	0			
Alaria	(3)	0.62	0.006	0		
Lessoniopsis	(4)	0.60	0.005	0.002	0	
Macrocystis	(5)	0.65	0.011	0.011	0.009	0

The *Costaria* sequence data were incorporated into a eukaryotic small-subunit rDNA phylogeny created by Sogin *et al.* (1986). *Costaria* clustered with the chlorophyll a + c bearing *Ochromonas* and the aquatic fungus *Achlya*, and was distinct from chlorophyll a + b bearing plants and a dinoflagellate (Fig. 6). These results support Cavalier-Smith's (1981) proposal that classification should be based on fundamental cell structure rather than mode of nutrition.

Ongoing Kelp rDNA Studies

Presently, we are sequencing kelp small-subunit rDNA directly following amplification by the polymerase chain reaction (PCR). Sequence data of this conserved region will allow us

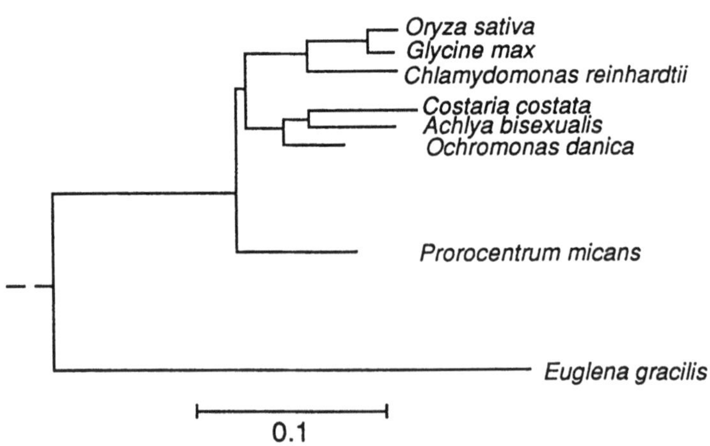

Fig. 6. Phylogeny of plants bearing chlorophyll a + b, chlorophyll a + c and an oomycete on the basis of divergence of small-subunit rRNA sequences (from Bhattacharya & Druehl 1988)

to construct phylogenies at the genus, family and order levels and perhaps suggest dates of divergence. Our objectives are to develop detailed phylogenies for the genus *Laminaria*, for all genera of the Laminariales, and for the majority of orders of the Phaeophyta.

Preliminary results suggest that there is little divergence at the genus and family level. For example, on the basis of 648 bp sequenced, members of the Alariaceae diverged by 0.005, members of the Lessoniaceae by 0.009, and the Lessoniaceace and Alariaceae diverged by 0.002 - 0.011 (Table 5).

CONCLUSIONS

Molecular DNA techniques have been successfully employed in the study of kelp intergeneric relationships. The recent application of gene amplification followed by direct sequencing will further increase the resolution and ease of these studies. Data derived from these investigations will significantly augment morphological, cytological, life cycle and biochemical characteristics which have been employed to estimate seaweed phylogenetic relationships.

As is often the case, data from new techniques not only aid in resolving old questions but also introduce new sets of problems. For example, how do we equate molecular DNA divergence to the presently employed taxonomic hierarchy? Do the single discovered IGS restriction site differences distinguishing the three *Alaria* species (Fig. 4) constitute reason to recognize only one species? Does the low degree of divergence between *Lessoniopis* (Lessoniaceae) and *Alaria* (Alariaceae) (Table 5) invalidate these families? How should we interpret the tight clustering of an aquatic fungus, a chrysophyte and a kelp (Fig. 6)? Does this cluster represent a Division of plants? If we consider the ultrastructure of these organisms, this may be a reasonable conclusion (Cavalier-Smith 1981).

The rDNA cistron is a widely studied system which promises powerful resolution of intergeneric relationships. To what extent should we rely on this system? In our studies, we have obtained compatible results from investigation of rDNA and cpDNA molecules. For example, both systems suggest very low interspecific deviation and both have thrown doubt on the family organization of the Laminariales.

ACKNOWLEDGEMENTS

The studies reviewed in this paper are, to a great extent, the result of thoughtful and patient nurturing by members of the Simon Fraser University Institute for Molecular Biology and Biochemistry, particularly Drs A. Beckenbach, D.L. Baillie and Ms K.A. Beckenbach.

Funding has been provided by the Simon Fraser University President's Research Fund and Natural Sciences and Engineering Research Council, Canada.

REFERENCES

Bhattacharya D, Druehl LD (1988) Phylogenetic comparison of the small-subunit ribosomal DNA sequence of *Costaria costata* (Phaeophyta) with those of other algae, vascular plants and oomycetes. J Phycol 24: 539-543

Bhattacharya D, Druehl LD (1989) Morphological and DNA sequence variation in the kelp *Costaria costata* (Phaeophyta). Mar Biol 102: 15-23

Bhattacharya D, Druehl LD, Baillie DL (1989) Population analysis of the kelp *Costaria costata* (Phaeophyta) using a polymorphic ribosomal DNA probe. Pl Syst Evol (in press)

Bhattacharya D, Druehl LD (1990) Restriction enzyme analysis of variation and taxonomy in the kelp genus *Laminaria*. Proc Int Seaweed Symp 13 (in press)

Black WC, McLain DK, Rai KS (1989) Patterns of variation in the rDNA cistron within and among world populations of a mosquito, *Aedes albopictus* (Skuse). Genetics 121: 539-550

Bolton JJ, Germann I, Lüning K (1983) Hybridization between Atlantic and Pacific representatives of the Simplices section of *Laminaria* (Phaeophyta). Phycologia 22: 133-140

Cavalier-Smith T (1981) Eucaryotic kingdoms seven or nine? Biosystems 14: 461-481

Cosson J, Oliveri R (1982) First results concerning the interspecific and intergeneric hybridization possibilities in Laminariales from the Channel Coast. C R Acad Sc Paris 295: 381-384

Druehl LD, Kemp L (1982) Morphological and growth responses of geographically isolated *Macrocystis integrifolia* populations when grown in a common environment. Can J Bot 60: 1409-1413

Fain SR (1986) Molecular evolution in the Laminariales: restriction analysis of chloroplast DNA. Ph.D. thesis, Simon Fraser University, Burnaby, Canada

Fain SR, Druehl LD, Baillie DL (1988) Repeat and single copy sequences are differentially conserved in the evolution of kelp chloroplast DNA. J Phycol 24: 292-302

Flavell RB, Bennett MD, Smith JB, Smith DB (1974) Genome size and the proportion of repeated nucleotide sequence DNA in plants. Biochem Genet 12: 257-269

Flavell RB, O'Dell M, Sharp P, Nevo E, Beiles A (1986) Variation in the intergenic spacer of ribosomal DNA of wild wheat, *Triticum dicoccoides*, in Israel. Mol Biol Evol 3: 547-558

Gerard VA (1987) Hydrodynamic streamlining of *Laminaria saccharina* Lamour. in response to mechanical stress. J Exp Mar Biol Ecol 107: 237-244

Hori H, Osawa S (1987) Origin and evolution of organisms as deduced from 5S ribosomal RNA sequences. Mol Biol Evol 4: 445-472

Lim B-L, Kawai H, Hori H, Osawa S (1986) Molecular evolution of 5S ribosomal RNA from red and brown algae. Jap J Genet 61: 169-176

Mroz L (1989) Ribosomal DNA assessment of genetic relationships among *Alaria* Greville (Phaeophyta) stands from Vancouver Island and vicinity. MSc. thesis, Simon Fraser University, Burnaby, Canada

Natvig DO, Jackson DA (1987) Random-fragment hybridization analysis of evolution in the genus *Neurospora*: The status of four-spored strains. Evolution 41: 1003-1021

Nei M, Li W-H (1979) Mathematical model for studying genetic variation in terms of restriction endonucleases. Proc Nat Acad Sci USA 76: 5269-5273

Norton TA, Mathieson AC, Neushel M (1981) Morphology and the environment. Bot Monogr 17: 421-451

Palmer JD (1987) Chloroplast DNA evolution and biosystematic uses of chloroplast DNA variation. Am Nat 130: S6-S29

Rose AM, Baillie DL, Candido EPM, Beckenbach KA, Nelson D (1982) The linkage mapping of cloned restriction fragment length differences in *Caenorhabditis elegans*. Mol Gen Genet 188: 286-291

Saghai-Maroof MA, Soliman KM, Jorgensen RA, Allard RW (1984) Ribosomal DNA spacer-length polymorphisms in barley: Mendelian inheritance, chromosomal location, and population dynamics. Proc Natl Acad Sci USA 81: 8014-8018
Saito Y (1972) On the effects of environmental factors on morphological characteristics of *Undaria pinnatifida* and the breeding of hybrids in the genus *Undaria*. In: Contributions to the systematics of benthic marine algae of the North Pacific, Abbott IA, Kurogi M (eds) Japanese Soc. Phycology, Kobe, p 117-130
Sanbonsuga Y, Neushul M (1978) Hybridization of *Macrocystis* (Phaeophyta) with other float-bearing kelps. J Phycol 14: 214-224.
Sapienza C, Doolittle WF (1982) Unusual physical organization of the *Halobacterium* genome. Nature 295: 384-389
Saunders GW, Druehl LD (1989) Alariaceae or Lessoniaceae; traditional morphological and molecular approaches to the taxonomy of *Lessoniopsis littoralis* (Laminariales, Phaeophyta). (Abst.) 13th Intern. Seaweed Symp. Vancouver, Canada
Schaal BA, Leverich WJ, Nieto-Sotelo J (1987) Ribosomal DNA variation in the native plant *Phlox divaricata*. Mol Biol Evol 4: 611-621
Schreiber E (1930) Untersuchen über Parthenogenesis, Geschlechtsbestimmung und Bastardierungsvermögen bei Laminarien. Planta 12: 331-353
Sogin ML, Elwood HJ, Gunderson JH (1986) Evolutionary diversity of eukaryotic small-subunit rRNA genes. Proc Nat Acad Sci USA 83: 1383-1387
Stam WT, Bot PVM, Boele-Bos SA, van Rooij JM, van den Hoek C (1988) Single-copy DNA-DNA hybridizations among five species of *Laminaria* (Phaeophyceae): phylogenetic and biogeographic implications. Helgol Meeresunters 42: 251-267
Sundene O (1964) The ecology of *Laminaria digitata* in Norway in view of transplant experiments. Nytt Mag Bot 11: 83-107
Svendsen P, Kain M (1971) The taxonomic status, distribution, and morphology of *Laminaria cucullata* sensu Jorde and Klavestad. Sarsia 46: 1-22
Young MW (1979) Middle repetitive DNA: A fluid component of the *Drosophila* genome. Proc Nat Acad Sci USA 76: 6274-6278

EVOLUTIONARY BIOGEOGRAPHY OF THE NORTH ATLANTIC ANTITHAMNIOID ALGAE

A. Athanasiadis*
University of Oslo
Biological Station Dröbak
N-1440 Dröbak
Norway

Abstract - Distributions of the North Atlantic representatives of the genera *Antithamnion*, *Antithamnionella*, *Balliella*, *Pterothamnion* and *Scagelia* are assessed based on re-examination of types and other collections. Ten species are recognized that fall within two distinct floristic elements; a warm temperate element comprising at least eight species of *Antithamnion*, *Antithamnionella*, *Balliella*, and *Pterothamnion*, and an Arctic-cold temperate element comprising only two species of *Scagelia* and *Antithamnionella*. Members of these floristic groups exhibit disjunct distributions or overlap within a narrow transitional area, in the eastern Atlantic Ocean extending from the northern British Isles to Trondheim.

The genera of the warm temperate element also occur in the Indo-Pacific Ocean, and have several endemic species in both regions. The occurrence in the North Atlantic Ocean may be explained by the westward migration of the ancient Tethyan flora into the young Atlantic and Mediterranean basins, and the subsequent isolation of these regions from the Indo-Pacific Ocean.

The two Arctic-cold temperate species are also present in the North Pacific Ocean (although only *Scagelia* occurs in the Arctic), and between Alaska and Baja California they co-occur with warm and cold temperate antithamnioids of the Pacific flora. The origins of these species remain speculative, although the presence of several closely related taxa (sister species?) suggests a Pacific origin for these two species.

INTRODUCTION

The term 'antithamnioid' is usually applied to species of Ceramiales with an *Antithamnion*-like thallus; it is a name of convenience without taxonomic rank or other precise circumscription. In the present paper, it accommodates the North Atlantic species of the genera *Antithamnion* Nägeli, *Antithamnionella* Lyle, *Balliella* Itono & Tanaka, *Pterothamnion* Nägeli, and *Scagelia* Wollaston that in a recent taxonomic account have been regarded as belonging to a single group, the tribe Antithamnieae (Moe & Silva 1980). Elsewhere I will give reasons for excluding *Balliella* from the Antithamnieae, and considering it as the potential sister taxon of

* Present address: University of Gothenburg, Department of Marine Botany, Carl Skottsbergs Gata 22, S-413 19 Gothenburg, Sweden.

NATO ASI Series, Vol. G 22
Evolutionary Biogeography of the
Marine Algae of the North Atlantic
Edited by D. J. Garbary and G. R. South

this group. In this paper I present a synopsis of the distributional data of the North Atlantic antithamnioid species and their immediate relatives[1] in order to discuss problems related to the recognition of potential subgroups (e.g. genera), centers of origin of taxa, and patterns in species distributions. Yet, such items are clearly interwoven with the evolutionary history of the whole tribe Antithamnieae, and require knowledge of this group on a worldwide basis.

While we currently lack this knowledge in the form of a comprehensive monograph, this account of North Atlantic antithamnioids is based on investigations that have been conducted with such a synthetic approach using the European members of this group as a starting point.

The warm temperate element

Eight species are included in this floristic element and these have their main distribution in parts of the Central Atlantic Ocean and/or in the Mediterranean Sea. With the single exception of *Balliella cladoderma* that inhabits deeper waters (20 to at least 95 m depth), all other species grow in the upper sublittoral zone (usually to a depth of c. 20 m).

ANTITHAMNION CRUCIATUM (C. AGARDH) NÄGELI (Fig. 1). This is the most commonly encountered species of *Antithamnion* in the North Atlantic Ocean. On European coasts, it occurs between Bergen (western Norway) and the Canary Isles, extending its distribution into the Mediterranean and the Black Seas (Athanasiadis 1986, 1987). On American coasts, it is reported between Onslow Bay, North Carolina (Schneider 1974), and Placentia Bay, Newfoundland (Whittick & Hooper 1977), as well as the Bermudas (Collins & Hervey 1917). Records from the central Atlantic Ocean (Barbados, Senegal, Cape Verde Islands) are based on older literature (see Taylor 1960, Price *et al.* 1986) and are in need of confirmation, as are records from the southern hemisphere (Harvey 1848). Despite these reservations, culture experiments showed that Mediterranean and Scandinavian isolates not only tolerate, but even grow luxuriantly at 25° C (Athanasiadis 1986), which should allow successful dispersal and colonization of the species in parts of the central and southern Atlantic Ocean.

Records from the Arctic waters (De Toni 1903, with reservations) are based on Kjellman's speculations on the identity of *Antithamnion pusillum* (Ruprecht) Kjellman (1877). These records should be discounted, since material of *Callithamnion pusillum* (type in LE)[2] was re-

---

[1]Here and elsewhere in the text, the term immediate (or closest) relatives applies to pairs of species with a high degree of overall similarity. It is assumed that such pairs are also sister taxa in the phylogenetic sense, which might well be the case, although this is not always demonstrated.

[2] Herbarium abbreviations follow Holmgren *et al.* (1981).

examined and was found to belong to a distichous variety of *Scagelia pylaisaei* (Montagne) Wynne. The northernmost locality for *A. cruciatum* is therefore at Bergen, but this record is also in need of confirmation[3].

Morphological variation in *A. cruciatum* is considerable and new observations confirm

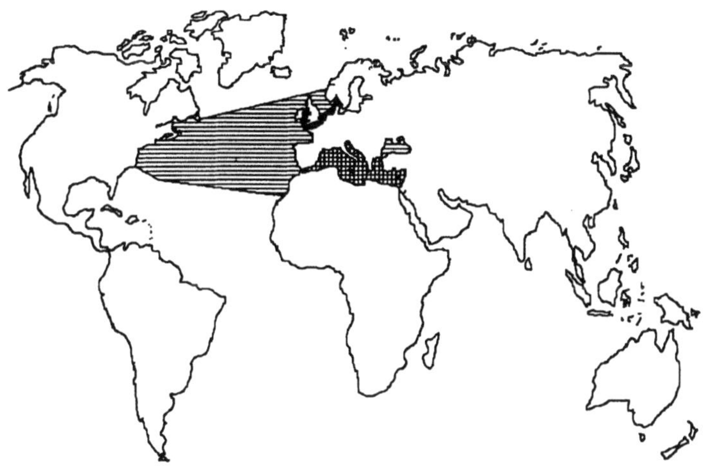

Fig. 1. *Antithamnion cruciatum*: North Atlantic, Mediterranean and Black Seas (all marked areas). Var. *scandinavicum*: NE. Atlantic (shadowed area). *Antithamnion heterocladum* and *Antithamnion tenuissimum*: Mediterranean Sea (crossed area)

Fig. 2. *Antithamnion densum*: NE. Atlantic, Peru, North Chile, Baja California to Prince William Sound, Patagonia, Tristan da Cunha, South Africa, and Japan Sea to South Korea (horizontally hatched area)

---

[3] It has not been possible to locate material of this collection in BG or O.

previous reports of specimens with occasional development of three branches[4] per axial cell (Harvey 1848) or with a pronounced distichous-opposite ramification of branches (Børgesen 1930). The latter feature that distinguishes Canarian plants suggests that the closest relatives of *A. cruciatum* are tropical and subtropical Atlantic-Mediterranean taxa, such as *A. ogdeniae* Abbott[5] and *A. compactum* (Grunow) Schiffner (Grunow 1867, type in HBG) that exhibit a similar branching pattern.

At the northern end of its distribution in southern Scandinavia, the type variety *cruciatum* becomes rare and is replaced by the variety *scandinavicum* that, in addition to its ecotypic adaptation to brackish waters, shows similar temperature responses in culture like Mediterranean isolates of the species (10° to 25° C) (Athanasiadis 1986). Variety *scandinavicum* is usually glandless and field plants may be confused with the Mediterranean entity *Antithamnion tenuissimum* (Hauck) Schiffner. This is the case with a Swedish record of *A. tenuissimum* (Waern 1961), that after re-examination of the original material (herb. Waern, Univ. of Uppsala) was found to belong to var. *scandinavicum* [6].

ANTITHAMNION DENSUM (SUHR) HOWE (Fig. 2). The occurrence of this species in the North Atlantic Ocean is so far based on two collections from the islands of Glénans and the Beg en Fry (North Brittany, L' Hardy-Halos 1968, as *A. defectum* Kylin). Material from the Glénans (Herb. L'Hardy-Halos, Laboratoire de Phycologie marine et de Morphogenese, Le Mans) was made available for comparison with authentic specimens of *A. defectum* Kylin (1925) from Friday Harbor (type and coll. in LD, and GB), the type material of *Callithamnion densum* Suhr (1840) from Peru (MEL), and other collections of this species from Tristan da Cunha (Baardseth 1941, as *Antithamnion* sp., coll. in O), South Argentina (San Julian, coll. in herb. Surbayrole, GB), North Chile [Tocopilla; Levring (1960); coll. in GB], South Alaska (Wynne 6263 & 5105-B, coll. in MICH), and South Africa [Namibia, coll. in MICH, as *A. leptocladum* (Montagne) Wynne (1986)]. All these specimens were found to be morphologically similar with respect to the distichous-opposite branching of the thallus, the adaxial ramification of branches, the development of gland cells on up to 7-celled branchlets, and the development of lateral axes from axial cells with suppression of the opposite branch.

---

[4]Throughout this paper the term (whorl-) branch is used to define a lateral filament of limited growth.

[5] *Antithamnion ogdeniae* Abbott (1979) appears to be a taxonomic synonym of *Callithamnion decipiens* J. Agardh (1842) from Mediterranean France.

[6] *Antithamnion cruciatum* var. *scandinavicum* Athanasiadis (1986) appears to be a taxonomic synonym of *Callithamnion cruciatum* var. *villosum* Kützing (1861) from Calvados.

According to Yoshida (1981a) and Wollaston (1972a), *A. sparsum* Tokida from Japan, and *A. setaceum* Gardner and *A. pygmaeum* Gardner from the Pacific North America should also be considered as heterotypic synonyms of *A. defectum*. These data suggest that *A. densum* is distributed in the warm and cold temperate waters of both hemispheres, while culture studies have shown that Pacific isolates can grow at temperatures between 10° and 20° C (Garbary *et al.* 1988) and reproduce between 16° and 19° C (Boo & Lee 1983).

While the North Atlantic record is regarded as an introduction into this region, the closest relative of *A. densum* appears to be *A. kylinii* Gardner that has been recorded between southern British Columbia and Baja California (Dawson 1962; Wollaston 1972a; Lindstrom & Gabrielson 1989), as well as in Peru [Callao (Howe 1914, as *A. densum*)]. The overlapping distribution of these apparently closely-related species suggests that *A. densum* might have its origin in the warm temperate waters of the Pacific Ocean.

ANTITHAMNION HETEROCLADUM FUNK & *ANTITHAMNION TENUISSIMUM* (HAUCK) SCHIFFNER (Fig. 1). These two species have been sporadically reported from the North Atlantic Ocean, the former from Azores (Ardré pers. comm.), and the latter from Atlantic France (L'Hardy-Halos 1968) and the Dutch coast (Stegenga & Mol 1983). However, the records of *A. tenuissimum* from the Atlantic coast of Europe are questionable and may be based on plants of var. *scandinavicum* that is common in this region (Athanasiadis 1988), while the Azorean record of *A. heterocladum* needs to be confirmed. Both species occur commonly in the Mediterranean Sea and their overlapping distribution and form variation (*A. heterocladum* may exhibit sub-equal branches and *A. tenuissimum* lateral axes replacing a branch in a pair) suggests that they probably have a common ancestor. Despite a strong morphological similarity, hybridization attempts have indicated intersterility (Athanasiadis unpubl. observ.).

If *A. densum* is regarded as an alien species that was introduced to European coasts from the southern Atlantic Ocean, the remaining North Atlantic-Mediterranean members of *Antithamnion* belong to the section *Cruciatae* J. Agardh (1892: 21) and are endemic, which suggests that they might also be monophyletic. This hypothesis should be based on the recognition of at least one autapomorphy for these four taxa (Hennig 1966:90), and the formation of pedicellate tetrasporangia provides such a character. The inter-relationships of this complex can now be analysed cladistically, using other morphological characters and the remaining species of the sect. *Cruciatae* as potential sister taxa (outgroups) to resolve character evolution (Ridley 1986:61). Eight widely accepted morphological characters were selected that define, through unique combinations, each member of the group (Table 1). Selection of sister taxa was limited to species defined by the same eight characters. Three species satisfied this criterion: *A. antillanum* Børgesen (distributed in the tropics of the Atlantic and the Pacific Oceans; Athanasiadis 1985a), *A. gracilentum* (Harvey) J. Agardh (distributed in the subtropics of the North and South Pacific Ocean; Wollaston 1968), and *A. ogdeniae* (distributed in the Mediterranean Sea and the tropics of the Atlantic Ocean; Athanasiadis 1985a).

Using the computer program MacClade (Maddison & Maddison 1987), three, five-taxon cladograms were produced, each based on a different outgroup). These three trees represent the simplest phylogenetic interpretations of the data and have the fewest number of homoplasies

Table 1. Data matrix for the four North Atlantic-Mediterranean taxa of *Antithamnion* section *Cruciatae* and their three potential sister taxa (outgroups). All characters unordered with one state condition: absent (0) or present (1). Coding based on data given by Athanasiadis (1983, 1985a, and 1986, for *A. antillanum, A. cruciatum* var. *cruciatum* and var. *scandinavicum, A. heterocladum, A. ogdeniae*, and *A. tenuissimum*), Kützing [1861, for var. *villosum* (= var. *scandinavicum*)], and Wollaston (1968, for *A. gracillentum*)

	Taxon						
Character	ogden[1]	anti[1]	graci[1]	cruci	scan	hetero	tenui
1. Gland cells	1	1	1	1	1	0	0
2. Monostichous ramification	0	0	0	0	0	1	1
3. Short branchlets (bearing gland cells or not)	1	1	1	1	1	0	0
4. Distichous-alternate ramif.	1	1	1	0	0	0	0
5. Distichous-opposite ramif.	1	0	1	1	1	0	0
6. Dimorphic branches	0	1	0	0	0	1	0
7. Pedicellate tetrasporangia	0	0	0	1	1	1	1
8. Development of axes singly	0	0	1	1	1	0	1

[1]outgroup taxa

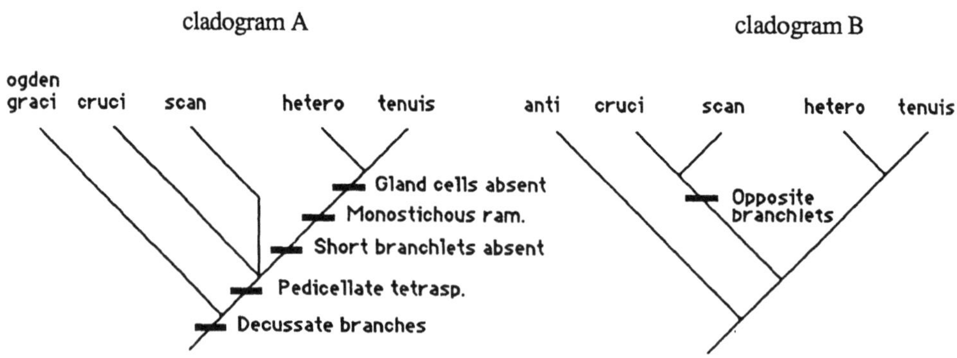

Fig. 3. Cladistic inter-relationships of the four North Atlantic-Mediterranean taxa of *Antithamnion* section *Cruciatae*, as shown by the most parsimonious phylogenetic trees and using *A. ogdeniae* (cladogram A), *A. gracilentum* (cladogram A), and *A. antillanum* (cladogram B) as outgroups. Horizontal bars indicate synapomorphies of characters that are coded in Table 1

(i.e. fewest number of evolutionary changes). When *A. gracilentum* was the outgroup, the shortest phylogenetic tree required eight steps and lacked homoplasies (Fig. 3, cladogram A). When *A. ogdeniae* was the outgroup (Fig. 3, cladogram A), the shortest cladogram required nine steps and there was one homoplasy (character 8). When *A. antillanum* was the outgroup (Fig. 3 cladogram B), the shortest cladogram had ten steps and there were two homoplasies (characters 6 and 8).

All these phylogenetic reconstructions suggest a monophyletic origin for the Mediterranean species *A. tenuissimum* and *A. heterocladum*, defined by no less than three synapomorphies. The fact that this monophyly is supported by evolutionary reversals (loss or reduction of characters 1, 2, and 3) suggests that these species have rather recently evolved, presumably after the Messinian salinity crisis that took place in the Late Miocene (see Joosten & van den Hoek 1986: 211). A common ancestry of var. *scandinavicum* and var. *cruciatum*, while not precluded by the first two phylogenetic trees (cladogram A), was supported by the third one (cladogram B) on the basis of one synapomorphy (character 5).

Using the assumptions that sister taxa merit the same rank and that *A. heterocladum* and *A. tenuissimum* are distinct species (the latter supported by biosystematic data), all phylogenetic trees support the elevation of var. *scandinavicum* to specific rank. Although the cladograms support such a classification scheme, it seems appropriate to withhold making taxonomic changes until further hybridization tests have been performed between the various taxa.

ANTITHAMNIONELLA ELEGANS (BERTHOLD) PRICE & JOHN (Fig. 4). This originally Mediterranean species is currently considered conspecific with several later described taxa from widely separated localities in the tropics and the subtropics (Cormaci & Furnari 1988). The proposed synonyms include *Aa breviramosa* (Dawson) Womersley & Bailey from California, Pacific Central America, the Solomon Islands, South Africa (Anderson & Stegenga 1989), North Carolina, and Curaçao (Stegenga & Vroman 1987), *A. sublittorale* Setchell & Gardner from California and southern Japan, and *A. tristicum* Joly & Yamaguishi and *A. ternatum* Joly & Cordiero from Brazil (Sao Paolo, Schneider 1984). In the North Atlantic Ocean, *Aa elegans* is recorded only from the Iberian Peninsula (Ardré 1970, Gallardo *et al.* 1985).

Examination of *Aa australis* from Tristan da Cunha (Baardseth 1941; type in O) showed that this species should also be regarded as a synonym of *Aa elegans*. It is also likely that further comparative studies with congeners from South Australia and South Africa (for example *Aa tasmanica*; Wollaston 1968; Stegenga 1986) will further extend the list of synonyms of this species. It should be noted, however, that the synonyms proposed previously are based on studies of preserved material, and until comparative studies of living plants and hybridization studies have been carried such synonymy should be regarded with some doubt.

With its worldwide distribution and genotypic variation, *Aa elegans* would appear to be a primitive member of a genus that currently includes several narrowly circumscribed species such as *Aa sarniensis* and *Aa spirographidis*.

ANTITHAMNIONELLA SARNIENSIS LYLE (Fig. 5). North Atlantic records extend as far as the Dutch coast (Stegenga & Mol 1983, as *Aa spirographidis*), including the southern coast

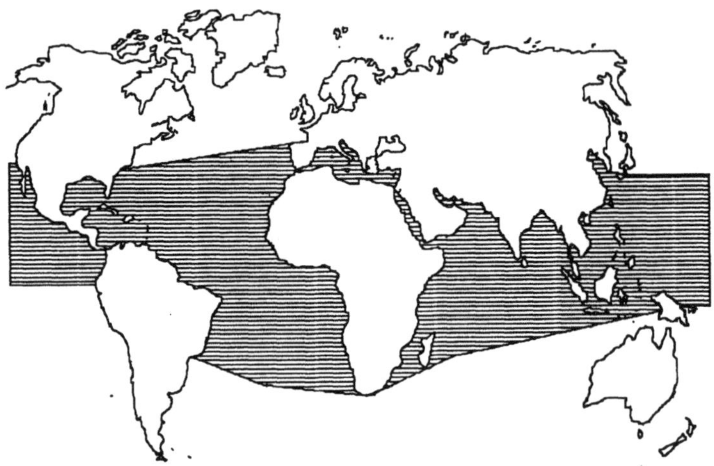

Fig. 4. *Antithamnionella elegans*: South California, Baja California, Pacific Costa Rica, Galapagos, Brazil, Tristan da Cunha, S. Africa, Ghana, Tongo, Canary Isles, Curaçao, North Carolina, Atlantic Spain, Mediterranean and Red Seas, Zanzibar, Mauritius, Solomon Isles, Hawaii, South Japan (horizontally hatched area)

Fig. 5. *Antithamnionella sarniensis*: NE. Atlantic (shadowed area). *Antithamnionella spirographidis*: NE. Atlantic, western Mediterranean, southern Australia, Japan Sea to South Korea, and California to Prince William Sound (shadowed and horizontally hatched areas)

of the British Isles (Lyle 1922, type in BM, and pers. observ.), Brittany (L'Hardy-Halos 1968, and pers. observ.), and Portugal (Ardré 1970). Huxley's (1978) record of *Aa spirographidis* from the Clyde Sea area might also belong to this species.

Aa sarniensis might be conspecific with *Antithamnionella ternifolia* (Hooker & Harvey) Lyle described from Cape Horn, and later reported from more northern latitudes on the coast of Patagonia (Mendoza 1969; coll. in GB), and from the Macquarie Islands in the southern Pacific (Ricker 1987). Other taxa that merit consideration as potential synonyms of this species are *Antithamnionella seriata* Baardseth (1941) from Tristan da Cunha (type in O) and *Antithamnion adnatum* (J. Agardh) J. Agardh (1892) from New Zealand (type and coll. in LD and GB).

Distinctive features of *Aa sarniensis* are the whorl-branching of the thallus with whorls of (2)3 or 4 branches per axial cell, and the isodiametric periaxial cells cells that are generally smaller than contiguous branch cells. Individual branches are usually simple, or occasionally ramified (particularly when surrounding carposporophytes), suggesting a close relationship of this species with both *Aa elegans* and *Aa spirographidis*.

ANTITHAMNIONELLA SPIROGRAPHIDIS (SCHIFFNER) WOLLASTON (Fig. 5). Atlantic records are restricted to the southern coast of the British Isles (Westbrook 1934, Guiry 1978), Brittany (L'Hardy-Halos 1968), the western Mediterranean (Feldmann-Mazoyer 1941) and the north Adriatic Seas (Schiffner 1916; type and coll. in B and BM). Extra-European records are based on material from southern Australia (Wollaston 1968), and the North Pacific where several locally described taxa (*A. glanduliferum* Kylin, *A. miharae* Tokida, *A. scrippsiana* Dawson, and *A. tenuissimum* Gardner) have recently been proposed as synonyms (Lindstrom & Gabrielson 1989). On the basis of this broad synonymy, *Aa spirographidis* would appear to have both a warm and cold temperate distribution in the North Pacific, occurring between Prince William Sound and Baja California (Lindstrom & Gabrielson 1989) and between the Russian coast of the Japan Sea (Perestenko 1980) and South Korea (Boo & Lee 1983). The European distribution of the species, however, is restricted to warm temperate coasts, and overlaps with the distribution of the two other warm temperate congeners. Whether or not a complex of several distinct entities is also involved in the Pacific region remains to be investigated using biosystematic approaches.

The closest relative of *Aa spirographidis* appears to be *Aa sarniensis*, a taxon regarded as a synonym by many authors in the past (e.g. Parke & Dixon 1976). Hybridization studies of Atlantic isolates have indicated sterility barriers between these species that can also be distinguished based on morphological features (L'Hardy-Halos 1968, 1986).

BALLIELLA CLADODERMA (ZANARDINI) ATHANASIADIS (Fig. 6). The occurrence of this species in the North Atlantic Ocean is based on Azorean material collected by the late Professor Tore Levring during his visit, possibly in the 60's (Athanasiadis 1985a). In the

Mediterranean it commonly grows in deep water habitats between 20 and at least 95 m depth (Seoane-Camba 1969), where water temperature is relatively stable throughout the year (around 12^0 to 15^{0}C).

The genus currently accommodates some eight species that are distributed in the tropics and the subtropics of the world [Pacific Mexico (Dawson 1962), Galapagos Islands (Young 1981), Curaçao (van den Hoek 1978), Puerto Rico (Ballantine & Wynne 1986), eastern Australia (Huisman & Kraft 1984), Indian Ocean (Wollaston 1984), southern Japan (Itono & Tanaka 1973), and western Australia (Huisman 1988)]. Most species are known from deep

Fig. 6. *Balliella cladoderma*: Mediterranean Sea and Azores (horizontally hatched area)

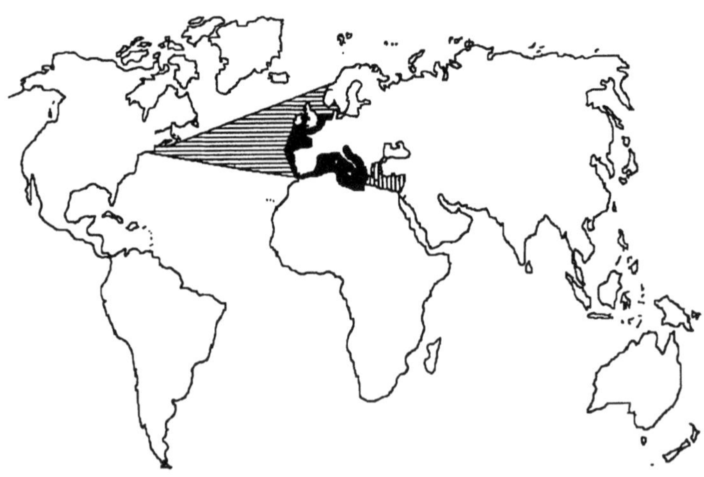

Fig. 7. *Pterothamnion plumula*: North Atlantic and western Mediterranean (shadowed and horizontally hatched areas). *Pterothamnion crispum*: NE. Atlantic and Mediterranean Sea (shadowed and vertically hatched areas)

water habitats of the southern Pacific and Indian Oceans, suggesting that the tropical and warm temperate waters of these regions have been the center of speciation of the genus. *Balliella cladoderma*, although endemic to the Atlantic-Mediterranean flora, appears to be closely related to the recently described *B. hirsuta* from western Australia (Huisman 1988).

Apart from a characteristic habitat, species of *Balliella* share several other features not found in other antithamnioid genera, e.g. the development of spherical gland cells connected distally to the mother cell, and cortication that, at least in *B. cladoderma*, becomes pseudoparenchymatous. Such features are apparently plesiomorphous, since they also characterize the genus *Warrenia* Harvey that is considered to be the most primitive member of the Ceramiaceae (Moe & Silva 1980). Other evidence that supports the exclusion of *Balliella* from the Antithamnieae will be presented elsewhere.

PTEROTHAMNION PLUMULA (ELLIS) NÄGELI (Fig. 7). Considerable confusion surrounds the circumscription of this taxon, causing problems in assessing its geographic distribution. Biosystematic studies of European and Scandinavian isolates support the recognition of at least four genotypically distinct varieties of a single species. Two of these varieties (the typical distichous and the tristichous-tetrastichous) are distributed between Atlantic France and Trondheim (L'Hardy-Halos 1968, Rueness & Rueness 1975, Sundene 1975, Rueness 1978). A third entity, originally described by Strömfelt (1888) as *A. plumula* var. *spinescens*, appears to be restricted to the exposed coasts of western Norway and the Faroes (coll. in KMN and GB), while a fourth undescribed variety with a more lax thallus has been located in the North Adriatic Sea (coll. in GB and L). Crosses between the three European entities have indicated completely interfertility, and these results suggest that *Pt. plumula* has a pronounced southern distribution, reaching its northern boundaries at Deep Cove Island, Nova Scotia (Hooper & Whittick 1975), and Trondheim on the west coast of Norway (Printz 1926). Further floristic investigations between Trondheim and Troms (a coastal area that remains little known) might extend the northern distribution of the species, but it is absent in the Troms and Finnmark areas (Jaasund 1965) and in Spitsbergen (Jenneborg pers. comm.). The southern Atlantic boundaries reach New Jersey (South & Tittley 1986) and the Iberian Peninsula (Ardré 1970), while the species also penetrates into the Baltic as far as Travemünde (Pankow 1971).

Mediterranean records of the distichous variety *plumula* are in need of re-investigation, in light of Sundene's (1975) observations of a sexually isolated entity with intermediate morphology between *Pt. plumula* and *Pt. crispum*. Records attributed to *A. plumula* var. *bebbii* (Reinsch) Feldmann should also be treated with caution, since Reinsch's (1875) original illustrations of *Callithamnion bebbii* from the Adriatic seem to be representative of *Pt. crispum* (cf. Wollaston 1980).

There are several records of *Pt. plumula* from the southern hemisphere [Macquarie Islands (Ricker 1987)], North Pacific [southern Japan (Itono 1977), and Pacific Mexico (Dawson

1962, 1966)], but none of these is based on a comparative study with European material. On the other hand, older records of this species from southern Australia [referred to as *Callithamnion plumula* var. *investiens* by J. Agardh (1876:24, type in LD)] have been segregated into a new genus and species, *Glandothamnus ramulentus* Wollaston (1980; type in MEL), whereas Okamura's (1916-23) record of *A. plumula* from southern Japan has been recognized as a new species of *Platythamnion* [*Pl. yezoense* Inagaki (1950)].

Apart from this complex of varietal, specific, and generic concepts of taxa that seem to be well accommodated in a single genus (*Pterothamnion*), the closest relatives of *Pt. plumula* appear to be species from the southern hemisphere. These relatives include *Pt. minutissimum* (Levring) Ricker (1987; type in GB) from Juan Fernandez and *Pt. simile* from Kerguelen (coll. in LD and GB), rather than the Atlantic-Mediterranean species *Pt. crispum* that was included previously within the varietal range of *Pt. plumula.*

PTEROTHAMNION CRISPUM (Fig. 7). The distribution of this species is restricted to warm temperate European coasts, between the Netherlands (Stegenga & Mol 1983) and the eastern Mediterranean (Athanasiadis 1985b), and the Sea of Marmara (coll. in WU). Extra-European records are unknown, although several, more recently, described taxa from the southern hemisphere show a close morphological relationship. These taxa include *C. plumula* var. *investiens* and *G. ramulentus* from southern Australia, *Pt. antarcticum* (Kylin) Moe & Silva (1980; isotype in GB) from the Antarctic Peninsula, as well as some North Pacific species referred to the genus *Platythamnion*, e.g. *Pl. villosum* Kylin (1925; type and coll. in LD and GB) and *Pl. recurvatum* Wollaston (1972b), the latter taxon also recorded from South Africa (Stegenga 1986).

Pterothamnion crispum was previously classified in *Platythamnion* J. Agardh, a taxon whose generic distinction from *Pterothamnion* Nägeli has been questioned (Whittick 1980; Athanasiadis 1985b). The immediate relatives of both *Pt. plumula* and *Pt. crispum* appear, however, to be taxa from the southern and northern Pacific, suggesting the occurrence of two separate phylogenetic lines: one encompassing species with (originally) a strictly distichous branching pattern (e.g. *Pt. simile*, *Pt. minutissimum*, and *Pt. plumula*), and a second lineage encompassing species with a permanent tetrastichous ramification (e.g. *Platythamnion heteromorphum* (J. Ag.) Setchell & Gardner (type in LD), *Platythamnion pectinatum* Kylin (type and coll. in LD, GB, and priv. coll.), and *Platythamnion crispum* (Ducluzeau) Feldmann]. If permanent tetrastichous branching is the derived condition, then the appearance of similar ramification in certain varieties of *Pt. plumula* could be explained as a case of convergence.

The Arctic-cold temperate element

Of the two species included in this floristic element, only *S. pylaisaei* occurs in the Arctic Ocean. Both species appear to be conspecific with, or closely related to taxa from the North Pacific Ocean.

ANTITHAMNIONELLA FLOCCOSA (MÜLLER) WHITTICK (Fig. 8). In the western Atlantic Ocean this species occurs between Massachusetts (South & Tittley 1986) and the Godthåbsfjord (64^0 26.5'N, 50^0 18' W) in western Greenland (Pedersen pers. comm.). The southern boundary on European coasts reaches the Clyde Sea-area in the northern British Isles (Huxley 1978) and Ireland (Guiry 1978). On the Scandinavian coast it has been collected between Bergen (priv. coll.) and Finnmark (Jaasund 1965, coll. in GB), while its northernmost record is represented by a single sublittoral collection off Bjørnøya (leg. Jenneborg, coll. in GB). *Aa floccosa* is also recorded from Iceland (Strömfelt 1886), the Faroes (Lyngbye 1819; as *Callithamnion plumula* var. *pusilla*, type and coll in C and LD), and the Barents Sea (Zinova 1955). The Atlantic distribution of the species is clearly cold temperate, and restricted to marine waters. European records refer to sterile plants or tetrasporophytes, except for a single unconfirmed report of 'sporocarps' from Scotland (Foslie 1890:45). Gametophytes and carposporophytes have, however, been recorded from Newfoundland (Whittick 1980), while carpogonial and spermatangial plants have also been observed in culture material originating from Bergen (pers. coll.).

Aa floccosa can be accommodated within the circumscription of *Aa pacifica* (Harvey) Wollaston, a similarly cold (to warm) temperate species found between the Aleutian Isles and Baja California (Lindstrom & Gabrielson 1989). Pacific plants referred to this taxon exhibit wider morphological variation including individuals with rare development of gland cells, decussate arrangement of branches along the bearing axis, and curving laterals [var. *uncinata* (Gardner) Wollaston (1972a)]. Moreover, in a recent collection of material from San Juan (U.S.), plants had ramified branches (particularly when surrounding carposporophytes) and tetrasporangia on uni- or multicellular pedicells. Sexual reproduction in *Aa pacifica* has also been more frequently reported (Daines 1913; Kylin 1925; Capt 1930; Wollaston 1972a; and pers. observ.).

If *Aa floccosa* and *Aa pacifica* are conspecific, their immediate relative becomes *Aa glandulifera* (Kylin) Wollaston and its potential synonyms from the North Pacific Ocean (see Wollaston 1972a), as well as *Aa alternans* Ricker (1987) from the Macquarie Islands in the South Pacific. From these distributions the Pacific Ocean may be recognized as a possible center of origin of *Aa floccosa*, and the North Atlantic Ocean may be a marginal area of its range. Despite this disjunct distribution, Scandinavian plants of *Aa floccosa* have not differentiated to a great extent from their Pacific counterparts, and have maintained certain primitive antithamnioid features, such as the strictly distichous-opposite ramification of the thallus and the successively divided tetrasporangia. This neoteny becomes particularly obvious when the warm temperate congeners and *Aa floccosa* (that meet only at the margin of their distribution, the Clyde Sea-area of the northern British Isles) are compared.

SCAGELIA PYLAISAEI (MONTAGNE) WYNNE (Fig. 8). This is a pronounced boreal species reported under various names from localities in the North Atlantic, North Pacific, and Arctic Oceans, including the White, Okhotsk and Murman Seas (Hansen & Scagel 1981; Wynne 1985). Its southern boundaries in the Atlantic Ocean reach the southern Baltic (Pankow 1971), the Clyde Sea-area of the northern British Isles (Huxley 1978), and New Jersey (South & Tittley 1986). In the North Pacific Ocean, it has been reported as far south as La Jolla, California (Wollaston 1972a, as *S. occidentale* (Kylin) Wollaston) and Nemuro on the eastern coast of Hokkaido (Yoshida 1981b, as *S. corallina* (Kjellman) Yoshida).

Scandinavian plants were described as a complex of at least three genetically isolated varieties (Sundene 1962). This view was based on biosystematic studies in which sexual reproduction, while commonly recorded in plants from the western Atlantic and North Pacific Oceans (Hansen & Scagel 1981; Kylin 1925; West & Norris 1966; Wollaston 1972a; and pers. observ. of coll. in UC and priv. coll.), was not observed in European and Scandinavian waters, except for a single unpublished record from Rogaland fjord, Norway (Åsen pers. comm.; coll. in KMN). Recent re-investigations of Scandinavian isolates and extensive examination of herbarium and liquid preserved collections originating from localities on the Swedish west coast, Newfoundland, Greenland, Spitsbergen, and Novja Zemlja (coll. in GB,

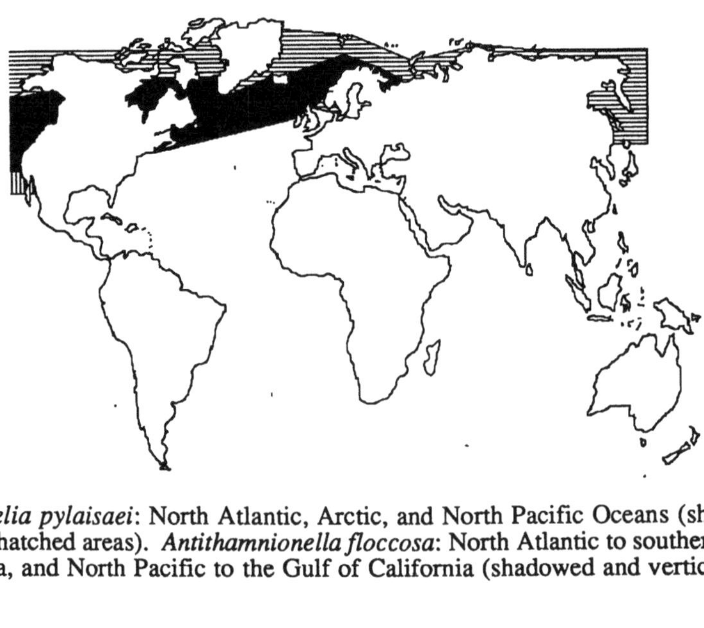

Fig. 8. *Scagelia pylaisaei*: North Atlantic, Arctic, and North Pacific Oceans (shadowed and horizontally hatched areas). *Antithamnionella floccosa*: North Atlantic to southern Greenland and Bjornoya, and North Pacific to the Gulf of California (shadowed and vertically hatched areas)

BG, KMN, C, LE, and priv. coll.), have confirmed Sundene's conclusions on the occurrence of at least three North Atlantic varieties, reproducing chiefly by apomeiotic tetrasporangia and fragmentation. An active process of differentiation was also demonstrated in a sexually reproducing strain from Bergen showing spontaneous mutations during tetrasporogenesis (Athanasiadis & Rueness unpubl. observ.).

If *S. pylaisaei* is regarded as a monophyletic assemblage of several entities, a central question becomes: what taxon should be considered as its immediate relative? On the one hand, *Scagelia* shares certain features with *Pterothamnion*, including the type of tetrasporangial division and the intercalary position of carposporophytes on the thallus. On the other hand, the form variation of *Scagelia* in culture shows features of *Antithamnionella*, such as the development of carpogonial branches on 2-celled laterals (Athanasiadis unpubl. observ.). All three genera have in common the type and position of gland cells and the development of branched spermatangial mother cells, but chromosome counts have indicated that Scandinavian *Scagelia* shares the haploid number of species of *Antithamnionella* (32-40; Rao 1960, Magne 1986, Athanasiadis unpubl. observ.) and not of *Pterothamnion* (24±2; Magne 1964, Athanasiadis 1985b).

A possible sister taxon of *Scagelia* is the monotypic *Scageliopsis* Wollaston (1980) from southern Australia (type in ADU). Despite its disjunct (antipolar) distribution, *Scageliopsis* shares many features with *Scagelia*. A notable exception to this similarity is the pseudodichotomous ramification of whorl-branches that also characterizes some endemic genera from the southern hemisphere that are closely related to the Antithamnieae, e.g. *Perithamnion* J. Agardh and *Tetrathamnion* Wollaston (1968).

The distichously branched variety of *Scagelia pylaisaei*, recognized as var. *boreale* by some authors, shows a superficial similarity with *Pt. plumula*. This similarity has resulted in misidentifications or interpretations of an infraspecific relationship between these taxa (Gobi 1878; Kjellman 1883). While the possibility of hybridization between *Scagelia* and *Pterothamnion* cannot be precluded, these genera have disjunct centers of distribution. Using the frequency of occurrence of sexual reproduction as an indicator, the center of distribution for *Scagelia* appears to be the cold temperate regions of the North Pacific and western Atlantic Oceans, contrasting with *Pterothamnion* which belongs to the warm temperate Atlantic-Mediterranean flora. Scandinavian and European populations of *Scagelia pylaisaei* are commonly sterile, and should therefore be considered as being at the periphery of the range.

With its high degree of morphological variation and broad affinities to several antithamnioid genera, *Scagelia* would appear to be a rather primitive member of the tribe Antithamnieae.

DISCUSSION

Distributions of species are based on identification of specimens and concepts of taxa. As a result, distributions are based on changeable criteria that may not be universally recognized. When such data concern minute ceramiaceous algae, the problems obviously increase and a critical approach must be applied before published information should be compiled and used as fact.

The distributions presented in this paper are largely based on examination of authentic collections and, compared to the information presented in the latest catalogue of the North Atlantic marine algae (South & Tittley 1986), give a far more detailed picture on the occurrence and distribution of the North Atlantic antithamnioids.

Species delimitation is here strictly based on the results of breeding tests (Sundene 1975; Rueness & Rueness 1975; Rueness 1978, Athanasiadis 1985b; L'Hardy-Halos 1986; Athanasiadis unpubl. observ.), that support the recognition of rather narrowly circumscribed units, such as *Pt. plumula* vs. *Pt. crispum*, *Aa sarniensis* vs. *Aa spirographidis*, and *A. tenuissimum* vs. *A. heterocladum*. It should be noted, however, that apparent sterility barriers might result from a variety of causes including failure to obtain functional gametangia because of suboptimal breeding conditions in culture. For example, cytological studies of cultured plants have shown that carpogonia may sometimes lack nuclei, and that auxiliary cells may fail to develop, despite presumed fertilization as indicated by spermatangial attachment and the development of the protruding connecting cell.

Ten North Atlantic antithamnioid species are recognized here and their geographic distribution coincides with two climatically defined regions in the North Atlantic Ocean (Michanek 1979; van den Hoek 1984; Joosten & van den Hoek 1986). These species form two distinct floristic elements: a warm temperate element with tropical affinities comprising at least eight species of the genera *Antithamnion*, *Antithamnionella*, *Balliella*, and *Pterothamnion*, and an Arctic-cold temperate element comprising only two species of the genera *Antithamnionella* and *Scagelia*. The distinction between these two elements is reflected by the marginal overlapping, or disjunct distributions of their members. Warm temperate species that penetrate into the cold temperate waters of the eastern Atlantic Ocean (as far as Trondheim) comprise only certain infraspecific taxa that apparently are undergoing speciation. On the other hand, the southern distribution of the two boreal species is rather limited, since they do not penetrate into the warm temperate waters of the North Atlantic further than New Jersey, the Clyde Sea-area of the British Isles and western Ireland.

The warm temperate element shows a greater diversity of taxa, that toward the southern parts of their range meet with several other antithamnioids of the Mediterranean and central Atlantic floras. This floristic element expands to a total of eighteen species and includes: *Acrothamnion butleriae* (Collins) Kylin [in the tropical central Atlantic and central Pacific

Oceans (Collins 1901, as *Antithamnion*; Wollaston 1977)], *Acrothamnion preissii* (Sonder) Wollaston [in the tropics and subtropics of the Pacific Ocean and recently recorded from the Mediterranean Sea (Cinelli & Sartoni 1969)], *Antithamnion antillanum* Børgesen [from the tropical central Atlantic and central Pacific Oceans (Athanasiadis 1985a)], *Antithamnion compactum* (Grunow) Schiffner [only known from Gibraltar (Grunow 1867, as *Sporacanthus*)], *A. heterocladum* Funk, *Antithamnion ogdeniae* Abbott [from the Central Atlantic Ocean and in Mediterranean Sea (Athanasiadis 1985a)], *A. nipponicum* Yamada [from the subtropical Pacific Ocean and recently recorded from Mediterranean France (Verlaque & Riouall 1989)], *Antithamnion piliferum* Cormaci & Furnari (1987; a Mediterranean endemic), *A. tenuissimum* (Hauck) Schiffner, and *Balliella pseudocorticata* (Dawson) Young [from the tropical central Atlantic and central Pacific Oceans (Dawson 1962, as *Antithamnion*; van den Hoek 1978, as *Bacothamnion curassavicum*; Huisman & Kraft 1984)].

Some of these species may be recent introductions to the Mediterranean and Atlantic floras, whereas the origins of other, widely recorded species from the Atlantic (or the Mediterranean Sea) and the Pacific Oceans remains speculative. The majority of taxa included in this floristic element seems to represent true endemics of the Atlantic-Mediterranean flora, and this high diversity may have resulted from geographic isolation associated with the westward migration of the Tethyan pantropical flora into the young Atlantic and Mediterranean basins, and its later isolation from the Indo-Pacific (van den Hoek 1984, Joosten & van den Hoek 1986). The hypothesis of a Tethyan ancestor of the present-day warm temperate Atlantic-Mediterranean antithamnioids, is supported by the fact that all species of this floristic element belong to cosmopolitan genera with even higher diversity in the tropical and subtropical Pacific Ocean where several other endemic taxa occur, particularly in the southern hemisphere. The present distributional data suggest that the Pacific Ocean is the center of origin of the tropical and warm temperate antithamnioids and that the North Atlantic Ocean is a marginal area, where certain genera dispersed and subsequently diversified.

The Arctic-cold temperate element comprises only two species that also occur (or have their closest relatives) in the North Pacific. Therefore, the present distribution of this element should be considered as circumboreal, and its origins should be traced in the evolutionary history of each of the included taxa.

Antithamnionella floccosa shows affinities with Pacific rather than Atlantic congeners. Assuming a monophyletic origin from a Pacific ancestor, its current distribution in the North Atlantic Ocean is attributable to migration (presumably via the Arctic Ocean). Alternatively, the assumption of an Atlantic ancestor would require the occurrence of Atlantic sister taxa that later became extinct, as well as a second speciation event in the Pacific region giving rise to the taxa here considered as immediate relatives of this species.

The same evolutionary scenario could be applied to *Scagelia pylaisaei*, which also shares most of its features with Pacific taxa. This assumes, however, that the similarities of certain varieties of this species to *Pterothamnion plumula* result from convergence or hybridization.

Undoubtedly, the origin of the Arctic-cold temperate element remains speculative. An apparent inconsistency in the present interpretation is that, while both boreal species seem to have areas of the North Pacific and western Atlantic Oceans as current centers of distribution, *Aa floccosa* shows a marked differentiation in the North Pacific Ocean, whereas *S. pylaisaei* shows a pattern with differentiation in the eastern Atlantic. Whether or not this difference is based on the separate biogeographic origins of the two species remains to be investigated.

ACKNOWLEDGEMENTS

I express my sincere thanks to the curators of the herbaria ADU, B, BM, BG, C, GB, HBG, KMN, L, LD, LE, MEL, MICH, O, PC, UC, and WU for generous loans of material and comments on authentic collections. Additional material from private collections was kindly provided by M. Th. L'Hardy-Halos (Laboratoire de Phycologie marine et de Morphogenese, Le Mans), Wm. F. Farnham (Portsmouth Polytechnic), F. Magne (Univ. Pierre et Marie Curie), P.-M. Pedersen (Univ. of Copenhagen), J. Rueness (Univ. of Oslo), and M. Waern (Univ. of Uppsala). I also thank two anonymous referees, A. Whittick (Memorial Univ. of Newfoundland), and D. Garbary (Univ. of St. Francis Xavier) for critically reading versions or parts of this paper and making valuable comments. Financial support to carry out this study was provided by the Swedish Natural Science Research Council through a post-doctoral scholarship for the years 1988 and 1989 at the Univ. of Oslo, Norway. The NATO Science Committee provided funds that enabled me to attend and present a version of this paper at the Advanced Research Workshop (Ref. ARW 931/88) entitled 'Evolutionary Biogeography of the Marine Algae of the North Atlantic Ocean', held in St. Andrews, New Brunswick, Canada, in September 1989.

REFERENCES

Abbott IA (1979) Some tropical species related to *Antithamnion* (Rhodophyta, Ceramiaceae). Phycologia 18: 213-217

Agardh JG (1842) Algae maris mediterranei et adriatici,... Fortin, Mason et Cie, Paris

Agardh JG (1876) Species Genera et Ordines Algarum...Volumen Tertium:... Epicrisis systematis Floridearum. T. O. Wiegel. Lipsiae

Agardh JG (1892) Analecta Algologica. Lunds Univ Årskr 28: 1-182

Anderson RJ, Stegenga H (1989) Subtidal algal communities at Bird Island, eastern Cape, S. Africa. Bot Mar 32: 299-311

Ardré F (1970) Contribution a l' étude des algues marines du Portugal. I- La flore. Port Acta Biol (B) 10: 1-423

Athanasiadis A (1983) The life history of *Antithamnion heterocladum* (Rhodophyta, Ceramiales) in culture. Bot Mar 26: 153-157

Athanasiadis A (1985a) North Aegean marine algae. I. New records and observations from the Sithonia Peninsula, Greece. Bot Mar 28: 453-468

Athanasiadis A (1985b) The taxonomic recognition of *Pterothamnion crispum* (Rhodophyta, Ceramiales), with a survey of the carposporophyte position in genera of the Antithamnieae. Br Phycol J 20: 381 - 389

Athanasiadis A (1986) A comparative study of *Antithamnion tenuissimum* and three varieties of *A. cruciatum*, including var. *scandinavicum* var. nov. (Rhodophyceae). Nord J Bot 6: 703 - 709

Athanasiadis A (1987) A survey of the seaweeds of the Aegean Sea with taxonomic studies on species of the tribe Antithamnieae (Rhodophyta). Goterna, Kungälv

Athanasiadis A (1988) The status and typification of *Antithamnion cruciatum* var. *pumilum* and its relationship to *A. cruciatum* var. *scandinavicum* (Rhodophyta). Br Phycol J 23: 395-398

Baardseth E (1941) The marine algae of Tristan da Cunha. Norske Videnskaps-Akademi, Oslo, No 9

Ballantine DL, Wynne MJ (1986) Notes on the Marine Algae of Puerto Rico. I, Additions to the Flora. Bot Mar 29: 131-135

Boo SM, Lee IK (1983) Life history and hybridization of *Antithamnion sparsum* Tokida (Rhodophyta, Ceramiaceae) in culture. Kor J Bot 26: 141-150

Børgesen F (1930) Marine algae from the Canary Islands, especially from Teneriffe and Gran Canaria. II. Rhodophyceae. Part III. Ceramiales. Dan Biol Medd 9: 1-159

Capt L 1(930) The morphology and life history of *Antithamnion*. Publ Puget Sound Biol Stat 7: 369-389.

Cinelli F, Sartoni G (1969) *Acrothamnion* J. Ag. (Rhodophyta, Ceramiaceae): genere algale nuovo per il mare Mediterraneo. Pubbl Staz Napoli 37: 567-574

Collins FS (1901) The algae of Jamaica. Proc Amer Acad Arts Sci 37: 231-270

Collins FS, Hervey AB (1917) The algae of Bermuda. Proc Amer Acad Arts Sci 53: 1-195

Cormaci M, Furnari G (1988) *Antithamnionella elegans* (Berthold) Cormaci et Furnari (Ceramiaceae, Rhodophyta) and related species, with the description of two new varieties. Phycologia 27: 340-346

Cormaci M, Furnari G (1987) *Antithamnion piliferum* sp. nov. (Ceramiaceae, Rhodophyta) from eastern Sicily (Mediterranean Sea). Crypt Algol 8: 223-232

Daines LL (1913) Comparative development of the cystocarps of *Antithamnion* and *Prionitis*. Univ Cal Publ Bot 4: 283-302

Dawson EY (1962) Marine red algae of Pacific Mexico. Part 7. Ceramiales: Ceramiaceae, Delesseriaceae. Allan Hancock Pac Exped 26: 1-207

Dawson EY (1966) New records of marine algae from the Gulf of California. J Arizona Acad Sci 4: 55-66

De Toni JB (1903) Sylloge algarum... Vol. IV. Sect. III. Padova

Feldmann-Mazoyer G (1941) Recherches sur les Céramiacées de la Méditerranée occidentale. Alger

Foslie M (1890) Contribution to knowledge of the marine algae of Norway. I East-Finmarken. Tromsöpostens printing office, Tromsö

Gallardo T, Gomez Garreta A, Ribera MA, Alvarez M, Conde F (1985) A preliminary checklist of Iberian benthic marine algae. Real Jardin Botanico, Madrid

Garbary D, Belliveau D, Irwin R (1988) Apical control of band elongation in *Antithamnion defectum* (Ceramiaceae, Rhodophyta). Can J Bot 66: 1308-1315

Gobi C (1878) Die Algenflora des Weissen Meeres. Mém Acad Imp Sci St-Pétersburg, ser 7 26: 1-92

Grunow A (1867) Algen. pp. 1-104, 11 pls. In: Fenzl, E (ed.), Reise seiner Majestät Fregatte Novara um die Erde. Botanischer Theil I Band Wien

Guiry M. 1978. A consensus and bibliography of Irish seaweeds. Bibliotheca Phycol 44. Cramer, Vaduz

Hansen GI, Scagel RF (1981) A morphological study of *Antithamnion boreale* (Gobi) Kjellman and its relationship to the genus *Scagelia* Wollaston (Ceramiales, Rhodophyta). Bull Torrey Bot Club 108: 205-212
Harvey WH (1848) Phycologia Britannica. Pls 145-216. Reeve, London
Hennig W (1966) Phylogenetic systematics. Translated by D. D. Davis & R. Zangerl. University of Illinois Press, Urbana
Hoek C van den (1978) Marine algae from the coral reef of Curaçao, Netherlands Antilles. I. Three new and one rarely observed species from the steep fore-reef slope. Aquat Bot 5: 47-61
Hoek C van den (1984) World-wide latitudinal and longitudinal seaweed distribution patterns and their possible causes, as illustrated by the distribution of Rhodophytan genera. Helgol Meeresunters 38: 227-257
Holmgren PK, Keuken W, Schofield EK (1981) Index Herbariorum. Pt. 1. The Herbaria of the World. Regnum Veg 106: 1-452
Hooper RG, Whittick A (1975) *Antithamnion plumula* (Ellis) Thur. in Le Jol. (Rhodophyceae: Ceramiaceae) in eastern Canada. Nat Can 102: 603-604
Howe MA (1914) The marine algae of Peru. Mem Torrey Bot Club 15: 1-185
Huisman JM (1988) *Balliella hirsuta* sp. nov. (Ceramiaceae, Rhodophyta) from Rottnest Island, western Australia. Phycologia 27: 456-462
Huisman JM, Kraft GT (1984) The genus *Balliella* Itono & Tanaka (Rhodophyta: Ceramiaceae) from eastern Australia. J Phycol 20: 73-82
Huxley R (1978) A brief guide to the British species of *Antithamnion* Nägeli. Br Phycol Soc Newsletter 11: 15-18
Inagaki K (1950) Some marine algae from the central Pacific coast of Japan. J Jap Bot 25: 20-26
Itono H (1977) Studies on the ceramiaceous algae (Rhodophyta) from southern parts of Japan. Bibliotheca Phycol. 35. Cramer, Vaduz
Itono H, Tanaka T (1973) *Balliella*, a new genus of Ceramiaceae (Rhodophyta). Bot Mag Tokyo 86: 241-252
Jaasund E (1965) Aspects of the marine algal vegetation of North Norway. Bot Gothoburg 4: 1-174
Joosten AMT, Hoek C van den (1986) World-wide relationships between red seaweed floras: a multivariate approach. Bot Mar 29: 195-214
Kjellman FR (1883) Norra Ishafvets algflora. Vega-Exped Vetenskapl Arbeten Vol 3
Kjellman FR (1877) Ueber die Algenvegetationen des Murmanschen Meeres an der Westküste von Nowaja Semlja und Wajgatsch. Nova Acta Reg Soc Sc Ups Ser III 12:1-86
Kylin H (1925) The marine red algae in the vicinity of the Biological Station at Friday Harbor, Wash. Lunds Univ Årsskrift N F Avd 2 21(9): 1-87
Kützing FT (1861) Tabulae Phycologicae... XI. Nordhausen
Levring T (1960) Reports of the Lund University Chile expedition 1948-49. 39. Contributions to the Marine Algal Flora of Chile. Lunds Univ Årsskrift N F Avd 2 56(10): 1-85
L'Hardy-Halos MT (1968) Les Ceramiaceae (Rhodophyceae, Florideae) des côtes de Bretagne: I. Le genre *Antithamnion* Nägeli. Rev Algol 9: 152-183
L'Hardy-Halos MT (1986) Observations on two species of *Antithamnionella* from the coast of Brittany. Bot Mar 29: 37-42
Lindstrom SC, Gabrielson PW (1989) Taxonomic and distributional notes on northeast Pacific Antithamnieae (Ceramiales, Rhodophyta). Jap J Phycol 37: 221-235
Lyngbye HC (1819) Tentamen Hydrophytologiae Danicae... Hafniae
Lyle L (1922) *Antithamnionella*, a new genus of algae. J Bot 60: 346-350
Maddison WP, Maddison DR (1987) [Manual of] MacClade, version 2.1, for the Apple Macintosh™. Harvard University
Magne F. 1964. Recherches caryologiques chez les Floridées (Rhodophycées). Cah Biol Mar 5: 461-671
Magne F. 1986. Anomalies du développement chez *Antithamnionella sarniensis* (Rhodophyceae, Ceramiaceae). II: Nature des individus issus des tétraspores. Crypt Algol 7: 215-229

Mendoza ML (1969) Estudio sistematico y ecologico de las Ceramiaceae (Algae-Rhodophyta) de Puerto Deseado Provincia de Santa Cruz (Argentina). Darwiniana 15: 287-362
Michanek G (1979) Phytogeographic provinces and seaweed distribution. Bot Mar 22: 375-391
Moe RL, Silva PC (1980) Morphological and taxonomic studies on the Antarctic Ceramiaceae (Rhodophyceae). II *Pterothamnion antarcticum* (Kylin) comb. nov. (*Antithamnion antarcticum* Kylin). Br Phycol J 15: 1-17
Okamura K (1916-1923) Icones of Japanese Algae IV. Kazamashobo, Tokyo
Pankow H (1971) Algenflora der Ostsee. I. Benthos (Blau-, Grün-, Braun- und Rotalgen). Gustav Fischer Verlag, Jena
Parke M, Dixon PS (1976) Check-list of British marine algae - third revision. J Mar Biol Ass UK 56: 527-594
Perestenko LP (1980) 'Algae from the Bay of Peter the Great'. Izd. Akad. Nauk SSSR. Leningrad [in Russian with Latin names of taxa and descriptions of new species]
Price JH, John DM, Lawson GW (1986) Seaweeds of the western coast of tropical Africa and adjacent islands: a critical assessment. IV. Rhodophyta. Bull Br Mus Nat Hist (Bot) 15: 1-122
Printz H (1926) Die Algenvegetation des Trondhjemsfjordes. Skr Norske Vidensk Akad I Mat Nat Kl 5: 1-274
Rao CSP (1960) Cytology of red algae. pp. 37-45. In: Proceedings of the symposium on algology. Kachroo, New Delhi
Reinsch PF (1875) Contributiones ad Algologiam et Fungologiam. Vol. 1. T. O. Wiegel, Lipsiae
Ricker RW (1987) Taxonomy and biogeography of Macquarie Island seaweeds. British Museum (Natural History), London
Ridley M (1986) Evolution and Classification. The reformation of cladism. Longman, London and New York
Rueness J (1978) Hybridization in red algae. In: Irvine DEG, Price JH (eds), Modern approaches to the taxonomy of red and brown algae. Academic Press, London and New York, p. 247-262
Rueness J, Rueness M (1975) Genetic control of morphogenesis in two varieties of *Antithamnion plumula* (Rhodophyceae, Ceramiales). Phycologia 14: 81-85
Schiffner V (1916) Studien über Algen des adriatischen Meeres. Wiss Meeresunters Abt Helgol 11: 129 -181
Schneider GW (1974) North Carolina marine algae. III. A community of Ceramiales (Rhodophyta) on a glass sponge from 60 meters. Bull Mar Sci 24: 1093-1101
Schneider GW (1984) Studies on *Antithamnionella*, *Callithamniella* and *Calloseris* (Rhodophyta, Ceramiales) from North Carolina, USA. Phycologia 23: 455-464
Seoana-Camba J (1969) Algas bentónicas de Menorca en los herbarios Thuret-Bornet y Sauvageau del Muséum National d'Histoire Naturelle de Paris. Inv Pesq 33: 213-260
South GR, Tittley I (1986) A checklist and distributional index of the benthic marine algae of the N. Atlantic Ocean. Huntsman Marine Laboratory and British Museum (Natural History), St Andrews and London
Stegenga H (1986) The Ceramiaceae (excl. *Ceramium*) (Rhodophyta) of the South West Cape Province, South Africa. Bibliotheca Phycol 74. Cramer, Vaduz
Stegenga H, Mol I (1983) Flora van de Nederlandse zeewiren. Konink Neder Natur Veren Nr 33
Stegenga H, Vroman M (1987) Notes on some Ceramiaceae (Rhodophyta) from Curaçao especially those from the exposed northeast coast. Blumea 32: 397-426
Strömfelt HFG (1886) Om Algvegetationen vid Islands Kuster. D. F. Bonniers, Göteborg
Strömfelt, HFG (1888) Algae novae quas ad litora Scandinaviae indagavit. Notarisia Anno III 9: 381-384
SuhrJN von (1840) Beiträge zur Algenkunde. Flora 18: 273-282
Sundene O (1962) Reproduction and morphology in strains of *Antithamnion boreale* originating from Spitsbergen and Scandinavia. Skr.Norske Vidensk-Akad Oslo I Mat-Naturv Klasse Ny Serie No 5: 1-19

Sundene O (1975) Experimental studies on form variation in *Antithamnion plumula* (Rhodophyceae). Norw J Bot 22: 35-42
Taylor WR (1960) Marine algae of the eastern tropical and subtropical coasts of the Americas. University of Michigan, Ann Arbor
Verlaque M, Riouall R (1989) Introduction de *Polysiphonia nigrescens* et d' *Antithamnion nipponicum* (Rhodophyta, Ceramiales) sur le littoral Méditerranéen Francais. Crypt Algol 10: 313-323
Waern M (1961) Tillägg till Sveriges rödalgsflora. Svensk Bot Tidskr 55: 234-235
West JA, Norris RE (1966) Unusual phenomena in the life histories of Florideae ion culture. J Phycol 2: 54-57
Westbrook MA (1934) *Antithamnion spirographidis* Schiffner. J Bot 72: 65-68
Whittick A (1980) *Antithamnionella floccosa* (O. F. Müller) nov. comb.: a taxonomic re-appraisal of *Antithamnion floccosum* (O. F. Müller) Kleen (Rhodophyta: Ceramiaceae). Phycologia 19: 74-79
Whittick A, Hooper RG (1977) The reproduction and phenology of *Antithamnion cruciatum* (Rhodophyta: Ceramiaceae) in insular Newfoundland. Can J Bot 55: 520-524
Wollaston EM (1968) Morphology and taxonomy of southern Australian genera of Crouanieae Schmitz (Ceramiaceae, Rhodophyta). Aust J Bot 16: 217-417
Wollaston EM (1972a) *Antithamnion* and related genera occurring on the Pacific coast of North America. Syesis 4: 73-92
Wollaston EM (1972b) The genus *Platythamnion* J. Ag. (Ceramiaceae, Rhodophyta) on the Pacific coast of North America between Vancouver, British Columbia, and southern California. Syesis 5: 43-53
Wollaston EM (1977) Morphology and taxonomy of the genus *Acrothamnion* J. Ag. (Ceramiaceae, Rhodophyta). Bull Jap Soc Phycol 25, Suppl (Mem Iss Yamada): 385-393
Wollaston EM (1980) Descriptions of two new genera, *Scageliopsis* and *Glandothamnus* (Ceramiaceae, Rhodophyta), including five previously undescribed species from southern Australia. Pac Sci 34: 109-127
Wollaston EM (1984) Species of Ceramiaceae (Rhodophyta) recorded from the International Indian Ocean Expedition, 1962. Phycologia 23: 281-299
Wynne MJ (1985) Concerning the names *Scagelia corallina* and *Heterosiphonia wurdemannii* (Ceramiales, Rhodophyta). Crypt Algol 4: 81-90
Wynne MJ (1986) Report on a collection of benthic marine algae from the Namibian coast (southwestern Africa). Nova Hedwigia 43: 311-355
Yoshida T (1981a) Note on *Antithamnion sparsum* Tokida (Rhodophyta, Ceramiaceae). Jap J Phycol 29: 47-50
Yoshida T 1981b. Observations on *Antithamnion miharae* Tokida and *A. corallina* Kjellman (Rhodophyta, Ceramiales) from the east coast of Hokkaido, Japan. J Fac Sci Hokkaido Univ Ser. V 12: 173-182
Young DN (1981) Taxonomic observations on eastern Pacific *Antithamnion* species (Rhodophyta: Ceramiaceae) described by E. Y. Dawson. Proc Biol Soc Wash 94: 94-100
Zinova AD (1955) 'Determinations of the red algae of the northern SSSR Seas'. Acad. Science, CCCP, Moskva [in Russian with Latin names of taxa]

DISTRIBUTION AND EVOLUTION OF NON-CORALLINE CRUSTOSE RED ALGAE IN THE NORTH ATLANTIC

Christine A. Maggs
Division of Environmental and Evolutionary Biology
School of Biology and Biochemistry
The Queen's University of Belfast
Belfast BT7 1NN
Northern Ireland

Abstract - The non-coralline (fleshy) red algal crusts of the North Atlantic are taxonomically diverse. They comprise approximately equal numbers of species with isomorphic or apomictic life histories, and those which are life-history phases of erect algae. The isomorphic crusts, in particular, are of biogeographic interest because their distributional patterns are generally not obscured by recent man-mediated events.

Seven crustose genera lacking an erect life-history phase occur in the North Atlantic: *Rhodophysema* and *Rhodophysemopsis* (Rhodophysemataceae; *Hildenbrandia* (Hildenbrandiaceae); *Cruoria* (Cruoriaceae): *Haemescharia* (Haemeschariaceae); *Plagiospora* (?Gloiosiphoniaceae); and *Peyssonnelia* (Peyssonneliaceae). In the Rhodophysemataceae, greater morphological diversity in the North Pacific, relative to the North Atlantic, is consistent with a North Pacific origin. Members of the Hildenbrandiaceae, which exhibit unique morphological features, occur in temperate waters of both hemispheres, reflecting their ancient origins. The Cruoriaceae, and the possibly related Haemeschariaceae, are endemic, to the North Atlantic and North Atlantic-Arctic Ocean, respectively. *Plagiospora*, a monotypic genus of uncertain affinities, is an amphi-North Atlantic endemic. *Peyssonnelia*, the largest genus, is of Tehyan origin. The high species diversity seen in the tropics and the Mediterranean decreases northwards on both sides of the the Atlantic. The only species presently in subarctic regions, *P. rosenvingii*, does not appear to be closely related to either the tropical or Mediterranean *Peyssonnelia* species. Instead, *P. rosenvingii* resembles NE Pacific species such as P. pacifica, and may have entered the North Atlantic via the Bering Strait during a relatively warm period. It reproduces only by monosporangia, perhaps because sexual reproduction was lost during migration through an unfavourable environment.

Algae with crustose life-history phases are represented in the North Atlantic by members of eleven families in three orders. *Atractophora* (Naccariaceae) is monotypic and known only from a small area of the NE Atlantic. *Ahnfeltia* (Ahnfeltiaceae) is circumpolar in both hemispheres, but this may result from relatively recent dispersal as South American material of *A. plicata* cannot be distinguished from northern hemisphere plants. Plastid DNA analysis in *Gymnogongrus* (Phyllophoraceae) has linked the two uncoupled life-history phases of an unidentified species that occurs in New England, Nova Scotia, and Northern Ireland, and it may assist in determining origins.

NATO ASI Series, Vol. G 22
Evolutionary Biogeography of the
Marine Algae of the North Atlantic
Edited by D. J. Garbary and G. R. South

INTRODUCTION

The non-coralline, or fleshy, crustose red algae of the North Atlantic Ocean consist of two discrete, taxonomically diverse sets of species linked only by gross thallus morphology. First, there are algae that undergo isomorphic life histories (Table 1), in which both the haploid gametophytic and the diploid tetrasporophytic phases are crustose. Crusts with *Rhodophysema*-type life histories are incluUded in Table 1. In these, carpogonia formed in sori on crustose gametophytes develop, after fertilization, directly into tetrasporangia and regenerative stalk cells. An approximately equal number of North Atlantic species (Table 2) exhibit heteromorphic life histories, in which only the tetrasporophytic phase is a crust and the gametophytes are erect, blade-like, filiform or compressed axes.

Why are crustose algae, and those with crustose phases, of biogeographic interest? First, the present distribution patterns of isomorphic crusts are less likely to reflect recent dispersal by man's activities than are those of the majority of red algae. Secondly, most species grow slowly, requiring several months to years to reach reproductive maturity. Thirdly, the crustose growth habit restricts spore dispersal as the spores are less likely to enter the water column and, when crusts are damaged by mechanical disturbance, only the parts remaining attached regenerate: detached fragments do not normally survive. Finally, and perhaps most significantly, many crusts do not grow epiphytically or on artificial substrata. Algae that grow on seagrasses or on larger buoyant seaweeds can be transported easily over long distances. Although a crustose tetrasporophytic thallus attached to a ship's hull would provide an ideal means of dispersal, in practice crusts are not generally found in this type of habitat. The only obvious natural vector for long-distance dispersal of crustose seaweeds is by ice-rafting of pebbles (Hommersand 1986). There are no known examples of introduced crustose red algae. A complete contrast is seen in many members of the Ceramiaceae, which grow and reproduce rapidly, fragment and regenerate easily, favour artificial substrata, and are often epiphytic. They have been introduced repeatedly to new areas, presumably by human activity. In some cases, such as that of two *Antithamnionella* species thought to be introduced to Atlantic Europe (L'Hardy-Halos 1985), their origin is now difficult to trace.

A second set of attributes of isomorphic crusts, their stress tolerance, is also of potential biogeographic importance (Wilce 1971). Many crustose species inhabit marginal environments, such as sand-scoured bedrock, and they are highly resistant to unfavourable conditions. Cells in the lower layers of crustose thalli are densely packed with starch grains and other storage materials (see, for example, Cabioch & Giraud 1982). This large nutrient reserve must contribute to survival during long periods of burial by sand, or overgrowth by competing organisms. In the high Arctic, *Haemescharia polygyna* is one of the few species that survives and perennates under ice cover during the dark winters (Wilce & Maggs 1989). Non-coralline

Table 1. North Atlantic crustose red algae with isomorphic, apomictic and *Rhodophysema*-type life histories

Order / Family	**Species**
Palmariales	
Rhodophysemataceae	*Rhodophysema elegans* (P. et H. Crouan ex J. Agardh) P. Dixon
	Rhodophysemopsis hyperborea (Rosenvinge) Masuda[2]
Hildenbrandiales	
Hildenbrandiaceae	*Hildenbrandia crouanii* J. Agardh
	Hildenbrandia rubra (Sommerfelt) Meneghini
Gigartinales	
Cruoriaceae	*Cruoria cruoriaeformis* (P. et H. Crouan) Denizot
	Cruoria pellita (Lyngbye) Fries
Haemeschariaceae	*Haemescharia hennedyi* (Harvey) Wilce et Maggs
	Haemescharia polygyna Kjellman
Peyssonneliaceae	*Peyssonnelia armorica* (P. et H. Crouan) Weber van Bosse in Børgesen
	Peyssonnelia atropurpurea P. et H. Crouan
	Peyssonnelia boudouresquei Yoneshigue
	Peyssonnelia conchicola Piccone et Grunow
	Peyssonnelia coriacea J. Feldmann
	Peyssonnelia dubyi P. et H. Crouan
	Peyssonnelia harveyana J. Agardh
	Peyssonnelia immersa Maggs et L. Irvine
	Peyssonnelia inamoena Pilger
	Peyssonnelia polymorpha (Zanardini) Schmitz
	Peyssonnelia rosenvingii Schmitz in Rosenvinge
	Peyssonnelia simulans Weber van Bosse in Børgesen
	Peyssonnelia squamaria (S.G. Gmelin) Decaisne
	Peyssonnelia stoechas Boudouresque et Denizot
?Gloiosiphoniaceae[2]	*Plagiospora gracilis* Kuckuck
Familial and generic placement uncertain	Undescribed *Gainia*-like species (Sears & Wilce unpubl.)

[1]Provisional placement of Rhodophysemataceae (see text).
[2]Family attribution provisional (see text).

crusts also penetrate to great depths in the tropics, as exemplified by the record of *Peyssonnelia* sp. from 189 m (at less than 0.001% of surface irradiance) on a seamount off Bermuda (Littler *et al.* 1985). At the opposite environmental extreme, *Hildenbrandia rubra* flourishes in intertidal caves and rock pools at a salinity less than 5 % in the Baltic (Waern 1952). Specialized survival mechanisms of crustose red algae include cuticle shedding and antibiotic activity (in *Peyssonnelia* species: Maggs 1983, Maggs & Cheney 1990), that prevent

Table 2. North Atlantic red algae with crustose life history phases, indicating the name under which the crustose phase was previously known

Order Family	Species	Crustose phase
Ahnfeltiales		
Ahnfeltiaceae	*Ahnfeltia plicata* (Hudson) Fries	*Porphyrodiscus simulans* Batters
Gigartinales		
Calosiphoniaceae	*Schmitzia hiscockiana* Maggs et Guiry	*S. hiscockiana* crust
Furcellariaceae	*Halarachnion ligulatum* (Woodward) Kützing	*Cruoria rosea* (P. et H. Crouan) P. et H. Crouan
Solieriaceae	*Turnerella pennyi* (Harvey) Schmitz	*Cruoria arctica* Schmitz in Rosenvinge
Gymnophloeaceae	*Platoma bairdii* (Farlow) Kuckuck	*P. bairdii* crust
	Schizymenia dubyi (Chauvin ex Duby) J. Agardh	*Haematocelis rubens* J. Agardh
Petrocelidaceae	*Mastocarpus stellatus* (Stackhouse in Withering) Guiry	*Petrocelis cruenta* J. Agardh
Phyllophoraceae	*Phyllophora traillii* Holmes ex Batters	*Erythrodermis allenii* Batters
	Gymnogongrus devoniensis (Greville) Schotter	*G. devoniensis* crust[1]
	Gymnogongrus ?patens (Goodenough et Woodward) J. Agardh[2]	*G. ?patens* crust
	Gymnogongrus sp.[3]	*Gymnogongrus* sp. crust
Sphaerococcaceae	*Sphaerococcus coronopifolius* Stackhouse	*Haematocelis fissurata* P. et H. Crouan[4]
Gloiosiphoniaceae	*Gloiosiphonia capillaris* (Hudson) Carmichael ex Berkeley	*Cruoriopsis hauckii* Batters
Kallymeniaceae	*Meredithia microphylla* (J. Agardh) J. Agardh	*M. microphylla* crust[5]
Bonnemaisoniales		
Naccariaceae	*Atractophora hypnoides* P. et H. Crouan	*Rhododiscus pulcherrimus* P. et H. Crouan
	Unknown	*Rhododiscus* sp.[6]

[1]Crust of *G. devoniensis* observed in carpospore cultures from Biarritz, SW France (Ardré 1978).
[2]Provisional identification of cultures from S Portugal.
[3]"*G. ?devoniensis* " from Nova Scotia, Canada, in Maggs (1988); it may represent an undescribed species (see text).
[4]See Maggs & Guiry (1982).
[5]See Guiry & Maggs (1984).
[6]With irregularly divided tetrasporangia: see text and Maggs (1988).

overgrowth by competitors. Crusts are highly resistant to grazing by virtue of their ability to regenerate from small remnants, in contrast to the destruction of the holdfast of an erect alga. Finally, several species are capable of reproducing apomictically (Maggs 1988) so that conditions unfavourable for sexual reproduction, including extreme rarity, will not lead to extinction of the population.

Crustose red algal life-history phases have some of the same survival advantages as isomorphic crusts, such as large reserves of storage materials, and resistance to mechanical damage and grazing; this also applies to erect algae with large crustose holdfasts. Refractive "gland cells" in the *Cruoria* phases of *Turnerella pennyi* and *Halarachnion ligulatum* (and in their basal holdfasts) are a possible grazing deterrent. Although the contents of these cells have not been examined chemically, halogenated terpenoids have been associated with gland cells in some red algae (Norris & Fenical 1982). A "bet-hedging" life history strategy (Slocum 1980) is of considerable survival value for some algae with heteromorphic life histories. Spores of the more resistant phase, the crust, can give rise either to the expected erect gametophytes or instead recycle the crustose phase. Environmental influences on the proportion of progeny following each developmental pattern permit the species to respond to changing environmental conditions. This phenomenon has been observed in four species belonging to three different orders (Maggs 1988), and is likely to be even more widespread.

It is reasonable to suggest, therefore, that the present-day distribution of crustose red algae provides potentially useful information on where the species, genera and families evolved. Their resistance to extinction has enabled them to survive previous biological crises such as the period of mass extinctions at the Cretaceous-Tertiary boundary (Alvarez *et al.* 1980); they would probably be among the few survivors of a nuclear winter.

The most recent summary of information on the taxonomy and distribution of crustose red algae is that of Denizot (1968). The main emphasis in the present review is on crusts with

Table 3. Pairs of closely related taxa in *Rhodophysema* and allied genera present in the North Atlantic and North Pacific, and North Pacific species with no close North Atlantic relative

North Atlantic	North Pacific
Rhodophysema elegans (P. et H. Crouan ex J. Agardh) P. Dixon et Abbott	*Rhodophysema minus* Hollenberg
Rhodophysema georgii Batters	*Rhodophysema georgii* Batters
Rhodophysemopsis hyperborea (Rosenvinge) Masuda	*Rhodophysemopsis laminariae* Masuda
Audouinella spetsbergensis (Kjellman) Woelkerling	*Audouinella spetsbergensis* (Kjellman) Woelkerling
None	*Rhodophysema nagaii* Masuda
None	*Rhodophysema odonthaliae* Masuda et Ohta
None	*Pseudorhododiscus nipponicus* Masuda

isomorphic life histories, including all species found in the North Atlantic. The distribution of algae with erect phases is generally better known, particularly those in the genera discussed by Lindstrom (1987) (*Turnerella, Furcellaria*) and Guiry & Garbary (1990) (*Gymnogongrus, Phyllophora, Mastocarpus*). This paper draws together information from taxonomic and life history studies to examine the geographic distribution of crustose red algae in relation to the origin of the North Atlantic seaweed flora.

CRUSTS WITH ISOMORPHIC AND *RHODOPHYSEMA*-TYPE LIFE HISTORIES

Rhodophysemataceae

South & Tittley (1986) list three species of *Rhodophysema* for the North Atlantic Ocean: *R. elegans, R. georgii* and *R. feldmannii* Cabioch (1975)[1]. *Rhodophysema feldmannii* is, however, now thought to be more correctly placed near to *Audouinella spetsbergensis* (Irvine & Guiry 1983; Saunders *et al.* 1989). Of the two currently recognized North Atlantic members of the genus, *R. elegans* is the only truly crustose species. Although the anatomy of *R. georgii* is initially similar to that of *R. elegans,* a large-celled "medulla" is formed, facultatively, by fusion and subsequent expansion of cells in neighbouring filaments (Cabioch 1975; Saunders & Bird 1989); mature plants are globular or hemispherical.

Rhodophysema elegans is exceptional among the crustose red algae. In complete contrast to the "typical crusts" discussed above, *R. elegans* exhibits many of the characteristics of an ephemeral or weedy species. It frequently occurs on artificial substrata such as glass and plastics, as well as on live mollusc shells, and sometimes appears on the glass walls of aquaria. Two life-history patterns have been documented in the North Atlantic (Saunders *et al.* 1989). In the sexual life history, haploid gametophytic crusts are usually monoecious; tetrasporangia, with regenerative stalk cells, develop directly from fertilized carpogonia. The second life history involves the formation of apomictic bisporangia and diploid spermatia by diploid crusts. Both meiotic tetraspores and apomictic bispores develop rapidly in culture and reproduce prolifically. Tetraspore cultures from Helgoland, Canada and Ireland formed tetrasporangial sori in 12-15 weeks (Kornmann & Sahling 1983; Saunders *et al.* 1989). Bisporangial crusts from Newfoundland, Nova Scotia and Ireland reproduced within 10 weeks.

---

[1]Species authorities are given in the text only for species not included in the tables.

Rhodophysema elegans is widely distributed in the North Atlantic, from Iceland to France (the type locality), in Greenland, and from Arctic Canada to Massachusetts (South & Tittley 1986). There are no apparent morphological differences between populations in the western and eastern Atlantic, which both exhibit the two life histories. As only monoecious isolates were obtained, no crossing experiments were possible. In *Palmaria palmata* (Linnaeus) O. Kuntze, failure of meiosis in hybrids indicates the development of transatlantic sibling species (van der Meer 1987). Partial incompatibility between Canadian and European isolates of *R. elegans*, therefore, might exist despite the lack of morphological differentiation.

Rhodophysema elegans has also been reported to occur in the North Pacific from Washington State to S California, Japan and the Aleutian Is. A recent comparative study of North Pacific with North Atlantic material (Saunders *et al.* 1989) concluded that two separate North Pacific taxa had been confused under the name *R. elegans*. Crusts from California and British Columbia epiphytic on seagrasses and smaller algae (including the subject of DeCew & West's (1982) life-history study) were apparently identical to *R. odonthaliae* (Masuda & Ohta 1981) from Japan. The second North Pacific crust, from Japan and the Aleutian Is., and from Washington to S California, was considered by Saunders *et al.* (1989) to be closely similar to European *R. elegans*. It differed in certain morphological features, however, and the name *R. minus* Hollenberg et Abbott was available for this taxon.

The genus *Rhodophysema* is confined apparently to the northern hemisphere. *Rhodophysema georgii* occurs in both the North Atlantic (NW Europe and Newfoundland to New York) and North Pacific (British Columbia to Oregon and Japan) (Irvine & Guiry 1983; Hawkes & Scagel 1986). There is some doubt, however, whether North Pacific and North Atlantic populations are identical, as no detailed taxonomic comparison has been made (Saunders & Bird 1989). As in *R. elegans*, a sexual life history may occur parallel to the apomictic life history (Saunders & Bird 1989). The only *Rhodophysema* species reported to have a tropical distribution is *R. africana* John et Lawson (1974), a name which has not been validly published as the description lacked an accompanying illustration (ICBN Article 39.1: Greuter *et al.* 1988). This alga, known only from Gabon and Angola, shows few *Rhodophysema*-like features (Lawson & John 1982) and requires re-examination.

Uncertainty over the systematic position of *Rhodophysema* leads to some difficulties in assessing the distribution of related taxa. Although it has usually been placed in the Palmariaceae, Irvine & Guiry (1983) and Hawkes & Scagel (1986) advocated the erection of a separate family of the Palmariales for *Rhodophysema*, and the Rhodophysemataceae has now been proposed (Saunders & McLachlan 1989). The ordinal position of this family is in doubt, owing to the current debate regarding the demarcation between Palmariales and Acrochaetiales (see Hawkes & Scagel 1986; Garbary & Gabrielson 1987, Pueschel 1989; Saunders & McLachlan 1989). The genera probably most closely related to *Rhodophysema* are

Pseudorhododiscus Masuda and *Rhodophysemopsis* Masuda. These are both crustose, and known only from the northern hemisphere. *Pseudorhododiscus nipponicus* from Japan was placed in a genus separate from *Rhodophysema* because tetrasporangia are immersed in the thallus, rather than being formed in superficial sori (Masuda 1976b). Its life history shows many features of the *Rhodophysema* type, although carpogonia have not as yet been observed. The pit-plug structure of *P. nipponicus* (Pueschel 1989) is closely similar to that of *Rhodophysema* species (Pueschel & Cole 1982). Masuda (1976a) based *Rhodophysemopsis* on a new species, *R. laminariae* from Japan, and included *Cruoriopsis hyperborea* Rosenvinge from Greenland and Ellesmere Is, Arctic Canada (Rosenvinge 1910, 1926). The vegetative anatomy of *R. laminariae* differs markedly from that of *Rhodophysema.* Cell fusions are entirely lacking and the erect filaments are loosely embedded in mucilage. Pit-plug ultrastructure has not been examined, but morphological observations are consistent with a life history of the *Rhodophysema*-type. *Halosacciocolax kjellmanii* Lund is another possibly palmarialean taxon of uncertain affinities, that resembles *Rhodophysema* in some morphological and life-history features although it is hemiparasitic rather than crustose. Pit plugs are ultrastructurally more like those of the host species *Devaleraea ramentacea* (Linnaeus) Guiry (Jónsson & Chesnoy 1988) than species of *Rhodophysema* (Pueschel & Cole 1982). The morphology of the spermatangial filaments is also quite different from those of *Rhodophysema. Audouinella spetsbergensis* is probably closely related to *Rhodophysema* (G.W. Saunders, pers. comm.).

Rhodophysema and its relatives show an exclusively northern hemisphere distribution and are likely to be of boreal origin. With the exception of *Pseudorhododiscus*, which has no known Atlantic representative, pairs of closely related or identical taxa occur in the North Pacific and North Atlantic Oceans (Table 3). This circumboreal distribution pattern has been noted in several red algal groups including the foliose Palmariaceae (Lindstrom 1987). Lindstrom (1987) and Lindstrom & South (1989) have suggested that the greater numbers of endemic genera in the North Pacific imply that this area is more likely to have been the source of species or genera with amphioceanic distributions, as in some invertebrate groups (Lüning 1985). The greater number and morphological diversity of Pacific *Rhodophysema*-like species is likewise consistent with a North Pacific origin.

Hildenbrandiaceae

In the North Atlantic there are two clearly-defined species of *Hildenbrandia*, viz, *H. rubra* and *H. crouanii*, separable on the basis of their tetrasporangial morphology. Two other species have also been reported from the North Atlantic. The description from Portugal of var.

lusitanica of the Californian species *H. occidentalis* Setchell in Gardner (Ardré 1959) is probably based on material of *H. crouanii*. Ardré (1959) did not compare the Portuguese material with *H. crouanii*, mistakenly believing *H. crouanii* to have oblique, rather than transverse, tetrasporangial cleavages. A complete taxonomic revision of this genus based on the selection of valid taxonomic criteria is a prerequisite for establishing the distribution of *Hildenbrandia* species.

The serious systematic problems in *Hildenbrandia* are due, in part, to our lack of understanding of the life history. Apart from "gemmae" in the freshwater species *H. rivularis* (Liebman) J. Agardh, the only known reproductive structures are tetrasporangia. Attempts to determine karyologically whether these are meiotic were inconclusive (Maggs and G. Saunders, unpubl. data). Pueschel (1988a) suggested that tissue sloughing in *H. rubra* might represent the shedding of a sessile tetrasporophyte formed on the perennial gametophyte.

The reported distribution of *H. rubra* (Denizot 1968, as *H. prototypus* Nardo) includes the Atlantic, Pacific and Indian Oceans and the Mediterranean but this must be treated with caution. *Hildenbrandia crouanii* is known only from the North Atlantic. It occurs from Norway to the British Isles (South & Tittley 1986) and in Greenland (Denizot 1968), but may also have been recorded from southern Europe as *H. canariensis*. In the western Atlantic it has recently been found in New Brunswick (South *et al.* 1988) and Nova Scotia, Canada, and is probably more widespread.

Species of *Hildenbrandia* occur in temperate regions of both hemispheres (Denizot 1968). A new order, Hildenbrandiales Pueschel et Cole (1982), was proposed on the basis of the the pit-plug structure in *Hildenbrandia*. It is supported by the unique mode of erosive tetrasporangial sorus development (Pueschel 1982) and direct formation of secondary pit connections (Pueschel 1988b), otherwise found only in the Corallinales. Tentative placement of *Apophlaea*, consisting of two species endemic to New Zealand, in the Hildenbrandiaceae (Denizot 1968) has now been confirmed by Pueschel (1989). *Apophlaea* species have thalli with an encrusting base and branching erect axes that reproduce only by tetrasporangia. Their distribution, in conjunction with their anatomy, suggests that Hildenbrandiales is one of the oldest red algal orders.

Cruoriaceae

Two *Cruoria* species, *C. cruoriaeformis* and *C. pellita*, are currently included in the Cruoriaceae (Maggs & Guiry 1989); the remaining crusts previously included in this genus represent life-history phases of algae from other families. Evidence from field collections indicates that both species undergo a monoecious isomorphic life history, but neither has been grown in culture. *Cruoria cruoriaeformis* forms thalli 1-2 cm in diameter on subtidal, loose-lying, live coralline algae (maerl). A much less usual habitat is crustose coralline algae on pebbles. *Cruoria pellita* is a much larger crust, often reaching 30 cm across on sublittoral bedrock in the British Isles, and is also common on pebbles. Both species are perennials and reproduce throughout the year by tetraspores and carpospores; sporelings and young plants are frequently observed near to mature crusts.

Cruoria cruoriaeformis is known only from the west and south-west coasts of the British Isles and from NW France, except for one record from Corsica which requires confirmation (Maggs & Guiry 1989). Its distribution within this range is very patchy and it is limited northwards in the British Isles by factors other than the occurrence of maerl. *Cruoria cruoriaeformis* is found in N Wales and the Isle of Man where the winter sea surface temperatures are about 7°C. The maerl beds of W Scotland are located in areas where the winter temperature is below 7°C (temperature data in Craig 1959; maerl bed positions in Maggs & Guiry 1989), indicating that winter temperature probably defines the northern boundary for this species. *C. cruoriaeformis* is one of the few crustose algae that seems vulnerable to extinction at present, as its habitat is so restricted and as maerl is a commercially-harvested resource. *Cruoria pellita* has been reported from Norway to Tangier, from the Baltic (Dixon & Irvine 1977), and in Iceland and west Greenland (South & Tittley 1986); clearly it has much wider environmental tolerances than *C. cruoriaeformis*. *Cruoria* species seem to be absent from the western North Atlantic. This is another example of a common distribution pattern thought to result from extinctions during Pleistocene migrations (Lüning 1985). Southwards displacement during glacial periods, and subsequent northerly recolonization of temperate species, was prevented by unfavourable substrata along the coasts of the Carolinas.

Cruoria, the only genus of the Cruoriaceae, is endemic to the northeastern Atlantic Ocean and it is therefore likely that the Cruoriaceae evolved in the North Atlantic rather than in the Arctic. There is a considerable degree of differentiation between the two species in terms of vegetative structure and reproductive development, *C. pellita* being more complex in several features. This suggests a long-evolutionary separation for these species, and may indicate that *Cruoria* originated in a temperate environment, since the morphologically simpler species, *C. cruoriaeformis*, is intolerant of low temperatures. Although the geographic ranges of the two species partly overlap, they do not share the same habitat, and hence could have evolved sympatrically.

The family perhaps most closely related to the Cruoriaceae is the Haemeschariaceae (M.H. Hommersand, pers. comm.), which may also be of North Atlantic origin. *Haemescharia* was included in the Cruoriaceae by Denizot (1968), and Vinogradova & Jacovleva (1989) consider this to be an appropriate position. Wilce & Maggs (1989) separated the Haemeschariaceae from the Cruoriaceae on the basis of the pre- and post-fertilization development and the irregularly cruciate intercalary tetrasporangia of the former, in contrast to the laterally-borne zonate tetrasporangia in the Cruoriaceae.

Haemeschariaceae

The genus *Haemescharia* has recently been reinstated for *H. hennedyi* and *H. polygyna*, both previously known as *Petrocelis* (Wilce & Maggs 1989; Vinogradova & Jacovleva 1989). These two species are very similar, in both vegetative and reproductive morphology, differing principally in the larger dimensions of most structures in *H. polygyna*, but also in minor anatomical details.

H. hennedyi is distributed in the north-eastern Atlantic from Iceland and S Norway to the south coast of England (Fletcher & Irvine 1982), and also occurs in the White Sea (Vinogradova & Jacovleva 1989). In the western Atlantic it is known only from the Bay of Fundy, Nova Scotia (Wilce & Maggs 1989). *Haemescharia hennedyi* grows typically on subtidal pebbles, but also emerges into the lower intertidal on bedrock and *Laminaria* stipes (Fletcher & Irvine 1982). In the British Isles, thalli are usually present in small numbers, while in the Bay of Fundy they are abundant just below extreme low tide level. A monoecious, isomorphic life history is indicated by field collections and the development of tetrasporophytes in carpospore cultures (Fletcher & Irvine 1982). *Haemescharia hennedyi* is winter-fertile and evidence of a short-day response in gametangial production was found in cultures from England (Maggs unpubl.). Crusts from Canada were morphologically indistinguishable from European material, but monoecious reproduction precluded crossing experiments.

Haemescharia polygyna is known from Arctic areas of Canada, Greenland and Siberia (Wilce & Maggs 1989; Vinogradova & Jacovleva 1989), and the species may be circumpolar in distribution. It is entirely subtidal, growing most abundantly at depths of 3-10 m and penetrating to at least 46 m on cobbles, boulders and shells. Based on field observations, *H. polygyna* appears to have a monoecious isomorphic life history; reproduction takes place in late summer to autumn.

The two *Haemescharia* species have non-overlapping distributions. *Haemescharia polygyna* is one of the very few arctic endemics (Wilce 1990), whereas *H. hennedyi* is restricted to cold-temperate areas of the North Atlantic. The Bay of Fundy population of *H. hennedyi* is probably a relic of a more extensive distribution of this species in the North Atlantic, prior to Pleistocene cooling of the ocean. The larger size of thalli and anatomical structures in *H. polygyna,* in comparison to those of *H. hennedyi,* may be an example of increased body size in relation to increasing latitude, as documented in some invertebrate groups (Lüning 1985). The Arctic species might be polyploid, in common with northern forms of many higher plants (Stebbins 1971). The family Haemeschariaceae is an Arctic-North Atlantic endemic that may be related to the Cruoriaceae. The considerable differences between these families, however, in contrast to the striking similarity of the two *Haemescharia* species, suggest a long evolutionary divergence. The lack of apparent differentiation between Canadian and European *H. hennedyi* populations, despite their reproductive isolation (perhaps since the Greenland-Scotland ridge sank about 25 million years ago, Lüning 1985) may indicate slow evolution of morphological characters. The absence of either family from the North Pacific and the distributions of all four *Haemescharia* and *Cruoria* species within the North Atlantic suggest a North Atlantic origin, with the subsequent speciation of *H. polygyna* in the Arctic.

Peyssonneliaceae

Peyssonnelia is the largest crustose genus of the Rhodophyta (Denizot 1968), with at least 54 described species (Maggs & Irvine 1983), and should be of considerable biogeographic interest. There are serious taxonomic problems in this genus, however, and published records must be treated with great caution. The present analysis is confined to the areas in which this genus has been best studied, from the Canadian Arctic to North Carolina, Greenland, and from N Norway to Portugal. A total of 14 *Peyssonnelia* species occurs within this region. There is a clear pattern of decreasing diversity northwards, on both sides of the North Atlantic (Fig. 1). Five species are found in North Carolina, but only one from Cape Cod northwards. On European coasts, high diversity exists over a longer stretch of coastline, six species occurring on the Atlantic coasts of the Iberian Peninsula, and five species in SW Britain and Ireland. Three of these penetrate to W Scotland and the Shetland Is, and only a single species is present in S Norway.

The five *Peyssonnelia* species recorded from North Carolina (Schneider & Reading 1987) have a tropical or Mediterranean distribution. *Peyssonnelia conchicola* was described from the Red Sea, and has been reported from various tropical locations including the Caribbean and Indo-Pacific (Schneider & Reading 1987). *Peyssonnelia inamoena*, previously known in North Carolina as *P. rubra* var. *orientalis* Weber-van Bosse, is widely distributed in the tropics, and also occurs in the Mediterranean. *Peyssonnelia simulans*, described from the Virgin Is, was considered by Denizot (1968) to be doubtfully distinct from *P. conchicola*, but Schneider & Reading (1987) reported that the two species were distinguishable in North Carolina. *P. stoechas*, which was described from the Mediterranean on the basis of a single sterile specimen (Boudouresque & Denizot 1975), has now been collected in North Carolina, also in a non-reproductive state (Schneider & Reading 1987). Schneider & Reading (1987, p 176) described a new species, *P. atlantica*, from Cape Lookout jetty, on living oysters at 4-6 m depth.

Peyssonnelia atlantica was thought to differ from *P. boudouresquei* Yoneshigue (1984, p 136) from Brazil, which also grows on living molluscs and barnacles, because it has parallel, rather than polyflabellate, basal layer filaments. This appears to be based on a misreading of Yoneshigue's description of parallel rows of cells in the basal layer. Comparison of material collected on live mollusc shells at Radio Island jetty, North Carolina, with the protologues of both *Peyssonnelia* species indicates that *P. atlantica* and *P. boudouresquei* are synonymous.

No species of *Peyssonnelia* are listed from North Carolina to Massachusetts by South & Tittley (1986). On the south coast of Cape Cod *P. dubyi* is common on shell bottoms at 5-12 m, reproducing from June to November (Sears & Wilce 1975). It does not extend into N Massachusetts (R.T. Wilce, pers. comm.) and has not been found in appropriate habitats in Nova Scotia. *Peyssonnelia dubyi* (including *P. codana*) is widely distributed in the Mediterranean (Boudouresque & Denizot 1975) and in the north-eastern Atlantic to S Norway (Fig. 1). Denizot (1968) considered that reports of this species from other areas, with the possible exception of Dawson's (1952) record from New Mexico, were dubious. Dawson's specimens in the US National Herbarium do not represent *P. dubyi*, differing markedly from it

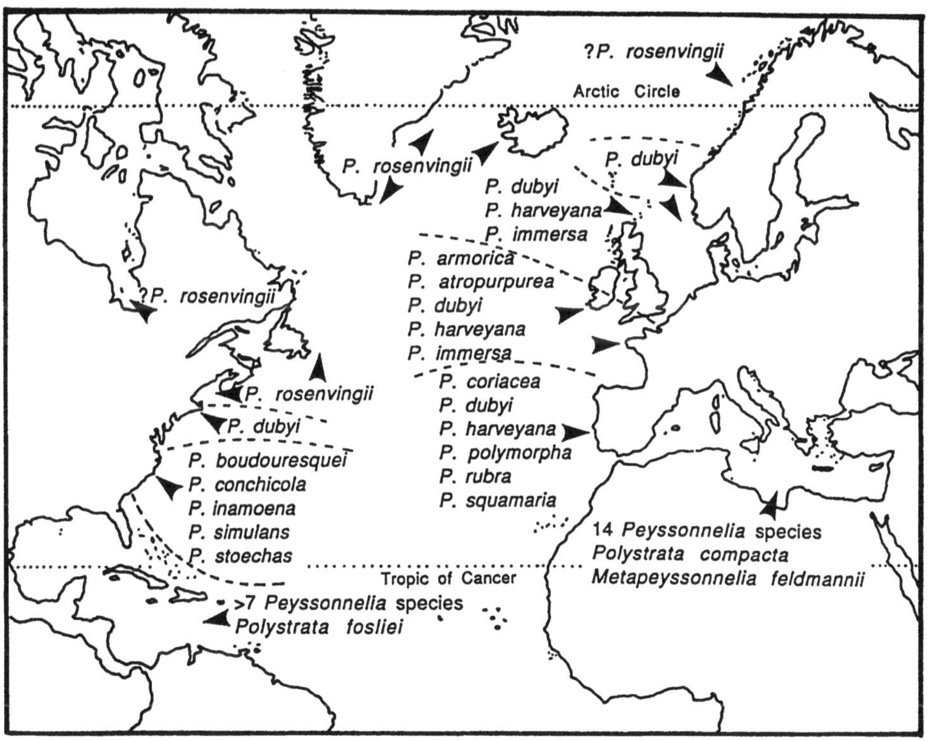

Fig. 1. Peyssonneliaceae of the North Atlantic, showing distribution of species of *Peyssonnelia* and of the genera *Metapeyssonnelia* and *Polystrata*

in such features as the multicellular rhizoids and calcified thallus. However, specimens in the herbarium of the British Museum (Natural History) from Tasmania, Australia, are closely similar to *P. dubyi* , and the distribution of *P. dubyi* may be wider than Denizot suggested.

Peyssonnelia rosenvingii is the only species present from N Massachusetts to Labrador (R.T. Wilce, pers. comm.). It was described from Greenland and also occurs in Iceland. *Haematostagon balanicola* Strømfelt (1886) from N Norway may be conspecific with *P. rosenvingii* (Denizot 1968), and requires further investigation. *Peyssonnelia rosenvingii* has not been found north of Labrador, but *P. johansenii* Howe from James Bay, Hudson Bay, is similar (R.T. Wilce, pers. comm.). Although Denizot (1968) felt that *P. rosenvingii* was only dubiously distinct from *P. harveyana*, Maggs (1983) and Maggs & Bird (unpubl.) have found that these species are quite distinct. The supposed tetrasporangia of *P. rosenvingii* (e.g. Edelstein *et al.* 1970) are actually monosporangia (Maggs & Bird, unpubl.), and neither tetrasporangia nor cystocarps have been observed in numerous freshly collected and herbarium specimens of this species. In the North Atlantic, *P. rosenvingii* is easily recognizable by the presence of conspicuous calcified nodules in the thallus. There are two pairs of species that share this characteristic: *P. caulifera* Okamura from Japan and *P. capensis* Montagne from South Africa. Although considered conspecific by Denizot (1968), these two species are readily distinguished based on anatomy (Yoshida 1975). Both differ from *P. rosenvingii* in their multicellular rhizoids, vegetative anatomy and growth habit. Two species from the NE Pacific are closer to *P. rosenvingii* in general morphology and, like *P. rosenvingii*, have unicellular rhizoids. *P. meridionalis* Hollenberg et Abbott has been reported from Oregon to central California; *P. pacifica* Kylin is known from the Aleutian Is and Alaska to Baja California, Mexico. These taxa are very difficult to separate, and require further taxonomic study. The relationship between *P. rosenvingii* and *P. meridionalis/pacifica* is difficult to evaluate because of the lack of any reproductive structures other than monosporangia in *P. rosenvingii*.

In the eastern North Atlantic, the *Peyssonnelia* flora of Portugal and W Spain (Fig. 1) has no species in common with North Carolina. All the Iberian species are of exclusively Mediterranean-eastern North Atlantic distribution, with the exception of *P. rubra,* which has been collected in Bermuda (Denizot 1968). *Peyssonnelia squamaria* and *P. coriacea* are very closely related, and are distinguished from all other members of the genus by their complete lack of calcification. *Peyssonnelia dubyi* and *P. harveyana* extend northwards to Brittany and SW British Isles. Of the three additional species that occur in this region, *P. armorica* (= *Cruoriella armorica* P. et H. Crouan) and *P. immersa* are known from the Mediterranean. Thus, their apparent absence from Portugal may be the result of under-sampling, particularly in the sublittoral zone. *Peyssonnelia atropurpurea* is endemic to this area, although it may be related to *P. magna* Ercegovic from the Mediterranean. There is a rapid decrease in the number of species in the genus northwards in the British Isles and European mainland, as first *P. atropurpurea* and

P. armorica, then *P. immersa* and *P. harveyana*, reach their northern limits. *Peyssonnelia dubyi* is the most widespread species in Europe, being found northwards to Trondhjemsfjord; it is also the most tolerant of such unfavourable environmental conditions as low salinity, and can survive in the western Baltic.

The Peyssonneliaceae is one of the few red algal families with a fossil record throughout the Cenozoic and late Cretaceous (Wray 1977, as Squamariaceae). The Peyssonneliaceae may have evolved from "ancestral Squamariaceans" that formed calcareous tropical reefs in the Permian Palaeotethys (Flügel 1977). These, in turn, may have been descendants of the late Palaeozoic "phylloid algae" (Wray 1977), but diagenetic changes and the lack of remains in the Triassic and early Jurassic, obscure lineages. The present-day distribution of the Peyssonneliaceae reflects their Tethyan origin, the greatest diversity being seen in the Indo-Pacific, Malaysia, Caribbean Sea and adjacent eastern Pacific Ocean (Hommersand 1986).

Traces of this pattern are discernable in the North Atlantic. The North Carolina *Peyssonnelia* species are of clear tropical affinities, and this is emphasized by the conspecificity of the only reported endemic, *P. atlantica*, with a Brazilian species. Unfortunately, the genus is incompletely known in the Caribbean, and Schneider & Reading (1987) emphasize that a study of *Peyssonnelia* in the Caribbean is long overdue. Numerous species are expected to occur in this area as seven were recorded in a single study (Weber-van Bosse 1916).

The large numbers of *Peyssonnelia* species on Atlantic European coasts adjacent to the Mediterranean, are clearly linked to the high diversity of Peyssonneliaceae in the Mediterranean, where a total of 14 species are present, in addition to two other genera, *Polystrata* Heydrich and *Metapeyssonnelia* Boudouresque, Coppejans et Marcot (1976). The type species of *Polystrata*, *P. dura* Heydrich, is from New Guinea and Indonesia; *P. fosliei* (Weber-van Bosse) Denizot occurs in Indonesia and the Caribbean, and *P. compacta* is known only from the Mediterranean (Denizot 1968; Searles & Schneider 1987). An unidentified *Polystrata* species has recently been reported from Bermuda (Searles & Schneider 1987). Only tetrasporangia are known in *Polystrata* and it was placed in the Peyssonneliaceae (Denizot 1968) on the basis of apparent homology of the downgrowing filaments with those of *Peyssonnelia squamaria*. Reproduction is entirely unknown in *Metapeyssonnelia*, based on *M. feldmannii* from the Mediterranean. Its vegetative anatomy is very unusual in that the margins of older plants are reported to undergo a "metamorphosis" from a *Peyssonnelia*-like structure to a *Polystrata*-like structure (Boudouresque *et al.* 1976). *Metapeyssonnelia* is allied to the Peyssonneliaceae by aragonite calcification and the presence of rhizoids (Boudouresque *et al.* 1976). The fourth peyssonneliacean genus, *Ramicrusta* Zhang et Zhou (1981), occurs only in the Indo-Pacific. It differs markedly from the other genera in forming cell fusions, and Zhang & Zhou (1981) also report secondary pit connections in material from China. Apart from the secondary pit connections, *Ramicrusta nanhaiensis* Zhang et Zhou appears identical to *Peyssonnelia calcea*

Heydrich (1897, p 10). *Peyssonnelia calcea* undoubtedly belongs to *Ramicrusta*, but further investigation is necessary to show whether it is conspecific with *Ramicrusta nanhaiensis*.

The northwards reduction in numbers of *Peyssonnelia* species in the North Atlantic reflects the tropical origins of the genus. *Peyssonnelia dubyi* is the only species tolerant of cold-temperate conditions, and is also the only non-tropical amphi-Atlantic species. Its presence in Massachusetts may be a relic of a more widespread distribution in the North Atlantic prior to glacial cooling. The morphology and subarctic distribution of *P. rosenvingii* suggest an origin unlike that of the other North Atlantic species. The species most similar to it occur in the NE Pacific, and it is possible that *P. rosenvingii* entered the North Atlantic via the Bering Strait during a relatively warm period. As *P. rosenvingii* reproduces only by monosporangia, sexual reproduction may have been lost during this migration through and into an unfavourable environment.

The numerous *Peyssonnelia* species distributed in most areas of the world show great morphological diversity, presumably a reflection of the group's antiquity. Theoretically, correlation of morphological characters with present-day distribution should provide insights into the evolutionary diversification of the genus. However, meaningful subdivision of *Peyssonnelia* into subgenera or evolutionary lines is very difficult. Subdivisions based on vegetative characters cannot be aligned with those based on reproductive characters. For example, the mucilaginous texture of a group of species was used until recently to characterize the genus *Cruoriella* (Yoneshigue 1984); other aspects of their vegetative and reproductive morphology indicate, however, that these species were individually more similar to species of *Peyssonnelia* than to each other (Maggs & L.M. Irvine, unpubl.). Lines of evolution in the Peyssonneliaceae seem to have been obscured by character reversals. This unsatisfactory taxonomic situation may be improved by further systematic study, or perhaps it may be amenable to the approach taken by Hommersand (1990) of examining the distribution of specific morphological trends.

Plagiospora gracilis

Plagiospora is a monotypic genus of uncertain affinities based on *P. gracilis* from Helgoland. Kylin (1956) included *Plagiospora* in the Cruoriaceae, and it was later placed, provisionally, in the Gloiosiphoniaceae by Irvine (1983), on the basis of its resemblance to crustose tetrasporophytes of *Gloiosiphonia* species.

Plagiospora gracilis grows on subtidal pebbles at depths of 2-22 m, forming crusts usually less than 1.5 cm, but occasionally up to 4.5 cm across. It is widely distributed in the North Atlantic, from Sweden to S England, and from Newfoundland to Massachusetts (Maggs unpubl.). Material from the eastern and western North Atlantic is morphologically indistinguishable. Until recently, tetrasporangia were the only reproductive structures known for *P. gracilis*. In the British Isles these are formed from late November to April, with a few sporangia present until June. In eastern Canada, reproduction is initiated earlier, in late August, and peaks in winter. Gametangia have been observed very rarely and only at one locality in Ireland, in very inconspicuous, superficial sori on the dioecious gametophytes. There was no evidence of post-fertilization development, although mature gametangia were collected over a three-month period. In cultures from Ireland, tetraspores gave rise directly to tetrasporophytes for three successive generations, indicating that tetrasporangia were mitotic.

Plagiospora gracilis is limited to cool-temperate regions of the North Atlantic. In the British Isles, tetrasporangial production was initiated when the water temperature dropped below 10°C in early winter, at a daylength of about 8.5 h; a short-day reproductive response at 10°C was demonstrated in culture (Maggs, unpubl.). In Massachusetts, *P. gracilis* grows at sites with a summer bottom temperature of 17°C (Sears & Wilce 1975). It is, therefore, likely that the most southerly records of *P. gracilis*, from S England, do not represent its actual southern limit. In NW France, winter temperatures below 10°C (Lüning 1985) would permit formation of tetrasporangia; likewise, the presence of *P. gracilis* in S Newfoundland suggests that it could occur further north in Scandinavia than presently recorded. Reports of tetrasporangia occurring in summer in eastern North America (e.g. Sears & Wilce 1975) are in contrast to the reproductive periodicity of this species in the British Isles. Although summer temperatures of about 10°C occur in the deep sublittoral of Cape Cod, the daylength would be expected to preclude reproduction. It is possible, therefore, that British Isles and North American populations are of different photoperiodic ecotypes.

The discovery of gametangia in *P. gracilis* has clarified its taxonomic position to some extent: the superficial sori preclude classification of *Plagiospora* in the Cruoriaceae as advocated by Kylin (1956). An alternative family placement has not yet been found, and Maggs (unpubl.) considers that a new gigartinalean family must be proposed for *Plagiospora*. The pit plugs lack cap layers, which is consistent with assignment to the Gigartinales (Pueschel 1989). The life history of *Plagiospora* seems to be in the final stages of becoming entirely apomictic, with only residual production of non-functional gametangia. No closely related family can be identified at present, with the possible exception of the Gloiosiphoniaceae, which appears to be boreal in origin (Hommersand 1986). The disjunct distribution of *P. gracilis* probably represents relics of a once-widepread distribution in the North Atlantic, prior to the widening of the North Atlantic and the sinking of the Greenland-Scotland ridge (Lüning 1985).

An undescribed North American endemic crust

An undescribed *Gainia*-like crust (J.R. Sears & R.T. Wilce, unpubl.), found from S Newfoundland to New Jersey, is the only crustose species known to be endemic to the temperate coasts of North America. This is an unusual distribution pattern, exhibited by only 20 other North Atlantic red algae (South 1987). Discussion of the phylogenetic relationships and evolutionary history of this species must await its description and assignment to a genus.

ALGAE WITH CRUSTOSE LIFE-HISTORY PHASES

Ahnfeltia plicata

Ahnfeltia plicata is widespread in the cold temperate North Atlantic, where it is the only representative of the genus, and is also circumpolar in both hemispheres (Maggs *et al.* 1989). Until recently, it was thought to occur in the North Pacific from Alaska to California, and across the Aleutian island chain to Sakhalin and Japan. A revision of the genus (Maggs *et al.* 1989) showed that the southern limit of *A. plicata* in the NE Pacific is northern Alaska; it is replaced in southern Alaska by *A. fastigiata* (Postels et Ruprecht) Makienko. The two species differ anatomically in such vegetative and reproductive characters as the position of elongate carposporophytes: solely on old axes of *A. plicata,* whereas, in *A. fastigiata* they are globular and formed only on young tissue. The crustose *Porphyrodiscus* phase of *A. fastigiata* is unknown in the field, but in culture, the *Porphyrodiscus* phases of the two species are quite distinct. *Ahnfeltia fastigiata* has larger tetrasporangia, more regular cell rows and fewer cell fusions.

Maggs *et al.* (1989) suggested that the distribution of *Ahnfeltia* species was consistent with a boreal origin and long evolutionary divergence of the two northern hemisphere species, *A. plicata* and *A. fastigiata*. It is interesting to note that their crustose phases are quite distinct, as are the erect thalli. *Ahnfeltia plicata* may have spread more recently to the southern hemisphere, perhaps during the Tertiary cooling of equatorial waters (van den Hoek 1982), since South American collections are indistinguishable from boreal ones. In South America, *A. elongata* Montagne, which is not clearly differentiated from *A. plicata,* appears to be a relatively recent segregate with a very restricted distribution. *Ahnfeltia* was previously included in the Phyllophoraceae on the basis of its vegetative anatomy; a taxonomic revision (Maggs &

Pueschel 1989) placed it in a new monotypic family and order. *Ahnfeltia* shows many primitive characteristics and may be among the oldest extant florideophyte genera.

Gymnogongrus sp.

North Atlantic species of *Gymnogongrus* are a heterogeneous assemblage, exhibiting several different life-history patterns (Ardré 1978; Guiry & Garbary 1990; Maggs 1990). Three species have potentially heteromorphic life histories with crustose phases (Table 2). One of these is found from Newfoundland to Massachusetts, and in Strangford Lough, N Ireland. It was provisionally identified (Maggs 1988) as *G. ?devoniensis*, but research in progress (S.E. Douglas, C.A. Maggs & C.J. Bird, unpubl.) shows that it differs from *G. devoniensis* at the type locality (Devon, England) and may represent an undescribed species. The erect phase reproduces mitotically by apomictic carpospores that have never given rise to crustose plants in culture. A crustose alga that resembles *Gymnogongrus* tetrasporophytes grows at the same sites, forming apomictic bispores or tetraspores. Using plastid DNA restriction fragment length analysis (Parsons *et al.* 1990), the two forms have been shown, unequivocally, to be morphological phases of the same species in which the life history has become uncoupled. Guiry & Garbary (1990) have suggested that this species is likely to have originated from the Pacific via a vicariant event. In this context, it is interesting to note similarities in morphology and life-history patterns between the North Atlantic alga and *G. leptophyllus* J. Agardh from the NE Pacific (DeCew & West 1981) that may indicate a close phylogenetic relationship.

Atractophora hypnoides

Atractophora hypnoides is endemic to a small area of the NE Atlantic, from Brittany (drift plants recorded from Biarritz) to W Scotland. It grows on unstable and artificial substrata, and has a complex life-history relationship with a small crust that is found on maerl and pebbles. The erect phase is relatively common in the southern part of its range, but is extremely rare in N Ireland and unknown in W Scotland, where the species is present only as the crustose tetrasporophyte. The ability of *A. hypnoides* to reproduce both phases from spores of the crustose phase is clearly important to the survival of this species in Scotland and N Ireland (Maggs 1988). Hommersand (1986) suggested that *Atractophora* and various other algae of simple construction are of Tethyan origin. The heteromorphic life history of *A. hypnoides* may have contributed to its long persistence despite a very narrow range of environmental tolerance in the gametophyte. In N Ireland, a putative crustose segregate of this species occurs on pebbles at a single locality, reproducing mitotically by tetraspores (Maggs 1988). It is quite common within a very limited area but there is no evidence that it has spread, which provides evidence of the limited dispersal ability of some crustose algae.

CONCLUSIONS

Once the serious taxonomic problems have been overcome, crustose algae can provide much information of biogeographic interest. Names in the literature, however, often appear to be random identifications; almost all published records must be checked before they can be interpreted. Analysis of the taxonomically revised crustose red algae of the cool-temperate North Atlantic shows that the number of genera endemic to this area has been underestimated. Lindstrom (1987) listed four endemic genera, to which *Cruoria*, *Plagiospora* and *Atractophora* can be added; *Haemescharia* is an Arctic cold-temperate North Atlantic endemic. Although no higher-level taxa were known as endemics (Lindstrom 1987), the Cruoriaceae is a North Atlantic endemic, and the Haemeschariaceae is endemic to the Arctic-boreal North Atlantic. Recognition of a separate family for *Plagiospora* would increase to three the number of endemic families.

ACKNOWLEDGMENTS

I am very grateful for helpful discussions with M.D. Guiry, M.H. Hommersand, L.M. Irvine, G.W. Saunders and R.T. Wilce. R.T. Wilce also supplied unpublished information. Material of *Peyssonnelia boudouresquei* collected in North Carolina by D. F.Kapraun was kindly loaned to me.

REFERENCES

Alvarez LW, Alvarez W, Asaro F, Michel HV (1980) Extraterrestrial cause for the Cretaceous-Tertiary extinction. Science 208: 1095-1108

Ardré F (1959) Un intéressant *Hildenbrandtia* du Portugal. Rev Algol NS 4: 227-237

Ardré F (1978) Sur les cycles morphologiques du *Gymnogongrus crenulatus* (Turn.) J. Ag. et du *Gymnogongrus devoniensis* (Grev.) Schott. (Gigartinales, Phyllophoracées) en culture. Rev Algol NS 13: 151-176

Boudouresque C-F, Denizot M (1975) Révision du genre *Peyssonnelia* (Rhodophyta) en Méditerranée. Bull Mus Hist Nat Marseille 35: 7-92

Boudouresque C-F, Coppejans E, Marcot J (1976) Un nouveau genre de Peyssonneliaceae, *Metapeyssonnelia* (Rhodophyta). Phycologia 15: 283-288

Cabioch J (1975) Le *Rhodophysema feldmannii* nov. sp. et les *Rhodophysema* (Rhodophycées, Cryptonémiales?) de la région de Roscoff. Botaniste 55: 105-118

Cabioch J, Giraud G (1982) La structure hildenbrandioïde, strategie adaptive chez les Floridées. Phycologia 21: 307-315

Craig RE (1959) Hydrography of Scottish coastal waters. Mar Res 2: 1-30

Dawson EY (1952) Marine red algae of Pacific Mexico. 1. Bangiales to Corallinaceae subf. Corallinoideae. Allan Hancock Pac Exped 17: 1-171

DeCew TC, West JA (1981) Life histories in the Phyllophoraceae (Rhodophyta: Gigartinales) from the Pacific coast of North America. I. *Gymnogongrus linearis* and *G. leptophyllus*. J Phycol 17: 240-250

DeCew TC, West JA (1982) A sexual life history in *Rhodophysema elegans* (Rhodophyceae): a re-interpretation. Phycologia 21: 67-74

Denizot M (1968) Les algues floridées encroûtantes à l'exclusion des Corallinacées. M. Denizot, Paris

Dixon PS, Irvine LM (1977) Seaweeds of the British Isles. Volume 1. Part 1. Introduction, Nemaliales, Gigartinales. British Museum (Natural History), London

Edelstein T, Chen L, McLachlan L (1970) Investigations of the marine algae of Nova Scotia. VIII. The flora of Digby Neck Peninsula, Bay of Fundy. Can J Bot 48: 621-629

Fletcher RL, Irvine LM (1982) Observations on the crustose red alga *Petrocelis hennedyi*. Bot Mar 25: 601-609

Flügel E (1977) Environmental models for upper Paleozoic benthic calcareous algal communities. In: Flügel E (ed) Fossil algae. Recent results and developments. Springer-Verlag, Berlin, p 314-343

Garbary DJ, Gabrielson PW (1987) Acrochaetiales (Rhodophyta): taxonomy and evolution. Crypt Algol 8: 241-252

Greuter W, Burdet HM, Chaloner WG, Demoulin V, Grolle R, Hawksworth DL, Nicolson DH, Silva PC, Stafleu FA, Voss EG, McNeill J (1988) International code of botanical nomenclature. Koeltz Scientific Books, Königstein

Guiry MD, Maggs CA (1984) Reproduction and life history of *Meredithia microphylla* (J. Ag.) J. Ag. (Kallymeniaceae, Rhodophyta) from Ireland. Giorn Bot Ital 118: 105-125

Guiry MD, Garbary DJ (1990) A preliminary phylogenetic analysis of the Phyllophoraceae, Gigartinaceae and Petrocelidaceae (Rhodophyta) in the North Atlantic and North Pacific. In: Garbary D, South, GR (ed) Evolutionary biogeography of the marine algae of the North Atlantic, p 265-290

Hawkes MW, Scagel RF (1986) The marine algae of British Columbia and northern Washington: Division Rhodophyta (red algae), class Rhodophyceae, order Palmariales. Can J Bot 64: 1148-1173

Heydrich F (1897) Corallinacéae, insbesondere Melobesieae. Ber Deutsch Bot Ges 15: 403-420

Hoek, C van den 1982. Phytogeographic distribution groups of benthic marine algae in the North Atlantic Ocean. A review of experimental evidence from life history studies. Helgol Meeresunters 35: 153-214

Hommersand MH (1986) The biogeography of the South African marine red algae: a model. Bot Mar 29: 257-270

Hommersand MH (1990) Biogeography of the marine red algae of the North Atlantic Ocean. In: Garbary D, South GR (ed) Evolutionary biogeography of the marine algae of the North Atlantic, p 349-410

Irvine LM (1983) Seaweeds of the British Isles. Volume 1 Rhodophyta. Part 2A Cryptonemiales (*sensu stricto*) Palmariales, Rhodymeniales. British Museum (Natural History), London

Irvine LM, Guiry MD (1983) Palmariales. In: Irvine LM (ed) Seaweeds of the British Isles. Volume 1 Rhodophyta. Part 2A Cryptonemiales (*sensu stricto*) Palmariales, Rhodymeniales. British Museum (Natural History), London

John DM, Lawson GW (1974) Observations on the marine algal ecology of Gabon. Bot Mar 17: 249-254

Jónsson S, Chesnoy L (1988) *Halosacciocolax kjellmanii*, parasite arctique de *Devaleraea ramentacea* (Palmariales, Rhodophyta): Organisation et rapports hôte-parasite. Bull Soc Bot Fr Lettres Bot 135: 211-227

Kornmann P & Sahling P-H (1983) Meeresalgen von Helgoland: Ergänzung. Helgol Meeresunters 36: 1-65

Kylin H (1956) Die Gattungen der Rhodophyceen. CWK Gleerups, Lund

Lawson GW, John DM (1982) The marine algae and coastal environment of tropical west Africa. Nova Hedwigia 70: 1-455

L'Hardy-Halos M-T (1985) Les Céramiacées (Rhodophycées, Floridées) des côtes de Bretagne. II. Particularités biologiques de l'*Antithamnion sarniense* (Lyle) G. Feldmann et de l'*Antithamnion spirographidis* Schiffner. Rev Cytol Biol Veg Bot 8: 89-116

Littler MM, Littler DS, Blair SM, Norris JN (1985) Deepest known plant life discovered on an uncharted seamount. Science 227: 57-59

Lindstrom SC (1987) Possible sister groups and phylogenetic relationships among selected North Pacific and North Atlantic Rhodophyta. Helgol Meeresunters 41: 245-260

Lindstrom SC, South GR (1987) Evidence of species relationships in the Palmariaceae (Palmariales, Rhodophyta) based on starch gel electrophoresis. Crypt Bot 1: 32-41

Lüning K (1985) Meeresbotanik. Verbreitung, Ökophysiologie und Nutzung der marinen Makroalgen. Georg Thieme Verlag, Stuttgart

Maggs CA (1983) A phenological study of two maerl beds in Galway Bay, Ireland. PhD thesis, National Univ Ireland, Galway

Maggs CA (1988) Intraspecific life history variability in the Florideophycidae (Rhodophyta). Bot Mar 31: 465-490

Maggs CA (1990) Taxonomy of phyllophoroid algae: the implications of life history. Proc Int Seaweed Symp 13 (in press)

Maggs CA, Cheney DP (1990) Competition studies of marine macroalgae in laboratory culture. J Phycol 26: 18-24

Maggs CA, Guiry MD (1982) The life history of *Haematocelis fissurata* Crouan frat. (Rhodophyta: Sphaerococcaceae). Br Phycol J 17: 235 (abstract)

Maggs CA, Guiry MD (1989) A re-evaluation of the crustose red algal genus *Cruoria* and the family Cruoriaceae. Br Phycol J 24: 253-269

Maggs CA, Irvine LM (1983) *Peyssonnelia immersa* sp. nov. (Cryptonemiales, Rhodophyta) from the British Isles and France, with a survey of infrageneric classification. Br Phycol J 18: 219-238

Maggs CA, Pueschel CM (1989) Morphology and development of *Ahnfeltia plicata* (Rhodophyta): proposal of Ahnfeltiales ord. nov. J Phycol 25: 333-351

Maggs CA, McLachlan JL, Saunders, GW (1989) Infrageneric taxonomy of *Ahnfeltia*. J Phycol 25: 351-368

Masuda M (1976a) Taxonomic notes on *Rhodophysemopsis* gen. nov. (Rhodophyta). J Jap Bot 51: 175-187

Masuda M (1976b) On a new red algal genus *Pseudorhododiscus*. Acta Phytotax Geobot 27: 123-132

Masuda M, Ohta M (1981) Taxonomy and life history of *Rhodophysema odonthaliae* sp. nov. (Rhodophyta). Jap J Phycol (Sôrui) 29: 15-21

Meer JP van der (1987) Experimental hybridization of *Palmaria palmata* (Rhodophyta) from the northeast and northwest Atlantic Ocean. Can J Bot 65: 1451-1458

Norris JN, Fenical W (1982) Chemical defense in tropical marine algae. In: Rützler K, Macintyre IG (eds) The Atlantic barrier reef ecosystem at Carrie Bow Cay, Belize, 1: Structure and communities. Smithsonian Institution Press, Washington, p 417-431

Parsons TJ, Maggs CA, Douglas, SE (1990) Plastid DNA restriction analysis links the heteromorphic phases of an apomictic red algal life history. J Phycol (in press)

Pueschel CM (1982) Ultrastructural observations of tetrasporangia and conceptacles in *Hildenbrandia* (Rhodophyta): Hildenbrandiales). Br Phycol J 17: 333-341

Pueschel CM (1988a) Cell sloughing and chloroplast inclusions in *Hildenbrandia rubra* (Rhodophyta, Hildenbrandiales). Br Phycol J 23: 17-23

Pueschel CM (1988b) Secondary pit connections in *Hildenbrandia* (Rhodophyta, Hildenbrandiales). Br Phycol J 23: 25-32
Pueschel CM (1989) An expanded survey of the ultrastructure of red algal pit plugs. J Phycol 25: 625-636
Pueschel CM and Cole KM (1982) Rhodophycean pit plugs: an ultrastructural survey with taxonomic implications. Amer J Bot 69: 7-3-720
Rosenvinge LK (1910) On the marine algae from north-east Greenland. Med Grønland 43:93-133
Rosenvinge LK (1926) Marine algae collected by Dr. H.G. Simmons during the 2nd Norwegian arctic expedition in 1898-1902. Rep 2nd Norw Arctic Exped *Fram* 37: 1-40
Saunders GW, Bird CJ (1989) Consideration of life history, morphology and taxonomy in *Rhodophysema georgii* Batters (Rhodophyta, Palmariales). Br Phycol J 24: 63-71
Saunders GW, McLachlan J.L. (1989) [1990] Taxonomic considerations of the genus *Rhodophysema* and the Rhodophysemataceae fam. nov. (Rhodophyta, Florideophycidae). Proc Nova Scotia Inst Sci 39: 19-26
Saunders GW, Maggs CA, McLachlan J.L. (1989) Life history variation in *Rhodophysema elegans* (Rhodophyta) from the North Atlantic and crustose *Rhodophysema* species from the North Pacific. Can J Bot 67: 2857-2872
Schneider CW, Reading RP (1987) A revision of the genus *Peyssonnelia* (Rhodophyta, Cryptonemiales) from North Carolina, including *P. atlantica* new species. Bull. Mar. Sci. 40: 175-192.
Searles RB, Schneider CW (1987) Observations on the deep-water flora of Bermuda. Hydrobiologia 151/152: 261-266
Sears JR, Wilce RT (1975) Sublittoral, benthic marine algae of southern Cape Cod and adjacent islands: seasonal periodicity, associations, diversity, and floristic composition. Ecol Monogr 45: 337-365
Slocum CH (1980) Differential susceptibility to grazers in two phases of an intertidal alga: advantages of heteromorphic generations. J Exp Mar Biol Ecol 46: 99-110
South GR (1987) Biogeography of the benthic marine algae of the North Atlantic Ocean - an overview. Helgol Meeresunters 41: 273-282
South GR, Tittley I (1986) A checklist and distributional index of the benthic marine algae of the North Atlantic Ocean. Huntsman Marine Laboratory and British Museum (Natural History), St Andrews and London
South GR, Tittley I, Farnham WF, Keats DW (1988) A survey of the benthic marine algae of southwestern New Brunswick, Canada. Rhodora 90: 419-451
Stebbins GL (1971) Chromosomal evolution in higher plants. Addison-Wesley Publ Co, Reading
Strømfelt HFG (1886) Om Algvegetationen vid Islands Kuster. Göteborgs Kgl. Vet. Akad., Uppsala
Vinogradova KJ, Jacovleva TA (1989) On the systematic position of the species *Petrocelis hennedyi* and *P. polygyna* (Rhodophyta, Cruoriaceae). Bot Zh Leningrad 74: 744-753 (in Russian)
Waern M (1952) Rocky-shore algae in the Öregrund Archipelago. Acta Phytogeog. Suecica 30: 1-298
Weber-van Bosse A (1916) Rhizophyllidaceae, Squamariaceae. In: Børgesen F (ed) The marine algae of the Danish West Indies III Rhodophyceae. Dansk Bot. Ark 3: 128-144
Wilce RT (1971) Some remarks on the benthic chrysophytes and the fleshy red and brown crusts. In: Riser NW, Carlson AG (eds) Cold water inshore marine biology - some regional aspects. Northeastern University, Boston, p 17-25
Wilce RT (1990) Role of the Arctic as a Bridge between the Atlantic and Pacific Oceans. In: Garbary D (ed) Evolutionary biogeography of the marine algae of the North Atlantic, p 323-348
Wilce RT, Maggs CA (1989) Reinstatement of the genus *Haemescharia* (Rhodophyta, Haemeschariacae fam. nov.) for *H. polygyna* and *H. hennedyi* comb. nov. (= *Petrocelis hennedyi*). Can J Bot 67: 1465-1479

Wray JL (1977) Late Paleozoic calcareous red algae. In: Flügel E (ed) Fossil algae. Recent results and developments. Springer-Verlag, Berlin, p. 167-176

Yoneshigue Y (1984) Flore marine de la région de Cabo Frio (Brésil) 4. Sur une espèce nouvelle du genre *Peyssonnelia* (Cryptonemiales: Rhodophyta). Vie Milieu 34: 133-137

Yoshida T (1975) On the structural characteristics of *Peyssonnelia caulifera* Okamura (Rhodphyceae, Squamariaceae) from Japan. Bull Jap Soc Phycol 23: 1-7

Zhang D, Zhou J (1981) *Ramicrusta*, a new genus of Peyssonneliaceae (Cryptonemiales, Rhodophyta). Oceanol. Limnol. Sinica 12: 538-544

A PRELIMINARY PHYLOGENETIC ANALYSIS OF THE PHYLLOPHORACEAE, GIGARTINACEAE AND PETROCELIDACEAE (RHODOPHYTA) IN THE NORTH ATLANTIC AND NORTH PACIFIC

Michael D. Guiry* and David J. Garbary†

*Department of Botany
University College Galway
The National University of Ireland
Galway
Republic of Ireland

†Department of Biology
St. Francis Xavier University
Antigonish
Nova Scotia
Canada B2G 1C0

Abstract - A preliminary cladistic analysis, using 71 characters and 64 taxa of carrageenan-containing algae from the Atlantic and Pacific, indicates that the Phyllophoraceae, Gigartinaceae and Petrocelidaceae are a monophyletic group of red algae. The cladogram is consistent with present familial concepts. Despite the loosely filamentous construction of most entities referred to the Gigartinaceae and the Petrocelidaceae, it appears that the Petrocelidaceae and the Phyllophoraceae may be more closely related, particularly regarding the formation of a sterile cell on the carpogonial branch, and the frequent production of tetrasporangia on crustose tetrasporophytes that are similar in construction. At the generic level, many entities are poorly defined, with too much emphasis being placed on the form of the thallus as a generic character. Within the Gigartinaceae, precise studies of carposporophyte development and the mode of formation of tetrasporangia are necessary before a final generic scheme can be proposed. Within the Phyllophoraceae, the genus *Phyllophora* appears to be polyphyletic, and *Phyllophora traillii* is referred to the genus *Erythrodermis* as *Erythrodermis traillii* (Holmes ex Batters) comb. nov. The recent removal of the type species of *Ahnfeltia* from the Phyllophoraceae means that species of *Gymnogongrus*-like Phyllophoraceae previously referred to *Ahnfeltia*, should be removed to another genus. The cladogram is consistent with the notion that species of Gigartinaceae, Petrocelidaceae and Phyllophoraceae presently in the Atlantic arose through vicariant speciation events distinguishing: (1) the north Atlantic and north Pacific (via the Arctic); (2) separating the North Pacific and North Atlantic (via the Tethys); and (3) distinguishing the warm eastern Pacific and warm western Atlantic (on either side of Central America).

INTRODUCTION

In the Gigartinales, four families (Chondriellaceae, Phyllophoraceae, Petrocelidaceae and Gigartinaceae) form a closely related complex. All genera of these families, except perhaps

NATO ASI Series, Vol. G 22
Evolutionary Biogeography of the
Marine Algae of the North Atlantic
Edited by D. J. Garbary and G. R. South

for *Chondriella*, contain greater or lesser amounts of carrageenans, sulphated hydrocolloids that form gels with a range of viscosities and that are of considerable economic importance (McCandless *et al.* 1982, 1983). Compared to most other families of the Gigartinales, their representatives show an almost monotonous regularity in the structure and development of the female reproductive organs: the diploid nucleus is transferred directly from the carpogonium to the nearby supporting/auxiliary cell and gonimoblast filaments develop inwards, outwards, or both. Variations in life history patterns (West & Hommersand 1981), have generated much interest in this complex. In addition, the Phyllophoraceae and Gigartinaceae are widely distributed and, in many cold- and warm-temperate regions, representatives of the Gigartinaceae form extensive, often dominant, intertidal and/or subtidal stands.

Distinctions between the Phyllophoraceae and Gigartinaceae have become increasingly blurred in recent years. Characters such as the loose, filamentous appearance of the medullary cells (Gigartinaceae) versus a compact, 'pseudoparenchymatous' appearance (Phyllophoraceae) and the presence or absence of an involucre ('pericarp') around the developing carposporophyte, have not stood up to close examination (see Mikami 1965; Guiry *et al.* 1984). There has been much controversy over the number of genera that should be recognized in the Gigartinaceae (Kim 1976; Silva 1979; Guiry 1984; Guiry *et al.* 1984); similar difficulties with generic limits are apparent in the Phyllophoraceae (Silva 1979; Masuda 1983).

The present study represents a preliminary attempt to use phylogenetic systematics (cladistics) in an analysis of the familial, generic and specific relationships within the complex, to try to clarify taxonomic and biogeographic relationships. Characters of vegetative and reproductive morphology and biochemical features are used. Cladistic analysis is based on the concept that only shared, derived character states provide evidence of phylogenetic relationship; by grouping taxa on the basis of such shared characters (synapomorphies), a hierarchy of nested sets of character states that represents the relationships of the taxa can be produced (Nelson & Platnick 1981; Wiley 1981). Cladistic analysis is a relatively recent arrival on the scene of red algal systematics: Gabrielson & Garbary (1987) presented a cladistic analysis of the orders of red algae; Lindstrom (1988) and Kraft (1989) analyzed the tribe Dumontieae (Dumontiaceae) and the genera of the Liagoraceae, respectively. Cladistic analysis can be sued to recognize past speciation events, and the relationship of these events with earth history. Garbary (1987) and Lindstrom (1987) have discussed algal biogeography, and the potential application of vicariance biogeography in seaweed studies; in the present paper, we evaluate the distribution and evolution of the Gigartinaceae, Petrocelidaceae and Gigartinaceae from this perspective.

Table 1. Species of Chondriellaceae, Gigartinaceae, Phyllophoraceae and Petrocelidaceae used in this analysis (NW= north west, NE=north east, SW= south west, SE= south east). Many putative species of *Iridaea* and *Rhodoglossum* from the Pacific and Atlantic (see South & Tittley 1986) are not included. Type species are indicated (T)

Family **Genus** and species	Atlantic	Pacific	Code
Chondriellaceae			
Chondriella *pusilla* Levring (T)		SE	CHON
Gigartinaceae			
Chondrus *crispus* Stackhouse (T)	NW NE		CCRI
C. elatus Holmes		NW	CELA
C. giganteus Yendo		NW	CGIG
C. ocellatus Holmes		NW	COCE
C. pinnulatus (Harvey) Okamura		NW	CPIN
C. verrucosus Mikami		NW	CVER
C. yendoi Yamada et Mikami in Mikami		NW	CYEN
Gigartina *acicularis* (Roth) Lamouroux	NE		GIAC
G. canaliculata Harvey		NE	GICA
G. corymbifera (Kützing) J. Agardh		NE	GICO
G. exasperata Harvey et Bailey		NE	GIEX
G. harveyana (Kützing) Setchell et Gardner		NE	GIHA
G. intermedia Suringar		NW	GIIN
G. leptorhynchos J. Agardh		NE	GILE
G. pistillata (S.G. Gmelin) Stackhouse (T)	NE SE		GIPI
G. spinosa (Kützing) Harvey		NE	GISP
G. teedii (Roth) Lamouroux	NE SE SW		GITE
G. tenella Harvey		NW	GITN
G. tepida Hollenberg		NE	GITP
G. johnstonii Dawson		NE	GIJO
G. volans (C. Agardh) J. Agardh		NE	GIVO
Iridaea *cordata* (Turner) Bory (T)		SE	ICOR
I. crispata Bory		SE	ICRI
I. splendens (Setchell et Gardner) Papenfuss		NE	ISPL
Rhodoglossum *lanceolatum* (Harvey) J. Agardh (T)		SW	RHOD
Phyllophoraceae (Z)			
Ahnfeltia *concinna* J. Agardh		NW	ACON
A. furcellata Okamura		NW	AFUR
A. gracilis (Yamada) Yamada et Mikami		NW	AGRA
A. yamadae (Segawa) Mikami		NW	AYAM
Besa *papillaeformis* Setchell (T)		NE	BPAP
B. stipitata Hollenberg et Abbott		NE	BSTI
Ceratocolax *hartzii* Rosenvinge	NE NW		CHAR

Table 1. cont. on following page

Table 1. cont.

Gymnogongrus catenatus Yendo		NW	GYCA
G. chiton (Howe) Silva et DeCew in Silva		NE	GYCH
G. crenulatus (Turner) J. Agardh	NE		GYCR
G. crustiforme Dawson		NE	GYCF
G. devoniensis (Greville) Schotter	NE NW		GYDV
G. divaricatus Holmes		NW	GYDI
G. flabelliformis Harvey		NW	GYFL
G. griffithsiae (Turner) Martius (T)	NE NW SW		GYGR
G. leptophyllus J. Agardh		NE	GYLE
G. linearis (C. Agardh) J. Agardh		NW	GYLI
G. martinensis Setchell et Gardner		NE	GYMA
G. paradoxus Suringar		NW	GYPX
Ozophora clevelandii (Farlow) Abbott (T)		NE	OCLE
O. latifolia Abbott		NE	OLAT
O. norrisii Abbott		NE	ONOR
Petroglossum pacificum Hollenberg (T)		NE	PPAC
P. parvum Hollenberg		NE	PPAR
P. undulatum Schneider et Searles	NW		PUND
Phyllophora crispa (Hudson) P.S. Dixon (T)	NE		PCRI
P. heredia (Clemente) J. Agardh	NE		PHER
P. pseudoceranoides (S.G. Gmelin) Newroth et A.R.A. Taylor	NE NW		PPSE
P. sicula (Kützing) Guiry et L. Irvine	NE		PSIC
P. traillii Holmes ex Batters	NE NW		PTRA
P. truncata (Pallas) A.D. Zinova	NE NW		PTRU
Schottera nicaeensis (Lamouroux ex Duby) Guiry et Hollenberg (T)	NE SE	SE	SNIC
Stenogramme interrupta (C. Agardh) Montagne ex Harvey (T)	NE	NE NW SE	STIN
Petrocelidaceae (Y)			
Mastocarpus stellatus (Stackhouse) Guiry (T)	NE NW		MSTE
M. pacificus (Kjellman) Perestenko		NE NW	MPAC
M. jardinii (J. Agardh) J. West		NE	MJAR
M. papillatus (C. Agardh) Kützing		NE SE?	MPAP
Mastocarpus sp. ined. Masuda		NW	MSPE

MATERIALS AND METHODS

Entities

Species of the Chondriellaceae, Phyllophoraceae, Petrocelidaceae and Gigartinaceae used in the phylogenetic analysis are shown in Table 1 together with their nomenclatural authorities and generalized distributions. Each species is assigned a four-letter code that is used in the data matrix (Table 3) and cladogram (Fig. 1). The type species of all genera have been included except for that of *Ahnfeltia, A. plicata* (Hudson) Fries, which was transferred recently to a new family and order (Maggs & Pueschel 1989), together with some other species of the genus (Maggs *et al.* 1989).

Species for analysis were chosen on the basis of the availability of information, but in most cases a full data set was not available. The family Chondriellaceae, which is thought to be closely related to the Gigartinaceae in terms of structure and reproduction (West & Hommersand 1981), was included to provide out-group analysis, and in the cladogram presented in Fig. 1, *Chondriella pusilla*, the only known species assigned to the family, has been chosen as the putative sister group; this species is known only from the Juan Fernandez Islands (south eastern Pacific), and only two small collections have been made. Most of the entities chosen for analysis are from the north Atlantic or north Pacific, and details of their structure, reproduction, and biochemistry are available in the following publications: Hollenberg (1943, 1945), Kylin (1956), Dawson (1961), Mikami (1965), Hollenberg & Abbott (1968), Abbott (1969), Schotter (1968), Newroth (1971a, 1971b, 1972),West (1972), Guiry & Hollenberg (1975), Polanshek & West (1975, 1977), Kim (1976), Schneider & Searles (1976), Magruder (1977), Dixon & Irvine (1977), West *et al.* (1977, 1978), Masuda *et al.* (1979, 1987), DeCew & West (1981), Masuda (1981, 1983, 1987), Masuda & Kurogi (1981), West & Hommersand (1981), Barwell *et al.* (1982), McCandless *et al.* (1982, 1983), West & Guiry (1982), Blunden *et al.* (1982), Guiry & West (1983), Barwell & Guiry (1984), Guiry (1984), Guiry *et al.* (1984), Sciuto *et al.* (1989). Some species, such as *Ahnfeltia gigartinoides* J. Agardh, have not been included because of confusion as to what these species represent (DeCew & West 1981, p. 248). In addition, some characters were scored on the basis of unpublished information from personal communications or from examination of material.

In an unpublished Ph.D. thesis, Leister (1977) concluded that the type species of the genus *Iridaea, I. cordata* (Turner) Bory, was a south eastern Pacific species (type locality: Isla de las Estados, Argentina) rather than a north eastern Pacific species (see also Silva 1979; Scagel *et al.* 1989, p. 295), and that the species in the north Pacific previously known as *I. cordata* (Abbott 1972; Abbott & Hollenberg 1976) should be known as *I. splendens* (Setchell et Gardner) Papenfuss. Leister (1977) presented a detailed and thorough analysis of the structure and reproduction of two species of *Iridaea* (*I. cordata* and *I. crispata* Bory) from the south eastern Pacific. A number of reproductive features used by Leister, including the mode of formation, the subsequent and relative development of the involucral filaments around the developing carposporophyte, and the mode of development of the tetrasporangial initials, have not been used in the present analysis, largely because reliable information on these features has not been published for most other species of Gigartinaceae. Nevertheless, any future reorganization of the genera of the Gigartinaceae will have to take these features into account.

For the sake of completeness, *Rhodoglossum lanceolatum* (Harvey) J. Agardh, the type species of the genus *Rhodoglossum*, also has been included. The concept of this genus

Table 2. Characters and character states used in phylogenetic analysis, their weightings (w), and consistency indices (CI)

Character Number. Character: states. Weighting (w). Consistency Index.

1. Thallus length: <5 mm; 5-15 mm; 15-150 mm; >150 mm. w=1. CI = 0.18
2. Thallus shape: flattened; compressed; terete; hemispherical. w=1. CI = 0.21
3. Branching: unbranched; dichotomous; otherwise. w=1. CI = 0.17
4. Tufted: yes; no. w=1. CI = 0.10
5. Substratum: not sandy; sandy. w=1. CI = 0.14
6. Habitat: intertidal; subtidal. w=1. CI = 0.25
7. Channeling: absent; present; at base. w=2. CI = 1.0
8. Number of bifurcations: absent; 1-5; >5. w=1. CI = 0.13
9. Second order branching: not pinnate; pinnate. w=2. CI = 0.20
10. Stipes: absent; <25% of thallus ; >25% of thallus. w=2. CI = 0.25
11. Holdfast: disc; rhizome; crust; other. w=2. CI = 0.43
12. Margins: not dentate; dentate. w=2. CI = 0.50
13. Spines: absent; present. w=2. CI = 0.33
14. Midribs: present; absent. w=2. CI = 1.00
15. Thallus wall: thick; thin. w=3. CI = 1.00
16. Cortical cells: 4 or less in a file; >4. w=3. CI = 1.00
17. Transition between cortex and medulla: sharp; gradual. w=3. CI = 0.33
18. Subcortical pit connections: present; absent. w=3. CI = 0.25
19. Medullary cells: 30 μm or less in diameter; >30 μm. w=3. CI = 0.14
20. Medullary cell length: breadth: > 3; 3 or < 3. w=3. CI = 0.25
21. Smaller cells interspersed in medulla: absent; present. w=3. CI =
22. Proliferous branching: absent; surface and margins; margins only. w=1. CI = 0.11
23. Reattachment of fronds from apices: absent; present. w=1. CI = 0.33
24. Carrageenan type in sporophyte: lambda; iota; otherwise. w=3. CI = 1.00
25. Carrageenan type in gametophyte: kappa; iota ; otherwise. w=3. CI = 0.67
26. Hordenine: absent; present. w=3. CI = 0.33
27. *Trans*-4-hydroxystachydrine: absent; present. w=3. CI = 1.00
28. Parasitic: no; yes. w=2. CI = 1.00
29. Epiphytic: never; sometimes. w=2. CI = 1.00
30. Phenology of reproduction: summer; winter; year round. w=2. CI = 0.33
31. Tetrasporangial sorus: sunken; protruding. w=4. CI = 0.24
32. Tetrasporangia: non-accessory; accessory; both. w=4. CI = 0.67
33. Tetrasporangial width: 25 μm or less; >25 μm. w=4. CI = 0.50
34. Tetrasporangial formation: singly; chains <4; chains of 4 or more. w=4. CI = 0.67
35. Tetrasporangial chains: branched; unbranched; not applicable.w=4. CI = 0.50
36. Tetrasporangial ontogeny: from cortical cells; from medullary cells; intermediate. w=4. CI = 0.33
37. Tetrasporangial sori: grouped; formed singly. w=4. CI = 0.12
38. Tetrasporangial sorus distribution: irregular; linear. w=4. CI = 1.00
39. Tetrasporangia in peltate proliferations: yes; no. w=4. CI = 1.00
40. Tetrasporangial sori erosive: no; yes. w=4. CI = 1.00
41. Superficial tetrasporangial sori formed on: both surfaces; one surface; absent.w=4. CI = 0.44
42. Tetrasporangial pore formed: no; yes. w=4. CI = 0.33
43. Tetrasporangia sori formed on: mature plants; juvenile plants. w=4. CI = 1.00

Table 2. cont. on following page

Table 2. cont.

44. Tetrasporangial location: on crusts or tetrasporoblasts; on erect plants; neither. w=4. CI = 0.33
45. Formation of reproductive structures: monoecious and dioecious; monoecious; dioecious. w=2. CI = 0.33
46. Spermatangial sori; in special proliferations; not in special proliferations. w=4. CI = 1.00
47. Spermatangial location: in pits; superficial. w=4. CI = 0.50
48. Spermatangial shape: cylindrical; spherical. w=4. CI = 0.50
49. Spermatangial width: 4 μm or less; > 4 μm. w=4. CI = 1.00
50. Carpogonial branches in special proliferations: yes; no. w=4. CI = 0.25
51. Number of cells in carpogonial branch: 3 ; 4. w=4. CI = 0.50
52. Sterile cell(s) on carpogonial branch: yes; no.w=4. CI = 1.00
53. Gonimoblast development: inwards; inwards and outwards; outwards only. w=4. CI = 0.33
54. Sterile filaments interspersed between carposporangia: yes; no. w=4. CI = 1.00
55. Carposporangia in chains of 2-4: yes; no. w=4. CI = 1.00
56. Carposporophytes on midrib line: yes; no. w=4. CI = 1.00
57. Carposporophytes in proliferations: yes; no. w=4. CI = 1.00
58. Carposporophytes: present; absent. w=4. CI = 0.33
59. Carpostomes: absent; one; >one. w=4. CI = 0.29
60. Tetrasporoblast: absent; present. w=4. CI = 0.33
61. Cystocarps in catenate chains: no; yes. w=4. CI = 0.17
62. Enveloping tissue around carposporophyte: absent; rudimentary; well developed. w=3. CI = 0.50
63. Sterile filaments on supporting cell: no; yes. w=4. CI = 0.17
64. Margins thickened: no; yes. w=2. CI = 1.00
65. Thallus width: <5 mm; 5-50 mm; >50 mm. w=1. CI = 0.15
66. Tetrasporangia formed: terminally in cortex; otherwise. w=4. CI = 1.00
67. Papillae: not applicable; simple; branched. w=2. CI = 1.00
68. Papillae: not applicable; straight; curved. w=2. CI = 1.00
69. Papillae: not applicable; terete; compressed. w=2. CI = 1.00
70. Intercalary divisions of medullary cells during carposporophyte development: no; yes. w=4. CI = 0.33
71. Intercalary divisions during tetrasporocyte development: no; yes. w=4. CI = 0.50

is very confused (see discussions in Baardseth 1941 and Parkinson 1981), and recent studies (K. Edyvane and H.B.S. Womersley pers. comm.) have indicated that the genus is confined to Australia and New Zealand.

It should be noted that many species, particularly in *Iridaea*, *Gigartina* and '*Rhodoglossum*' were not included in this preliminary analysis; it is recognized that inclusion of these species might modify some of our taxonomic and biogeographic conclusions.

Characters

Seventy-one characters were chosen for analysis (Table 2). A majority of these characters is based on features of reproduction, but characters based on aspects of thallus dimensions, vegetative structure, branching patterns, habitat, phenology and biochemistry also were incorporated when available. Characters were not coded according to any preconceived notion of a transformation series for this preliminary analysis. Consistency indices and weightings for each character are included in Table 2. A consistency index is a measure of how well a cladogram fits the data that generated it, and is equal to the total number of character states in the character matrix (Table 3), divided by the actual number of character state transitions in the cladogram times 100. For a particular character, the consistency index is an expression that represents a relative measure of the amounts of parallelisms and reversals for that character in the cladogram.

As can be seen from Table 2, a great diversity of characters has been used. Characters 1 and 65 set arbitrary parameters for blade length and width. Character 2 deals essentially with the shape of the thallus whilst character 3 is concerned with branching pattern. Characters 4, 5 and 6 are largely habitat features. Character 7 deals with channeling, but is treated as being present only when the thallus is strongly channeled. Character 8 is the presence or absence and number of bifurcations, and character 9 is the presence or absence of pinnate branching. Stipe presence or absence and relative size are treated in character 10. The type of attachment structure and its extent is treated as character 11. Marginal dentation and the presence or absence of spines are considered in characters 12 and 13. A distinction is made between papillae, which are proliferations formed specially for the purpose of producing female reproductive structures (as in species of *Mastocarpus*), and spines, which are surface protuberances that are not exclusively formed for the production of reproductive structures.

Character 14 is an autapomorphy of *Phyllophora crispa* (formation of tetrasporangia in peltate proliferations). Character 15 was found to be difficult to score, in that wall thickness is a relative feature. Characters 16 and 17 are related; species with more than four cells in a cortical file appear to have a gradual transition between cortex and medulla. Similarly, species with more than four cells in a cortical file tend to have obvious secondary pit connections in the subcortex (character 18), whereas those with less than four tend not to have such connections (or this information is not reported). Characters 19 and 20 attempt to differentiate between a 'loose medulla with a filamentous appearance' and a 'compact medulla with a "pseudoparenchymatous" appearance'. Difficulties with terminology are apparent here: all multicellular red algae are essentially of a filamentous construction and 'pseudoparenchymatous' is a relative term that cannot be defined with any accuracy. (Compare, for example, the medullary dimensions given by Masuda (1987) for Japanese species of *Gymnogongrus*.) A medullary filament diameter of less than 30 μm and a length: breadth ratio of more than three have been chosen, somewhat arbitrarily, to distinguish 'filamentous' from 'pseudoparenchymatous'. Masuda (1987) has pointed out that smaller cells often occur interspersed between larger medullary cells in some Japanese species of *Gymnogongrus*, and, although this feature has been scored as character 21, details are not available for other species. Character 23

Table 3. Data matrix for the 64 taxa used in the analysis (identified by a four-letter code; see Table 1) and 71 character states (see Table 2)

CODE	5	10	15	20	25	30	35	40	45	50	55	60	65	71
CHON	1210?	00101	?0011	11?00	010??	??0??	00002	0??10	?001?	????1	010??	110?0	01000	0000?0
CCRI	20110	00201	00011	11000	00000	00002	01120	10010	20012	11101	01001	11000	00001	100000
CELA	30110	00101	00011	11000	020??	??00?	01120	10010	2001?	????1	01001	11000	01?01	100010
CGIG	30110	00101	00011	11000	000??	??00?	01120	10010	20012	11101	01001	11000	01102	100000
COCE	20110	00201	00011	11000	00000	??002	01120	10010	20012	11101	01001	11000	01101	100000
CPIN	31110	00111	00011	11011	020??	??00?	011?0	10010	20012	11101	01101	11000	10001	100000
ISPL	30010	?0001	00011	11101	00000	??00?	01120	11010	00012	110?1	01001	110?0	02102	100010
ICOR	30010	?0001	00011	11101	000??	??00?	00121	01010	01012	110?1	01001	11000	01102	100010
ICRI	30010	?0001	00011	11101	020??	??00?	01120	11010	01012	110?1	01001	11000	02102	100010
RHOD	30210	10101	00011	11000	020??	??00?	00121	01010	00011	11??1	01001	11010	02101	100001
CVER	30110	02101	00011	11010	00000	??00?	01120	10010	2001?	????1	01001	11000	01101	100000
CYEN	30200	00101	00011	11000	000??	??00?	01120	10010	00012	11??1	01001	11000	00002	100000
GIAC	22100	00211	10011	11000	00100	00001	0012?	00010	00012	11111	01001	11000	02?00	1000?0
GICA	31200	00111	10011	11000	00100	??00?	0????	?1010	0?012	11???	????1	110?0	02?00	1000?0
GICO	30210	?0101	01111	11000	00000	??00?	00???	01010	0?012	11???	010?1	110?0	02002	1000?0
GIEX	30200	?0101	01111	11000	00000	??00?	00???	?1010	0?01?	?????	????1	110?0	0??02	1000?0
GIHA	30201	00101	01111	11000	020??	??00?	00???	?1010	0?012	?????	????1	110?0	0??01	1000?0
GIIN	2220?	00111	?0011	11000	001?0	??00?	?????	?????	????2	11111	01001	11000	02000	1000?0
GILE	32210	00111	00111	11000	01000	??00?	00???	???1?	??012	?????	????1	110?0	02?01	1000?0
GIPI	32111	00211	00011	11000	010?0	0000?	02120	10011	00011	11111	01001	11000	02?00	100010
GISP	30210	00111	01111	11000	010??	??00?	0????	0??10	0?012	11??1	??0?1	11000	0??01	1000?0
GITE	31110	10111	00111	11000	000??	0?002	00120	00010	01012	11111	01001	11000	02101	100010
GITN	22200	00111	00011	11000	00???	??00?	?????	??10	0?012	11???	??01	11000	02?00	1000?0
GITP	21100	00111	1001?	11000	001??	??00?	?????	???1?	0?01?	1????	????1	110?0	0??00	1000?0
GIJO	31200	00111	10111	11000	001?0	??000	00120	01010	0?012	11111	01001	11000	02101	1000?0
GIVO	30111	10110	00011	11000	020??	??00?	0????	0?01?	0?012	11??1	01001	11000	02?02	1000?0
MSTE	30100	01201	00010	11000	01000	11002	00102	00010	10000	11100	00001	11010	00001	1121?0
MPAC	20100	01101	00010	11000	01000	01002	00102	00010	10000	11100	00001	11010	00001	121110
MJAR	20110	01101	00010	11000	01000	01002	00?02	00010	10000	11100	00001	11010	00011	1211?0
MPAP	20100	01201	00010	11000	01000	01002	00102	00010	10000	11100	00001	11010	00001	1212?0
MSPE	30100	01201	00010	11000	010?0	01002	00102	00010	10000	11100	00001	11020	00001	112110
ACON	2210?	00200	0001?	11?01	000?1	??0??	10021	00010	10002	11001	00001	11020	00?00	100000
AFUR	221??	?02?0	0001?	11001	000??	??0??	?????	?????	???0?	????1	??001	110?0	00?00	100000
AGRA	1220?	00100	2001?	1100?	000??	??0??	?????	?????	???0?	????1	00?01	110?0	00?00	100000
AYAM	2110?	00111	0001?	11011	120??	??0??	?????	?????	???0?	????1	00001	110?0	00?01	100000
BPAP	0201?	00000	2001?	?10?1	?00?1	??00?	?????	?????	?????	?????	?0??1	11000	0??00	1000?0

Table 3. cont. on following page

Table 3. cont.

BSTI	12010	10000	2001?	11011	?00??	??00?	?????	?????	?????	?????	???01	11000	00?00	1000?0	1000?0
GYCA	2110?	?0211	0001?	11010	120??	??0??	?????	?????	???02	11001	00001	11020	10000	1000?0	1000?0
GYCH	30100	00201	0001?	11001	?0001	??00?	10?21	01010	?002?	????1	????1	11101	00?00	1000?0	1000?0
GYCR	20101	00201	20011	10?01	?00??	0?001	10021	00010	00022	11??1	?0001	11101	00?00	1000?0	1000?0
GYCF	1011?	00101	0001?	1??0?	?2001	??00?	?????	?????	???0?	11???	??0?1	110?0	10?00	1000?0	1000?0
GYDV	20100	00201	00011	11001	?00??	??002	?????	?????	???0?	???01	1?001	11020	10?00	1000?0	1000?0
GYDI	21100	00211	20011	10011	020??	??00?	?????	?????	???02	11001	00001	11020	10100	1000?0	1000?0
GYFL	2010?	00201	20011	11011	02001	??002	10021	0?010	10002	11001	00001	11020	10100	1000?0	1000?0
GYGR	22101	00201	20011	11001	?00?1	0?00?	10021	01010	10?22	11??1	0?1?1	11101	?0?00	1000?0	1000?0
GYLE	20101	00201	00011	110??	?2001	??00?	00021	0001?	10002	11001	0?001	11020	00?00	1000?0	1000?0
GYLI	30101	00202	20011	110??	?00??	??00?	00021	00010	10002	11??1	??0?1	11020	00?00	1000?0	1000?0
GYMA	30100	00201	0001?	?1?1?	?10??	??00?	?????	?????	???0?	????1	????1	110?0	0??00	1000?0	1000?0
GYPX	3210?	00201	0001?	11001	020?1	1?0??	10021	000??	???02	11001	00001	11020	10100	1000?0	1000?0
PPAC	20110	20201	00011	00011	020??	??00?	?????	?????	????2	11???	??001	11000	00?00	100000	100000
PPAR	202?0	10101	10011	00011	02???	??00?	10?21	01010	00012	11??0	??001	11010	00?00	100000	100000
PUND	202??	00201	00011	00011	010??	??0??	?????	?????	????1	11?0?	??01	11000	00?01	1000?0	1000?0
ONOR	30110	10201	0001?	000??	?10?2	??00?	?????	?????	????2	01???	?????	100?0	00?01	1000?0	1000?0
OCLE	30110	10202	0001?	00011	?20??	??00?	10021	01010	?0012	010??	?????	100?0	00?01	1000?0	1000?0
OLAT	30110	10201	0001?	000??	?20??	??00?	?????	?????	????2	01???	?????	100?0	00?01	1000?0	1000?0
PCRI	20110	00101	00001	00111	010?2	1?011	10021	00000	0?012	00001	0?0?1	10000	00001	1000?0	1000?0
PHER	20110	10202	00011	00111	?00??	??00?	?0021	010??	00012	0????	??0?1	1?000	00?01	1000?0	1000?0
PPSE	20110	00202	00011	00111	000?1	0?002	10021	01010	00012	0000?	000?1	10000	00001	1000?0	1000?0
PSIC	20101	00102	00011	00111	000??	??001	10021	01010	0001?	?????	????1	1??0	???01	1000?0	1000?0
PTRA	20110	00101	00011	00111	000??	??001	10021	00010	10002	00000	00001	10000	00100	100000	100000
PTRU	30110	10202	00011	00111	00021	??002	10021	00010	00021	10000	001?1	1?101	00001	1000?0	1000?0
SNIC	20110	00101	10011	00111	021?1	??001	10011	00110	00112	11??1	10010	11000	00001	100000	100000
STIN	30110	00201	00011	00111	00001	??000	10011	00010	00012	11??1	000?1	01000	10101	100000	100000
CHAR	03210	10000	30011	00011	010??	??102	00021	01010	00000	10??1	0?101	11101	00000	1000?0	1000?0

deals with the ability of fronds to reattach, an important ecological feature of many species of *Gigartina*.

Carrageenan types are treated as characters 24 and 25, the relevant information being taken largely from McCandless *et al.* (1982, 1983); however, we have treated all the hybrid carrageenans listed in McCandless's papers as being of one type of carrageenan or another, in other words, kappa-iota hybrids are treated as being kappa and iota-kappa hybrids as iota. Additional information on carrageenans was also provided by Gretz (pers. comm.). Dragendorff-positive compounds (organic substances that produce coloured complexes with potassium bismuth iodide solution: Blunden *et al.* 1982) are treated in characters 26 and 27.

Only a single putative parasite (character 28), *Ceratocolax hartzii*, is known in the complex and this is found exclusively on *Phyllophora truncata*. Epiphytism and phenology are treated as characters 29 and 30, respectively, but information was found to be very patchy in both of these categories. Tetrasporangial features are treated in characters 31 through 44 and

66, while the development of male, female and carposporophyte (or carpotetrasporophyte) structures are dealt with as characters 45 through 63, 70 and 71. Two autapomorphies, the occurrence of carposporophytes in a midrib-like line (*Stenogramme interrupta*), and the formation of tetrasporangial sori in a linear series on juvenile plants (North Atlantic *Schottera nicaeensis*), are included amongst these characters.

Thickening of the margins of the vegetative thallus is treated as character 64. Morphological features of Petrocelidacean papillae are included in characters 67 through 69. Character 70 was the most difficult character to score, as it was often not clear whether intercalary divisions of medullary cells are occurring in the development of carposporophytes, and this is an important reproductive character that needs further study. Character 71 is an autapomorphic character found only in *Rhodoglossum lanceolatum* (Guiry unpubl. obs. on plants from Victoria, Australia in the herbarium of the University of Melbourne). The data matrix for the 64 taxa and 71 characters is shown in Table 3.

After an initial series of analyses using unweighted charcaters, it was apparent that a number of characters, particularly those relating to plant habit and ecology, were highly homoplasious. Additionally, because of the greater importance of reproductive and internal anatomical features for generic and familial taxonomy in this algal complex, a system of character weighting was devised as an alternative to discarding characters which are often used for species level taxonomy. In this weighting system, characters relating to ecological charcaters or plant habit were weighted 1; characters associated with external blade morphology were weighted 2; characters dealing with internal anatomical features were weighted 3; and characters dealing with reproduction were weighted 4 (see Table 2).

Analysis

An overall cladistic analysis of the 64 taxa and 71 characters was carried out on a Macintosh SE30 microcomputer using MacClade® version 2.1 (March, 1987, © W.P. Maddison & D.R. Maddison). The program PAUP version 3.0 (Phylogenetic Analysis Using Parsimony) was not used because processing time on a Macintosh SE30 was impossibly long with the number of taxa utilized in our study. A cladogram for all taxa and characters is shown in Fig. 1.

RESULTS AND DISCUSSION

Familial considerations

The four families analyzed in this study form the carpogonium terminally on carpogonial branches borne on supporting cells that are thought to be derived from ordinary vegetative cells of the subcortex. These supporting cells act as generative auxiliary cells after fertilization. Tetrasporangia are cruciately divided in all species. As shown in Fig. 1, the only character that separates the Chondriellaceae from the other three families is the occurrence of tetrasporangia singly in a terminal position in the cortex. The next major separation is that of the Gigartinaceae and Petrocelidaceae/Phyllophoraceae into two well-defined assemblages. The only absolute character (with a consistency index of 1), that could be considered highly conserved in a evolutionary sense is the occurrence of a sterile cell or cells on the second cell of the carpogonial branch (see Masuda 1987). Such a sterile branch is absent in all species of Gigartinaceae for which this information is available (Table 3). It appears to be absent also in the Chondriellaceae (Levring 1941). The occurrence of a sterile cell on the second cell of the carpogonial branch in species of *Mastocarpus* (the only genus currently referred to the Petrocelidaceae), suggests a closer relationship between the Phyllophoraceae and the Petrocelidaceae, than between the Gigartinaceae and Petrocelidaceae. This does not appear to be correlated with any features of the cortical cells, as it occurs in species with long files of cortical cells (Petrocelidaceae and some species of Phyllophoraceae) and those with short files (some Phyllophoraceae). The occurrence of a sterile cell, cells or branches on the *supporting* cell of the carpogonial branch (character 63) — a feature frequently mentioned by authors as if it had some significance — does not appear to have any phylogenetic significance as these sterile cells are very irregular in their occurrence (CI =0.17).

The only species referred to the Phyllophoraceae that have more than three cells in the carpogonial branch are *Schottera nicaeensis* and *Gymnogongrus devoniensis*. These observations were made by Schotter (1968), and it is not entirely clear from his illustrations whether or not one of these cells represent a sterile branch, because the pit connections are not shown. It is likely that all three families typically have three-celled carpogonial branches, with four-celled carpogonial branches being rare, or even absent. Two-celled carpogonial branches are sometimes mentioned by authors (e.g. Dixon & Irvine 1977), but these may represent immature three-celled branches.

Formation of tetrasporangia on crustose plants (*Petrocelis*-like and *Erythrodermis*-like) is a feature that unites the Petrocelidaceae and the Phyllophoraceae, and this feature is absent in the Gigartinaceae *sensu stricto*. As pointed out by DeCew & West (1977), crustose tetrasporophytes in the Phyllophoraceae and Gigartinaceae are rather similar in construction and are suggestive of a link between the families. Maggs (1989), however, has remarked on some

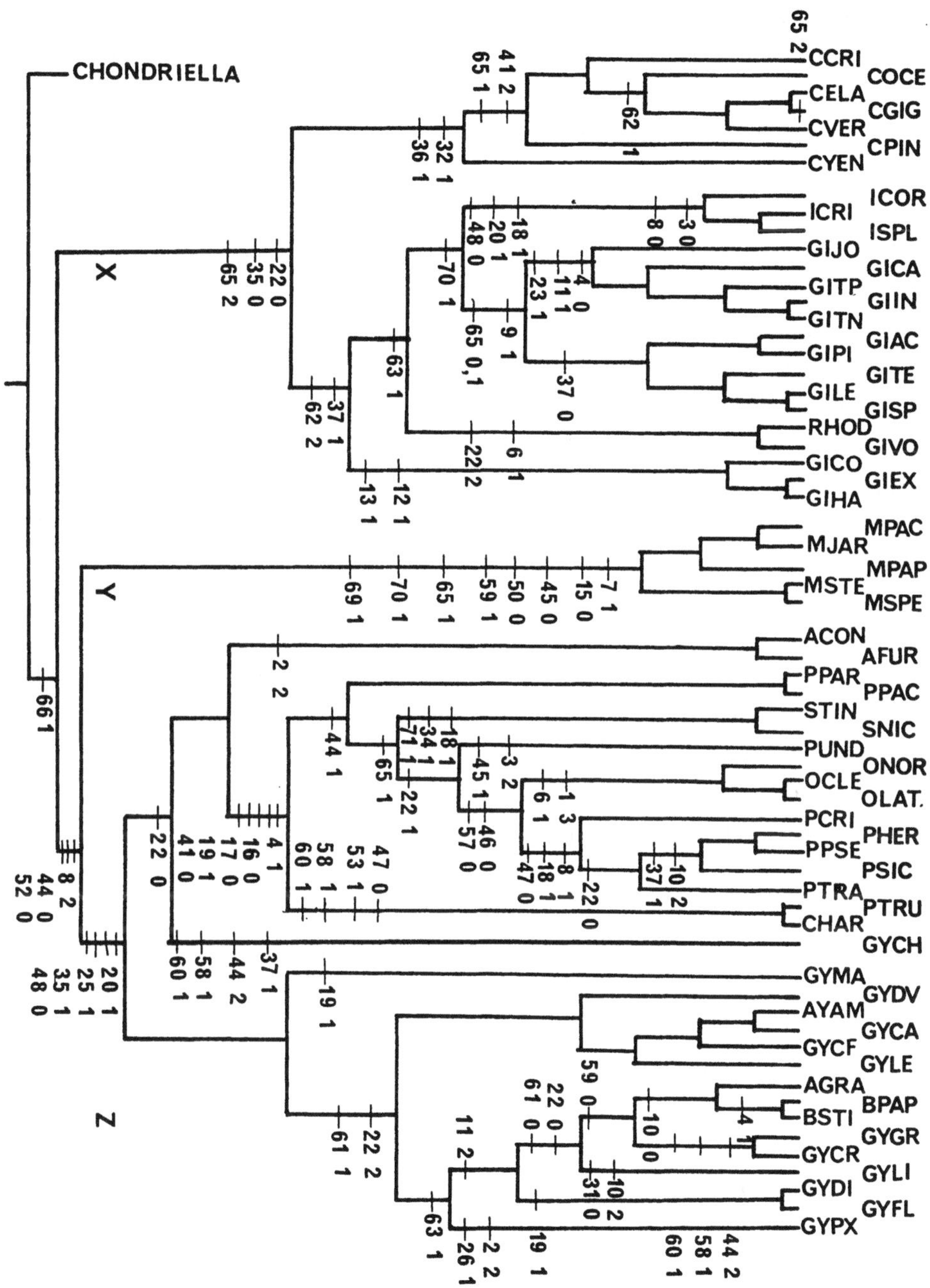

Fig. 1. Cladogram showing hypothesis of relationships among selected North Atlantic and North Pacific species of Gigartinaceae (X), Petrocelidaceae (Y) and Phyllophoraceae (Z). For abbreviations, and details of characters and data see Tables 1-3. Synapomorphies indicated on the tree indicate character number and state. Note: apomorphies are not indicated for all terminal taxa or clades

differences in construction between the tetrasporophytes of various Phyllophoracean algae. The occurrence of the Dragendorff-positive compound hordenine may also unite the Petrocelidaceae and Phyllophoraceae (Barwell *et al.* 1982), but few species have been tested, and some species tested from both families do not contain this compound.

An involucre ('pericarp', 'pericarpium proprium') surrounding the developing carposporophyte is completely absent in the Phyllophoraceae and Petrocelidaceae (character 62). Mikami (1965) has reported intercalary divisions (character 70) in the medullary cells of *Mastocarpus pacificus* (as '*Gigartina pacifica*') and another species of *Mastocarpus* (as '*Gigartina mammillosa*'; see Masuda *et al.* 1987, 1989), associated with the formation of the carposporophyte. Similar intercalary divisions have been reported in *Chondrus nipponicus* Yendo (as '*C. crispus*'), *C. elatus*, and possibly *C. yendoi*. This feature may indicate a relationship between some western Pacific species of *Chondrus* and the genus *Mastocarpus*. *Chondriella pusilla* develops a rudimentary involucral network (and has been scored as such), but further studies are necessary. Most species of the Gigartinaceae (species of *Gigartina*, *Iridaea* and *Rhodoglossum*) form an involucre of varying extent (Kim 1976), but *Chondrus crispus* (the type species from the Atlantic), and some other *Chondrus* species from the north western Pacific do not. Some other species currently referred to the genus *Chondrus* from Japan form a rudimentary involucre in the early stages of carposporophyte development (Mikami 1965, Brodie 1989), but this may not have the same ontogeny as that found in species of *Gigartina*.

Tetrasporophytes of all three families generally contain lambda-carrageenans or their derivatives; the Phyllophoraceae has, however, iota-carrageenans or their derivatives in the gametophytes, whereas the Petrocelidaceae and Gigartinaceae have kappa-carrageenans or their derivatives in the gametophytes. No information is available on carrageenan types in the Chondriellaceae. Tetrasporangial diameter (character 33) and spermatial shape (character 49) are similar in the Petrocelidaceae and the Gigartinaceae, although there are exceptions to the spermatial shape character in the latter family.

The Petrocelidaceae is well characterised by several characters including the presence of: channeling, a thick thallus wall, papillae, one carpostome, and intercalary divisions of medullary cells during carposporophyte development.. Channeling of the thallus has been reported in a number of species of Gigartinaceae, but in all these instances this seems to be a variable character[1]. A very characteristic feature of species of *Mastocarpus* is the widespread occurrence of populations that contain apogamic or monoecious individuals, the carpospores of which may give rise to further erect axes, without the formation of a tetrasporophyte (West *et al.* 1983). In the data matrix, the occurrence of such populations has been arbitrarily scored as 'monoecious and dioecious'. Such individuals and life histories occur in all species of

[1]For example, *Gigartina volans* from the eastern Pacific, which is only channeled at the base of the thallus.

Mastocarpus reported to date. Carpospores from a 'narrow form' of *Gymnogongrus leptophyllus* from California produced what appear to be apogamic individuals in culture (DeCew & West 1981), but there is some uncertainty as to the identity of these plants. Such life histories have not been reported in the Gigartinaceae, although Kim (1976) reported several instances of apomeiosis in the Gigartinaceae.

A unique feature of the Petrocelidaceae is the occurrence of carpogonial branches in specially-formed papillae. This is a somewhat controversial matter, as many terete or slightly compressed species of *Gigartina* are regarded as being 'papillate'. In *Mastocarpus* species, male plants do not form papillae; papillae only are produced on the female plants, and carpogonial branches are confined to these. Papillae may be formed on the concave surfaces of the blades, or from the margins and the apices, but carpogonial branches and carposporophytes do not occur in the blade proper. Supposedly papillate species of *Gigartina*, such as the type species, *G. pistillata*, form female reproductive structures (and spermatangia in the monoecious species) in short, tertiary branches that cease growth; the carposporophyte thus appears to be papillate. These short, fertile branches also occur on males (in dioecious species), and in tetrasporophytes of *G. pistillata*. Spines are surface features that occur on plants of all reproductive states; on female plants spines may bear carposporophytes, but these may also occur in the thallus proper. Carpostomes have been reported frequently for species of *Mastocarpus*. Masuda *et al.* (1989) have reported multiple carpostomes in a species from Japan, whereas other species of the genus are said to have only one.

The Gigartinaceae is characterised by tetrasporangia occurring in branched tetrasporangial files (Fig. 1), whereas in the Phyllophoraceae the files always are unbranched. Reports in the literature of unbranched rows of tetrasporangia occurring in the Gigartinaceae need reinvestigation, but all species referred to the family, that have been examined to date, have branched files of tetrasporangia (Hommersand & Guiry unpubl. obs.). It is generally necessary to examine the developing tetrasporocyte files to be certain of the development pattern. In the type species of *Rhodoglossum*, *R. lanceolatum* from Australia, tetrasporangia develop from the products of intercalary divisions of inner cortical cells; the tetrasporangia ultimately are borne in unbranched chains, a feature unique to this taxon in the Gigartinaceae. The spermatia of the Phyllophoraceae tend also to be > 5 μm in diameter and spherical or ovoid rather than < 5 μm in diameter and cylindrical as in the Petrocelidaceae and Gigartinaceae.

Generic implications

GIGARTINACEAE. The genus *Chondrus*, with six species confined to the north western Pacific and a possible further rarely recorded and poorly known species, *Chondrus platynus* (C. Agardh) J. Agardh, in the colder waters of the north Pacific, is well represented in the north Pacific basin. Reports of *C. crispus* from Japan are of another species (*C. nipponicus* Yendo; Guiry & Masuda in Lüning *et al.* 1987) that has a warm-water distribution, mainly on the west coast of Japan. Morphological and breeding data (Brodie, Guiry & Masuda unpubl. data) show that none of the western Pacific species referred to the genus *Chondrus* are related to Atlantic *C. crispus* at the species level. At the genus level, it is likely that *C. yendoi*, *C. pinnulatus*, *C. verrucosus*, *C. giganteus* and *C. elatus* are not closely related to *C. crispus* (Brodie, Guiry & Masuda, unpubl. data). This complex of species (and genera) is, however, connected in the cladogram (Fig. 1), largely on the basis of the development of tetrasporangia as accessory branches of medullary cells in these species. This type of development, however, also occurs in *Iridaea crispata* from South America (Leister 1977) and, to a greater extent, in *I. splendens* from Pacific North America (Kylin 1928 as *I. cordata*). Thus, even though it appears that there is a complex of species of *Chondrus* in the North Pacific, that could conceivably have given rise to *C. crispus*, we are quite uncertain at present as to the phylogenetic relationships between these species. In Japan, the only species referred to the genus *Chondrus*, with a cold-water distribution pattern, is *C. pinnulatus*. This species is unrelated to *C. crispus* in that it has medullary filaments with an almost pseudoparenchymatous appearance. The possibility remains, however, that there are entities in the colder waters of the north Pacific, such as *C. platynus*, that are sister species of *C. crispus*.

Of the species of *Gigartina* that occur in the north Atlantic basin, *G. pistillata* has no close relatives in the north Pacific. The species reaches its northern limit on the south coast of Ireland and does not occur in the western Atlantic. More or less continuous records of its distribution exist along the coast of Africa to South Africa, and its generic affinities appear to lie with a complex of species in South Africa, Australia and New Zealand (Hommersand pers. comm.; see Hommersand 1986). In the cladogram (Fig. 1), *G. pistillata* and *G. acicularis* are the most closely related of the species of *Gigartina* in the north Atlantic. This relationship is more apparent than real, however, since most of the reproductive details for these species need clarification. *Gigartina teedii* and *G. acicularis* are species with warm-water distributions, reaching their northern limit on European coasts on the west coast of Ireland (Guiry & Cunningham 1984; Guiry & Maggs 1985). There are indications from breeding data (Guiry *et al.* 1987) that *G. teedii* may be generically related to a complex of species passing under the name *G. intermedia* in the western Pacific. Breeding data (Guiry 1984; Guiry *et al.* 1987) also show that isolates of this species from Brazil, Italy, France, England and Ireland are completely interfertile and do not interbreed with the morphologically similar *G. johnstonii* from the Gulf of California, and with various strains of *G. intermedia* from Japan and Korea. The cladogram (Fig. 1) also indicates a relationship between *G. teedii* from the Atlantic, *G. leptorhynchos* from the eastern Pacific, and *G. tenella* from Japan; this relationship is probably tenuous, as the connection has been established on the basis of features of gross morphology, rather than on any reproductive similarity. The relationships shown between *G. tenella, G. intermedia, G. canaliculata* and *G. johnstonii* are poorly resolved, due to a substantial lack of information on the reproductive features of these plants. All of these latter species are warm-water species that are probably endemic to the north Pacific. *Gigartina exasperata, G. harveyana* and *G. corymbifera* are three very closely related, blade-like, spine-forming species occurring in the eastern Pacific that are, at best, difficult to distinguish (see Kim 1976; Scagel *et al.* 1989). These latter three species are poorly known reproductively, but it is likely that they are representative of a generic complex endemic to the north eastern Pacific.

Iridaea cordata, I. crispata and *I. splendens* appear from the cladogram to be closely related with the latter two species being more related to each other than to *I. cordata*. Unfortunately, it has not been possible to incorporate details of the development of carposporophyte and tetrasporangia (Leister 1977) in the cladograms, and only a few taxa within the *Iridaea* complex have been included in the analysis. It is clear, nevertheless, that the North Pacific species *I. splendens* has no close relatives in the North Atlantic.

PETROCELIDACEAE. *Mastocarpus* is well defined as a genus.

PHYLLOPHORACEAE. *Stenogramme interrupta* is known from Japan and Korea, and from Prince William Sound in Alaska south to Baja California; it also is found on the Galapagos Islands (Scagel *et al.* 1989) and in New Zealand. In the Atlantic, it is known from the west coast of Ireland south to Morocco and the Cape Verde Islands. *Schottera nicaeensis* is known from the west coast of Scotland south to Portugal, but has not been reported from the northwestern Atlantic. It has been reported recently from Namibia (Wynne 1986), Natal (Norris & Aken 1985), Australia (Lewis & Kraft 1979), and Chile (Santelices *et al.* 1989). Although there are considerable disjunctions in the distributions of *Stenogramme* and *Schottera*, they do appear to be widespread and warm-water in their origins. These are well-defined genera, largely on the basis of autapomorphies.

Petroglossum pacificum and *P. parvum* are known only from the eastern Pacific, and previously were thought to occur only in California and Mexico; however, Santelices *et al.* (1989) recently reported *P. pacificum* from Chile. *Petroglossum parvum* is in need of reinvestigation, to establish its precise generic relationship with *P. pacificum*, the type species of the genus. *Petroglossum undulatum* is known from a single collection at a depth of 28 m off North Carolina, and Fig. 1 suggests a relationship between this species and *Ozophora*. Plants are monoecious (in contrast to the other two species of *Petroglossum*), and carposporophyte development appears to be accompanied by extensive elongation of the surrounding cortical files (Schneider & Searles 1976 fig. 15); this feature is not typical of the Pacific species, but is reminiscent of the type of development found in *Stenogramme interrupta* (Kylin 1928, 1956). Little is known of the early stages of development of the gonimoblast, and tetrasporangial plants are unknown. It is possible that *Petroglossum undulatum* represents an undescribed genus endemic to the Atlantic.

Ozophora latifolia, *O. clevelandii* and *O. norrisii* are northeastern Pacific endemics which reach their northern limit of distribution in northern Washington. Species of *Ozophora* are very similar in structure, and in some aspects of their reproduction, to species of *Phyllophora*, particularly *P. crispa* (the generitype). Kylin (1956) was not convinced that *Phyllophora*, *Ozophora*, and *Petroglossum* were distinct genera, and considered the latter two genera to be synonyms of the first. *Ozophora* and *Petroglossum* are distinguished from *Phyllophora* by the absence of spermatangia in pits; in *Ozophora* the spermatangia are borne in specially formed, cordate leaflets on the surfaces of the blades (Abbott 1969). However, *Ozophora clevelandii* does have, in common with *Phyllophora crispa*, a characteristic type of proliferous branching, the occurrence of reproductive structures in specialised leaflets or papillae. McCandless *et al.* (1982) reported that the carrageenans of *P. crispa* and *O. norrisii* were not typical of those found in other Phyllophoraceae. The generic status of *Phyllophora*, *Ozophora* and *Petroglossum* needs reinvestigation.

The variation in reproduction and life history patterns in *Phyllophora* species in the North Atlantic suggests that these species should be assigned to different genera. *Phyllophora crispa* is probably monotypic. *Phyllophora pseudoceranoides* and *P. heredia* clearly are not related to *P. crispa* in their general thallus structure and in the formation of tetrasporangia directly on the thallus surface (instead of in peltate leaflets), rather they appear to be related to one another. *Phyllophora sicula*, with its expanded crustose base and formation of tetrasporangia in centrally placed sori on unmodified blades, is probably representative of a further genus, and is quite unrelated to either *P. crispa* or *P. pseudoceranoides*. Male and female gametangia and the carposporophyte are unknown in *P. sicula*; its placement in the genus *Phyllophora* is only a matter of necessity. Maggs (1989) recently showed that *Phyllophora traillii* has a heteromorphic life history, in which the tetrasporangia are formed in crustose plants previously known as *Erythrodermis allenii* Batters (see also Newroth & Markham 1972). The formation of spermatangia in shallow pits on specially formed leaflets, and the general structure of the thallus are the only reasons for placing this species in the genus

Phyllophora, and we propose that this species be placed in the genus *Erythrodermis*[2]. Similarly, the unusual life history and reproduction of *Phyllophora truncata* sets this species apart from other species of the genus, as is clearly shown in the cladogram in Fig. 1; this species also should be referred to a separate genus.

The type species of *Gymnogongrus*, *G. griffithsiae*, is common in the Atlantic, being found on both sides of the North Atlantic and in the western South Atlantic. This species forms a tetrasporoblast similar to that found in *Phyllophora truncata* and *Gymnogongrus crenulatus*, but is terete or slightly compressed, and arises from an extensive crustose base. *Gymnogongrus crenulatus* also arises from an expanded crustose base, but the blade is distinctly flattened, and the medullary cells are more 'pseudoparenchymatous'. Cordeiro-Marino & Poza (1981) have reported that this species forms gonimoblast filaments that develop inwards to form an internal cystocarp. The resulting carposporangia are said to germinate *in situ* to form filaments that grow outwards to form an external pustule (tetrasporoblast) that ultimately forms tetrasporangia. This mode of development is quite different from anything reported in other tetrasporoblast-forming species of the genus (see Schotter 1968, Ardré 1978, Masuda 1987) and needs to be reinvestigated. There is no doubt, however, that tetrasporoblast-forming species of gymnogongroid and phyllophoroid algae need to be generically separated from those that form internal carposporophytes.

Biogeographical implications

FAMILIAL COMPLEX. The Phyllophoraceae, Petrocelidaceae and Gigartinaceae constitute a monophyletic group that underwent major radiation in temperate conditions in the North Pacific. All families have their greatest species diversity in the Northern Hemisphere, and in the Phyllophoraceae and Petrocelidaceae there are few Southern Hemisphere taxa. The cladogram offers no explanation of the biogeographical origin of families, although Hommersand (1986) suggested that the complex evolved in the North Tethyan Ocean. See Hommersand (1990, this volume) for a different perspective on the taxonomy and biogeography of this complex of families.

It is of interest that all pairs of identifiable sister species (except for *Iridaea crispata* and *I. splendens* that are not closely related) have highly overlapping distributions within the biogeographic regions identified in Table 1. This supports the notion of sympatric or quasi-sympatric speciation proposed by Pielou (1978), that differs from the allopatric speciation model usually assumed in most speciation events. On the other hand, it might be difficult to

[2]*Erythrodermis traillii* (Holmes ex Batters) comb. nov. (BASIONYM: *Phyllophora traillii* Holmes ex Batters (1890, p. 334, reprint p. 114).

resolve many recent speciation events, in temperate regions with north/south aligned coasts, as being allopatric, because of numerous migrations, and subsequent floristic intermingling imposed by Pleistocene glaciations. In addition, Japan, which has most of the species from the western Pacific considered in this study, is oceanographically very complex, and the resulting complexity in patterns of vicariant speciation need to be studied in greater detail.

GIGARTINACEAE. The complex of species presently included in the genus *Chondrus* arose in the northwestern Pacific and diversified there. Our cladogram suggests a vicariant event separating *C. crispus* in the Atlantic, and the *C. ocellatus*, *C. giganteus*, and *C. verrucosus* clade in the western Pacific. The genus is apparently absent in the eastern Pacific[3], although *Chondrus platynus*, not included in our analysis, is present in the Bering Sea, and may be the sister species to *C. crispus*. Of the western Pacific species of *Chondrus*, *C. ocellatus*, *C. nipponicus*, *C. elatus* and *C. pinnulatus* are distributed southwards from Japan to China and Korea (Lee & Kang 1986); the other species are apparently endemic to Japan.

Within the *Iridaea/ Rhodoglossum/ Gigartina* complex, *Gigartina exasperata, G. harveyana* and *G. corymbifera* form a clear-cut group. These species are closely related (see Kim 1976), and are different from the remainder of the complex in that they have spines and are blade-like. Their relationship with the remainder of this group remains unresolved. The relationship between *Rhodoglossum lanceolatum* and *Gigartina volans* is undoubtedly artifactual. Our cladogram suggests that *Iridaea* arose in the Southern Hemisphere, and that some members of the complex migrated into the Northern Pacific. It remains to be demonstrated whether or not the North American species of *Iridaea* arose by a single vicariant event, or by a number of vicariant events.

Several vicariant events are implicated in the remaining *Gigartina* species that we studied. One such event apparently distinguishes Pacific species (the GIJO-GITN clade) from a primarily Atlantic clade (GIAC-GISP). The former species group shows further vicariance in distinguishing eastern and western Pacific species (*G. tepida* in the east and *G. intermedia*, *G. tenella* in the west). The clade GIAC to GISP are all warm water species with some entities occurring on either side of central America (e.g. Baja and Venezuela). The variety of apparent vicariant events supports an ancient origin for this complex.

PETROCELIDACEAE. At least three species of *Mastocarpus* (Petrocelidaceae) occur in the North Pacific, whereas only one is known from the North Atlantic. Although *Mastocarpus* is confined to the North Atlantic, the genus is not confined to the North Pacific, as a species

[3]Scagel *et al.* (1989) report that *Chondrus ocellatus* Holmes is distributed from the Aleutian Is. south to Oregon, but note that Kim (1976) re-examined specimens on which some of the records were based and concluded that these specimens were probably referable to *Iridaea cornucopiae* Postels et Ruprecht. It is likely therefore that the genus *Chondrus* does not occur in Pacific north America.

occurs in Chile (West *et al.* 1983). *Mastocarpus papillatus, M. pacificus* and *M. stellatus* are circumboreal in distribution. The cladogram (Fig. 1) does not indicate either a North Pacific or North Atlantic origin for this genus. A vicariant event appears to have given rise to *M. stellatus* in the Atlantic, and to *Mastocarpus* sp. ined. on the east coast of Honshu, Japan (Masuda *et al.* 1987, 1989). West *et al.* (1983) reported that some strains of these two entities were interfertile. Further studies are needed to determine if *Mastocarpus* sp. represents a recent introduction into Japan.

PHYLLOPHORACEAE. In the clade ACON to GYCH, the first three species or species pairs comprise monophyletic lines in the northeastern Pacific (GYCH), northwestern Pacific (ACON, AFUR) and northeastern Pacific (PPAR, PPAC), implicating ancient vicariant events. This is supported further by the fact that *Stenogramme* and *Schottera* are more or less cosmopolitan in distribution. The phyllophoroids (*Phyllophora /Ozophora* complex) would appear to have arisen from a common ancestor in either the Atlantic or Pacific. A vicariant event occurred, followed by speciation in each ocean, giving rise to the *Ozophora* species in the Pacific, and the North Atlantic *Phyllophora* species (except the clade *P. truncata* and *Ceratocolax hartzii*) in the Atlantic. Judging from its disjunct distribution (see Newroth 1971c) in the cold waters of the North Atlantic and in the Black Sea, *Phyllophora truncata* appears to be a very ancient species that is not closely related to any other species of *Phyllophora*. *Phyllophora sicula* is a warm-water species that is also found in the western Mediterranean.

Similar to the phyllophoroid clade discussed above, the complex GYMA to GYPX also shows an apparent early radiation in the northeastern Pacific (*Gymnogongrus martinensis*), northwestern Pacific (*G. paradoxus*) and in the Atlantic (*G. devoniensis*). Different vicariant events are apparent at successive levels in the tree. Within the GYDV to GYLE clade, there are two vicariant events: the first distinguishing the Atlantic *Gymnogongrus devoniensis* from the Pacific AYAM to GYLE clade. One species of the *Gymnogongrus devoniensis* complex is circumboreal in the Atlantic[4], and cold-water in distribution, and is likely to have originated from the Pacific via a vicariant event; the group AYAM to GYLE suggests a speciation event in the eastern Pacific (*Gymnogongrus leptophyllus*), and then a vicariant event separating *G. crustiforme* in the eastern Pacific, from a clade comprising *Ahnfeltia yamadae* and *G. catenatus* in the western Pacific. The sequence described above is repeated in the AGRA to GYCR clade, with an initial division separating Atlantic and Pacific species, and a subsequent event distinguishing eastern and western Pacific species. Although the two clades GYDV-GYLE and

[4]C.A. Maggs has informed us that *Gymnogongrus devoniensis* from the British Isles includes at least two species, one with a southerly distribution (south western England and Ireland) and one with a more northerly distribution (north eastern Ireland). The latter entity is also found in the eastern Atlantic in cold water and records of *Gymnogongrus crenulatus* from this area refer to the latter, undescribed species (see also Maggs 1988).

AGRA to GYCR show congruent evolutionary patterns, the vicariant events distinguishing Atlantic and Pacific species may be different. One of the species presently included in the *Gymnogongrus devoniensis* complex is cold water in distribution, and its relationship with Pacific species may be through the Arctic. Other species belonging to the *G. devoniensis* complex, and *G. crenulatus* and *G. griffithsiae*, are more warm-water in distribution, and the relationship with the Pacific may have been via the Tethys Sea.

CONCLUSIONS

Our cladistic analysis of 64 taxa from the Atlantic and Pacific, using *Chondriella pusilla* as the out-group, indicates that the Phyllophoraceae, Gigartinaceae and Petrocelidaceae are a monophyletic complex of red algae, that probably correspond to three natural families. Despite the loosely filamentous construction of entities referred to the Gigartinaceae and the Petrocelidaceae, it seems likely that the latter family and the Phyllophoraceae, are more closely related, based on the formation of a sterile cell on the carpogonial branch and in the production — in at least some entities — of tetrasporangia on crustose tetrasporophytes that are of similar construction. At the generic level, many entities are defined poorly with too much emphasis being placed — consciously or unconsciously — on the form of the thallus as a generic character. Within the Gigartinaceae, precise studies of carposporophyte development, and the mode of formation of tetrasporangia, are necessary before a final generic scheme can be proposed. Within the Phyllophoraceae, the genus *Phyllophora* appears to be polyphyletic and at present is held together by a single character: the formation of spermatangia in pits. With the removal of the type species of *Ahnfeltia* from the Phyllophoraceae (Maggs & Pueschel 1989), species of *Gymnogongrus*-like Phyllophoraceae, previously referred to *Ahnfeltia*, should be removed to another genus. Further studies of *Besa papillaeformis*, the type species of the genus *Besa*, are necessary to decide if this would be a suitable place to refer these species[5], although this seems unlikely. The status of *Gymnogongrus* itself is in some doubt: *G. griffithsiae*, the type species, forms tetrasporoblasts, and species with internal cystocarps should not be placed in this genus.

[5]In a footnote DeCew & West (1981) stated 'Re-examination of the type material of *Besa papillaeformis* Setchell and Gardner, the type of the genus, reveals that the tetrasporophyte and gametophyte are isomorphic (DeCew, unpublished observations). We believe that *Besa* belongs to the Gigartinaceae due to the filamentous nature of the medulla'. Nevertheless, no further details have been published and McCandless, West & Guiry (1982) state that the tetrasporophytes are unknown in the genus *Besa*. Carrageenan chemistry (McCandless et *al.* 1982) and the occurrence of a sterile branch on the carpogonial branch (Kim 1976) suggest that *Besa* belongs in the Phyllophoraceae.

Although a full data set is not available, it seems likely that *Chondrus crispus*, *Mastocarpus stellatus* and *Gymnogongrus devoniensis* arrived in the Atlantic from the North Pacific, both areas having related species with cold-water affinities. *Gigartina* species in the North Atlantic probably had a warm-water origin. It is also likely that at least some *Phyllophora* species in the North Atlantic are closely related to *Ozophora* species in the North Pacific. The cladogram suggest vicariant events at several levels within different clades, distinguishing monophyletic groups of Pacific and Atlantic species and between species occurring in the eastern and western Pacific. Our cladogram is consistent with the notion that species of Gigartinaceae, Petrocelidaceae and Phyllophoraceae presently in the Atlantic arose through vicariant speciation events: (1) distinguishing the North Atlantic and North Pacific (via the Arctic); (2) separating the North Pacific and North Atlantic (via the Tethys); and (3) distinguishing the warm eastern Pacific and warm western Atlantic (on either side of Central America).

ACKNOWLEDGMENTS

We are very grateful to the following persons for helpful discussions and permission to use unpublished data: J.A. Brodie, K.S. Edyvane, M. Gretz, M.H. Hommersand, C.A. Maggs and H.B.S. Womersley. P.W. Gabrielson made numerous helpful comments on the manuscript. This study was carried out while MDG held the James Chair in Pure and Applied Science at St Francis Xavier University and was supported by a Natural Sciences and Engineering Research Council of Canada operating grant to DG.

REFERENCES

Abbott IA (1969) Some new species, new combinations, and new records of red algae from the Pacific coast. Madroño 20: 42-53

Abbott IA (1972, '1971') On the species of *Iridaea* (Rhodophyta) from the Pacific coast of North America. Syesis 4: 51-72

Abbott IA, Hollenberg GJ (1976) Marine algae of California. Stanford University Press, California

Ardré F (1978) Sur les cycles morphologiques du *Gymnogongrus crenulatus* (Turn.) J. Ag. et du *Gymnogongrus devoniensis* (Grev.) Schott. (Gigartinales, Phyllophoracées) en culture. Rev Algol NS 13: 151-176

Baardseth E. (1941) The marine algae of Tristan da Cunha. Results Norw Scient Exped Tristan da Cunha No 9: 1-173

Barwell CJ, Guiry MD (1984) Hordenine content of *Gigartina stellata* and Carrageen. Br Phycol J 19: 189 (abstract)

Barwell C, Farnham WF, Fletcher (1982) The occurrence of hordenine in European members of the Phyllophoraceae and Gigartinaceae. Br Phycol J 17: 229 [abstract]

Batters EAL (1890) A list of the marine algae of Berwick-on-Tweed. Hist Berwicksh Nat Club 12: 221-392 [reprint 1-172]

Blunden G, Gordon SM, McLean WFH, Guiry MD (1982) The distribution and possible taxonomic significance of quaternary ammonium and other Dragendorff-positive compounds in some genera of marine algae. Bot Mar 25: 563-567

Brodie JA (1989) Life histories, morphology and photoperiodic responses in Japanese *Chondrus* (Rhodophyta). Unpublished Ph.D. Thesis, National University of Ireland

Cordeiro-Marino M, Poza AC (1981) Life history of *Gymnogongrus griffithsiae* (Turner) Martius (Phyllophoraceae, Gigartinales). Proc Int Seaweed Symp 10: 155-161

Dawson EY (1961) Marine red algae of Pacific Mexico Part 4. Gigartinales. Pacif Nat 2: 191-343

DeCew TC, West JA (1977) A life history of *Ahnfeltia gigartinoides*: a possible link between the Phyllophoraceae and Gigartinaceae. J Phycol 13(Suppl.): 16(abstract)

DeCew TC, West JA (1981) Life histories in the Phyllophoraceae (Rhodophyta: Gigartinales) from the Pacific coast of north America. I. *Gymnogongrus linearis* and *G. leptophyllus*. J Phycol 17: 240-250

Dixon PS, Irvine LM (1977) Seaweeds of the British Isles Vol. 1 Rhodophyta Part 1 Introduction, Nemaliales, Gigartinales. British Museum (Nat. Hist.), London

Gabrielson PW, Garbary DJ (1987) A cladistic analysis of Rhodophyta: Florideophycidean orders. Br Phycol J 22: 125-138

Garbary D (1987) A critique of traditional approaches to seaweed distribution in light of the development of vicariance biogeography. Helgol Meeresunters 41: 235-244

Guiry MD (1984) Structure, life history and hybridisation of Atlantic *Gigartina teedii* (Rhodophyta) in culture. Br Phycol J 19: 37-55

Guiry MD, Cunningham EM (1984) Photoperiodic and temperature responses in the reproduction of north-eastern Atlantic *Gigartina acicularis* (Rhodophyta: Gigartinales). Phycologia 23: 357-367

Guiry MD, Hollenberg GJ (1975) *Schottera* gen. nov. and *Schottera nicaeensis* (Lamour. ex Duby) comb. nov. (= *Petroglossum nicaeensis* (Lamour. ex Duby) Schotter) in the British Isles. Br Phycol J 10: 149-164

Guiry MD, Maggs CA (1985) Notes on Irish marine algae—7 *Gigartina teedii* (Roth) Lamour. (Rhodophyta). Ir Nat J 21: 490-493

Guiry MD, West JA (1983) Life history and hybridization studies on *Gigartina stellata* and *Petrocelis cruenta* (Rhodophyta) in the north eastern Atlantic. J Phycol 19: 474-494

Guiry MD, Tripodi G, Lüning K (1987) Biosystematics, genetics and upper temperature tolerance of *Gigartina teedii* (Rhodophyta) from the Atlantic and Mediterranean. Helgol Meeresunters 41: 283-295

Guiry MD, West JA, Kim D-H, Masuda M. (1984) Reinstatement of the genus *Mastocarpus* Kützing (Rhodophyta). Taxon 33: 53-63

Hollenberg GJ (1943) New marine algae from southern California. II. Am J Bot 30: 571-579

Hollenberg GJ (1945) New marine algae from southern California. III. Am J Bot 32: 447-451

Hollenberg GJ, Abbott IA (1968) New species of marine algae from California. Can J Bot 46: 1235-1251

Hommersand MH (1986) The biogeography of the South African marine red algae: a model. Botanica marina 29: 257-270

Hommersand MH (1990) Biogeography of the marine red algae of the North Atlantic Ocean. In: Garbary DJ, South GR (eds), Evolutionary biogeography of the marine algae of the North Atlantic. Springer Verlag, Berlin, p 349-410

Kim D-H (1976) A study of the development of cystocarps and tetrasporangial sori in Gigartinaceae (Rhodophyta, Gigartinales). Nova Hedwigia 27: 1-146

Kraft GT (1989)*Cylindraxis rotundata* gen. et sp. nov. and its generic relationships within the Liagoraceae (Nemaliales, Rhodophyta). Phycologia 28: 275-304

Kylin H (1928) Entwicklungsgeschichtliche Florideenstudien. Acta Univ Lund NF, Avd 2 24(4): 1-127
Kylin H (1956) Die Gattungen der Rhodophyceen.Gleerups, Lund
Lee IK, Kang JW (1986) A check list of marine algae in Korea. Korean J Phycol 1: 311-325
Leister GL (1977) Taxonomy and reproductive morphology of *Iridaea cordata* (Turner) Bory and *Iridaea crispata* Bory (Gigartinaceae, Rhodophyta) from southern South America. Unpubl. Ph.D. Thesis, Department of Botany, Duke University
Levring T (1941).Die Meeresalgen der Juan Fernandez-Inseln. In Skottsberg C (ed) The Natural History of Juan Fernandez and Easter Island. Vol. 2, Almqvist & Wiksells Lund, p 601-670
Lewis JA, Kraft GT (1979) Occurrence of a European red alga (*Schottera nicaeensis*) in southern Australian waters. J Phycol 15: 226-230
Lindstrom SC (1988) The Dumontieae, a resurrected tribe of red algae (Dumontiaceae, Rhodophyta). Phycologia 27: 89-102
Lüning K, Guiry MD, Masuda M (1987) Upper temperature tolerance of north Atlantic and north Pacific geographical isolates of *Chondrus* species (Rhodophyta). Helgol Meeresunters 41: 297-306
Maddison WP, Maddison DR (1987) [Manual of] MacClade, version 2.1, for the Apple Macintosh TM. Harvard University
McCandless EL, West JA, Guiry MD (1982) Carrageenan patterns in the Phyllophoraceae. Biochem Syst Ecol 10: 275-284
McCandless EL, West JA, Guiry MD (1983) Carrageenan patterns in the Gigartinaceae. Biochem Syst Ecol 11: 175-182
Maggs CA (1988) Intraspecific life history variability in the Florideophycidae (Rhodophyta). Bot Mar 31: 465-490
Maggs CA (1989) *Erythrodermis allenii* Batters in the life history of *Phyllophora traillii* Holmes ex Batters (Phyllophoraceae, Rhodophyta). Phycologia 28: 305-317
Maggs CA, Pueschel CM (1989) Morphology and development of *Ahnfeltia plicata* (Rhodophyta): proposal of Ahnfeltiales ord. nov. J Phycol 25: 333-351
Maggs CA, McLachlan J, Saunders GW (1989) Infrageneric taxonomy of *Ahnfeltia* (Ahnfeltiales, Rhodophyta). J Phycol 25: 351-368
Magruder WH (1977) The life history of the red alga *Ahnfeltia concinna* (Rhodophyta, Gigartinales). Phycologia 16: 197-203
Masuda M (1981) Further observations on the life history of *Gymnogongrus flabelliformis* Harvey (Rhodophyta) in culture. J Fac Sci Hokkaido Univ Ser 5 12: 159-164
Masuda M (1983) The life history of *Ahnfeltia concinna* J. Agardh (Rhodophyta, Gigartinales) from Japan. Jap J Phycol 31: 180-189
Masuda M (1987) Taxonomic notes on the Japanese species of *Gymnogongrus* (Phyllophoraceae, Rhodophyta). J Fac Sci Hokkaido Univ Ser 5 14:39-72
Masuda M, Kurogi M (1981) The life history of *Gigartina ochotensis* (Ruprecht) Ruprecht (Rhodophyta) in culture. J Fac Sci Hokkaido Univ Ser 5 12: 165-171
Masuda M, DeCew TC, West JA (1979) The tetrasporophyte of *Gymnogongrus flabelliformis* Harvey (Gigartinales, Rhodophyta). Jap J Phycol 27: 63-73
Masuda M, Nabeshima Y, Yamamoto Y (1989) Taxonomic notes on *Mastocarpus* sp. (Rhodophyta) from Kii Peninsula. Nankiseibutu: Nanki Biol Soc 31: 1-6
Masuda M, West JA, Kurogi M (1987) Life history studies in culture of a *Mastocarpus* species (Rhodophyta) from central Japan. J Fac Sci Hokkaido Univ Ser 5 14: 11-38
Mikami H (1965) A systematic study of the Phyllophoraceae and Gigartinaceae from Japan and its vicinity. Sci Paps Inst Algol Res Fac Sci Hokkaido Univ 5: 181-285
Nelson G, Platnick N (1981) Systematics and Biogeography, Cladistics and Vicariance. Columbia University Press, New York
Newroth PR (1971a) Redescriptions of five species of *Phyllophora* and an artificial key to the north Atlantic Phyllophoraceae. Br Phycol J 6:225-230
Newroth PR (1971b) Studies on life histories in the Phyllophoraceae. I. *Phyllophora truncata* (Rhodophyceae, Gigartinales). Phycologia 10: 345-354
Newroth PR (1971c) The distribution of *Phyllophora* in the north Atlantic and Arctic regions. Can J Bot 49: 1017-1024

Newroth PR (1972) Studies on life histories in the Phyllophoraceae. II. *Phyllophora pseudoceranoides* and notes on *P. crispa* and *P. heredia* (Rhodophyceae, Gigartinales). Phycologia 11: 99-107

Newroth PR, Markham JW (1972). Observations on the distribution, morphology, and life histories of some Phyllophoraceae. Proc Int Seaweed Symp 7: 120-126

Norris, RE, Aken, ME (1985) Marine benthic algae new to South Africa. S Afr J Bot. 51: 55-65

Parkinson PG (1981) *Iridea*, Illuminating observations on the publication, typidication, orthography and history of the name *Iridea* Bory 1826 . . . Pettifogging Press, Auckland

Pielou, EC (1978) Latitudinal overlap of seaweed species: evidence for quasi-sympatric speciation. J Biogeogr 5: 227-238

Polanshek AR, West JA (1975) Culture and hybridization studies on *Petrocelis* (Rhodophyta) from Alaska and California. J Phycol 11: 434-439

Polanshek AR, West JA (1977) Culture and hybridization studies on *Gigartina papillata* (Rhodophyta). J Phycol 13: 141-149

Santelices B, Ramirez ME, Abbott,I A (1989) A new species and new records of marine algae from Chile. Br Phycol J 24: 73-82

Scagel RF, Gabrielson PW, Garbary DJ, Golden L, Hawkes MW, Lindstrom SC, Oliveira JC, Widdowson TB (1989) A synopsis of the benthic marine algae of British Columbia, southeast Alaska, Washington and Oregon. Phycol Contrib Dept Bot Univ Br Columbia, No. 3: 1-532

Schneider CW, Searles RB (1976) North Carolina marine algae. VII. New species of *Hypnea* and *Petroglossum* (Rhodophyta, Gigartinales) and additional records of other Rhodophyta. Phycologia 15: 51-60

Schotter G. (1968) Recherches sur les Phyllophoracées. Notes posthumes publiées par Jean Feldmann et Marie-France Magne. Bull Inst Océanogr Monaco 67(1383): 1-99

Sciuto S, Chillemi R, Morrone R, Patti A, Piattelli M. (1989) Dragendorff-positive compounds in some Mediterranean red algae. Biochem Syst Ecol 17: 5-10

Silva PC (1979) The benthic algal flora of central San Francisco Bay [and] Appendix A. Systematic catalogue of the benthic algal flora of San Francisco Bay. In Conomos T San Francisco Bay, the urbanized estuary. Pacific Division, American Association for the Advancement of Science, San Francisco, p 287-345

South GR, Tittley I (1986) A checklist and distributional index.of the benthic marine algae of the north Atlantic ocean. Huntsman Marine Laboratory and British Museum (Nat. Hist.), St Andrews and London

West JA (1972) The life history of *Petrocelis franciscana*. Br Phycol J 7: 299-308

West JA, Guiry MD (1982) A life history study of *Gigartina johnstonii* (Rhodophyta) from the Gulf of California. Bot Mar 25: 205-211

West JA, Guiry MD, Masuda M (1983) Further investigations of the genetic affinities and life history patterns of the red alga *Gigartina*. In Tseng CK (ed) Proceedings of the Joint China-U.S. Phycology Symposium. Quingdao, Nov. 16-21, 1981. Science Press, Beijing, p 137-166

West JA, Hommersand MH (1981) Rhodophyta: life histories. In Lobban CS, Wynne MJ (eds) The biology of seaweeds. Botanical monographs 17. Blackwell, Oxford, 133-193

West JA, Polanshek A, Guiry MD (1977) The life history in culture of *Petrocelis cruenta* J. Agardh (Rhodophyta) from Ireland. Br Phycol J 12: 45-53

West JA, Polanshek AR, Shevlin DE (1978) Field and culture studies on *Gigartina agardhii* (Rhodophyta). J Phycol 14: 416-426

Wiley EO (1981) Phylogenetics. Wiley, New York

Wynne, MJ (1986) Report on a collection of benthic marine algae from the Namibian coast (Southwestern Africa). Nova Hedwigia 43: 311-355

ALGAL PROVINCES IN THE NORTH ATLANTIC - DO THEY EXIST?

Ian Tittley*, Gordon L.J. Paterson*, P. John D. Lambshead* and G. Robin South#

Natural History Museum*
London SW7 5BD
U.K.

Huntsman Marine Science Centre#
St. Andrews
New Brunswick, EOG 2XO
Canada

Abstract - Biogeographical studies of the benthic algal flora of the North Atlantic Ocean were initially based on subjective assessments of patterning in species distribution. Subsequently, computer based (clustering and ordination) approaches have been used to attempt more objective appraisals. The present paper commences from a null hypothesis (that algal biogeographic provinces are not proven to exist), and analyses a larger number of data sets than hitherto used (1116 species from 33 floras). Two methods of analysis were used: an ecocladistic (Wagner tree) method (PAUP program), a phenetic ordination method (DECORANA program). Previous biogeographic analyses utilising cluster-methods identified apparently distinct algal provinces in the North Atlantic Ocean. The present ecocladistic analyses produced ecocladograms with no clear basal division of branches likely to indicate major biogeographical provinces, the main exception being the flora of the mid-Atlantic Azores island group, which was well separated from the continental European flora and showed affinities with the American flora. Branching at higher levels in the ecocladograms suggested a degree of floristic dissimilarity between east and west Atlantic, which can be related to the species poor nature of the latter. Ordination confirmed these results, clearly indicating a continuum of floristic relationships in the North Atlantic. A second series of analyses performed on a comparatively well known regional flora (comprising 33 local floras in the British Isles and adjacent areas, Faroes to the Channel Islands and northern France), traversing propounded biogeographical boundaries, revealed no clear biogeographical separation of the Faroes or Shetland Islands, but a continuum of floristic relationships. It is suggested therefore that the North Atlantic Ocean is a single biogeographical province, with perhaps a southern boundary excluding the Azorean flora.

INTRODUCTION AND BACKGROUND

Biogeographical studies on the benthic marine algal flora of the North Atlantic Ocean were initially based on subjective assessment of patterning in species distribution. Of the earliest workers in the subject, Børgesen & Jónsson (1905) investigated the distributions of marine algae using seventeen floras in the Arctic Sea (from the Bering Straits to Novaya Zemlia), the northern northeast Atlantic (from Novaya Zemlia to Scotland) and the northern northwest

NATO ASI Series, Vol. G 22
Evolutionary Biogeography of the
Marine Algae of the North Atlantic
Edited by D. J. Garbary and G. R. South

Atlantic (Maine to New Jersey). They proposed five main groupings (arctic; subarctic; boreal-arctic; cold-boreal; warm-boreal). Farlow (1881) identified three biogeographical areas on the New England coast of America. The flora to the north of Maine was considered to be "markedly arctic". Farlow identified Cape Cod as a biogeographical boundary between a cold-water flora to the north and a warm-water flora to the south.

More recently, Hoek & Donze (1967) and Hoek (1975) proposed a scheme of algal phytogeographical regions in the northern Atlantic Ocean. The latter study covered a larger area (south to the Caribbean and tropical Africa) than the present study and employed a clustering method to help identify such regions. In an appraisal of world-wide Rhodophyta distribution patterns, Hoek (1984) and Joosten & Hoek (1986) employed, respectively, agglomerative clustering and multivariate methods of analysis producing results which, as far as the north Atlantic is concerned, largely supported the earlier proposed phytogeographical regions.

A different approach in identifying biogeographical relationships was pursued by Lindstrom (1987); this involved a phylogenetic appraisal of members of the Dumontiaceae (Rhodophyta) using cladistics. Through analysis of a group of species with well-defined geographical distributions, taxonomic cladograms thus generated could also be used to suggest biogeographical relationships. From this analysis, Lindstrom (1987) concluded that the two sides of the Pacific Ocean were floristically closely related, and that the Pacific flora was more closely related to that of the northwestern Atlantic; the northeastern Atlantic was more distantly related.

Alvarez *et al.* (1988) produced a reassessment of Northern Atlantic seaweed biogeography that encompassed both the North Atlantic flora to the equator and the Mediterranean flora, a much larger area than covered by the present work. Their study, like the present, sought biogeographical patterning in the major taxonomic groups (Chlorophyta, Phaeophyta and Rhodophyta). An agglomerative clustering method was used, and the phytogeographic regions proposed broadly confirmed those identified by van den Hoek and co-workers.

Prud'homme van Reine (1988) assessed the relationships of the marine algal flora of the Azores to other floras in the Atlantic. The floras of 'areas' between the Congo River (tropical Africa) in the south, and north west England in the north, were compared using a clustering method. The Azores flora was shown to be more closely related to the Mediterranean and European floras than to that of Africa.

Working to the south of the area under investigation, Lawson (1978) appraised the distribution of seaweed floras in the tropical and subtropical regions of the Atlantic. Two methods of analysis were employed: clustering (Group Average Clustering) and ordination (Indicator Species Analysis, based on Reciprocal Averaging). Groupings of floras which were recognised were broadly interpreted in the sense of phytogeographical regions. In a subsequent paper (Lawson 1988), algal distribution patterns were reappraised but with additional data, and using a variant of Reciprocal Averaging (Decorana); reassessing from these ordinations, Lawson discussed the relationship between the species-rich western Atlantic flora and the species-poor eastern flora, the reciprocal of the situation in the northern North Atlantic.

South (1987), in an overview of the biogeography of the marine algae of the North Atlantic, presented a rationale for the present work, based on the recent collation of information in the form of a check-list and geographical index to the benthic marine algae of the North Atlantic (South & Tittley 1986). Using this resource, the present study poses the question: "Is the patterning of species distributions in the north Atlantic such that biogeographical provinces can be recognised ?" Hypotheses advanced by other workers are thereby tested. The data resource enabled this to be undertaken on both broad oceanic and regional levels.

MATERIALS AND METHODS

Our approach in this study has been to start from a null hypothesis which is that biogeographical provinces are not proven to exist in the North Atlantic; hitherto, many biogeographical appraisals have been extremely subjective in approach. The data used for this analysis comprised the 1116 species of Chlorophyta, Phaeophyta and Rhodophyta listed in South & Tittley (1986). The basic unit of information is the presence or absence of a species at a given site (no quantitative values were used). The broad geographical scope also follows that employed in South & Tittley (1986). The North Atlantic Ocean is, therefore, defined as ranging from the Azores to Spitzbergen in the east, and from Virginia, USA, to the eastern arctic Canada in the west; also included are Jan Mayen, Bear Island, Iceland and Greenland (see Fig. 1). The delimitation of the 33 locations is less than ideal, since largely politically determined, but with a degree of natural geographical bias. For data sources, see South & Tittley (1986).

A second tier of analysis was performed on a regional flora embracing six of the stations (one in part) used in South & Tittley (1986). The region comprised the British Isles and adjacent islands, and the continental coast (Figure 2). It was selected because the flora is well-known, and because the region traversed propounded biogeographical boundaries. Data were assembled for 516 species from 31 stations (mainly counties or equivalents; a complete station list and data sources are given in Table 1).

Two main numerical methods were used to analyse the algal distributions: (i) a cladistic (Wagner tree) method (PAUP v.2.4 program, Swofford 1985)), (ii) a phenetic ordination method (Detrended Correspondence Analysis, DCA or DECORANA, part of the Vespan II suite, Malloch 1988)).

Table 1. Sources of information for floras of the British Isles and adjacent areas

Area	Authors
Faroes	Irvine (1982)
Shetland	Irvine (1974, 1980)
Orkney	Wilkinson (1975)
St Kilda	Watling *et al.* (1970)
Hebrides	Norton (1972)
Mull	Price & Tittley (1978)
Colonsay	Norton *et al.* (1969)
Clyde	Clokie & Boney (1979)
Galloway	Burrows(1960), Wilkinson (1980)
I. o. Man	Knight & Parke (1931), Kain (1960), Burrows (1958)
N.Wales	Fletcher & Jones (1975, 1976), Jones *et al.* (1977)
Grampian	Wilkinson (1979)
Forth	Wilkinson & Tittley (unpubl.)
E. Borders	Norton (1976)
Durham	Edwards (1976)
Scarborough/S. Yorkshire	Perkins (1953)
Lincolnshire	Price *et al.* (1977)
Norfolk	Chapman (1937)
Kent	Tittley & Price (1977, 1978)
N.E.France	Tittley & Price (1978)
I.o.Wight/Solent	Kain (1958), Jones (1980), Farnham (1982)
Dorset	Burrows (1964), Whittaker (1980)
S.Wales	Wilkinson (1982)
Lizard	Price *et al.* (1980, 1981)
Scilly I.	Russell (1968)
Channel I.	Dixon (1961)
Cork	Guiry (1978)
Kerry	Guiry (1978)
Galway	Guiry (1978)
Mayo	Guiry (1978)

Numerical phylogenetic methods have been employed in the analysis of ecological data (Lambshead 1986; Lambshead & Paterson 1986). When cladistic methods are applied to ecological data, species are used as characters, and sites as species. Presence is considered as derived, and absence as primitive; absences can be coded as derived, when there is good historical evidence (cf. Legendre 1986). However, in most ecological studies such information is lacking, and it is not possible to determine whether joint absences are homologous (i.e. due to the same ecological factor excluding the species at two sites). One potential disadvantage in analysing ecological data, using the Wagner algorithm used in PAUP, is that it assumes that parallelisms (species present as the result of different factors), and reversals (the absence of a species which could have expected to be present), are equally likely. It is, therefore, possible that spurious groupings based on joint absences could be created; however, in practice the method is robust, and the patterns generated appear to be natural groupings sensu Mill (1872).

The output from an ecological analysis using a Wagner algorithm is an ecocladogram.

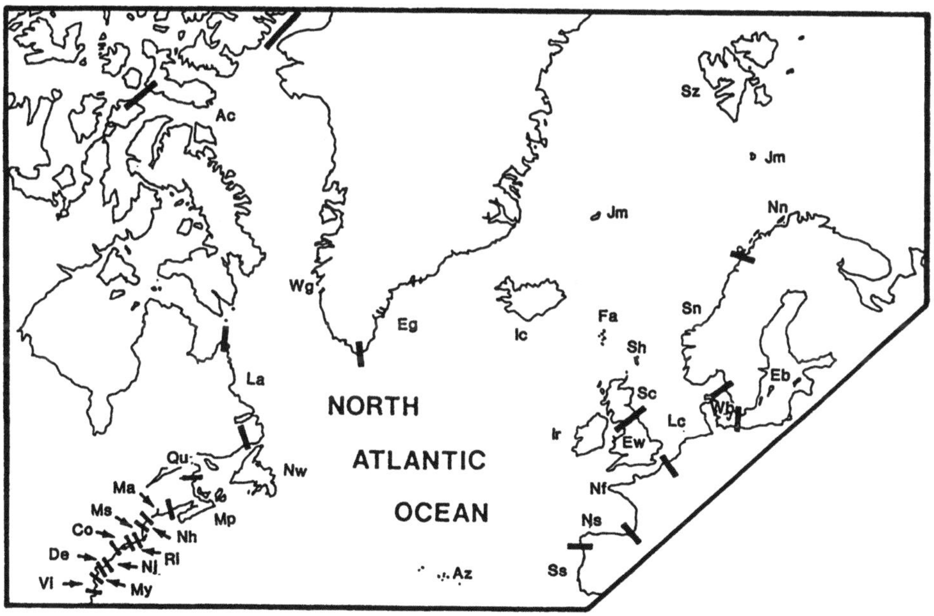

Fig. 1. The thirty-three floras used in the North Atlantic analyses.
Abbreviations used: Az, Azores; Ss, Southern Spain and Portugal; Ns, Northern Spain; Nf, Northern France; Ir, Ireland; Ew, England and Wales; Sc, Scotland; Sh, Shetland; Lc, Low Countries; Wb, Western Baltic; Eb, Eastern Baltic; Sn, Southern Norway; Nn, Northern Norway; Fa, Faroes; Jm, Jan Mayen and Bear Island; Sz, Spitzbergen; Ic, Iceland; Eg, Eastern Greenland; Wg, Western Greenland; Ac, Arctic Canada; La, Labrador; Nw, Newfoundland; Qu, Quebec; Mp, Maritime Provinces; Ma, Maine; Nh, New Hampshire; Ms, Massachusetts; Ri, Rhode Island; Co, Connecticut; Nj, New Jersey; De, Delaware; My, Maryland; Vi, Virginia

The latter are presented in two forms, a geometrically symmetrical diagram, and/or usually an asymmetrical diagram in which the branch lengths are scaled, according to the number of attributes (species) separating the nodes or characterising the terminal branches.

The ecocladograms were rooted, using the "Lundberg rooting" option in the program (Lundberg 1972); this means that the ecocladogram is rooted, at a point where a station containing only zeros (i.e. no species present) would be rooted. Opportunity was also taken to use the Azores as an outgroup, when our initial results indicated that it was a sister group to the remaining sites. The data often produced a large number of equally parsimonious ecocladograms, and these were appraised for similarity in pattern.

Ecological cladistics has advantages over phenetic methods in that it uses parsimony as an objective criterion to select between competing patterns generated. The ecocladogram selected is that which makes the least number of character changes, and gives the shortest tree length. Research suggests that numerical ecocladistic techniques can outperform phenetic methods in their power to derive meaningful patterns (Lambshead 1986; Lambshead & Paterson 1986; Paterson 1986).

Ordination, and especially the Decorana (DCA) program, is popular with ecologists and has been successfully used in the analysis of algal distributions (Lawson 1988). Ter Braak & Prentice (1988) reviewed methods used to assess species distributions in response to environmental gradients. They suggest that where gradients are large scale, and environmental parameters are not quantified, as in this study, the use of indirect gradient methods such as DCA can be effective. These methods are particularly useful in hypothesis-generation based on the patterns observed (Green 1980). DCA is a reasonably robust practical method, despite its disadvantages in certain circumstances (Minchin 1987; Pielou 1984).

It was not possible to analyse the algae as a single group, because of the limitations on matrix size used by PAUP; each of the main classes was therefore analysed separately. Since algal classes are known to have differing ecological responses, it makes sense to analyse them separately (cf Alvarez *et al.* 1988). Although other studies (e.g. Lawson 1978, 1988) have analysed combined algal classes, we were concerned that, in so doing, information might be lost. It was also necessary to break the Rhodophyta down further. Two schemes were used: (a) a random split into two groups, and (b) division based on more systematic groupings, with the Ceramiales being treated separately. The analyses undertaken are listed in Table 2.

In this study, our null hypothesis would be fully supported by the generation of maximally resolved ecocladograms; these are characterised by a single comb-like series of branches (pectination). This is irrespective of whether or not there is a recognisable sequence of floras represented by the branches. Appraisal of the geographical relationships within a sequence of branches may yield additional information. Similarly, a single cluster or string of floras (sites) in an ordination would also support the null hypothesis; interpretation of the arrangement of points in the ordination may yield additional information. If major floristic

Table 1. Sources of information for floras of the British Isles and adjacent areas

Area	Authors
Faroes	Irvine (1982)
Shetland	Irvine (1974, 1980)
Orkney	Wilkinson (1975)
St Kilda	Watling *et al.* (1970)
Hebrides	Norton (1972)
Mull	Price & Tittley (1978)
Colonsay	Norton *et al.* (1969)
Clyde	Clokie & Boney (1979)
Galloway	Burrows(1960), Wilkinson (1980)
I. o. Man	Knight & Parke (1931), Kain (1960), Burrows (1958)
N.Wales	Fletcher & Jones (1975, 1976), Jones *et al.* (1977)
Grampian	Wilkinson (1979)
Forth	Wilkinson & Tittley (unpubl.)
E. Borders	Norton (1976)
Durham	Edwards (1976)
Scarborough/S. Yorkshire	Perkins (1953)
Lincolnshire	Price *et al.* (1977)
Norfolk	Chapman (1937)
Kent	Tittley & Price (1977, 1978)
N.E.France	Tittley & Price (1978)
I.o.Wight/Solent	Kain (1958), Jones (1980), Farnham (1982)
Dorset	Burrows (1964), Whittaker (1980)
S.Wales	Wilkinson (1982)
Lizard	Price *et al.* (1980, 1981)
Scilly I.	Russell (1968)
Channel I.	Dixon (1961)
Cork	Guiry (1978)
Kerry	Guiry (1978)
Galway	Guiry (1978)
Mayo	Guiry (1978)

Fig. 2. Map of British Isles and adjacent areas indicating floras used in analysis

groups exist (rejecting our null hypothesis), we would expect these to be indicated by ecocladograms, with clusters of branches from major basal dichotomies. However, the presence of groupings in an ecocladogram would not necessarily refute our null hypothesis. Appraisal of the geographical relationships with branching sequences is important here, since there could be seemingly random arrangements of floras (sites) within and between groups of branches, which may deny simple explanation. Similarly, in an ordination, we would expect floristic groupings to form distinct clusters, but with the proviso that the members of such clusters relate geographically, or in terms of another environmental parameter.

RESULTS

For the ecocladistic analyses see the ecocladograms illustrated as Figs 3 - 5 and 7. Ordinations (first two axes) using Decorana are illustrated in Figs 6 and 8.

Area ecocladograms of the North Atlantic

PHAEOPHYTA. The basic structure of the two equally parsimonious ecocladograms produced in analysis 1 was similar, differing mainly in the arrangement of the ultimate branches; ecocladogram 1 (Fig. 3a) showed that the Maryland algal flora splits off from the rest of the Atlantic flora, and that the remaining floras could be further divided into two unequally-sized groups (branching marked node 1 in Fig. 3a): (i) the Azores and Iberian floras, together with those from Massachusetts to Virginia, and also the Eastern Baltic; (ii) the remaining floras of the North Atlantic. Branching pattern of the cladogram suggests that the latter comprises a series of North Atlantic Arctic floras, then two groups of floras (splitting at node marked 2 in Fig. 3a): (i) the floras of Newfoundland, Quebec, Maritime Provinces, Maine and New Hampshire, all arising from a single branch; (ii) those of Europe, north of Iberia.

The significance of the first branch of the cladogram in analysis 1 (Fig. 3a), separating the impoverished Maryland Phaeophyta flora, is questionable. The second branch from the base is more important, and suggests two main, south and north, components to the North Atlantic Phaeophyta flora. When spatial geographical relationships are appraised, the comb-like series of American floras suggests a clinal sequence of floristic relationships. The inclusion of the Eastern Baltic flora in the American group is difficult to explain, a seemingly spurious result; it is plainly an atypical flora, being species-poor and characteristic of a non-marine environment. The next series of branches (Arctic floras) are species-poor. The cluster of branches arising

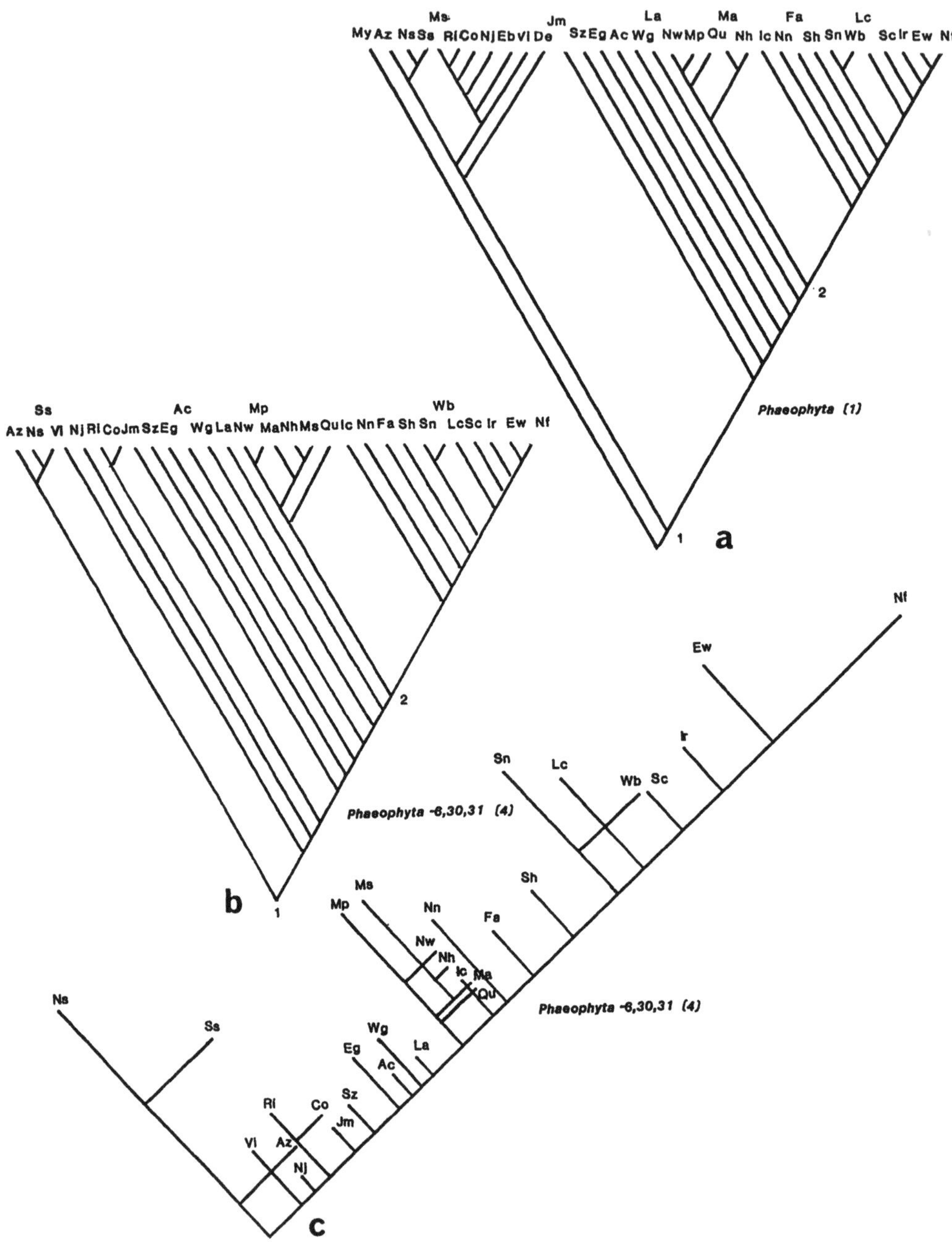

Fig. 3. Area ecocladogram for the North Atlantic: Phaeophyta; (a) analysis 1, cladogram 1; (b) analysis 2, cladogram 4; (c) analysis 2, scaled cladogram 4. Chlorophyta; (d) analysis 7, cladogram 11; (e) analysis 4, cladogram 30; (f) analysis 4, scaled cladogram 11. (For abbreviations see caption to Fig. 1)

Fig. 3 cont.

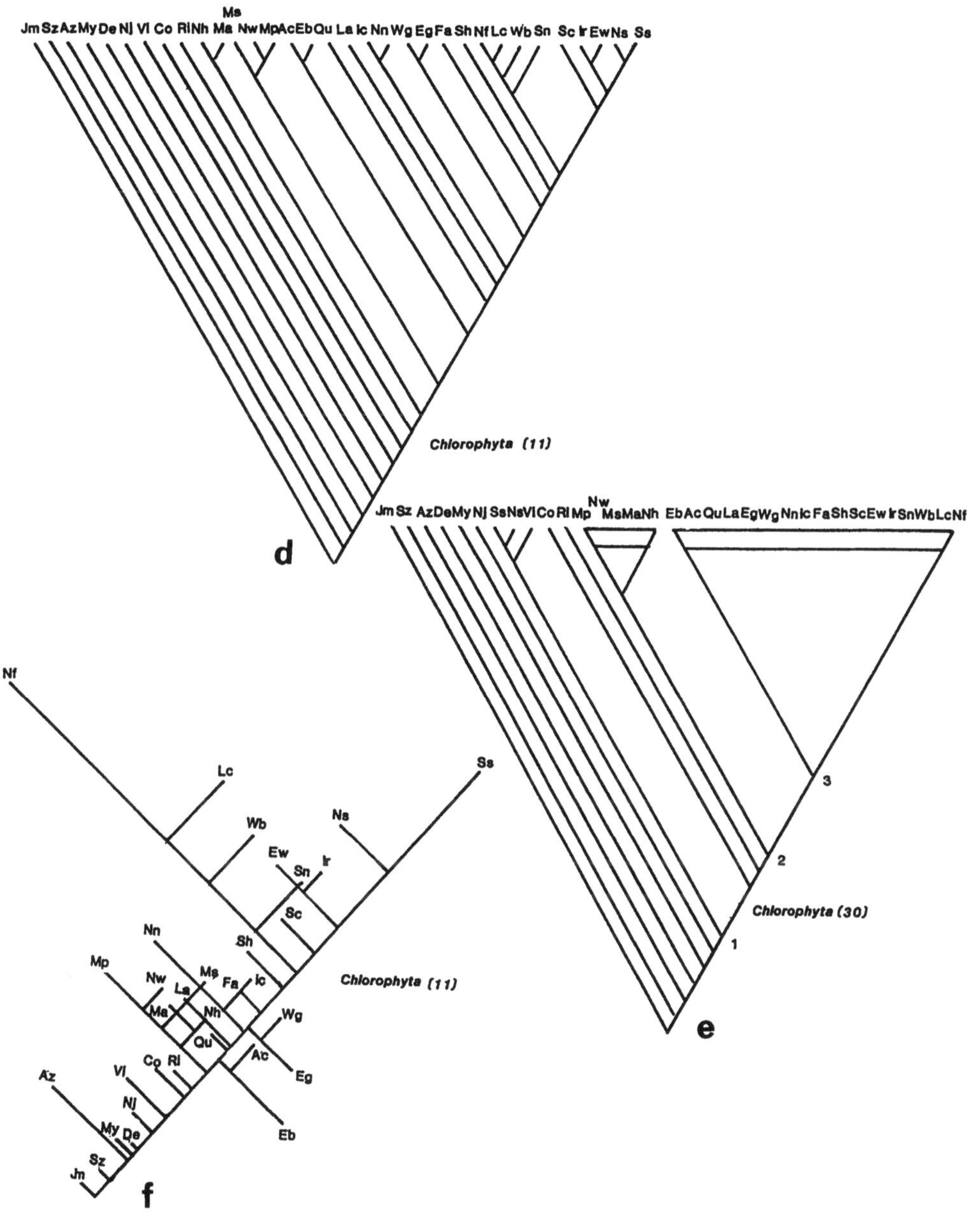

from one of the comb series (at the node marked 2 in Fig. 3a) indicates a small local grouping of northeastern American floras (Newfoundland to New Hampshire) and an east-west separation of Atlantic floras. A clear geographical (north to south) sequence was seen in the arrangement of European floras from Iceland to Northern France, and reflected a trend of species-poor to species-rich floras.

Since the original analysis indicated some questionable floristic relationships, the Atlantic flora was reinvestigated, without the seemingly atypical floras of Maryland, Delaware and the Eastern Baltic (analyses 2 and 3). These sites are deemed atypical since they are all strongly estuarine, species present being characteristic of such habitats.

The sixteen ecocladograms produced in analysis 2 differed little in basic structure, varying mainly in the ordering of terminal branches. Ecocladogram 4 (Figure 3b) separated the Phaeophyta floras of the Azores and Iberia from the North Atlantic flora (node marked 1 in Fig. 3b). There then followed a long comb series, comprising the geographically-close Connecticut to Virginia floras (the main difference from analysis 1) followed by the arctic floras. Higher level branching (node marked 2 in Fig. 3b) suggested two groupings: (i) northern northeast America (Newfoundland to New Hampshire); (ii) European floras (a series from Iceland to Northern France). From these results, clinal/overlapping floristic relationships could be inferred. When the cladogram (Fig. 3c) is scaled according to the number of characters (species) separating the dichotomies, only a small number separate the basal series, suggesting that these floras are similar; e.g., the eastern Greenland, Arctic Canada, western Greenland and Labarador branches are successively separated by only 5, 3 and 4 species.

The sixteen ecocladograms produced in analysis 3 (not illustrated), which were rooted by the Azores flora, were almost identical in structure to those described in analysis 2.

CHLOROPHYTA. Figure 3d shows ecocladogram 11 (one of 36 equally parsimonious trees produced by the program), in which floristic groupings were not so well defined. The first two branches separated Jan Mayen and Spitzbergen from the remaining Atlantic floras; the next in the comb series of branches is the Azores, which was followed by American floras. European floras occurred higher in the cladogram but were not separated from the American floras as by the Phaeophyta and Rhodophyta analyses (see below). The positioning of floras low in the ecocladogram reflected species-paucity.

Figure 3e shows an alternative ecocladogram (30) for the same analysis. The basic structure is similar to Fig. 3d, differing in that the Iberian floras were not positioned with the European, but with the American, floras (node marked 1 in the figure); the northern northeastern American sites formed a small group branching at the node marked 2. Node 3 formed a large grouping of sites spanning a wide geographical area from Quebec, the high Arctic (excluding Jan Mayen and Spitzbergen) and Europe to Northern France.

A scaled ecocladogram (Fig. 3f) indicated the closeness of middle level branching due to few separating characters and, therefore, only small differences between floras. The long branch representing the French flora reflected the large number of species recorded only from France; its high position in the cladogram emphasised the large number of features shared with other floras.

RHODOPHYTA. In analysis 5, ecocladogram 10 (Fig. 4a; 31 trees produced) separated the Maryland and Delaware floras from the remaining North Atlantic floras. The next grouping (at the node marked 1 in Fig. 4a) comprised those floras of Virginia, New Jersey and the Azores (well-separated from the European flora), suggesting a north-south division. A comb-like series of branches, representing the Eastern Baltic and high Arctic floras, precede the remaining North Atlantic Rhodophytan floras. The branching pattern suggests that the latter can be divided into two floristic groups (node marked 2 in Fig. 4a): (i) a north-western group (Labrador to Connecticut); (ii) a north-eastern group (western Greenland, northern Norway, Iceland to Iberia). These main groups correspond broadly to those identified in the Phaeophyta analysis; the comb series of branches suggests floristic continuity. An alternative cladogram (30, Figure 4b) presents a similar picture, but with a distinct Azores, Virginia, Maryland, Delaware, New Jersey grouping, arising from the basal branch. In this cladogram the western Greenland flora is included in the Arctic series.

In analysis 6 (with the Eastern Baltic, Delaware and Maryland floras removed) the branching pattern in ecocladogram 1 (Fig. 4c; fourteen trees produced) linked the Virginia and Azores Rhodophytan floras as a small group (node marked 1 in Fig. 4c) separate from the rest of the North Atlantic. The comb series of high Arctic floras formed sister floras to the remaining North Atlantic flora; as noted previously, the latter comprised two groupings, east and west Atlantic floras (separating at the node marked 2 in Figure 4c).

In analysis 7, the second subset of the Rhodophyta, the ecocladogram (not illustrated) suggests a clearer north-to-south floristic grouping. The impoverished Arctic flora, and also the Newfoundland flora formed a distinct group, separating at the base of the tree. Within the remainder first the Eastern Baltic then the Azores are separated off. Finally, the remaining sites group into a contiguous series corresponding to the European and American coasts.

In analysis 8 (Rhodophyta second subset with Eastern Baltic, Delaware and Maryland removed), ecocladogram 1 (of four produced; Fig. 4e) showed a long comb series of high Arctic and northern Canadian floras, preceding the remaining North Atlantic flora. The latter comprised two groupings (splitting at node marked 1 in Fig. 4e): one included the Azores with American floras as far north as the Maritime Provinces (Canada), the Azores flora being closest to the Virginia flora; the other comprised European floras from Iceland to Iberia.

Both main groupings were represented by a series of branches, suggesting clinal floristic relationships. The scaled ecocladogram (Fig. 4d) more clearly depicted the floristic relationships in the North Atlantic, particularly the similarity of the American floras.

Results of the analyses of orders corresponded to those from the random subsets and broadly confirmed the relationships previously identified. Ecocladogram 1 of analysis 9 (Ceramiales; Fig. 5a; 50 produced) typified the results, with the North Atlantic being divided into two unequal groups (node marked 1 in Figure 5a). The smaller comprised the

Azores, Eastern Baltic, Virginia and Maryland floras; the larger group comprised a series of high Arctic floras, with two main North Atlantic groupings (splitting at node marked 2 in Fig. 5a) which, in this case, comprised (i) the floras from Newfoundland to New Jersey, and (ii) the floras from Iceland to Iberia, both a comb series.

In analysis 10, fifty equally parsimonious ecocladograms were produced, and all

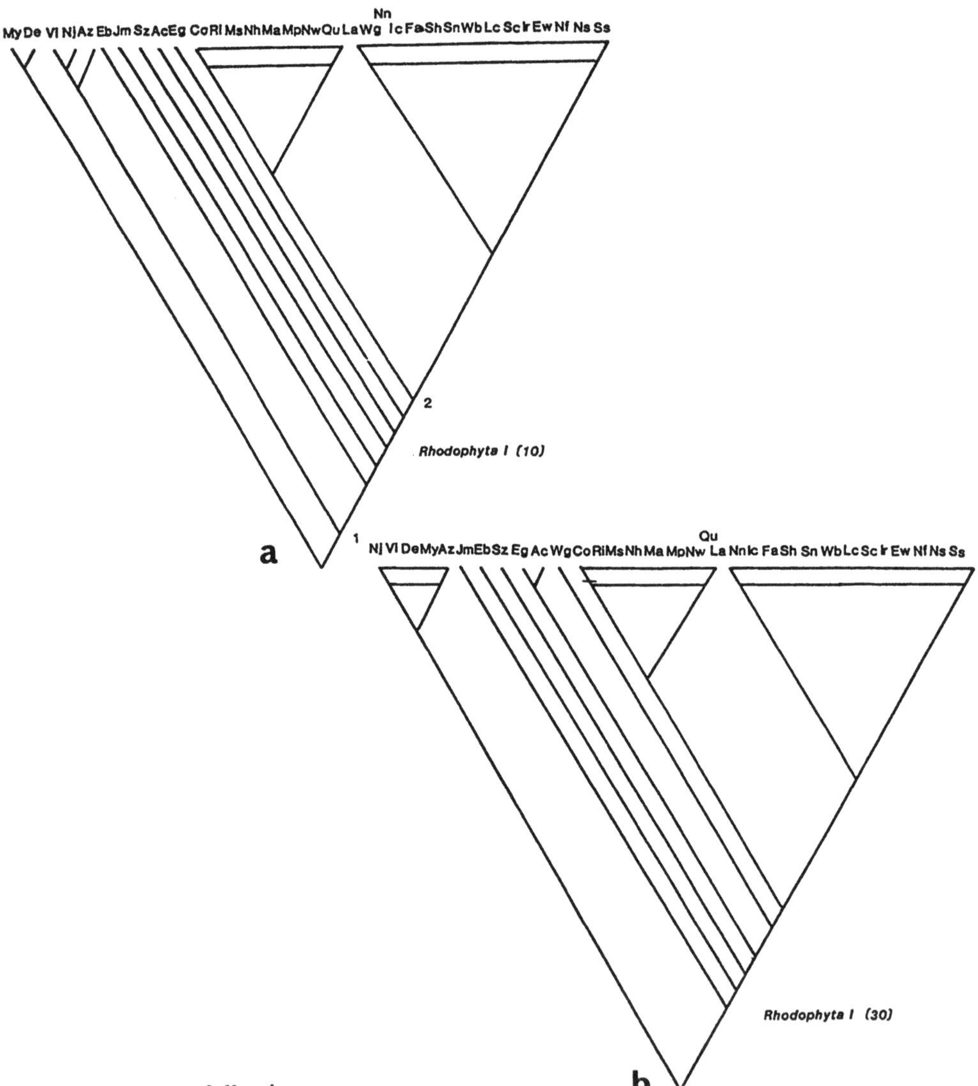

Fig. 4 cont. on following page

Fig. 4. Area ecocladogram for the North Atlantic: Rhodophyta; (a) analysis 5, cladogram 10; (b) analysis 5, cladogram 30; (c) analysis 4, cladogram 1; (d) analysis 4, scaled cladogram; (e) analysis 8, cladogram 1. (For abbreviations see caption to Fig. 1)

Fig. 4 cont.

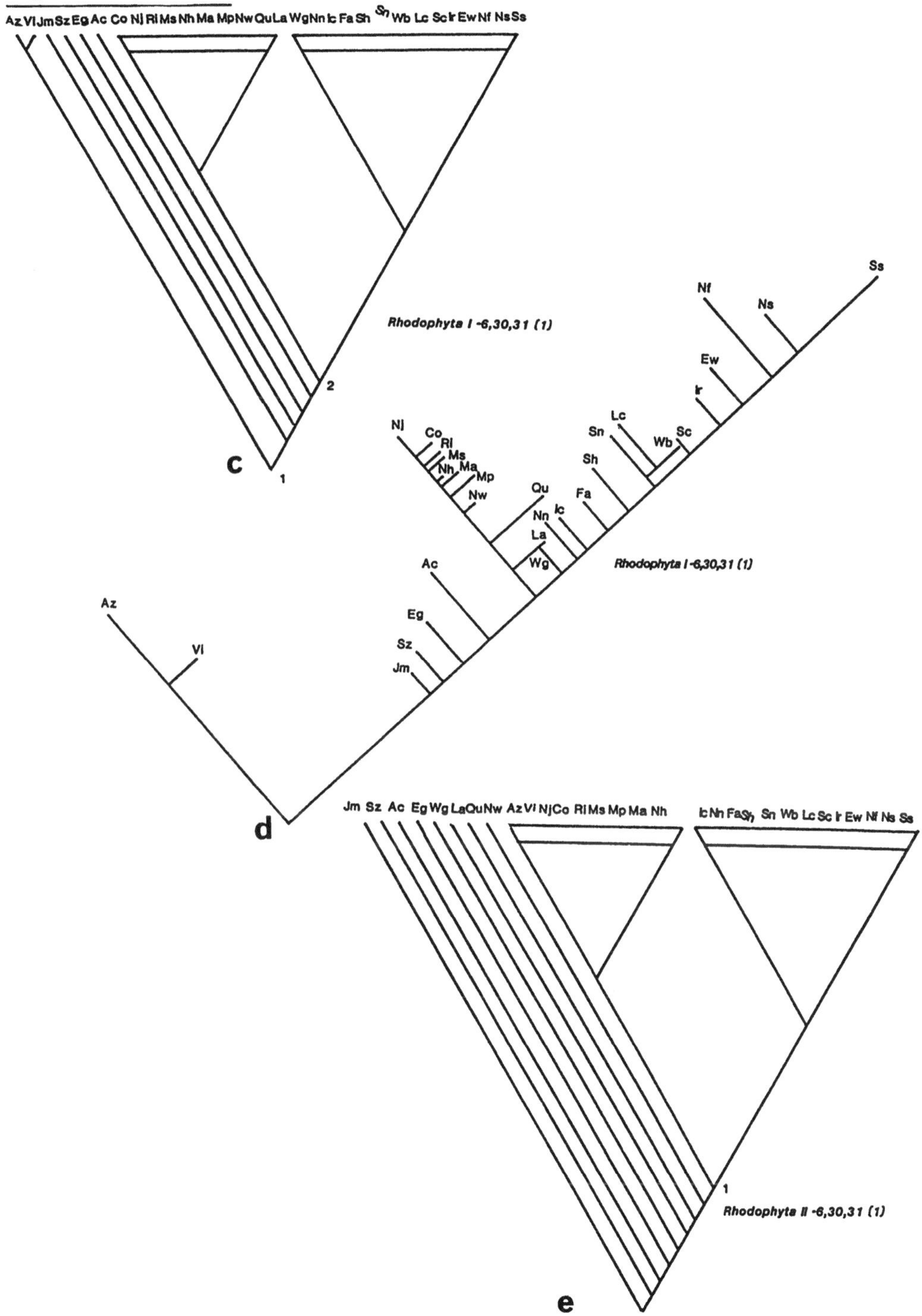
Az Vl Jm Sz Eg Ac Co Nj Rl Ms Nh Ma Mp Nw Qu La Wg Nn Ic Fa Sh Sn Wb Lc Sc Ir Ew Nf Ns Ss
Rhodophyta I -6,30,31 (1)
2
1
c
Ss
Nf
Ns
Ew
Ir
Lc
Sn
Wb
Sc
Sh
Nj
Co
Rl
Ms
Nh
Ma
Mp
Nw
Qu
Fa
Nn
Ic
La
Wg
Rhodophyta I -6,30,31 (1)
Ac
Az
Eg
Vl
Sz
Jm
d
Jm Sz Ac Eg Wg La Qu Nw Az Vl Nj Co Rl Ms Mp Ma Nh
Ic Nn Fa Sh Sn Wb Lc Sc Ir Ew Nf Ns Ss
1
Rhodophyta II -6,30,31 (1)
e

showed broadly similar branching patterns. Ecocladogram 50 (Fig. 5b) suggested that the North Atlantic flora comprised two unequally-sized floristic groupings (branching at node 1). The smaller encompassed the Azores, Virginia, Maryland, Delaware, New Jersey and Eastern Baltic floras; the larger group could be subdivided (more or less as described in analysis 9), but with a longer basal comb series of branches, encompassing the high Arctic and northern

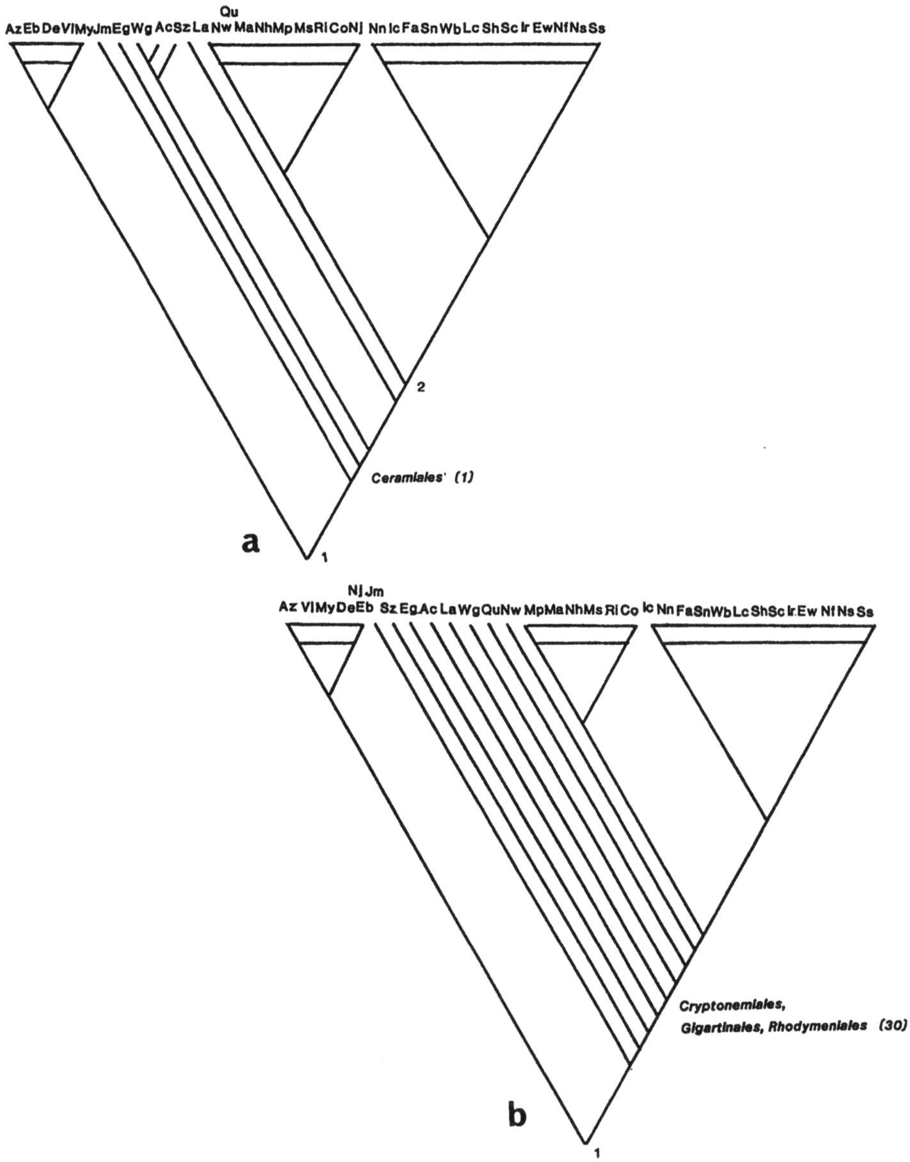

Fig. 5. Area ecocladogram for the North Atlantic: Rhodophyta; (a) analysis 9, cladogram 1; (b) analysis 20, cladogram 30. (For key to abbreviations see caption to Fig. 1)

Canadian floras, and a smaller north-eastern North American grouping. A variant (ecocladogram 25 - not illustrated) separated the Atlantic into two floristic groupings, but with the smaller group comprising the floras of Virginia, Maryland, Delaware and New Jersey; the rest of the Atlantic comprised the Azores flora, sister to the remaining flora.

Ordination analysis of the North Atlantic data

The plot of the first two axes of the Phaeophyta ordination (Analysis 14, Fig. 6a) showed two more or less linear arrangements of floras: one parallel to the axis 1, the other to axis 2; these met at a cluster of floras in the bottom left of the plot. The linear sequence along the first axis comprised a contiguous geographical south-to-north series of European floras, from the Azores on the right of the plot, to high Arctic sites at lower left; in the other direction the sequence was a contiguous north-to-south series of American floras from the high Arctic, to the Virginia, Maryland and Delaware floras at the top of the plot. This strongly suggested a clinally overlapping series of floristic relationships. The plot of axes 1 and 3 (not illustrated) maintained the linear arrangement (along axis 1), and separated only the Eastern Baltic flora on axis 3, underlining its dissimilarity from the other fully marine Atlantic floras. The overall trend in the ordination along axis 1, was from species-rich floras to species-poor (in the Arctic).

Ordination of the Chlorophytan floras (Analysis 15; Fig. 6b) was different from that of the Phaeophyta. The North Atlantic floras were largely grouped in a single cluster, suggesting (non-clinally) overlapping floristic relationships. Outliers were the species-poor floras of Arctic Canada (on axis 1), the Azores (on axis 2), Jan Mayen/Bear I. (axes 2 and 3) and Spitzbergen (on axis 3); the latter three formed the basal branches of the cladograms. It is interesting to note that the Eastern Baltic Chlorophytan flora did not appear to be distinct.

The ordination of Rhodophyta I (Analysis 16; Fig. 6c) mirrored that of the Phaeophyta (Analysis 14), with linear arrangements of European and American floras merging into the Arctic floras (lower left in the plot); the European sequence extended diagonally to axis 1, the American sequence along axis 2. In both cases, the most southerly sites were the furthest from the Arctic grouping. A plot of axes 1 and 3 maintained the linear arrangement of floras, but with the flora of Jan Mayen/Bear Island well separated along axis 3. As before, the main trend along axis 1 was from species-rich to species-poor sites. The overall results confirmed the clinal nature of floristic relationships in the North Atlantic.

The ordination of the second Rhodophyta subset (Analysis 17; Fig. 6d) showed the same broad patterns as the previous analysis (16), except that the American and European linear sequences were not so pronounced. The main difference in this ordination was the separation of the Azores flora from both principal sequences, along both axes 1 and 2, strongly

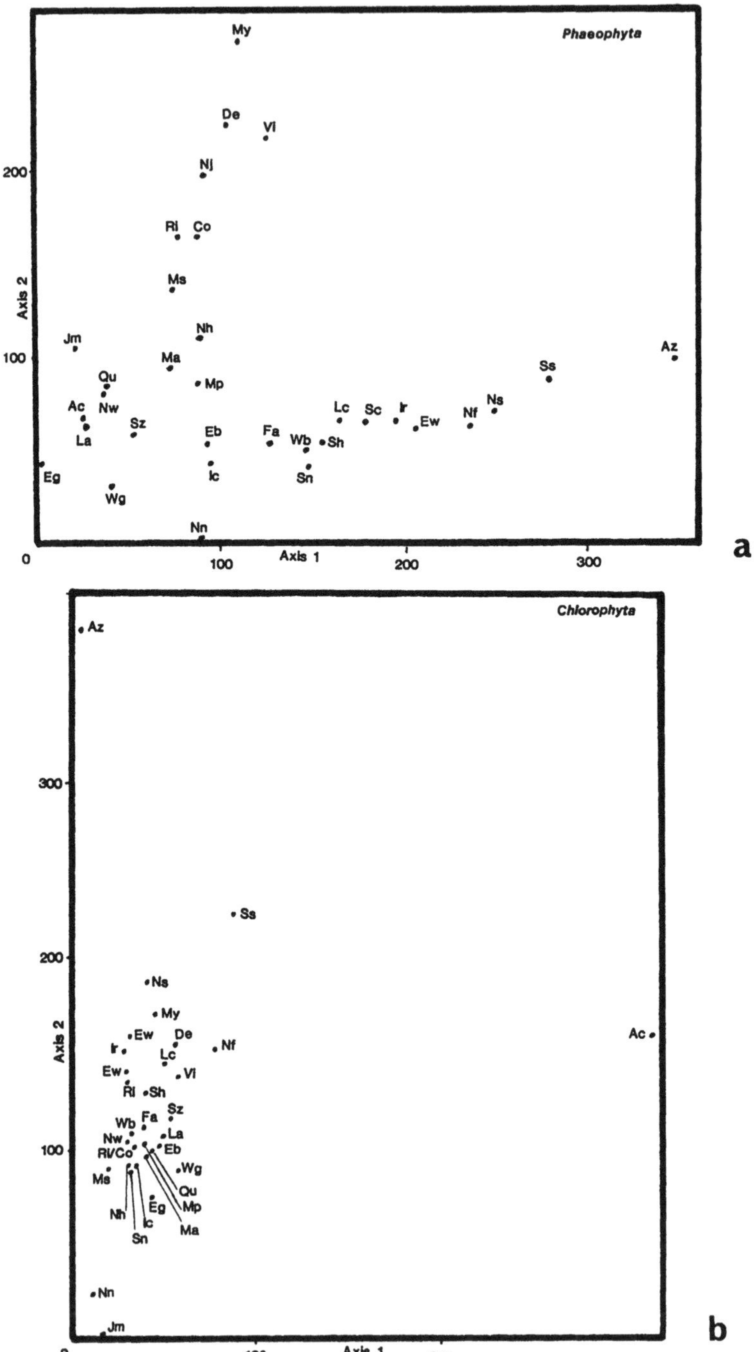

Fig. 6 cont. on following page

Fig. 6. North Atlantic ordinations; (a) analysis 14, Phaeophyta; (b) analysis 15, Chlorophyta; (c) analysis 16, Rhodophyta I; (d) analysis 17, Rhodophyta II. (For abbreviations see caption to Fig. 1)

Fig. 6 cont.

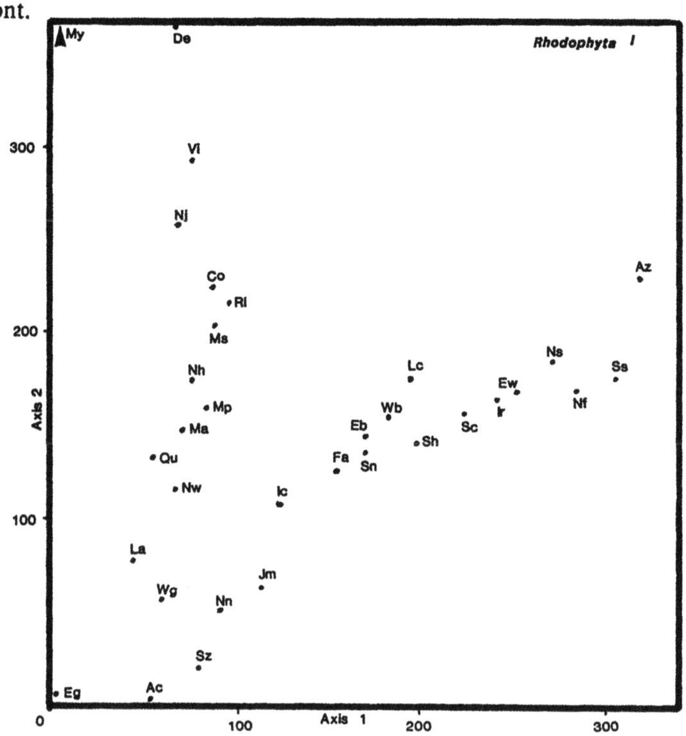

Rhodophyta I
My
De
Vi
Nj
Co
Rl
Ms
Nh
Mp
Ma
Qu
Nw
Ic
La
Wg
Nn
Jm
Sz
Eg
Ac
Fa
Eb
Sn
Wb
Lc
Sh
Sc
Ir
Ew
Ns
Nf
Ss
Az
Axis 2
Axis 1
0
100
200
300

c

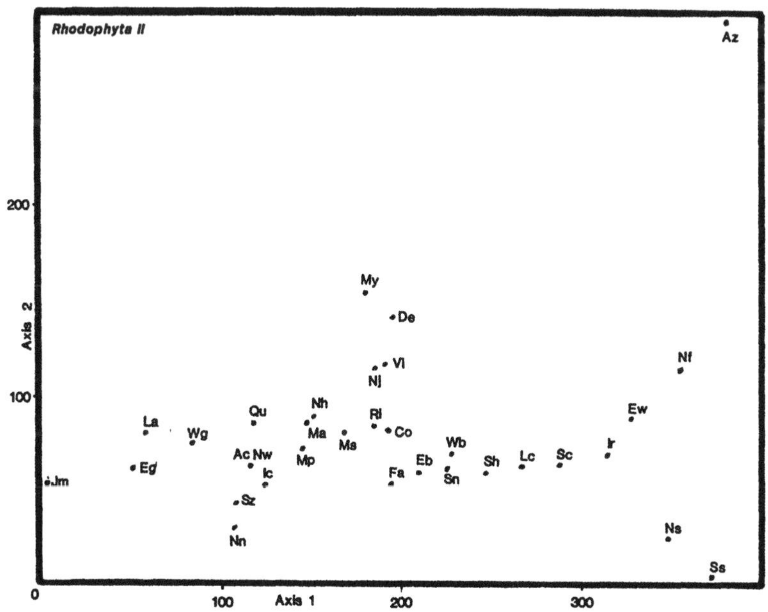

Rhodophyta II
Az
My
De
Vi
Nj
Nf
Ew
Ir
Qu
Nh
Rl
Ma
Co
Ms
La
Wg
Ac Nw
Ic
Mp
Eg
Jm
Sz
Nn
Wb
Eb
Fa
Sn
Sh
Lc
Sc
Ns
Ss
Axis 2
Axis 1
0
100
200
300

d

suggesting floristic discontinuity. The ordination of axes 1 and 3 again presented a more or less linear arrangement of floras, although the separation of the species-poor Jan Mayen flora cannot easily be explained.

The strong clinal series of floristic relationships outlined here is supported, in all four ordinations, by a spread of points (sites) along axes 1 and 2, exceeding two standard deviations and, in some cases, attaining almost four standard deviations.

Ecocladograms of the British Isles and adjacent floras

PHAEOPHYTA. In all fifty equally parsimonious ecocladograms produced in this analysis (11), groupings of floras were recognised. In ecocladogram 2 (Fig. 7a), the two basal branches represented the species-poor floras of Lincolnshire and South Wales. There then followed a short series of branches arising from a single branch (annotated "East" in Fig. 7a; node marked 1) representing the floras of Northern France (adjacent to Kent), Kent and Norfolk; the next two branches were the floras of Durham, and the Faroes and the Grampian Region. Successive branches represented: (i) Scottish floras (nodes marked 2, 3, 4); (ii) the floras of the south and west coasts of Britain (node marked 5). There was a small degree of geographical overlap between these floristic groupings. In other ecocladograms (e.g. 24, Fig. 7b), the same broad pattern of groupings was evident, but the Faroes flora (node marked 1) occurred at higher levels, often close to the south and west coast groupings (node marked 2). Occasionally, the east coast sites of Scarborough (Yorkshire) and Durham were associated with the Scottish sites. In all analyses, floras within the main groups were unordered. As in the North Atlantic results, the overall trend in branching reflected species-rich floras (high in the cladogram) to species-poor floras (basally).

RHODOPHYTA. The ecocladograms resulting from this analysis (Figures 7c, d) were similar in structure to those for the Phaeophyta: branching patterns revealed "East" coast, Scottish, south and west coast groupngs of floras. As before, the Scottish sites formed a series of branches (e.g. nodes marked 3, 4, 5, 6 in Fig. 7c) followed by a grouping of dichotomies arising from a single branch (e.g. node marked 7 in Fig. 7c). Again, there was little geographical arrangement within these groups; the floras of Lincolnshire and St Kilda formed the basal branches. In ecocladogram 5 (Fig. 7c), the Faroes flora was one of the basal branches (node marked 1), perhaps suggesting biogeographical separation. This was, however, contradicted by ecocladogram 10 (Fig. 7d), which arranged the Faroes flora at a higher level (node marked 1), nearer the Scottish group.

CHLOROPHYTA. Although the ecocladograms demonstrated three main floristic groups (Fig. 7e), these showed no obvious geographical arrangement; this result is consistent with the ordination analysis (see below). The branch representing the Faroes occurred high in the cladogram.

Ordination analysis of the British Isles and adjacent floras

The plot of axes 1 and 2 of the Phaeophyta (Analysis 18; Fig. 8a) contained a single large cluster of floras with a few outliers. Northern and eastern floras tended to occur to the left in the plot, western and southern sites to the right. Appraisal of axis 3 yielded no additional information. Ordination of the Chlorophyta (Analysis 19; Fig. 8b) again showed a single large cluster of floras with a few outliers; however, in this case, there was no positioning which

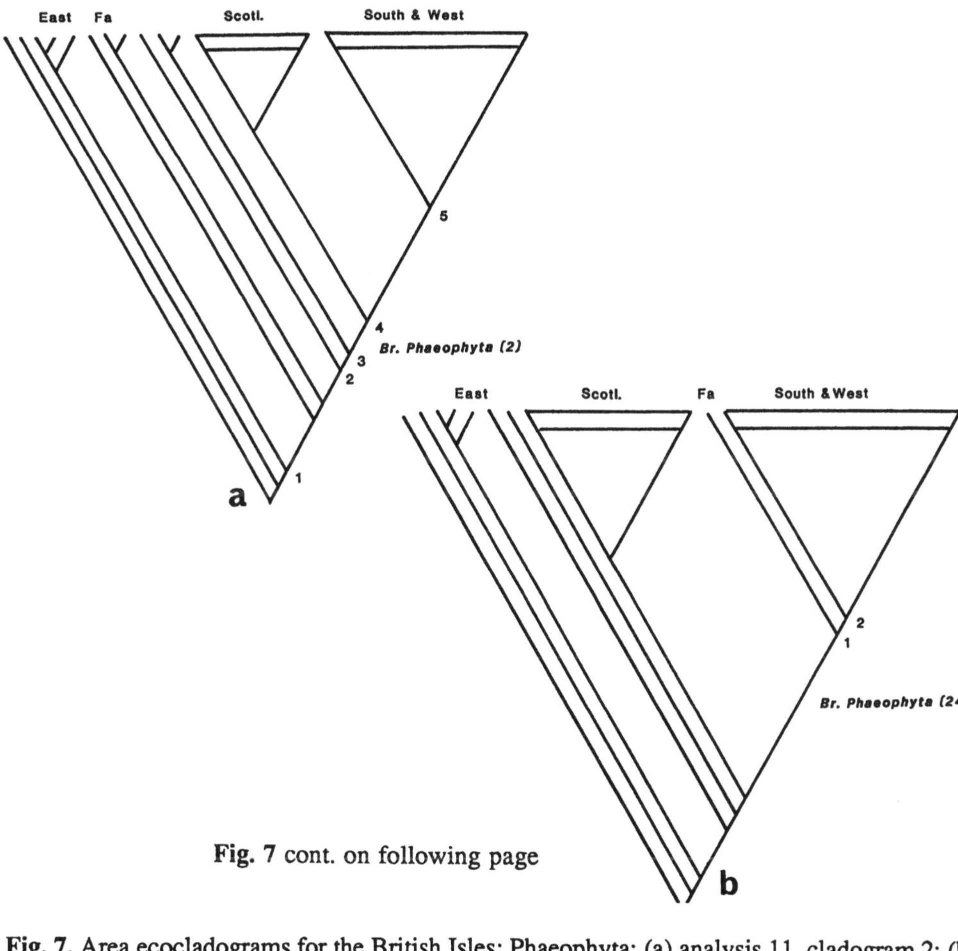

Fig. 7 cont. on following page

Fig. 7. Area ecocladograms for the British Isles: Phaeophyta; (a) analysis 11, cladogram 2; (b) analysis 11, cladogram 24. Rhodophyta: (c) analysis 12, cladogram 5; (d) analysis 12, cladogram 10. Chlorophyta; (e) analysis 13, cladogram 3

Fig. 7 cont.

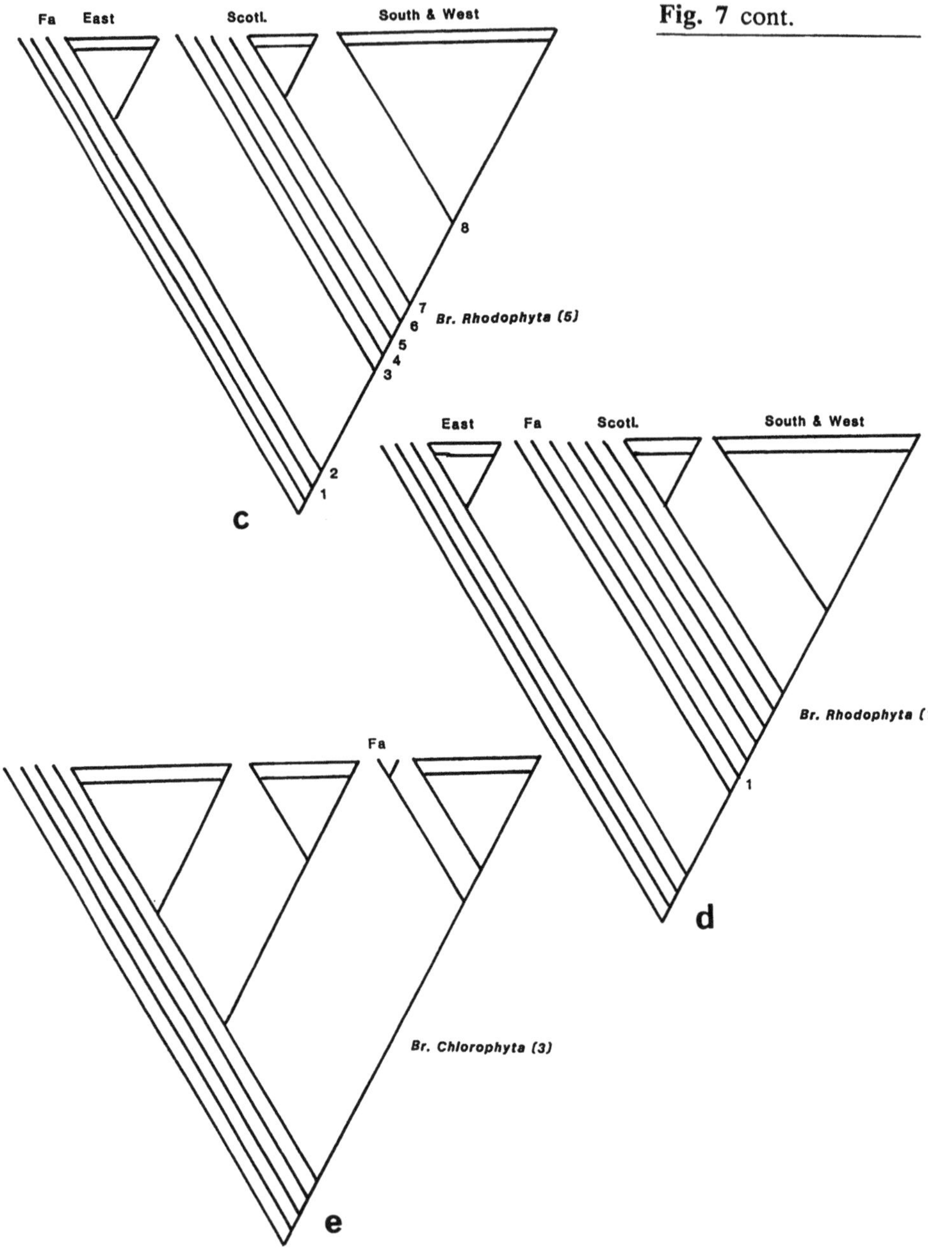

could clearly be related to geography. Appraisal of axis 3 yielded no further information. Ordination of the Rhodophyta (Analysis 20; Figure 8c) also showed a single large cluster of sites. As with the Phaeophyta, there was a geographical bias in the distribution of sites (floras) in the plot: northern and eastern floras were positioned to the right; southern and western floras to the left. The axis 3 yielded no additional information. These ordinations suggest no real biogeographical division between the Faroes and northern France, but simply a sequence of overlapping floras.

These ordinations contrast with those of the North Atlantic (see above), in that the spread of points (sites) along axes 1 and 2 falls within two standard deviations, suggesting similar and overlapping floras not strongly clinally related.

GENERAL DISCUSSION

Species impoverishment, determining the relationships of sites in both cladograms and ordinations, is a characteristic of many areas in the North Atlantic. As was mentioned previously, absences are not necessarily homologous, being attributable to a range of ecological factors; the latter, in the Arctic, are different from those affecting the floras of Maryland and Delaware, with both of these latter differing from those operating in the Eastern Baltic. It was interesting to note that the species-poor floras of the soft-substrate coastlines of the Low Countries and New Jersey, usually associated with their geographical nearest neighbours in both cladograms and ordinations.

That cladistic analyses produced more than one equally parsimonious cladogram in each case indicates considerable variability in floristic relationships. Much of this variation, suggested by the reordering of either basal comb branches of species-poor floras, or of ultimate branches, amounts to "noise" in the analysis and appears to be of little significance. The large number of possibilities expressed in the parsimonious trees reduces the likelihood of reading too much into individual trees. What were sought, in order to establish biogeographical discontinuities, were groupings based on consistent patterns in basal dichotomies.

The important points suggested by the consistency of branching patterns, that emerged from these analyses, were: (i) the separation of the Azores from the European floras; (ii) the association of the Azores flora with those of Virginia and adjacent areas; (iii) North Atlantic eastern and western groupings; (iv) clinal series of relationships among the latter two groupings; (v) a clinal series of high Arctic relationships which link eastern and western floristic groupings; (vi) local floristic groupings on the American coast. Most of these points could also be inferred from the ordination analysis.

Comparisons with previous appraisals

The problems of defining precise biogeographical boundaries reach back to the earliest investigations. Børgesen & Jónsson (1905) noted that, on both sides of the Atlantic, the Arctic

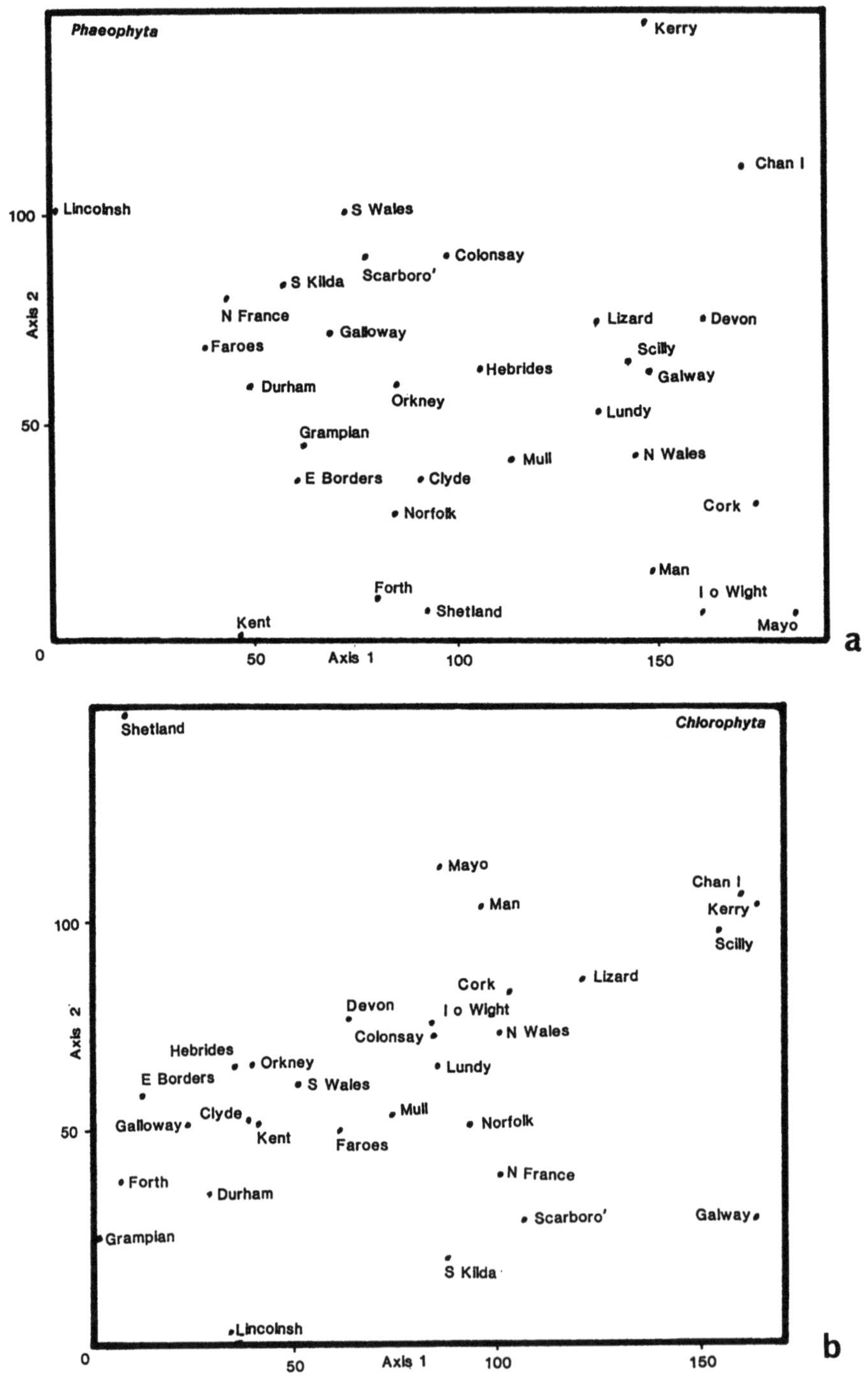

Fig. 8 cont. on following page

Fig. 8. British Isles Ordinations: (a) analysis 18, Phaeophyta; (b) analysis 19, Chlorophyta; (c) analysis 20, Rhodophyta

Fig. 8 cont.

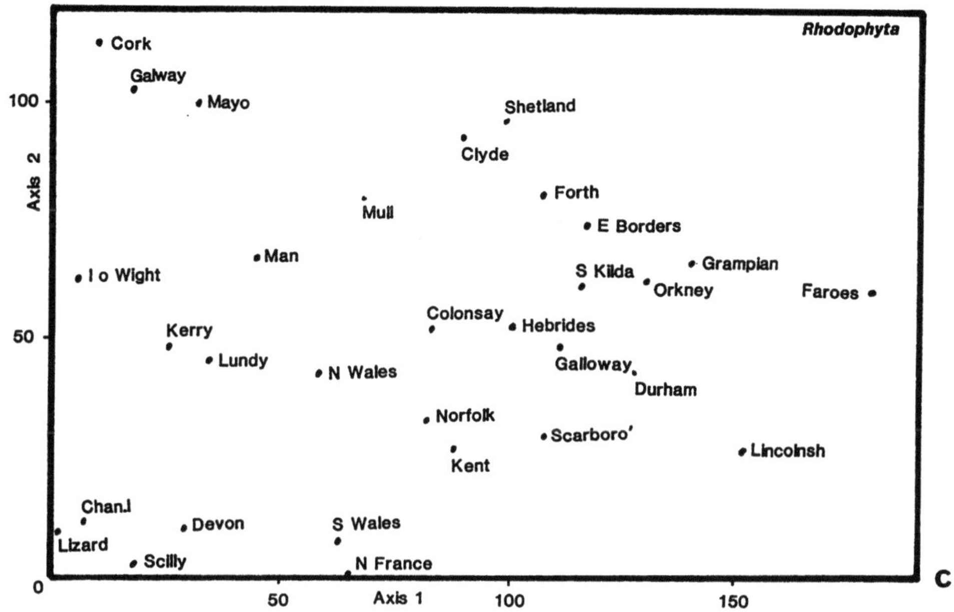

flora merged, without boundaries, into the boreal Atlantic flora. Although Farlow (1881) initially identified Cape Cod (on the American coast) as an important boundary, he subsequently pointed out that it could not be precisely defined, because some southern species occurred to the north, and some northern species occurred to the south. The central and northern of Stephenson & Stephenson's (1972) three biotic groups on the north east coast of America, showed markedly overlapping ranges, with component species occurring on either side of Cape Cod.

The present investigation supports the observations of earlier workers that there is a continuum of floristic relationships between high Arctic floras, and those to the south. Our results from ecocladistic analysis along the American coast suggest two possible (north and south) local floristic groupings. The evidence does not fix a boundary specifically at Cape Cod, but suggests a transition region. The ranges of site combinations depicted in the ecocladograms encompassed an area from Newfoundland, and occasionally even as far north as Labrador, to New Hampshire, Connecticut and occasionally as far south as New Jersey. Since these local floristic groupings were characterised by high level branching in the cladograms, they do not warrant major biogeographical recognition. Ordination emphasised the continuous nature of floristic relationships. We were unable to confirm, using either cladistic or ordination methods, the "Straits of Belle Isle marine benthic biogeographic boundary" to the north, between Newfoundland and Labrador (Hooper 1982; also inferred by Hoek 1975). Hooper considered this to be the boundary between the boreal and arctic provinces, because of a 35% decrease in species present in the flora from Newfoundland to

Labrador. On the basis of our analyses, we do not consider the Arctic, as currently understood, to be biogeographically distinct.

In an assessment of the algal phytogeography of the European Atlantic coasts, Hoek & Donze (1967) proposed "...one very obvious floristic discontinuity...along the coast of the British Isles..." and another between Arctic Europe and Spitzbergen. In a wider appraisal of North Atlantic biogeographic regions, Hoek (1975) identified two broad geographical groupings of sites (a clustering method was used to support the case for biogeographical regions): (i) the Canaries/North African, Mediterranean to Ireland floras; (ii) the Faroes to Spitzbergen floras. Within these groupings, five phytogeographic areas were identified. These results supported the earlier proposed biogeographical (floristic) discontinuities between Ireland and the Faroes, and between Arctic Europe and Spitzbergen. Alvarez *et al.* (1988), in a similar study, proposed boundaries slightly different to Hoek's (1975). An important boundary was identified as lying between the Shetlands and continental Europe (north of the Low Countries), and the British Isles; our results do not support this. That between Ireland and the Faroes could not be detected, in either North Atlantic or regional studies; the latter demonstrate overlapping floristic relationships in the area. Both the present, and an earlier analysis (Tittley *et al.* 1986), showed that the relatively species-poor Faroes flora had affinites with the species-poor flora of eastern Scotland. If there were a major discontinuity between the Faroes and British Isles floras, this should have been shown in the ecocladograms, by the Faroes flora forming a branch splitting off from other British local floras, and in the ordinations, by separation from the main cluster of British sites. One of the Rhodophytan ecocladograms (Fig. 7c) did suggest this; another (Fig. 7d) however, associated the Faroes with the Scottish Rhodophyta floras. In the Phaeophyta cladograms (Figs 7a and 7b), the consistently higher positioning of the Faroes branch suggested much stronger similarity between the Faroes and British floras than was previously thought; the same is true for the Chlorophyta. These results reflect the known trend of a decrease (more pronounced than in the Phaeophyta) in Rhodophytan species-diversity from south to north.

The evidence for a Norway-to-Spitzbergen discontinuity is stronger in the cladograms, with the species-poor Spitzbergen flora well separated from northern European floras and sister to other high Arctic floras. Whether this justifies a distinct biogeographic province is questionable. Seen from a whole North Atlantic perspective, the high Arctic floras relate more closely to one another, but also form part of a continuum of relationships, linking with the warmer water floras to the south; this is most clearly demonstrated in the ordinations. These high Arctic areas are floristically impoverished sites where endemicity is low.

One result from the Phaeophyta ecocladistic analyses strongly suggested a linked biogeographical division, on both eastern and western sides of the Atlantic. The association of the Azores and Iberian floras with an American grouping (Connecticut to Virginia, cf. Fig. 3a), initially appeared to support a division coming close to the boundary of the "cold temperate

Atlantic boreal region" as proposed by Hoek (1975). On the European coast a possible separation of the "western province of the cold temperate Atlantic boreal region" from the "Lusatian province of the warm temperate Mediterranean Atlantic region" (sensu Hoek) is suggested. The alternative ecocladogram in the analysis supports the latter, but without the American link. The Chlorophytan analyses were less clear-cut; the majority of ecocladograms (cf. Fig. 3d) placed the Iberian floras at the distal end at a series of branches, representing a cline of European floras, although alternative ecocladograms (cf. Fig. 3e) did link the Iberian flora with Virginia, in an American series of branches. Thus, the case for a biogeographical boundary is less clear here, and is weakened further by the results of the Rhodophytan analyses, which consistently placed the Iberian floras at the distal end of a clinal series of European floras. Ordination emphasised clinal floristic relationships along both European and American coasts, with no suggestion of a "cold temperate" boundary.

Appraisal of the Azores flora (Prud'homme van Reine 1988) showed that the flora shared species with America, Macaronesia, Eurafrica (Morocco to England) and the western Mediterranean; 10 species were endemic. Fewer species were shared with America, than with eastern Atlantic sites. A dendrogram of floristic similarities showed the Azores flora to be associated with a group of floras, ranging from Macaronesia and Morocco to England, and including the western Mediterranean (the other group comprised tropical, subtropical African and Cape Verdes floras). The former group could be further subdivided, with the Macaronesian flora a sister group to the Azores, Morocco - England, and western Mediterranean floras. The latter division placed the Azores flora sister to the rest. This confirms our results, which suggest a separation of the Azores and European floras. The possible affinity with the American flora suggested by our results could be, as Prud'homme van Reine pointed out, due to dispersal from America following the main Atlantic currents. However, it is apparent that the greater proportion of species on the Azores is shared with the western Atlantic flora (Prud'homme van Reine 1988), a feature less easily explained by the prevailing currents.

Interestingly, results of Alvarez *et al.* (1988) for the Chlorophyta depict relationships different to those for the Phaeophyta and Rhodophyta; this mirrors our results, using both ecocladistic and ordination methods, that showed floristic groupings to be less clearly identifiable. They attributed this to Chlorophycean cosmopolitanism, and in this we are agreed.

We have not attempted from our results to define any major biogeographical unit, save to note that the Azores flora in the ecocladograms was consistently separable from the European flora, and showed affinites with the American flora. Although we identified a separation of eastern and western Atlantic algal floras, in both sets of analyses they were inter-related by the floras of the Arctic region. Is the northern North Atlantic benthic marine algal flora essentially an American flora (common to both sides), with distributions facilitated by the major North Atlantic currents, but enriched, in Europe, with additional southern European/Mediterranean/African elements? These latter represent an important source of eastern

Atlantic endemicity; about 60% of species are endemic to the European coast, only 5% endemic to the American coast, and 35% common to both sides. The creation of biogeographical provinces in the North Atlantic has hitherto been based more on species absence, than on endemicity. We would suggest that endemicity is important in determining biogeographical provinces.

Problems with methodology

The difficulties in undertaking ecological-biogeographical analyses basically relate to the problems of sampling. The strength of the present analysis lies in the use of a larger number of sampling sites, within a geographical area smaller than that appraised by previous workers. Large-scale biogeographical units are best demonstrated through the clustering or grouping of many small units. The floristic discontinuity between Ireland and the Faroes identified by Hoek & Donze (1967), may not have emerged had Scottish sites been included in the analysis. The study took in only twelve sites and approximately 25% of the total flora. Appropriate sampling is important, and will significantly affect the results of an analysis, irrespective of method used, e.g. Lawson (1978) ordinated the Hoek & Donze (1967) data and produced results similar to theirs.

All authors have pointed out that achieving comparable sampling areas is also a major problem; our work, similarly, suffers in this respect. Some sites are too small (Delaware, Maryland); others (Eastern Greenland) too large. We would agree with Lawson (1988) that the ideal way to undertake these analyses would have been to use unit lengths of coastline, which would have eliminated the problems caused by Delaware and Maryland. We also have reservations about omitting sites from analyses; we should perhaps have combined them (the computer program did not allow for this, only omissions were possible). But what should be done about such ecologically very different areas as the Eastern Baltic? The availability of floristic data is still very patchy in the North Atlantic, despite the fact that the flora is better known than in other oceans; there are also many poorly-understood groups, making determination difficult and imprecise, with inevitable consequences for the present type of study.

Another problem relates to the interpretation of results, and the definition of the hierarchy of biogeographical regions, provinces and subordinate areas. The nomenclature used, hitherto, has also been neither clear nor consistent. Although dendrograms have been used to express floristic similarity, they appear not to have been used to help define biogeographical units. To have used specific cut-off levels would have been helpful in the uniform definition of a biogeographical unit. Hoek's (1975) dendrogram (Fig. 4 in his paper) could, for example, have been interpreted (applying the same cut-off similarity levels used elsewhere in the dendrogram) to show that his "cold temperate North Atlantic" and "Arctic" regions constitute a single unit. Similarly, from the Alvarez *et al.* (1988) dendrogram (Fig. 1 in their paper), an "Arctic region" was proposed on the grounds of the Arctic Canadian and Spitzbergen floras clustering together (despite the separation of the Greenland flora), but what the dendrogram actually shows is that their Arctic Canada, Spitzbergen, temperate USA and cold temperate Canadian floras are together more similar, and constitute a unit, together with the Greenland flora - again, suggesting no clear separation of Arctic and temperate floras. Another puzzle is that of why their subsequent analysis of biogeographical regions (combinations of floras) gave results at variance with their results for floras. This may have been a reflection of the different sampling units used. Dendrogram 6A (in their paper) could also be interpreted (using designated cut-off levels) to suggest that their North Atlantic regions constitute a single North Atlantic unit. These problems associated with clustering techniques have been discussed by Carney *et al.* (1983).

The present study has used ecocladistic methodology experimentally, to help define biogeographical relationships. That the results did not conflict seriously with those obtained by employing detrended correspondence analysis (which has previously been successfully used in biogeography) is encouraging for the future. Both methods suggested, overall, a clinal series of floristic relationships in the North Atlantic, and the linking of east and west Atlantic floras in the Arctic. Cladistic analysis, additionally, brought out the possible Azores - America relationship inferred in only one Decorana plot. It also suggested a stronger separation of the Azores and European floras, than was indicated by ordination.

CONCLUSIONS

Before algal biogeographical studies can constructively proceed much further, the problems relating to basic methodology, especially to sampling, must be addressed.

Our results show that there is little evidence for deviation from our null-hypothesis, that major biogeographical provinces are not recognisable in the northern North Atlantic Ocean. The Azores to Virginia floristic affinity may form an exception to this, possibly indicating a boundary in the extreme south of the area. Apart from an east-west separation of floras (south of the Arctic), the North Atlantic flora is characterised by a continuous pattern of overlapping and gradually changing floras. This facet of the situation has been at least tacitly recognised by all workers since biogeographical studies began in the North Atlantic Ocean.

ACKNOWLEDGEMENTS

We wish to thank J.H.Price (NHM), Dr P. Legendre (Université de Montreal) and an unnamed reviewer for critically reading the manuscript.

REFERENCES

Alvarez M, Gallardo T, Ribera MA, Gomez Gareta A (1988) A reassessment of Northern Atlantic seaweed biogeography. Phycologia 27: 221-233

Børgesen F, Jónsson H (1905) The distribution of the marine algae of the Arctic Sea and of the northernmost part of the Atlantic. In: Warming E (ed), Botany of the Faeröes based upon Danish investigations. Part III: Appendix. Gyldendalske, Copenhagen and Christiana, p I-XXVII

Burrows EM (1958) Sublittoral algal population in Port Erin Bay, Isle of Man. J Mar Biol Ass U K 37: 687-703

Burrows EM (1960) A preliminary list of the marine algae of the Galloway coast. Br Phycol Bull 2: 23-25

Burrows EM (1964) A preliminary list of the marine algae of the coast of Dorset. Br Phycol Bull 2: 364-368

Carney RS, Haedrich RL, Rowe GT (1983) Zonation of fauna in the deep sea. In: Rowe GT (ed) The Sea, vol 8. Wiley, New York, p. 371-398

Chapman VJ (1937) A revision of the marine algae of Norfolk. Bot J Linn Soc 51: 205-263

Clokie JJP, Boney AD (1979) Check-list of marine algae in the Firth of Clyde. Scot Fld Stud 1979: 3-13

Dixon PS (1961) List of marine algae collected in the Channel Islands during the joint meeting of the British Phycological Society and the Société Phycologique de France September 1960. Br Phycol Bull 2: 71-80

Edwards P (1976) An assessment of possible pollution effects over a century on the benthic marine algae of Co. Durham, England. Bot J Linn Soc 70: 269-305

Farlow MD (1881) Marine algae of New England and adjacent coast. Government Printing Office, Washington

Farnham WF (1982) Species list of the benthic algae of the Isle of Wight. Portsmouth Polytechnic, Portsmouth.

Fletcher A, Jones WE (1975) The second report of the Coastal Surveillance Unit. December 1975. University College of North Wales, Bangor

Fletcher A, Jones WE (1976) The third report of the Coastal Surveillance Unit. December 1976. University College of North Wales, Bangor

Green RH (1980) Multivariate approaches in ecology: an assessment of ecological similarity. Ann Rev Ecol Syst 11: 1-14

Guiry MD (1978) A consensus and bibliography of Irish seaweeds. Bibliotheca Phycologiuca 44: Cramer, Vaduz

Hoek C van den (1975) Phytogeographic provinces along the coasts of the Northern Atlantic Ocean. Phycologia 4: 317-330

Hoek C van den (1984) World-wide latitudinal and longitudinal seaweed distribution patterns and their possible causes, as illustrated by the distribution of Rhodophytan genera. Helgol Meeresunters 38: 227-257

Hoek C van den, Donze M (1967) Algal phytogeography of the European Atlantic coasts. Blumea 15: 63-89

Hooper RG (1982) The Straits of Belle Isle marine benthic biogeographic boundary. Br Phycol J 17: 233.

Irvine DEG (1974) The marine vegetation of the Shetland Isles. In: Goodier R (ed) The natural environment of Shetland. Nature Conservancy Council, Edinburgh, p 107-113
Irvine DEG (1980) Marine algae. In: Berry RJ, Johnson JL (eds) The Natural History of Shetland. Collins, London, p 267-272
Irvine DEG (1982) Seaweeds of the Faroes 1: The flora. Bull Br Mus Nat Hist (Bot) 10: 109-131
Joosten AMT, Hoek C van den (1986) World-wide relationships between red seaweed floras: a multivariate approach. Bot Mar 29: 195-214
Jones EB (1980) Marine micro-organisms and seaweeds in the Solent. In: Anon (ed) The Solent estuarine system An assessment of present knowledge. Natural Environmental Research Council, Swindon, p 64-72
Jones WE, Fletcher A, Bennell SJ, McConnell BJ, Mack Smith S, Mitchell J, Roberts CM (1977) The fourth report of the Coastal Surveillance Unit.... University College of North Wales, Bangor
Kain JM (1958) Observations on the litoral algae of the Isles of Wight. J Mar Biol Ass U K 37: 769-780
Kain JM (1960) Direct observations on some Manx sublittoral algae. J Mar Biol Ass U K 39: 609-630
Knight M, Parke MW (1931) Manx algae An algal survey of the south end of the Isle of Man. Proc Trans Liverpool Biol Soc 45: 1-155
Lambshead PJD (1986) Sub-catastrophic sewage and industrial waste contamination as revealed by nematode faunal analysis. Mar Ecol Progr Ser 29: 247-260
Lambshead PJD, Paterson GLJ (1986) Ecological analysis - an investigation of numerical cladistics as a method for analysing ecological data. J Nat Hist 20: 895-909
Lawson GW (1978) The distribution of seaweed floras in the tropical and subtropical Atlantic Ocean: a quantitative approach. Bot J Linn Soc 76: 177-193
Lawson GW (1988) Seaweed distribution patterns as revealed by ordination with reference to the Atlantic and southern Oceans. Helgol Meeresunters 42: 187-189
Legendre P (1986) Reconstructing biogeographic history using phylogenetic-tree analysis of community structure. Syst Zool 35: 68-80
Lindstrom SC (1987) Possible sister groups and phylogenetic relationships among selected North Pacific and North Atlantic Rhodophyta. Helgol Meeresunters 41: 245-260
Lundberg JG (1972) Wagner networks and ancestors. Syst Zool 21: 398-413
Malloch AJC (1988) Vespan II. A computer package to handle and analyse multivariate species data and handle and display species distribution data. University of Lancaster, Lancaster
Mill JS (1872) Book IV: of operations subsidiary to induction, a system of logic ratioinactive and inductive. In: Robson JM (ed) John Stuart Mill: Collected Works 8. Toronto University Press, Toronto
Minchin P (1987) An evaluation of the relative robustness of techniques for ecological ordination. Vegetatio 69: 89-107
Norton TA (1972) The marine algae of Lewis and Harris in the Outer Hebrides. Br Phycol J 7: 375-385
Norton TA (1976) The marine algae of the eastern border counties of Scotland. Br Phycol J 11; 19-27
Norton TA, McAllister HA, Conway E, Irvine LM (1969) The marine algae of the Hebridean island of Colonsay. Br Phycol J 4: 125-136
Paterson GLJ (1986) Aspects of the zoogeography of some benthic animals in the Rockall Trough. Proc R Soc Edinburgh 88B: 316-318
Perkins D (1953) Marine algae-seaweeds. pp. 67-80. In: Walsh GR, Rimington FC The natural history of the Scarborough district, vol 1. Scarborough Field Naturalists' Society, Scarborough
Pielou EC (1984) The interruption of ecological data: a primer on classification and ordination. Wiley & Sons, New York
Price JH, Hepton CEL, Honey SI (1980) The inshore benthic biota of the Lizard Peninsula, South West Cornwall. I. The marine algae: history; Chlorophyta; Phaeophyta. Cornish Stud 7: 7-37

Price JH, Hepton CEL, Honey SI (1981) The inshore benthic biota of the Lizard Peninsula, South West Cornwall. II. The marine algae: Rhodophyta; discussion. Cornish Stud 8: 5-36
Price JH, Tittley I (1978) Marine algae (excluding diatoms). Chapter 19. In: Jermy, AC, Crabbe JA (eds), The Island of Mull, a survey of its flora and environment... British Museum (Natural History), London, p 19.1-19.37
Price JH, Tittley I, Honey SI (1977) The benthic marine flora of Lincolnshire and Cambridgeshire: a preliminary survey. Naturalist, Hull 102: 3-20, 91-104
Prud'homme van Reine WF (1988) Phytogeography of seaweeds of the Azores. Helgol Meeresunters 42: 165-185
Russell G (1968) List of marine algae from the Isles of Scilly. Br Phycol Bull 3: 579-584
South GR (1987) Biogeography of the benthic marine algae of the North Atlantic Ocean - an overview. Helgol Meeresunters 41: 273-282
South GR, Tittley I (1986) A checklist and distributional index of the benthic marine algae of the North Atlantic Ocean. Huntsman Marine Laboratory and British Museum (Natural History), St Andrews and London
Stephenson TA, Stephenson A (1972) Life between tidemarks on rocky shores. W.H.Freeman and Co, San Francisco
Swafford DL (1985) Phylogenetic analysis using parsimony. PC version 2.4
Ter Brak CJF, Prentice IC (1988) A theory of gradient analysis. Adv Zool Res 18: 271-317
Tittley I, Price JH (1977) An atlas of the seaweeds of Kent. Trans Kent Fld Club 7: 1-80
Tittley I, Price JH (1978) The benthic marine algae of the eastern English Channnel: a preliminary floristic and ecological account. Bot Mar 212: 499-512
Tittley I, Price JH, Fincham AA, George JD (1986) The macrobenthos of Chalk and Greensand Shores in Southeastern England. A report for the Nature Conservancy Council. British Museum (Natural History), London
Watling R, Irvine M, Norton TA (1970) The marine algae of St. Kilda. Trans Proc Bot Soc Edinburgh 41: 31-42
Whittaker JE (1980) The Fleet Dorset- A seasonal study of the watermass and its vegetation. Proc Dorset Nat Hist Archaeol Soc 100: 73-99
Wilkinson M (1975) The marine algae of Orkney. Br Phycol J 10: 387-397
Wilkinson M (1979) Marine algae of the Grampian region of Scotland. Br Phycol J 14: 33-41
Wilkinson M (1980) The marine algae of Galloway. Br Phycol J 15: 265-273
Wilkinson M (1982) Marine algae from Glamorgan. Br Phycol J 17: 101-106

ROLE OF THE ARCTIC OCEAN AS A BRIDGE BETWEEN THE ATLANTIC AND PACIFIC OCEANS: FACT AND HYPOTHESIS

Robert T. Wilce
Department of Botany
University of Massachusetts
Amherst, Massachusetts 01003
U.S.A.

Abstract - The Arctic Ocean is recognized as a 'bridge' for marine biota between the Atlantic and Pacific Oceans. The present-day Arctic Ocean marine flora is essentially a depauperate extension of the N Atlantic flora, with few endemic species. It is 'stamped' by the environment in its adaptations to cold temperatures and low light conditions. Characteristics and geographic limits of the Arctic biome are discussed, as well as previous misconceptions of this biome. This flora contains five floristic groups: N Pacific, N Atlantic, Indo-Pacific, cosmopolitan and endemic species; these groups are based on phylogenetic relationships, distribution tracks, and presumed origins. The recognition of these origins is also based on assumptions of genome antiquity, and morphological and reproductive stasis, i.e., continuity of species traits from a pre-Cretaceous, possibly early Mesozoic, ancestry.

Trans-migration of the N Atlantic-N Pacific biota via the Arctic Ocean occurred periodically during the Tertiary, and was dependent primarily upon massive plate movements. During its evolution, periodic marine incursions, with accompanying biota, occurred from four ocean areas: N Pacific, present Caribbean via Cannonball Seaway, present Mediterranean via Turgai Strait, and the N Atlantic. Post-Pleistocene algal exchange between the N Pacific and N Atlantic is not evident. Post-Pleistocene species migration between the North Atlantic and Arctic Ocean is on-going and dependent only on temperature adaptation.

THE ARCTIC ENVIRONMENT AND ITS SEAWEED FLORA

Phycological literature contains little information concerning the attached algae actually from arctic marine environments. Much of this literature pertains to species from the subarctic and even cold temperate regions (Kjellman 1883; Farlow 1886; Rosenvinge 1893, 1933; Børgesen & Jónsson 1905; Taylor 1954; Wilce 1959). The physical and biological character of this environment, its attached flora and so-called "indicator" species have long been misunderstood (Taylor 1957; and others). One may easily be convinced of its arctic character in the presence of large masses of floe ice, formation of an ice foot, seasonally sporadic ambient water temperatures of -2 to 4^0C., and lists of species known to occur in high arctic areas. However, both floe ice and cold temperatures occur seasonally throughout the subarctic, the Gulf of Maine, and much of the Icelandic and the southern Svalbard coasts -- but these are not arctic marine environments. Species lists must be evaluated carefully before they are used as

NATO ASI Series, Vol. G 22
Evolutionary Biogeography of the
Marine Algae of the North Atlantic
Edited by D. J. Garbary and G. R. South

criteria to assign an arctic character to a marine environment. Most benthic algal species from truly arctic environments also occur along the subarctic and temperate coasts cited above; examples of these species are cited in Table 1. Species in Table 2 are also North Atlantic, all with arctic distribution, and most known from the subarctic, temperate areas of the northern Atlantic. One, *Climacosorus mediterranus*, is known from several high arctic areas, the temperate coast of Norway and from several areas of subtropical Mediterranean. All these taxa have enigmatic distribution patterns, and, as a result, are difficult to use as indicators of phytogeographic areas. Differences in the biota of arctic and subarctic shores and sublittoral are subtle --less so with the biota of the marine arctic and that of temperate areas. Floristic differences, patterns, species form and the structure and composition of species assemblages are all helpful in zone recognition.

Definition of the arctic environment

The origin and character of the marine waters can be considered equally important in zone assessment. Arctic marine areas are those influenced primarily by unmixed marine waters of polar origin or other waters of long residence in the polar sea (Dunbar 1968). These are waters of minimal ambient temperatures with a yearly average of -2 to 4 (+5) degrees C. Nevertheless, in some arctic environments, e.g., areas of river outflow, closed embayments, heads of fjords, and where tidal waters are trapped by arctic low shore boulder barriers, surface water temperatures fluctuate dramatically and may even exceed 5^0C. (Madsen 1940; pers. obs.).

Delineation of the arctic marine environment should not be based on inshore water temperatures since the physical characteristics of the water column are by nature so variable.

Table 1. Seaweed taxa with northern but not arctic distributions. For authorities of taxa listed in this and following tables, and in the text, see South & Tittley (1986) and Scagel *et al.* (1989)

CHLOROPHYTA	PHAEOPHYTA	RHODOPHYTA
Derbesia marina	*Ectocarpus* sp.	*Bangia* sp.
Enteromorpha (most)	*Coilodesme* sp.	*Dumontia contorta*
Cladophora sp.	*Papenfussiella* sp.	*Peyssonnelia* sp. (*rosenvingii*)
	P. plantaginea	*Fimbrifolium dichotomum*
	Laminaria digitata	*Corallina officinalis*
	L. nigripes	*Callithamnion* sp.
	Fucus vesiculosus	

Benthic zoologists and phycologists alike return to species occurrence which reflect time-averaged water temperature regimes, to suggest phytogeographic provinces and their boundariès (Briggs 1974; Hoek 1975; 1982a,b; South 1983, 1987). Even then, lines of delineation on this basis are sometimes difficult to apply except where dramatic phytogeographic barriers occur (e.g., Cape Cod, Massachusetts, Strait of Belle Isle, Newfoundland, Cape Dyer, Baffin Island, Simmons 1906; Madsen 1940; Lee 1973; Pielou 1977; Wilce 1977; South 1983, 1987). What may be an obvious phytogeographic province for littoral and shallow sublittoral species of one area may not be as apparent for other species of the deeper sublittoral in the same area (Steele 1974; South 1983; Wilce pers. obs.). For example, the littoral zones of Disko Bay and Pangnirtung Fjord must be considered subarctic

Table 2. Arctic/northern species of seaweeds with enigmatic distributions. Note: no Chlorophyta are included in this list

CHRYSOPHYTA	PHAEOPHYTA	RHODOPHYTA
Phaeosaccion collinsii	*Pilayella macrocarpa*	*Peyssonnelia rosenvingii*
	Climacosorus mediterranus	*Dumontia*-like red*
	Agarum cribrosum	*Callophyllus cristata*
		Kallymenia schmitzii
		Kallymeniaceous red#
		Turnerella pennyi

*Sachs H.

#Ellesmere Island; branched

Table 3. Species composition of sublittoral loose-lying algal populations in stressed arctic environments

CHLOROPHYTA	PHAEOPHYTA	RHODOPHYTA
Chaetomorpha linum	**Sorapion simulans*	*Phyllophora interrupta*
C. melagonium	**Petroderma maculiformis*	*Rhodomela confervoides*
Chlorochytrium schmitzii	**Pseudolithoderma* sp.	coralline crusts including
	Desmarestia viridis	**Lithothamnion glaciale*
	D. aculeata	

*present with shell and pebbles

environments. Both areas contain a number of algal species (e.g., *Fucus vesiculosus*, *Ascophyllum nodosum*, digitate *Laminaria*, *L. digitata* and *L. nigripes*, *Cladophora rupestris*, *Phymatolithon laevigatum* and *Peyssonnelia rosenvingii*) with continuous species tracts southward into the Gulf of Maine, Iceland and the Norwegian coasts (Taylor 1957; Caram & Jónsson 1972; Rueness 1977). These shore areas are hardly arctic in character although winter conditions suggest the arctic environment. The species cited above, along with a variety of littoral gastropods and other algae, are inhabitants of subarctic shores, but do not extend into the arctic on the Baffin coast or through the Vaigatt on the west Greenland coast (Madsen 1940; Vibe 1950; Feyling-Hanssen 1953; Petersen 1962; Wilce, pers.obs.). In contrast to the ameliorated subarctic shore environment, the sublittoral and its vegetation below 10 m on the Baffin Island and west Greenland coasts have a strongly arctic character, reflecting colder, and generally less variable, water temperatures. Even so, species occurrence and subtle differences in the structure of the sublittoral vegetation suggest subarctic biogeographic provinces. Both southern Baffin and most of west Greenland are influenced by the mixed warm water of the Irminger Current, which, though not easily detectable by the phycologist working in the field, demonstrates its presence by moderating the marine environment and influencing species

Table 4. Common denominator species: arctic survivors. List includes 9 Chlorophyta, 15 Phaeophyta and 17 Rhodophyta, for a total of 41 species

CHLOROPHYTA	PHAEOPHYTA	RHODOPHYTA
Chlorochytrium schmitzii	*Pilayella littoralis*	Dilsea integra
C. inclusum	*Petroderma maculiforme*	*Ahnfeltia plicata*
Endocladia viridis	*Sorapion simulans*	*Palmaria palmata*
Ulothrix flacca	*Lithoderma fatiscens*	*Ceratocolax hartzii*
Urospora wormskjoldia	*Pseudolithoderma* sp.	*Harveyella mirabilis*
Chaetomorpha linum	*Sphacelaria arctica*	*Turnerella pennyi*
C. melagonium	*S. plumosa*	*Clathromorphum compactum*
Rhizoclonium riparium	*S. caespitula*	*C. circumscriptum*
Blidingia minima	*Desmarestia aculeata*	*Leptophytum laeve*
	D. viridis	*L. foecundum*
	Chorda tomentosa	*Lithothamnion glaciale*
	Laminaria saccharina	*Phyllophora interrupta*
	Alaria esculenta	*Odonthalia dentata*
	Agarum cribrosum (?)	*Scagelia boreale*
		Polysiphonia arctica

occurrence. Sparse populations of *Mytilus edulis* and *Balanus balanoides* occur in the northern part of Melville Bay, west Greenland and suggest marine conditions different from those in the Thule District immediately to the north, or from those on the Canadian side of Baffin Bay at similar latitudes.

The arctic flora- a response to Kjellman

Despite misconceptions about the arctic marine environment and its seaweed flora, there is a very general picture of that province and its flora (Lund 1951, 1959; Lee 1966; Lee 1973, 1980). In this paper, I hope to clarify this picture with additional information as to what constitutes the Arctic Ocean flora, and the flora of its peripheral seas. This new approach should also broaden our understanding of floristic relationships among northern algal floras.

Our current misconception of the arctic flora has its roots with Kjellman (1883) who stated, that "In the by far the greater part of the Arctic Sea the littoral zone possesses no vegetation at all or an exceedingly meagre one." He continues, "In the Kara Sea traces of a littoral vegetation have been detected only in two places.", while, "The main mass of the vegetation may be said to be diffused over the sublittoral zone, but the vegetation found here and there is poor in species as well as individuals." According to Kjellman, "the scarcity of algae within the littoral zone and the upper part of the sublittoral in the greater parts of the Arctic Sea depends on the pernicious influence of ice."

Kjellman's story is a grim and largely inaccurate account of the Arctic vegetation in general and of the environmental factors responsible for the structure and composition of this flora. His conclusions on the paucity of Arctic Ocean vegetation are based, unfortunately, on his visits to physically harsh and geologically unfavorable localities that would preclude the existence of epilithic organisms of any alga or animal species. Conditions he describes and the scarcity of algae he records while aboard the Vega (1878-1880) suggest a most inhospitable environment for attached algae throughout the greater portion of the north Russian coast. This is the "stressful" arctic shore and shallow sublittoral environment, and represents most of the Arctic Ocean shores and shallow water coastal areas. These stressful habitats are present toward the heads of all northern fjords and also along the immense open coasts of the northern USSR and north Alaska. Such habitats have, in common, minimal solid substratum, and prolonged periods of maximum ice cover with associated ice abrasion and low light, low temperature and low salinity regimes. It is no wonder that Kjellman's picture of the Arctic Ocean environment was so strongly biased; he experienced primarily the stressful arctic environment. In the following account I compare the "stressful" Arctic Ocean environment and its flora, with the "optimal" arctic environment and its flora.

The features of the "stressful" environment described above all dictate, as Kjellman stated, a flora of few species and low biomass. In the most stressful habitats, shore populations are missing, or they may be reduced to stunted plants of *Urospora*, *Ulothrix*, *Pilayella* and a weak black band of the *Calothrix* association. This prokaryote dominated black band in the uppermost littoral is composed of 8-12 species, including *Calothrix crustacea*, *Pilinia rimosa*, and *Rufusiella fosliaeanus* as the most conspicuous members of this microflora. Similarly, the sublittoral vegetation is depauperate, and contains mostly unattached, loose-lying algal populations (Table 3; Kjellman 1883; Rosenvinge 1893; Just 1951; Lund 1959; Lee 1980; Wilce, pers. obs.).

With the addition of hard substratum to the most "stressed" habitats, species abundance and diversity increase. These substrata include the unusual occurrences of scattered outwash cobble and boulders along the north Alaskan and western Canadian coasts, either sparingly or in boulder fields. These ice-rafted gravels (and boulders), deposited ca 15-10,000 B.P., are lag deposits from the erratic-bearing Flaxman formation (Reimnitz *et al.* 1985; Roderick 1979), and form a relatively free, hard substratum in the otherwise sand-silt dominated habitat of these arctic coasts. The species listed in Table 4 are the arctic "survivors," a relatively larger group that occurs on solid substratum in the stressed arctic habitats, compared to similar habitats that lack hard substratum.

The remainder of the Arctic Ocean flora occurs in habitats moderately stressed or with "optimal" conditions and with slid substratum. The flora of this "optimum" arctic marine environment occurs mainly along the rocky coasts of northwest and east Greenland and a large part of the eastern Canadian archipelago (Lund 1959; Wilce, pers. obs.). A few species (Table 5), thus far known only from the arctic, add a touch of character to the monotony of this arctic ocean flora. The arctic flora is, in fact, the greater portion of the northwest Atlantic flora, gradually reduced in species with increased latitude. These remaining species, largely circum-arctic, have adapted variously to the arctic environment.

Adaptations to the arctic environment

Survival under what appears to be adverse environmental conditions means adaptation to one or more environmental factors. The arctic shallow water environment can be characterized by the following features: a long period of continuous and winter darkness, continuous low light regimes in most of the sublittoral, low ambient temperature, long periods of low salinity and complete annual ice cover for 8-10 months, or more (Wilce 1964; Dunbar 1968; Dunton 1986). Arctic shore environments appear even more hostile than the sublittoral. The range of shore temperatures may exceed 50°C. Shore biota remains alternatively frozen or thawed during most of each year, and beneath the ice foot that covers the mid- and upper shore, darkness or low light levels prevail (Bentham 1937; Petersen 1962). The shore environment during ice-free periods is inundated with fresh water from sea ice melt and land run-off (Wilce 1959). Ice scour occurs throughout the open season but it is especially evident during mid-summer break-up. Obviously, some arctic algal species survive this most rigorous environment. Modes of adaptation include physiological changes, the timing of reproductive and growth periods, minimization of vegetative development, an emphasis on early reproductive development and development of perennating, loose-lying, solely vegetative populations. The development of enzymes to permit periodic heterotrophic growth may be a part of arctic algal physiology, as an adaptation to low light levels, (Wilce 1966). Heterotrophic uptake has not been shown to occur in shallow water (<20m) kelp production (Chapman & Lindley 1980), but it has not been experimentally disproved for deep water (50-100 m +) arctic perennial forms (e.g., crusts, numerous unicellular species, and several small filamentous forms). If the plants in such populations are completely photoautotrophic, they likely have other adaptations to permit growth in an essentially dark environment.

We know few details of how arctic species adapt to their environment, and this information is applicable only to one or two species (Chapman & Lindley 1980; Dunton & Schell 1986). Probably most critical is temperature adaptation (Clausen 1951). While some North Atlantic species are on the very edge of their northward distribution in southern Labrador, others are at their current northern limit in central Baffin Island; still other North Atlantic species have a general circum-arctic distribution. We might hypothesize that in time some or all of the subarctic species will become residents of the arctic environment. Some species have demonstrably extended their ranges southward into temperate North Atlantic (e.g., *Laminaria longicruris*, *Agarum cribrosum*, *Papenfussiella callitricha*, *Turnerella pennyi* and

Table 5. Species of Rhodophyta and Phaeophyta that are arctic endemics. Note: there are no endemic chlorophyte species

PHAEOPHYTA	RHODOPHYTA
Thule brown crust	*Dumontia* -like axis
Ralfsia ovata	Kallymeniaceous red
Acrocystis groenlandica	*Haemescharia polygyna*
Symphyocarpus longsetus	
Phaeostroma endophyticum	
Jonssonia pulvinata	
Punctaria glaciale	
Platysiphon verticillatus	

Odonthalia dentata). Adaptation to temperature by seaweeds, perhaps all algae, as with all other forms of evolution, is progressive and cryptic (Clausen 1951; Futuyma 1986).

MAJOR FEATURES OF ARCTIC OCEAN GEOLOGICAL DEVELOPMENT AND FLORISTIC CONSEQUENCES

In this section I outline the physical development of the north polar sea and comment on the broad character of its seaweed floras from its conception through the Pleistocene. Figure 1 shows a general representation of the current North Polar Seas and peripheral marine areas, and those areas that I consider as 'arctic'.

The late Cretaceous to Paleocene Arctic Ocean

The older of the two deep water basins that comprise the Arctic Ocean, the Siberian-American basin, was a large open sinus of the North Pacific during the mid to late Cretaceous. Plate movements cut off this portion of the North Pacific to form a shallow inland sea (Scheffer 1967; Smith & Briden 1977; Churkin & Trexler 1981; Fujita *et al.* 1982). Further plate movements and the Losomonov Ridge orogeny formed the present recognizable Arctic Ocean. The late Cretaceous-Paleocene flora of the Arctic Ocean was North Pacific in origin, and cool temperate in character. It is likely to have also contained a large number of species currently referred to as "Cosmopolitan."

The Cannonball and Turgai Seaways

During the Paleocene the Arctic Ocean developed two large seaway connections with the world oceans, the Cannonball and Turgai Seaways (Fig. 2; Matsumoto 1980). Paleontological and geological evidence suggests that the Cannonball Seaway connected the Arctic ocean through northern Canada, south to the Gulf of Mexico. Molluscs, ostracod and foraminifera faunas from a current Arctic Ocean Paleocene site (Ocean Point) are similar to the fossil fauna of the Cannonball Seaway of North and South Dakota and believed to be of similar age (Stanley 1965; Cvancava 1966; Macbeth & Schmidt 1973; Carter *et al.* 1977; Niloleva 1977; Marincovich *et al.* 1985).

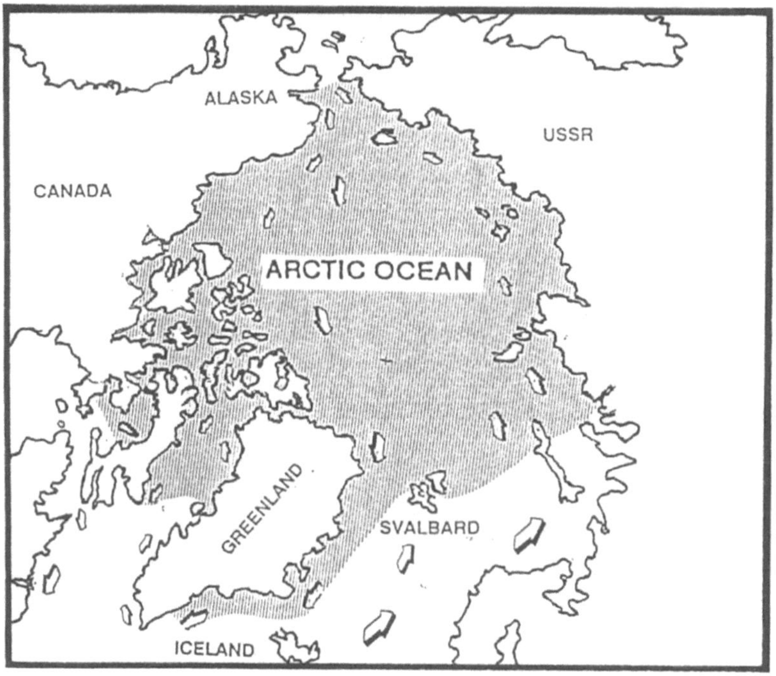

Fig. 1. The present Arctic Ocean showing primary currents

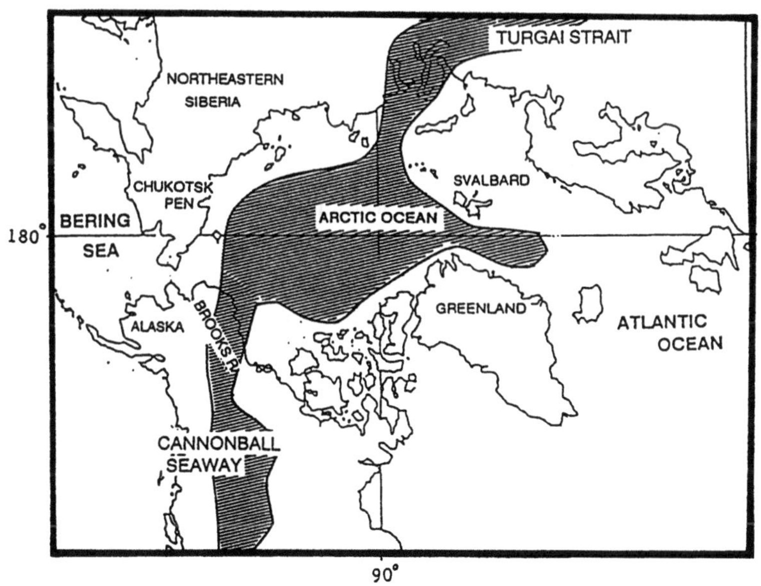

Fig. 2. The Arctic Ocean during the Paleocene-Eocene: 55-35 my B.P.

By the mid- to late Paleocene, a second seaway from the Arctic Ocean, the Turgai Strait, extended southward across Eurasia into a northern arm of the Tethys (Matsumoto 1980; Kitchell & Clark 1982). Thus, throughout much of the Paleocene and Eocene, the Arctic Ocean, with its initial Pacific biota, was connected to the world oceans by two large seaways (Pomerol 1982). Changes in the initial Pacific character of the Arctic Ocean flora resulted with additions of warm water taxa from an early Caribbean Sea and a northern arm of the Tethys Ocean (Marincovich *et al.* 1985).

The Oligocene-Miocene Arctic Ocean 'cul-de-sac'

Much later, probably in the early Oligocene (32-34 my B.P., Eldholme & Thiede 1980), and even into the Eocene (Kitchell & Clarke 1982), additions of Atlantic taxa to the Arctic Ocean flora and fauna occurred. This mix of species was a result of the opening of the Arctic Ocean into the North Atlantic. By this time both large seaways and the Pacific Ocean connections with the Polar Ocean had closed and an Oligocene-Miocene Arctic Ocean 'cul-de-sac' resulted (Fig. 3). Fossil silicoflagellates of Atlantic character from an Arctic Ocean core (Bukry 1984) and the reduced similarity of terrestrial vertebrates (McKenna 1975, 1983; Marincovich *et al.* 1985) in North American and European deposits, suggest an early sea-floor spreading between Greenland and Svalbard in the mid to late Eocene. Certainly, a massive exchange of Arctic Ocean and North Atlantic waters with representatives of their respective floras and faunas occurred through a Greenland-Norwegian Sea deep water connection by mid-Miocene (Eldholme & Thiede 1980; Young & McNeil 1982). Fossil evidence (Hopkins & Marincovich 1984) shows an Atlantic character to Arctic Ocean fauna during this time and into the Pliocene. Atlantic seaweeds were likely to have accompanied the mollusc, ostracod and silicoflagellate populations into the arctic basin, and one may also assume that elements of the Arctic Ocean flora invaded the North Atlantic. By the time of the Atlantic transgression (late Oligocene-Miocene) the northern and southern seaways leading from the polar basins (Fig. 2) were lost, owing to continental plate movements (Vinogradov 1967). However, warm water species brought into the Arctic Ocean via these seaways must have merged with the Miocene Atlantic species incursion. Evidence from terrestrial and marine fossils from this time suggest continued warm-cool temperate high arctic terrestrial and marine environments (MacNeil 1957; Hills & Matthews 1974; Matthews 1976; Hickey *et al.* 1983).

The Pliocene opening of the Arctic Ocean

The Miocene Arctic ocean 'cul-de-sac' was dramatically changed with the submergence of the Bering-Chukotsk Peninsula land bridge between 3-3.5 myu B.P. (Fig. 4). Considerable geological (Grim & McManus 1970; Eittreim *et al.* 1979; Vail & Hardenbol 1979) and paleontologic (Hopkins 1967; Einarsson *et al.* 1967; Gladenkov 1979; Repenning 1983; Hopkins & Marincovich 1984; Marincovich *et al.* 1985) evidence supports a massive movement of North Pacific Pliocene species into the Arctic and North Atlantic Oceans. Thus the late Pliocene, early Pleistocene Arctic Ocean flora and fauna probably contained its maximum diversity, with taxa from four different floras, i.e., Pacific, early Caribbean, late Tethys-Indian Ocean and Atlantic. In addition, species must have evolved from this Arctic Ocean stock from the late Cretaceous to the Pliocene, c 60 my. Fossil types from several Arctic Ocean basin sites and from elsewhere in the high arctic (Rosenkrantz 1970; Henderson *et al.* 1976; Estes & Hutchison 1980; Marincovich *et al.* 1985) suggest ambient Arctic Ocean water temperature ranged between 3-6°C. The Paleocene Arctic Ocean was open and ice-free, possibly even during winter darkness (Kitchell & Clark 1982). A Pliocene temperate marine environment in the polar basin seems indisputable, perhaps until well into the Pleistocene era. Northern hemisphere cooling and the formation of ice sheets in northern areas in the late Pliocene/early Pleistocene was followed by an ice covered Arctic Ocean by 700,000 B.P. (Hopkins & Marincovich 1984). Only North Atlantic water and elements of the current North Atlantic biota entered the much reduced Arctic Ocean during the last half of the Pleistocene. However, the origins and character of current Arctic Ocean flora must be based on the complex of a pre-Pleistocene Arctic Ocean flora. In fact, elements of this flora probably survived repeated Pleistocene glacial periods to ultimately form the North Atlantic "seed stock." From this "seed stock," the Arctic Ocean has been recolonized during the last 8-10,000 years.

The Pleistocene and post-Pleistocene Arctic Ocean

The number and duration of Pleistocene glacial and interglacial events and the extent of ice cover during these periods is not of consideration to this study. It is sufficient that, at least during the interglacial period of 125,000 years ago, Arctic Ocean water temperatures were possibly comparable to those of today (Troitsky 1974; Schytt *et al.* 1982). Warm interglacial periods permitted limited species exchange through a shallow Bering Strait and a much greater species exchange through a larger seaway into the North Atlantic. Cold periods converted the Arctic Ocean into another 'cul-de-sac', but one which was much smaller and of different origin than its Miocene counterpart (Figs 4-5). In fact, roughly 60% of the present Arctic Ocean surface area was exposed coast during Pleistocene glacial periods if one refers to the 100-110 m

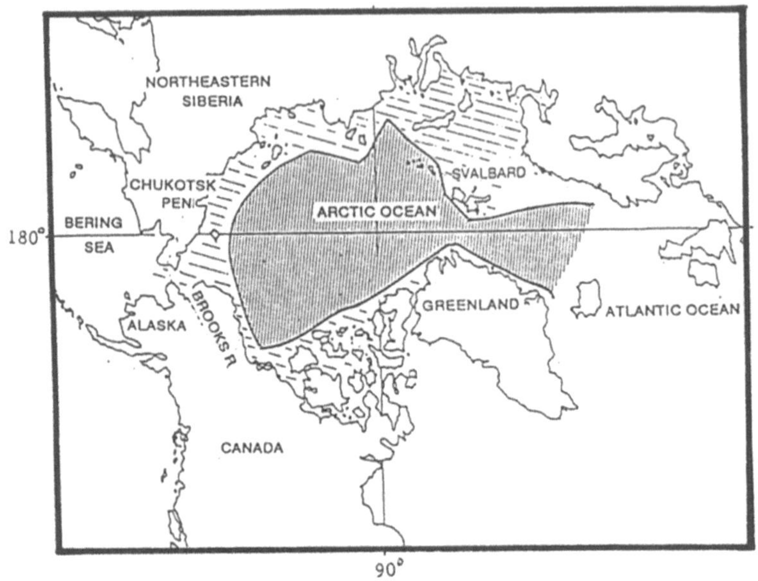

Fig. 3. The Oligocene-Miocene Arctic Ocean ' cul-de-sac': 34-32 my B.P.

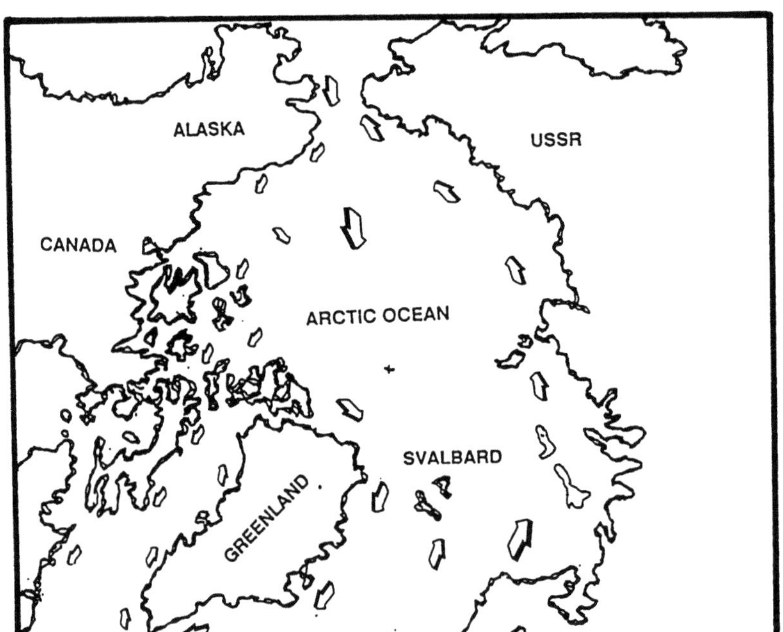

Fig. 4. The Pliocene-early Pleistocene Arctic Ocean: 3.5-1.5 my B.P.

bathometric line as the lower limit of seawater regression (Hopkins & Marincovich 1984). Of more importance to the phytogeographer interested in present-day Arctic Ocean seaweeds, is the extent of Pliocene and early Pleistocene species survival into the Holocene. What taxa remain from pre-Pleistocene Arctic Ocean "seed stock" and how can these species be identified?

Interpretation of Arctic Ocean cores suggest periods of massive environmental change, periods of extinctions, certainly of biotic displacement, all within the last 700,000 years (Herman & Hopkins 1980). The general scenario of intermittent glacial and interglacial climates is a phenomenon of the last one million years. The Arctic Ocean was perennially ice covered only during the last 0.7 my (Hopkins & Marincovich 1984; Stewart & England 1983; Schytt *et al.* 1982). Another interpretation based on Arctic Ocean cores, was that the north polar ocean was perennially covered for the last 3 my (Clark 1982). Despite a healthy controversy concerning the nature of Pleistocene Arctic Ocean environment and the extent to which it remained ice covered, a post-Pleistocene seaweed flora of considerable diversity remains. This

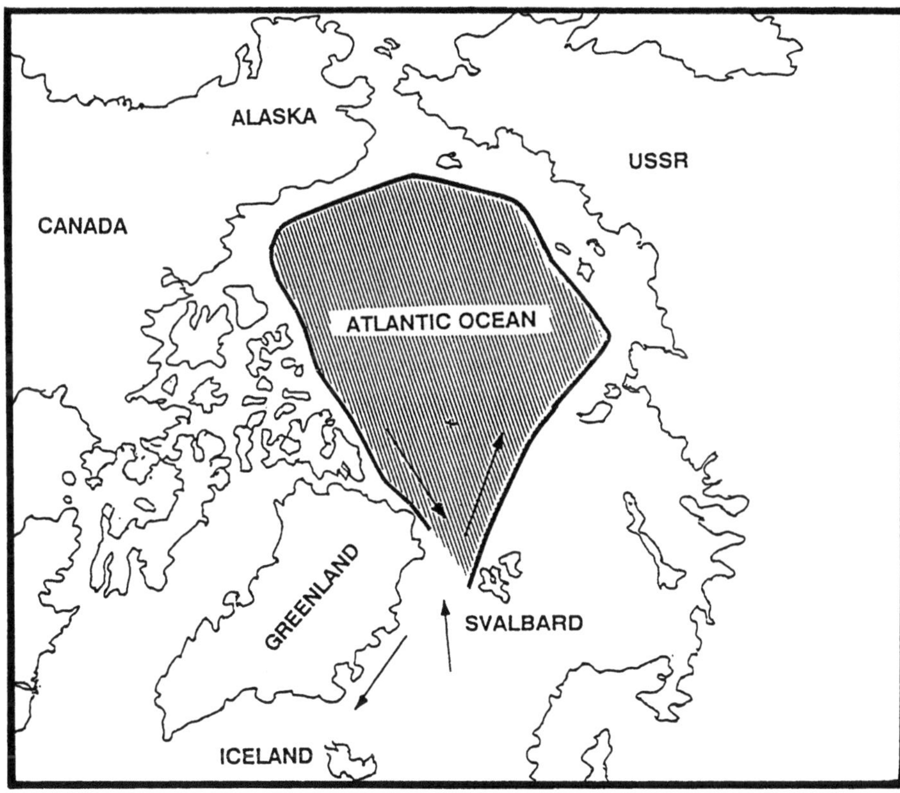

Fig. 5. The mi-to-late Pleistocene Arctic Ocean 'cul-de-sac': 1.5 my - 10,000 B.P. (one glacial peak representation)

flora can be identified and utilized in the interpretation of phylogenetic and biogeographic affinities. To date, conclusions concerning the character and relationships of the Arctic Ocean flora have been hypothetical, and based on shared relationships, mostly at the generic and family levels and, of course, on details of the geological history of the area. The North Atlantic marine flora has had the most impact on the flora of the Arctic Ocean. The following account, beginning with an interpretation of the western North Atlantic flora, outlines a working hypothesis which matches floristic-phylogenetic details with our current understanding of an historical development of the Arctic Ocean basin.

ORIGINS OF THE BENTHIC ARCTIC OCEAN FLORA

My first assumption is that the northwest Atlantic flora contains taxa associated with five different biogeographic origins. This assumption is based on current species distributions, and on morphological and reproductive similarities, which can be used to infer generic and familial affinities. Assignment of taxa into these five groups is entirely hypothetical. Fossil seaweed remains, by which relationships and even approximate dating can be made, are few. Vicariant relationships, the assessment of taxonomic affinity with geological history, must be one of our basic tools of interpretation (Garbary 1987). Use of cognitive information from a variety of disciplines together with details of geological history often suggest phylogenetic relationships of widely disjunct taxa, indeed, even floras. The most obvious example of such a relationship is seen in the occurrence of essentially all (>91%) of the western North Atlantic flora along northern European coasts (Wilce 1977). One must conclude from this that the western and eastern Atlantic elements of these two floras had, at one time, a common origin (Wilce 1973) prior to the Pleistocene. Elements of this North Atlantic flora were long resident in the Arctic basin and were merely "moved about" by Pleistocene climates. However, in addition to migrations, extinctions and repopulations, Pleistocene climates greatly and differentially modified the North Atlantic common stock along the two Atlantic coasts. The western North Atlantic flora probably consists of parental stock, and even species, from at least five different origins, possibly some of Cretaceous age and older. In this paper I consider these species as having one of the following biogeographic origins: "Cosmopolitan", "Pacific", 'Indo-Pacific", "Atlantic" and "Endemic".

The cosmopolitan species group

Cosmopolitan species, numerous in the western North Atlantic and elsewhere (Table 6), comprise the first of the five groups. These species are identified by their wide distribution throughout the world's oceans, and by their long and confusing distribution tracks. Many are broadly eurythermal, euryhaline and probably of ancient origin. A higher proportion of temperate and cold-adapted green algae belong to the cosmopolitan group, but a number of North Atlantic brown and red algae are included.

The Pacific species group

The Pacific group with shared North Atlantic-North Pacific distributions (Table 7) forms a second component of the western North Atlantic flora. Most are broadly disjunct, but a few have a continuous distribution via the Arctic Ocean. For example, Lindstrom (1987) cites only 14 species of red algae with a North Pacific-Arctic Ocean-North Atlantic distribution. One of Lindstrom's species, *Ceramium rubrum*, does not occur in the arctic. *C. rubrum* may occur in the southernmost Chuckchi Sea along the Alaskan coast which would suggest a subarctic environment. From that area (Cape Thompson, Alaska), Chihara (1967) cites twelve species, three of which (*Halosaccion glandiforme*, *Pterosiphonia bipinnata*, *Odonthalia kamstchatica*) are of Pacific Ocean origin. Nine of the twelve species have Pacific and Atlantic Ocean distributions, and seven of these also occur in the Arctic Ocean. About 10% of the North Pacific flora are members of the western North Atlantic flora, i.e., they occur from Cape Cod northward. In the case of many species with these distributions, the problem of either Pacific or Atlantic origin remains hypothetical. Species tracks may be read in both directions, and centers of distribution are vague for most species. Five genera (*Pelvetia*, *Pterosiphonia*, *Aglaothamnion*, *Plocamium*, *Catenella*) are particularly interesting and enigmatic in their known distributions. These genera, with their characteristic species, occur in the eastern European North Atlantic, southern Iceland and the North Pacific, but are absent in the northwest Atlantic and in the Arctic. The evolutionary history of these distinctive taxa, and their probable North Pacific center of distribution, suggest migration from west to east through a temperate north polar ocean, which necessitates species movement prior to 2 my BP. One must assume that other species also accompanied them into the Atlantic, but origins of these species are more difficult to identify. Despite this problem, geological events involving the development of the North Pacific, Arctic and North Atlantic Oceans, as well as ample fossil evidence, suggest such a migration (Gladenkov 1979).

It is apparent that species with shared North Atlantic-North Pacific occurrence are far fewer than genera with these shared distributions(c 74% genera vs c 10% species). Most North Atlantic genera are represented in the North Pacific with one or more sister species, e.g., the Pacific and Atlantic Turnerella species, and numerous others (Simmons 1906; Abbott & Hollenberg 1976; Lebednik 1977; Lindstrom 1987). In a consideration of Pacific-Atlantic origins, one must also attempt to account for the large number of North Pacific endemic genera

Table 6. Cosmopolitan species of seaweeds with North Atlantic distribution

CHLOROPHYTA	PHAEOPHYTA	RHODOPHYTA
Ulothrix sp.	*Pilayella littoralis*	*Audouinella purpurea*
Chaetomorpha linum	*Ectocarpus siliculosus*	*Nemalion helminthoides*
C. melagonium	*Giffordia* sp.	*Ahnefeltia plicata*
Rhizoclonium riparium	*Leathesia difformis*	*Ceramium rubrum*
Cladophora sp.	*Petalonia fascia*	*Antithamnion cruciatum*
Blidingia mimimum	*Scytosiphon lomentaria*	
Enteromorpha sp.	*Desmarestia viridis*	
Bryopsis plumosa	*Sphacelaria cirrosa*	
Derbesia marina	*Laminaria saccharina*	
Codium fragile		

Table 7. A list of putative Pacific species. These are conspicuous brown and red seaweeds with Atlantic-Pacific but not Arctic Ocean distributions. Note: numerous species of Chlorophyta belong to this category, but are not listed here

PHAEOPHYTA	RHODOPHYTA
Leathesia difformis	*Ceramium rubrum*
Colpomenia sinuosa	*Agardhiella tenera*
Sphacelaria sp.	*Polyides rotundus*
Laminaria nigripes	*Schizymenia dubyi*
Melanosiphon intestinalis	*Neoagardhiella baileyi*
Scytosiphon dotyi	*Champia parvula*
	Corallina officinalis
	Gelidium crinale

(Hommersand 1986). Lindstrom (1987) provides a list of red algal genera endemic to the North Pacific, and Abbott & Hollenberg (1976) cite numerous brown algal endemics including a diversity of kelp, e.g., *Hedophyllum, Costaria, Pleurophycus, Egregia, Eisenia, Pterygophora, Macrocystis, Nereocystis, Pelagophycus, Dictyoneuropsis, Dictyoneurum, Postelsia, Lessoniopsis*. The few endemic genera of red algae cited by Lindstrom, and the much larger number proportion of endemic kelp genera in the Pacific versus the Atlantic, raise many questions concerning the age of Pacific endemics. Why are some of these endemic taxa not represented in the North Atlantic? Were these old genera in the North Pacific and a part of the Pacific flora "trapped" with the formation of the Arctic Ocean, or did they enter the Pliocene Arctic Ocean and only later, in the Pleistocene, become eliminated from the Polar Ocean? The larger question, and seminal to species exchange through the Arctic Ocean, is why some or all of these genera are not represented in the North Atlantic.

The Indo-Pacific species group

A third group of northwest Atlantic species may be "Indo-Pacific" and of Tethyan origin, all with warm water affinities. These species, or their ancestral stock, may have entered the Arctic Ocean or the North Atlantic during the early Tertiary or late Cretaceous. Others entered the Arctic basin and ultimately migrated into the North Atlantic during the Miocene Pacific Ocean transgression (Fig. 3). Five taxa stand out from the remaining western north Atlantic seaweeds as possibly Tethyan in origin owing to their family relationships, and the absence of phylogenetically related stock in the North Pacific and North Atlantic Oceans. Doubtless others occur in the North Atlantic, but their lack of distinctiveness does not permit this association. These five species have narrow distributions but the ancestral stock at the generic and family levels point to an ancient origin either in the southern hemisphere or in warm water areas of the northern hemisphere. The five species suggested as having an Indo-Pacific origin are: *Papenfussiella callitricha*, *Turnerella pennyi*, *Kallymenia schmitzii*, *Callophyllis cristata*, and an undescribed additional member of the Kallymeniaceae.

Papenfussiella callitricha, a subarctic annual species of Chordariaceae, is the only northern hemisphere member of a genus with 5-6 other species, all known from warm water. Divergence of *P. callitricha* from the other members of the genus suggests both an ancient separation and an initial warm to cool temperate North Atlantic environment (South 1987). *P. callitricha* appears to have changed its physiology from a temperate species during pre-Pliocene-Pleistocene times to a Holocene subarctic-temperate species. One might ask if *P. callitricha* is once more invading the temperate environment, with a migration from central west Greenland to southern Newfoundland. Can we expect *P. callitricha* in the Gulf of Maine in the not too distant future?

Turnerella pennyi is a distinctive subarctic/arctic perennial and stands apart from the remainder of the western North Atlantic flora in its structure, primitive reproductive features (Knauss 1988) and in its relative isolation as a member of the Soleriaceae in the north Atlantic and Arctic Oceans. The sister species, *T. mertensiana*, occurs in Japan and the North Pacific, but not in the Arctic Ocean. Most members of the Soleriaceae occur in the Austral-Indo-Pacific, their center of distribution (Gabrielson & Hommersand 1982). Other members of the family are found monotypically in the North Pacific, but only *Neoagardhiella baileyi* stands apart as a single temperate western North Atlantic member of the Soleriaceae. Central to the problem of species origin are the questions of the phylogenetic affinities of *T. pennyi* and *T. mertensiana*, and that of the poor representation of the family in the North Pacific and North Atlantic Oceans. Movement of *T. pennyi* from the Pacific into the Atlantic seems plausible considering the close relationship of the two species and of *Turnerella* to the closely related northern Pacific *Opuntiella californica* (Perestenko 1976). However, one may easily question the origin of *Turnerella*, the direction of its movement in the northern hemisphere and the timing of Pacific and Atlantic species evolution and migration.

Three members of the Kallymeniaceae (i.e., *Kallymenia schmitzii*, *Callophyllis cristata* and an undescribed species) have structural and reproductive features and phylogenetic relationships sufficiently distinct to suggest an ancient warm water origin, and long isolation from the family stock. Representatives of the Kallymeniaceae, like the Soleriaceae, appear isolated, seemingly out-of-place in the Arctic Ocean and the northern portions of the western North Atlantic. The Kallymeniaceae have a southern hemisphere, temperate to warm water center of origin. Arctic Ocean-western North Atlantic species assigned to the Kallymeniaceae stand apart from remaining members of the northern flora. Little known *Kallymenia schmitzii*, from Ungava Bay, Ellesmere Island and central west Greenland, remains enigmatic in its life history, but its identity as a *Kallymenia* is confirmed by vegetative structure and female reproductive apparatus. The species track of *K. schmitzii* lies in the arctic and subarctic, entirely within the western North Atlantic. By contrast, a second species in the family, *Callophyllis cristata* has a truly broad species track, one of the few red algae with North Pacific-circum-Arctic and North Atlantic distribution. This species occurs in the cool temperate to high arctic region, with uniformly 0 to minus degree temperatures, and full to low light conditions. *C. cristata*, *Turnerella pennyi* and *K. schmitzii* are the last of the erect red algae to leave the vegetation with increased depth (Lund 1959; Sears & Cooper 1978; Wilce pers obs.). Considering the current broad distribution of *C. cristata*, it is interesting to speculate whether *C. cristata*'s pre-Pliocene origin in the Arctic Ocean was from the North Pacific or North Atlantic. Whatever the source of its dispersal into the Arctic Ocean, major changes in temperature adaptation in this species must have occurred on several occasions during the last 1.5-2.0 my B.P. While similar adaptation occurred in numerous Arctic and North Atlantic Ocean species, *C. cristata*, *K. schmitzii* and an other unknown species, remain distinctive members of the Kallymeniaceae in these boreal to arctic movements.

The unknown third kallymeniaceous plant has been collected twice in the Canadian arctic archipelago. Relationships are vague as only two vegetative plants were seen. The plants are erect, branched, terete in section with stellate cells in the medula and may reach 12 cm. Both plants were collected in late August, one in Grise Fjord, Ellesmere I., and the second near Cape Hatt, Baffin I. Both species were morphologically and structurally unlike all other red algae present in the Arctic and North Atlantic, other than those of the Kallymeniaceae. Evolution of this species from other members of the family in the Arctic Oceans seems remote; a better choice would be Pacific kallymeniaceous stock, or one of the more numerous Indo-Pacific species no longer present in the northern hemisphere.

The North Atlantic species group

Species now recognized as "North Atlantic" are perhaps most enigmatic in terms of their pre-North Atlantic origins. These species form a large percentage of the western North Atlantic flora, most of which are shared with the European seaweed flora. Moreover, these "Atlantic" species are of particular importance to this study as they constitute more than 90% of the species present in the current Arctic Ocean flora. Atlantic species are numerous; most occur in cool temperate environments, but many extend into the arctic with longer species tracks. While a number of "Atlantic" species occur in warm subtropical central Atlantic waters, it is generally agreed (Hoek 1982, 1984; South 1987; South & Tittley 1985; Searles 1984) that most north Atlantic taxa do not share a common origin with species of the Atlantic tropical and subtropical floras.

Ultimately, one must consider the origin of our "North Atlantic" species as only partially derived from a post-Pliocene movement of Arctic Ocean taxa into the north Atlantic (Fig. 4). That post-Pliocene flora, with its multiple origins (all warm water or cool temperate), merged with an existing cool temperate (Herman and Hopkins 1980) North Atlantic flora of unknown composition. This unknown flora is the enigmatic component of the North Atlantic Pleistocene vegetation and, indeed, of our present-day flora. What was the character of this North Atlantic flora that met, and ultimately merged, with the Pliocene Arctic Ocean flora? As suggested earlier, the northern hemisphere climate remained temperate with ambient ocean temperatures of 3-4°C until well into the Pleistocene. Phytogeographic provinces in the North Atlantic had already been established by a strong North Atlantic Drift to the east, though the position of the Drift, and consequently, of these phytogeographic provinces, was considerably further south than it is today. Were Pliocene North Atlantic taxa simply a small percentage of central western and eastern warm water Atlantic species? Were they species from the south which adapted to a cool temperate environment? The origins of that warm water Atlantic flora are also multiple and, in consequence, must be ancient. That Atlantic flora was largely isolated during or prior to the Tertiary, before the Arctic Ocean transgression which brought its additional species.

The Endemic species group

Endemics form the last group of western North Atlantic species. They are few in the western Atlantic and Arctic Ocean, more abundant in the eastern North Atlantic. In each instance, one may ask whether these are "new" species, products of post-Pleistocene evolution, or older? If the former is correct, species such as *Platysiphon verticillatus* and *Punctaria glaciale*, must have evolved as Pleistocene cold-adapted taxa, most of which have short distribution tracks, today's arctic endemics; if their origin is older (and this is likely), a warm to temperate water origin must be considered. Similar temperature regimes occurred during Pleistocene interglacials or in the pre-Pleistocene temperate marine environments. Several of these currently cold-adapted species (*Laminaria solidungula*, *L. longicruris*, *Agarum cribrosum*, *Omphalophyllum ulvaceum*, *Papenfussiella callitricha*, *Turnerella pennyi*, *Phyllophora interrupta*), to cite a few, have longer distribution tracks and extend into the temperate realm. Broad temperature and geographic ranges of these species suggest both temperature adaptations and migration to different phytogeographic provinces. Other, presently cold-adapted species, such as *P. verticillatus* and *P. glacialis*, have short distribution tracks and are known only from the eastern Canadian Arctic and Greenland. Regardless of the species age and distribution range, most of these species show a close generic relationship to other North Atlantic and North Pacific taxa.

Another group of endemics occurs in the eastern North Atlantic Ocean; these species must be considered differently. If one considers the origin of these endemics as Pacific, Indo-Pacific and pre-Pliocene Atlantic, and their current European occurrence as a function of Pleistocene events, their isolation in the Atlantic becomes understandable. A late Pliocene, early Pleistocene scenario might show a far more uniform North Atlantic marine flora with the current European endemics widely distributed along the cool temperate coasts, of both North America and Europe. Ultimately, differential Pleistocene effects along the North Atlantic American and European coasts, and the difference in post-Pleistocene temperature regimes on these coasts resulted in a large part of the current floristic disparity in the North Atlantic Ocean.

SYNTHESIS: THE MODEL AND ITS CONSEQUENCES

At this time, phycologists have a working knowledge of the floras of the North Atlantic, North Pacific and Arctic Oceans. The outline of the paleogeographic development of the Arctic Ocean and its peripheral ocean areas from the Cretaceous through the Pleistocene is known. Marine geologists, paleontologists, geochemists and paleobotanists have provided information concerning paleoclimates in the northern hemisphere and the probable timing of movements of marine faunas and floras. This evidence suggests that a marine environmental stasis existed for the entire duration of the pre-Pleistocene Arctic Ocean, and that the marine environment was cool temperate through the early Pleistocene. Fossil evidence suggests multiple origins for the pre-Pleistocene Arctic Ocean flora, with its members being either warm-water or cool temperate-adapted.

The evolutionary scenario outlined above has two necessary assumptions in terms of algal evolution: the first assumption is that most of our current North Atlantic, North Pacific, and Arctic Ocean floras must be considered ancient, with origins well before the Tertiary (for some taxa quite possibly into the Paleozoic). The second assumption, that has also been articulated for other organisms (Simpson 1953), is that for species of great age, there is a conservation of morphological and reproductive features and reproductive strategies. Populations of widely disjunct species, those that I have referred to as "cosmopolitan," repeatedly show these forms of genetic conservation, i.e., an apparent lack of change in species over time. In all probability, the same generalizations concerning form and reproductive strategies may be said for the remaining North Atlantic algal species.

During the Paleocene and earlier, the Arctic Ocean was a large sinus of the North Pacific Ocean, and throughout its pre-Pleistocene the Arctic Ocean contained species with separate origins, including the northern Pacific. The Arctic Ocean was formerly of great importance in the distribution of species from the North Pacific to the North Atlantic and from the North Atlantic into the North Pacific. Atlantic-Pacific exchange occurred first in the Pliocene and continued into the early to mid-Pleistocene. Thereafter, there was a variable exchange through the Bering Strait because of the occurrence and degree of interglacial warming. Few Pacific algae are present in the Chuckchi and Beaufort Seas, despite the present-day occurrence of Pacific water and some solid substratum of the Flaxman formation along the north Alaskan and central Canadian coasts (Dunton *et al.* 1982). The western Arctic Ocean flora and, indeed, the Arctic Ocean flora in general, are composed only of western North Atlantic species and a few species termed "endemic." One must assume, based on their absence along the north Alaskan and Beaufort Sea coasts, that North Pacific algae are not yet adapted to cold water. With continued interglacial warming and a possible reopening of the Bering-Chukotsk peninsula, the Arctic Ocean may once again develop its Pliocene biotic diversity. It may be stated that the Arctic Ocean has played an important role in the longitudinal distribution of species from west to east and east to west. The lack of North Pacific taxa in the Arctic Ocean, particularly in the Chuckchi and Beaufort Seas, suggests, however, that little species exchange has occurred since the last interglacial, or is presently occurring.

REFERENCES

Abbott IA, Hollenberg GJ (1976) Marine algae of California. Stanford Univ. Press, Stanford CA

Bentham R (1937) The ice-foot. Appendix III in Shackleton: Arctic Journeys, The story of the Oxford Univ. Ellesmere Land Exped. 1934/35, London
Børgesen F, Jónsson H (1905) The distribution of the marine algae of the Arctic Sea and of the northernmost part of the Atlantic. Bot of the Faeroes, 3: Appendix I-XXVIII Copenhagen
Briggs JC (1974) Marine Zoogeography. McGraw-Hill, New York
Bukry D (1984) Paleogene paleoceanography of the Arctic Ocean is constrained by middle or late Eocene age of USGS core F1-422: Evidence from silicoflagellates. Geology 12: 199-201
Caram B, Jónsson S (1972) Nouvel inventoire des algues marines de l'Islande. Act Bot Isl 1: 5-31
Carter LM, Repenning CA, Marincovich LN, Hayel JE, Hopkins DM, McDougall, Kristin, Naeser CW (1977) Gubik and pre-Gubik Cenozoic deposits along the Colville River near Ocean Point, North Slope, Alaska. US Geol Sur Circular 751-B: 12-14
Chapman ARO, Lindley JE (1980) Seasonal growth of (*Laminaria solidungula*) in the Canadian high arctic in relation to irradiance and dissolved nutrient concentrations. Mar Biol 47: 1-5
Chihara M (1967) Some marine algae collected at Cape Thompson of the Alaskan Arctic. Bull Nat Sci Mus Tokyo 19: 183-204
Churkin M Jr, Trexler JH Jr (1981) Continental plates and accreted oceanic terranes in the Arctic. In: Nairn AEM, Churkin M Jr, Stehli, FG (eds) The ocean basins and margins, 5. Plenum, New York, p 1-20
Clark DL (1982) Origin, nature, and world climate effect of Arctic Ocean ice cover. Nature 300: 321-325
Clausen J (1951) Evolution of plant species. Cornell Univ Press, Ithaca, New York
Cross WE, Wilce RT, Fabijan MF (1987) Effects of experimental releases of oil and dispersed oil on arctic near-shore macrobenthos. III. Macroalgae. Arctic 40(suppl 1): 211-219
Cvancava AV (1966) Revision of the fauna of the Cannonball Formation (Paleocene) of North and South Dakota. Contrib. Mus Paleontology, U Mich 20(10): 1-97
Dunbar MJ (1947) Note on the delimitation of the Arctic and Subarctic zones. Can Fld Nat 61: 12-14
Dunbar MJ (1968) Ecological development in polar regions. Englewood Cliffs, NJ, Prentice-Hall
Dunton KH, Schell D (1986) Seasonal carbon budget and growth of (*Laminaria solidungula*) in the Alaskan high arctic. Mar Ecol Prog Ser 31: 57-66
Dunton KH, Reimnitz E, Schonberg S (1982) An arctic kelp community in the Alaskan Beaufort Sea. Arctic 35: 465-484
Einarsson T, Hopkins DM, Doell RR (1967) The stratigraphy of Tjornes, northern Iceland, and the history of the Bering land bridge. Stanford Univ Press, p 312-325
Eittreim S, Grantz A, Whitney OT (1979) Cenozoic sedimentation and tectonics of Hope Basin, southern Chuckchi Sea. In: Sisson AL (ed) The relationship of plate tectonics to Alaskan geology and resources. Alaskan Geol Soc Symp, 6th, Anchorage, B1-B11
Eldholme O, Thiede J (1980) Cenozoic continental separation between Europe and Greenland. Palaeogeo Palaeoclim Palaeoecol 30: 243-259
Estes R, Hutchinson JH (1980) Eocene lower vertebrates from Ellesmere Island, Canadian Arctic Archipelago. Palaeogeo Palaeoclim Palaeoecol 30: 325-347
Farlow WG (1886) Notes on Arctic algae; based principally on collections made at Ungava Bay by Mr LM Turner. Proc Am Acad Arts Sci 31(NS13): 469-477
Feyling-Hanssen RW (1953) The barnacle (*Balanus balanoides*) in Spitsbergen, Norsk Polar-institutt, Skrifter nr 98
Fujita, Kazuya, Newberry JT (1982) Tectonic evolution of northeastern Siberia and adjacent regions. Tectonophysics 89: 337-357
Futuyma DJ (1986) Evolutionary Biology. Sinauer, Sunderland, MA
Gabrielson PW, Hommersand MH (1982) The Atlantic species of *Solieria* (Gigartinales, Rhodophyta): their morphology, distribution and affinities. J Phycol 18: 31-45
Garbary DG (1987) A critique of traditional approaches to seaweed distribution in light of the development of vicariance biogeography. Helgol Meeresunters 41: 235-244
Gladenkov YB (1979) Comparison of late Cenozoic molluscan assemblages in northern regions of the Atlantic and Pacific Oceans. Intn Geol Rev 21: 880-890

Grim MS, McManus DA (1970) A shallow seismic-profiling survey of the northern Bering Sea. Mar Geol 8: 293-320
Henderson G, Rosenkrantz A, Schiener EJ (1976) Cretaceous-Tertiary sedimentary rocks of west Greenland. In: Geology of Greenland, Gronl Geol Undersogelse 1976: 341-362
Herman Y, Hopkins DM (1980) Arctic oceanic climate in late Cenozoic time. Science 209: 557-562
Hickey LJ, West MR, Choi DK (1983) Arctic terrestrial biota: Paleomagnetic evidence of age disparity with mid-northern latitudes during the late Cretaceous and early Tertiary. Science 221: 1153-1156
Hills LV, Matthews Jr JV (1974) A preliminary list of fossil plants from the Beaufort Formation, Neighan Island, District of Franklin. Geol Surv Can Pap 74-1B: 224-226
Hoek C van den (1975) Phytogeographic regions in the northern Atlantic Ocean. Phycologia 14: 317-330
Hoek C van den (1979) The phytogeography of (*Cladophora*) (Chlorophyceae) in the northern Atlantic Ocean, in comparison to that of other algal species. Helgol Wiss Meeresunters 32: 374-393
Hoek C van den (1982) The distribution of benthic marine algae in relation to the temperature regulation of their life histories. Biol J Linn Soc 18: 81-144
Hoek C van den (1982) Phytogeographic distribution groups of benthic marine algae in the North Atlantic Ocean. Helgol Meeresunters 35: 153-214
Hoek C van den (1984) World-wide latitudinal and longitudinal seaweed distribution patterns and their possible causes, as illustrated by the distribution of Rhodophycean genera. Helgol Meeresunters 38: 227-257
Hoek C van den (1987) The possible significance of long-range dispersal for the biogeography of seaweeds. Helgol Meeresunters 41: 261-272
Hommersand MH (1986) The biogeography of the South African marine red algae: a model. Bot Mar 29: 257-270
Hooper RG, South, GR (1977) Distribution and ecology of (*Papenfussiella callitricha*) (Rosenv) Kylin (Phaeophyceae, Chordariaceae). Phycologia 16: 153-157
Hopkins DM (1967) Quaternary marine transgressions in Alaska. In: Hopkins DM (ed) The Bering land bridge. Stanford Univ Press, p 47-90
Hopkins DM, Marincovich L (1984) Whale biogeography and the history of the Arctic Basin. Groningen Rijksuniversitet, Netherlands, Works of the Arctic Center 8: 7-24
Just J (1951) Marine biological investigations of Jorgen Bronlund Fjord North Greenland. Medd Gronl 184(5): 1-42
Kitchell JA, Clark DL (1982) Late Cretaceous-Paleogene paleogeography and paleocirculation: evidence of North Polar upwelling. Palaeogeo Palaeoclim Palaeoecol 40: 135-165
Kjellman FR (1883) The algae of the Arctic Sea. Kongl Svensk Vetensk Akad Handl 20(5): 1-349
Knauss ME (1988) Reproductive development and systematics of some primitive families of Gigartinales (Rhodophyta). Ph.D. thesis, U North Carolina, Chapel Hill
Lebednik PA (1977) The Corallinaceae of northwestern North America. I. (*Clathromorphum*) Foslie emend Adey. Syesis 9: 59-112
Lee RKS (1973) General ecology of the Canadian Arctic benthic marine algae. Arctic 26: 32-43
Lee RKS (1980) A catalogue of the marine algae of the Canadian Arctic. Natl Mus Can, Nat Sci Publ Bot 9: 1-83
Lee RW (1966) Ecological study of the Mould Bay marine algae. MS thesis, U Mass, Amherst, MA
Lindstrom SC (1987) Possible sister groups and phylogenetic relationships among selected North pacific and North Atlantic red algae. Helgol Meeresunters 41: 245-260
Lund S (1951) Marine algae from Jörgen Brönlunds Fjord in eastern North Greenland. Medd Gronl 128(4): 1-26
Lund S (1959) The marine algae of East Greenland. I. Taxonomical Part. II. Geographic Distribution. Medd Gronl 156(1): 1-247; (2): 1-72
Macbeth JI, Schmidt RAM (1973) Upper Cretaceous foraminifera from Ocean Point, northern Alaska. J Paleont 47: 1047-1061

MacNeil FS (1957) Cenozoic megafossils of northern Alaska. US Geol Sur Prof Paper 294-C: 99-126

Madesen H (1940) A study of the littoral fauna of North Greenland. Medd Gronl 124(3): 1-24

Marincovich Jr L, Brouwers EM, Carter LD (1985) Early Tertiary marine fossils from northern Alaska: Implications for Arctic Ocean paleogeography and faunal evolution. Geology 13: 770-773

Matthews JR JV (1976) Insect fossils from the Beaufort Formation: geological and biological significance. Geol Sur Can Paper 76-1B: 217-227

Matsumoto T (1980) Inter-regional correlation of transgressions and regressions in the Cretaceous Period. Cretaceous Res 1: 359-373

McKenna MC (1975) Fossil mammals and early Eocene North Atlantic land continuity. Ann Missouri Bot Gard 62: 335-353

McKenna MC (1983) Cenozoic paleogeography of North Atlantic land bridges. In: Bott MHP, Saxov S, Talwani M, Thiede J (eds) Structure and development of the Greenland-Scotland Ridge. Plenum, New York, p 351-399

Nikoleva IA (1977) New ostracode species from the Paleogene of the Turgai Basin. Paleont Soc, Ann Yearbook Editorial Vol: 191-199

Petersen GH (1962) The distribution of (*Balanus balanoides*) L. and (*Littorina saxatilis*) Olivi, var (*groenlandica*) Mencke in northern West Greenland. Medd Gronl 159(9): 1-43

Perestenko LP (1976) The red algae of the far-eastern seas of the USSR. (*Turnerella*) Schmitz, (*Opuntiella*) Kylin (Solieriaceae, Gigartinales). Nov Syst Plant Non Vasc 13: 39-50 (in Russian)

Pielou EC (1977) The latitudinal spans of seaweed species and their patterns of overlap. J Biogeogr 4: 299-311

Pomerol C (1982) The Cenozoic Era, Tertiary and Quaternary. Ellis Harwood, Chichester,

Reimnitz Erk, Graves SM, Barnes PW (1985) Beaufort Sea coastal erosion, shoreline evolution, and sediment flux. US Geological Survey Open-File Rep 85-380, 66 p

Repenning CA (1983) New evidence for the age of the Gubik Formation, Alaskan North Slope. Quatern Res 19: 356-372

Rodeick CA (1979) The origin, distribution and depositional history of gravel deposits on the Beaufort Sea continental shelf, Alaska. US Geol Sur Open-File Rep 79-234: 1-87

Rosenkrantz A (1970) Marine upper Cretaceous and lowermost Tertiary deposits in West Greenland. Gronl Geol Under, Misc Paper No. 78: 406-453

Rosenvinge L Kolderup (1933) Marine algae from Kangerdlugssuak. Medd Gronl 104(8): 1-14

Rosenvinge L Kolderup (1893) Grönlands Havalger. Medd Gronl 3: 765-981

Rueness J (1977) Norsk Algenflora. Universitets-forlaget, Oslo

Scagel RF, Gabrielson PW, Garbary DJ, Golden L, Hawkes MW, Lindstrom SC, Oliveira JC, Widdowson TB (1989) A synopsis of the benthic marine algae of British Columbia, southeast Alaska, Washington and Oregon. Phycological Contribution No. 3, Department of Botany, University of British Columbia, Vancouver

Scheffer VB (1967) Marine mammals and the history of Bering Strait. In: Hopkins DM (ed) The Bering Land Bridge, Stanford U Press, California, p 350-363

Schytt V, Bostrom K, Hjort C (1982) Geoscience during the Ymer-80 expedition to the Arctic. Geol Foreningens I. Stockholm Forh 103: 109-119

Searles RB (1984) Seaweed biogeography of the mid-Atlantic coast of the United States. Helgol Meeresunters 38: 259-271

Sears JR, Cooper RA (1978) Descriptive ecology of offshore, deep-water benthic algae in the temperate western North Atlantic Ocean. Mar Biol 44: 309-314

Simmons HG (1906) Remarks about the relations of the floras of the Northern Atlantic, the Polar Sea and the Northern Pacific. Bot Zbl (Beih) 19: 149-194

Simpson GG (1953) The major features of evolution. Columbia Univ Press, New York

Smith AG, Briden JC (1977) Mesozoic and Cenozoic paleocontinental maps. Cambridge U Press, Cambridge

South GR (1983) Benthic marine algae. In: South GR (ed) Biogeography and ecology of the island of Newfoundland. W. Junk, The Hague, p 385-420

South GR (1987) Biogeography of the benthic marine algae of the North Atlantic Ocean - an overview. Helgol Meeresunters 41: 273-282

South GR, Tittley I (1986) A checklist and distributional index of the benthic marine algae of the North Atlantic Ocean. Huntsman Marine Laboratory and British Museum (Natural History), St. Andrews and London
Stanley EA (1965) Upper Cretaceous and Paleocene plant microfossils and Paleocene dinoflagellates and hystrichosphaerids from northwestern South Dakota. Bull Amer Paleont 49(222): 175-384
Steele DH (1975) Marine climate and the biogeography of the surface waters in the northwest Atlantic. Nat Can 102: 189-198
Stewart TG, England J (1983) Holocene sea-ice variations and paleoenvironmental change, northernmost Ellesmere Island, NWT, Canada. Arctic Alpine Res 15: 1-17
Taylor WR (1954) Cryptogamic flora of the Arctic II. Algae: Non-planktonic. Bot Rev 20(6,7): 363-399
Taylor WR (1957) Marine algae of the northeastern coast of North America. Univ Michigan Press, Ann Arbor
Troitsky SL (1974) Subarctic Pleistocene molluscan fauna. In: Herman Y (ed) Marine Geology & Oceanography of the Arctic Seas. Springer-Verlag, p 257-270
Vail PR, Hardenbol J (1979) Sea-level changes during the Tertiary. Oceanus 22: 71-80
Vibe C (1950) The marine mammals and the marine fauna in the Thule District (Northwest Greenland) with observations on ice conditions in 1939-41. Medd Gronl 150(6):
Vinogradov AP (1967) Atlas of the lithological paleogeographical maps of the USSR. Moscow, USSR Acad Sci & Ministry of Geol, 15 maps
Wilce RT (1959) The marine algae of the Labrador Peninsula and northwest Newfoundland (Ecology and Distribution). Nat Mus Can, Bull 158, Biol Ser No. 56:1-103
Wilce RT (1964) Studies of attached marine algae in northwest Greenland. Proc Intn Seaweed Symp 4: 280-287
Wilce RJ (1966) Heterotrophy in Arctic sublittoral seaweeds: An hypothesis. Bot Mar 10: 185-197
Wilce RT (1977) The North Atlantic benthic marine flora: origins and phytogeographic limits. J Phycol 13(suppl):73 (abstract only)
Young FG, McNeil DH (1982) Cenozoic stratigraphy of Mackenzie Delta, Northwest Territories, NTS 107B, 107C. Geol Sur Can, Open File Rep 872

BIOGEOGRAPHY OF THE MARINE RED ALGAE OF THE NORTH ATLANTIC OCEAN

Max H. Hommersand
Department of Biology
University of North Carolina at Chapel Hill
Chapel Hill, N.C. 27599-3280
U.S.A.

Abstract - The temperate marine red algal flora of the North Atlantic Ocean contains two principal components: a former Arctic flora, and the remnant of a mid- to late-Tethyan flora. The problem for the biogeographer is to determine how these floras were assembled and subsequently modified. My evaluation is based on an analysis of morphological characters that define family clusters, families, genera and infrageneric taxa among Rhodophyta. The following scenario is suggested by these observations.

Phylogenetic lines that evolved in the Tethyan Ocean in the Mesozoic became widely distributed, largely through plate tectonics and seafloor spreading. Their descendants presently have major centers of dispersal in the SW Pacific (Australia, New Zealand) and the NW Pacific (NE Asia and Japan). Concomitant with gradual cooling of the oceans beginning in Cretaceous and accelerated cooling since Oligocene, a flora evolved within the Tethyan Ocean composed mainly of new phylogenetic lines. With the gradual collapse of the Tethyan Seaway in the Neogene, a portion of that flora, including many archtypal species, was isolated in the NE Atlantic. The larger component in the western Tethys evolved through adaptation to high levels of predation by herbivores. Taxa that had evolved earlier retreated to temperate waters nearer the poles. One refugium for this older element was the Arctic Ocean which received taxa from the west (N Pacific), and from the southeast (via the Turgai Strait and west Siberian Sea). The completion of the Bering Land Bridge, possibly in Eocene, isolated the Arctic flora from the influence of the N Pacific. Extension of the mid Atlantic ridge system to the Arctic Ocean ultimately brought the Arctic flora into contact with the N Atlantic flora, probably in Oligocene. Some components of this flora underwent bipolar distribution in the Atlantic reaching South America, Antarctica and islands of the the West Wind Drift. Extinction of many N Atlantic taxa in the Neogene was followed by the addition of new elements that evolved in the Indo-W Pacific, migrated around South Africa, and were added either by amphitropical distribution along the coast of West Africa or through long range dispersal by way of the Caribbean Sea. A few N Pacific species appear to have reached the N Atlantic since the opening of the Bering Strait in Pliocene.

NATO ASI Series, Vol. G 22
Evolutionary Biogeography of the
Marine Algae of the North Atlantic
Edited by D. J. Garbary and G. R. South

INTRODUCTION

The North Atlantic Ocean has undergone major geological and climatic changes since its formation in the Mesozoic, changes that are reflected in the composition and floristic relationships of its seaweeds. This is nowhere more evident than in the case of the red algal flora of the NE Atlantic and the Mediterranean Sea, a region rich in endemic genera and species. Future taxonomic studies can only increase our sense of the uniqueness of this flora, as rhodophytan species from other areas, presently placed in genera based on European types, receive new names. Directed graphs depicting nearest neighbor relationships between regions, based on ordinations of genera by region (Joosten & van den Hoek 1986), dramatically illustrate the separateness of present-day temperate Atlantic/E Pacific and Atlantic/Indo-W Pacific red seaweed floras.

How is the complexity and special quality of the rhodophytan flora of the eastern North Atlantic to be understood? The only large body of information available at present for such a study comes from the comparative analysis of morphological characters. The approach we have taken has been to investigate, in so far as is possible, the ontogeny of vegetative and reproductive characters, and to make morphological comparisons between taxa based on characters that are expressed at comparable stages of development. Our observations, combined with the morphological evidence in the literature, form the basis for the biogeographic speculations put forward here.

A model of the biogeography of the algae of a region is no better than the accuracy of the phylogenetic hypothesis upon which it is based. There are few certainties in the classification of the Rhodophyta upon which such an hypothesis might be constructed. In my experience, the families recognized by Kylin (1956) have proved to be remarkably sound, even when their ordinal position is questioned. This is a great help in that most useful biogeographic comparisons involve relationships between families, or between clusters of genera within a family. The ultrastructural studies of Pueschel & Cole (1982) that led to the recognition of several new orders, also have aided greatly in clarifying rhodophytan taxonomy at a higher level. Gabrielson & Garbary (1987) have published a cladistic analysis of the orders of red algae incorporating the newest observations. The most recent event has been the establishment of two new orders of agarophytes, Ahnfeltiales (Maggs & Pueschel 1989) and Gracilariales (Fredericq & Hommersand 1989), which were removed from the Gigartinales. By far the largest remaining problem is the classification of the Gigartinales, sensu Kraft & Robins (1985), which still contains over forty families. I have found it necessary in this paper to cluster the families of Gigartinales into smaller groups that appear to be related phylogenetically, and which are seen to have biogeographic significance. These are treated either as separate families or as family clusters (complexes), according to my present opinion regarding their taxonomic identity and possible evolutionary relationships.

Evidence concerning the biogeography of the North Atlantic Rhodophyta are treated in this paper first by taxonomic group and then by region. All morphological evidence supporting a biogeographic affinity of a specific taxonomic group is given in the section on morphology and taxonomy, together with any speculations regarding possible centers of origin and pathways of distribution of specific taxa. Next, I have identified seven patterns of biogeographic relationships by region, which I believe are correlated with events in the geological and climatic history of the earth since the beginning of the Mesozoic (Tables 1-7). Each of these is analysed separately in the discussion section. Tables 1-7 also serve as an index, linking the floristic relationships described in the discussion, to taxa covered earlier in the treatment on morphology and taxonomy. Finally, the possible impact of large scale historical events and processes on the biogeography of the red algae of the North Atlantic Ocean are considered in the conclusions.

MORPHOLOGY AND TAXONOMY OF THE RHODOPHYTA IN RELATION TO THEIR BIOGEOGRAPHY

Rhodochaetales, Compsopogonales

These two orders form a phylogenetic line characterized cytologically by the absence of pit plug caps and cap membranes (Pueschel & Magne 1987; Scott *et al.* 1988) and dictyosomes not associated with mitochondria (Scott & Broadwater 1989). Reproductive bodies (spermatia, carpospores, monospores) are cut off, one at a time, by a curved wall, with part of the cytoplasm remaining with the parent cell (Magne 1989). The monotypic genus *Rhodochaete*, known from the Mediterranean and Caribbean Seas, appears to be a relict taxon confined to the N Atlantic; as such, it represents an important addition to the list of primitive taxa known only from this region. The Compsopogonales, in contrast, are widely distributed in freshwater and marine environments.

Bangiales

The Bangiales, consisting of a single family, Bangiaceae, with two genera, *Bangia* and *Porphyra*, is characterized cytologically by pit plugs with one cap layer lacking cap membranes

(Pueschel 1987), and a dictyosome-mitochondrial association like that found in the Florideophycidae (Scott & Broadwater 1990). The life history is heteromorphic. Spermatia and carpospores are formed in packets through successive perpendicular divisions of the parent cell, with monospores borne terminally, and conchospores produced in fertile cell rows in the filamentous Conchocelis stage (Cole & Conway 1980).

Both *Bangia* and *Porphyra* are elements of the cool-temperate floras of the Northern and Southern Hemispheres, and also range into the tropics. I expect that the Bangiales will ultimately prove to be an old assemblage, represented by distinct groups in the Northern and Southern Hemispheres that were separated initially during the opening of the Tethyan Ocean in the Mesozoic. *Bangia atropurpurea* is generally considered to be a cosmopolitan species, however, cytological and morphological studies (Cole *et al.* 1983; Sheath & Cole 1984) suggest that a complex of species will eventually be resolved, which may then be studied in the context of their evolutionary biogeography. Though only a few species of *Porphyra* are currently recognized from the Southern Hemisphere, the list is growing. For example, approximately a dozen species are now recognized in New Zealand (W. Nelson pers. comm.) where only four were previously recorded. Modern migrations of *Porphyra* into the tropics have, nevertheless, occurred. Kapraun and Lemus (1987) have shown that conchospore formation and release takes place under short day conditions, and is temperature dependent in *P. spiralis* var. *amplifolia* from Venezuela. Originally described from southern Brazil, this species appears to have retained its response to cool temperatures and short day conditions that characterize its ancestors.

A rich *Porphyra* flora is known from both the western and eastern N Pacific. Several species clusters from these regions are also represented by counterparts in the N Atlantic (Lindstrom pers. comm.). These could have resulted from modern distribution since the opening of the Bering Strait in Pliocene; however, I suspect that they reflect an older affinity extending back to the first half of the Cenozoic.

Acrochaetiales

At the present time the Acrochaetiales (Garbary & Gabrielson 1987) contains a single family, variously referred to as the Acrochaetiaceae Fritsch, the Rhodochortonaceae Nasr, or the Audouinellaceae Feldmann ex Woelkerling (Lee & Lee 1988). The order is considered to be archetypal of Florideophyceae (Garbary 1978). Classification at the family and generic level is in a state of flux, and the order may well prove to be polyphyletic. Following Stegenga (1985), I recognize the genera: *Acrochaetium*, *Audouinella*, *Calaconema*, *Rhodochorton*, and *Rhodothamniella*, within Acrochaetiaceae. So far as I have been able to discern from a tabulation of reports in the literature (Glazer *et al.* 1982; Saunders 1987), two of the genera listed above (*Colaconema* and *Rhodochorton*) contain R-phycoerythrin, with the rest containing B-phycoerythrin. B-phycoerythrin, also found in the Bangiales, should characterize the older assemblage. Recently, Saunders & McLachlan (1990) have transferred *Rhodophysema*, *Rhodophysemopsis* and *Halosacciocolax* to a new family, Rhodophysemataceae, which they placed provisionally in the Palmariales. Since the Rhodophysemataceae is also charcaterized by the presence of B-phycoerythrin, I have included it here under Acrochaetiales for purposes of comparison.

Photoperiodic control of the life history has been demonstrated in most genera belonging to the Acrochaetiales found at high latitudes (Dring 1989). I believe that the modern Acrochaetiales evolved a successful life history near the poles, through the incorporation of photoperiodic mechanisms for regulating vegetative growth and reproduction. In this regard they are like the Bangiales, and are forerunners of this behavior in succeeding families of Rhodophyta. With the exception of *Rhodothamniella*, all of the genera of Acrochaetiaceae that have been reported from the N Atlantic, are also known from the N pacific. The same is true of Rhodophysemataceae. I take this as evidence that all, except *Rhodothamniella*, originated in the Northern Hemisphere either in the North Pacific or Arctic Ocean. Most Acrochaetiaceae appear to have undergone bipolar distribution and are widely distributed in the Southern Hemisphere, including Antarctica.

Palmariales

The Palmariales, family Palmariaceae, possess R-phycoerythrin and are characterized cytologically by pit plugs with two cap layers and a cap membrane (Pueschel & Cole 1982). Spermatangia are borne in pairs. Tetrasporangia are formed by periclinal division of a vegetative cortical cell into an inner stalk cell and a large outer tetrasporangial initial. Division of the tetrasporangial initial is cruciate, and the inner stalk cell may enlarge and divide to produce a secondary tetrasporangial mother cell and a further stalk cell (Guiry 1974). The life cycle is biphasic with isomorphic male gametophytes and tetrasporophytes. Female gametophytes are microscopic, consisting of a crustose disc composed of cells that reach maturity in days with the production of terminal carpogonia. A fertilized carpogonium gives rise directly to the tetrasporophyte (van der Meer & Todd 1980). According to these criteria the Palmariaceae contains the genera *Palmaria* (including Leptosarca, see Ricker 1987), *Devaleraea*, *Halosaccion* and *Neohalosacciocolax*.

Using starch gel electrophoresis of enzymes, Lindstrom & South (1989) showed that the type species of *Palmaria*, *P. palmata*, from the N Atlantic, is distinct from all other species of *Palmaria*, *Devaleraea*, and *Halosaccion*. Calculated distances between N Atlantic *Palmaria* and the other taxa corresponded to 45.3 million years. While they found this separation date to be unrealistic, it fits perfectly with expections based on the model I am proposing here. In my opinion, the Palmariaceae evolved in the N Pacific and Arctic Oceans in the late Cretaceous. *Palmaria palmata* was probably isolated in the Arctic sometime during the Cenozoic and reached the N Atlantic concurrently with the opening of a passageway through the Barents Shelf, perhaps in Oligocene. Subsequently, *Palmaria* was distributed to the S Atlantic and to Antarctica and the islands of the West Wind Drift inOligocene or miocene. Ricker (1987) finds very little difference between either sporophytes or gametophytes of *P. decipiens* from antarctic and subantarctic waters and *P. palmata* from the N Atlantic. Relatively rapid softening of blades of *P. decipiens* following fixation in 4% formalin-seawater was the one distinction he noted.

Lindstrom & South (1989) found that the distance between *Palmaria mollis* (a solid, flattened species occurring in Pacific North America) and *Devaleraea ramentacea* (a hollow, cylindrical species found in both the N Pacific and N Atlantic) corresponded to a separation of 3.1 million years. This suggests that the two are closely related species, despite their obvious morphological dissimilarity. If so, *Devaleraea* must have been introduced from the N Pacific to the N Atlantic after the opening of the Bering Strait in the Pliocene.

Corallinales

I am not sufficiently familiar with this group to include it for discussion here.

Nemaliales

The Nemaliales are characterized by pit plugs with two cap layers and a cap membrane (Pueschel & Cole 1982; Pueschel 1989). Thalli are multiaxial, with sympodial growth of the axial filaments (Coomans & Hommersand 1990). A terminal or lateral carpogonial branch is present. Gonimoblasts develop directly from the carpogonium after fertilization, or from a derivative cell formed by the first division of the fertilized carpogonium. The order contains two families: Liagoraceae and Galaxauraceae. Kraft (1989) has analysed generic characters and summarized generic distributions of the Liagoraceae; *Trichogloea*, *Trichogloeopsis*, *Liagoropsis*, and *Liagora* (sensu lato) appear to be represented by species clusters distributed between the Indo-W Pacific and the Caribbean regions. Such a distribution pattern could have come about through eastward migration by way of the Alaskan Peninsula in the Cretaceous or Paleogene, a westward distribution through the Tethyan Ocean, or a westward distribution by way of southern Africa at a later date. *Helminthora* exhibits a Tethyan distribution represented by separate N Atlantic and austral Indo-W Pacific species clusters. *Nemalion helminthoides* could have originated in the N Pacific, and appears to have been distributed to the N Atlantic by way of the Arctic Ocean, either in the Paleogene, or more recently in Pliocene.

The *Scinaia* assemblage of the Galaxauraceae afford a better test of an Indo-Pacific origin, followed by an eastward migration of species clusters; indeed, such a pathway was originally proposed by Svedelius (1924) to account for the distribution of several species clusters within *Scinaia*. Whereas Svedelius proposed a pan-tropical distribution in Pliocene by way of the sinus of Panama, I envision a late Cretaceous to Paleogene distribution by way of the boreal N Pacific prior to the onset of Oligocene. Species clusters include: *Scinaia okamurae* (Japan)-*confusa* (Pacific N America)-*halliae* (Caribbean)-*furcellata* (Europe), with all but the last species formerly placed in *Pseudogloiophloea*; *Scinaia japonica* (Japan)-*johnstoniae* (Pacific N America)-*johnstoniae* (West Africa)-*complanata* (Atlantic); *Scinaia cottonii* (Japan)-*latifrons* (Pacific N America)-*imbricata* (Caribbean)-*cottonii* (West Africa). These all appear to be distinct from Indian Ocean species. New observations are needed to test this hypothesis.

The genus *Galaxaura* is more advanced than *Scinaia* in that spermatangia and cystocarps are formed in conceptacles. The subsidiary filaments borne on the hypogynous cell of the carpogonial branch are more highly branched, and an extensive fusion cell is formed after fertilization (Hommersand & Fredericq 1990). The tetrasporophyte is erect, and as fully developed as the gametophyte. Species of *Galaxaura* are distributed world-wide in tropical waters. I believe that *Galaxaura* is a modern genus that originated in the Indo-W Pacific, and was largely distributed in the Neogene. The absence of any but cosmopolitan species of *Galaxaura* in western Europe and the Mediterranean Sea, testify to this opinion.

Ahnfeltiales

The Anhfeltiales are agarophytes with pit plugs lacking cap layers or cap membranes. The life history is heteromorphic, with erect, branched male and female gametophytes and a crustose tetrasporophyte. Spermatangia are cut off transversely, and carpogonia are terminal. The carposporophyte develops directly from the carpogonium, fuses facultatively with gametophytic cells, and produces outwardly developed gonimoblasts with terminating carposporangia (Maggs & Pueschel 1989). The order consists of one family, one genus and three species (Maggs *et al.* 1989). Other species presently placed in *Ahnfeltia* belong in the Phyllophoraceae.

Of the three species of *Ahnfeltia*, the type species, *A. plicata*, is widespread in the N Atlantic, extending into the N Pacific from Cape Pierce, Alaska, to the Sea of Okhotsk, Siberia, and in the S Atlantic, from southern Argentina to the Strait of Magellan, the Antarctic Peninsula, and islands of the West Wind Drift. *A. fastigiata* is found only in the N Pacific from Hokkaido through Alaska to northern Baja California. The third species, *A. elongata*, is known only from south-central Chile. A comparison of species characters (Maggs *et al.* 1989) reveals that *A. plicata* and *A. elongata* are closely related, whereas *A. fastigiata* is separated by several prominent characters, including the absence of cortical growth rings, and the development of female sori on young, rather than mature axes.

The Ahnfeltiales must be a very old order, and *A. fastigiata* and *A. plicata* have surely been separated for a long time, perhaps since the late Cretaceous or early Paleogene. Because Ahnfeltiales is a relict group, it is impossible to assign a likely center of origin to the order. The assessment that *A. fastigiata* is the more primitive of the two species, favors a N Pacific origin. The distribution of *A. plicata* and its sister species, *A. elongata*, to the S Atlantic could have taken place during two separate cooling cycles in the Neogene. Displacement of *A. elongata* northward to the coast of south-central Chile is not unexpected. This distribution pattern is shared by many species endemic to southern South America. The presence of *A. plicata* in the boreal N Pacific could have resulted from migration across the Arctic since Pliocene, or from recent distribution through the agency of man.

Gracilariales

The Gracilariales are agarophytes characterized by pit plugs with parallel striations perpendicular to the ends of the plug core; cap membranes are present, but with cap layers probably absent (Pueschel 1989). In the Gracilariaceae the carpogonium terminates a two-celled filament flanked by two or more modified sterile filaments. Gonimoblasts develop directly from the carpogonium after fusion with cells of the sterile filaments, and bear carposporangia in chains within a cavity surrounded by a pericarp developed externally from cortical filaments (Fredericq & Hommersand 1989).

In my opinion the Gracilariaceae is an ancient family that originated in eastern Gondwanaland, and was subsequently divided into Northern and Southern Hemisphere components at a late stage, during the opening of the Tethyan Ocean in the Mesozoic. The southern group, composed of *Melanthalia* and *Curdiea* from Australia and New Zealand, are characterized by the presence of a broad apical meristem with multiaxial growth, in which secondary pit connections are formed late in the differentiation of the medulla, and are largely absent in the cortex. Gonimoblasts completely fill the cystocarp cavity, which does not appear to be formed schizogenously, and tetrasporangia are nemathecial, borne between raised cortical filaments. The northern group, containing *Gracilariopsis*, *Gracilaria* and *Hydropuntia* (including *Polycavernosa*), appears to be uniaxial, with secondary pit connections forming between cortical filaments near the thallus apex. The cystocarp cavity forms schizogenously and is not completely filled by the gonimoblasts, and tetrasporangia are superficial, scattered over the cortex (Frederic & Hommersand 1990). I envision that *Gracilariopsis* and some sections of *Gracilaria* were distributed from the NW Pacific to Pacific N America by way of the Alaskan Peninsula, either in late Cretaceous or in the Paleogene, ultimately reaching the Caribbean Sea, and possibly also Europe and the Mediterranean Sea. Other species clusters in *Gracilaria* may have a Tethyan distribution, while still other *Gracilaria* species, and species belonging to *Hydropuntia*, appear to have extended their range westward from the Indian Ocean to the western Atlantic and Caribbean Sea by way of southern Africa.

Gelidiales

Like the Graciariales and the Ahnfeltiales, the Gelidiales are agarophytes containing neutral agarose. Pit plugs have cap membranes with one cap layer, and parallel striations extending inward from the ends of the plug core (Pueschel 1989). Spermatangia are cut off transversely from the spermatangial mother cells. Carpogonial branches are intercalary, and give rise directly to gonimoblasts that fuse with cells of preformed nutritive filaments before initiating carposporangia (Hommersand & Fredericq 1988). Like the Gracilariales, the Gelidiales appear to have originated in eastern Gondwanaland, and to have become separated into two groups with the opening of the Tethyan Ocean. In this instance, however, the southern group now contains only the type species of *Pterocladia*, *P. lucida* (including *Pterocladiastrum*), from Australia and New Zealand, with the remaining genera and species having originated in the Northern Hemisphere. In *Pterocladia lucida* the nutritive filaments that sustain the developing gonimoblasts are derived from medullary cells situated across the base of the cystocarp; these are weakly differentiated, and stain poorly with hematoxylin (pers. obs.). The cystocarp is unilocular with ostioles formed only on one side. Other species currently placed in *Pterocladia*, related to *P. capillacea*, may be unilocular with ostioles present on only one side; however, the nutritive filaments are derived from the four basal cells of the third order filaments, borne on the innermost second order filaments, and form a central core around the central axis. The gonimoblast extends as a cylinder outside the central core, with which it becomes connected and bears carposporangia in chains (Fredericq & Hommersand, pers. obs.). Other sexually reproducing Gelidiaceae are like *Gelidium* in having well-differentiated nutritive filaments formed from the basal cells of third-order filaments on either side of a medial septum composed of second-order filaments that divide the cystocarp into two halves (bilocular condition), usually with ostioles present on both sides (Hommersand & Fredericq 1988). As in the Gracilariaceae, the present center of distribution of the northern group of Gelidiaceae is the northern Indo-W Pacific, where most genera occur. *Gelidium*, *Gelidiella* and the *Pterocladia capillacea* group are probably all polyphyletic, and are widely distributed. An understanding of the biogeography of the Gelidiaceae is not possible without a thorough taxonomic revision of the family.

Bonnemaisoniales

Most current systems place the Naccariaceae and Bonnemaisoniaceae in the order Bonnemaisoniales. In recognizing a close relationship between these two families, Kylin (1928, 1956) emphasized the feature that the hypogynous cell (cell directly below the carpogonium) produces a tuft of filaments ('nutritive tuft') whose contents are consumed soon after fertilization at an early stage of gonimoblast development, and that a 'typical' auxiliary cell is absent. The Naccariaceae presently contains three genera: *Atractophora*, (Atlantic Europe), *Naccaria*, (widespread in the Atlantic and Pacific Oceans), and *Reticulocaulis* (Hawaiian Islands). Chihara & Yoshizaki (1972) divided the Bonnemaisoniaceae into two groups: one containing *Delisea*, *Ptilonia* and *Leptophyllus* (primarily Southern Hemisphere), the other *Bonnemaisonia* and *Asparagopsis* (primarily Northern Hemisphere).

Several themes can be identified in the Bonnemaisoniales that have phylogenetic and biogeographic significance. Branching has evolved from seemingly random sequences of lateral initiation in *Atractophora* to highly patterned sequences in *Bonnemaisonia* and *Asparagopsis* in which the fate of a dividing cell is determined by its position and by the plane of cell division (Coomans & Hommersand 1990). With increased precision of vegetative branching, the location of the carpogonial branch and cystocarp becomes specified. Nutritional efficiency is achieved in the Bonnemaisoniales through the formation of secondary assimilatory filaments (Naccariaceae) or pericarps (Bonnemaisoniaceae), and associated modification of internal gametophytic tissues for nutrient processing and storage (Hommersand & Fredericq 1990). Finally, the prostrate *Hymenoclonium*-type tetrasporophytic phase has been replaced in the most advanced Bonnemaisoniaceae by a tetrasporophyte that exhibits bipolar germination, and expresses certain developmental features found in the gametophyte (see West & Hommersand 1981).

Atractophora exhibits the simplest type of vegetative growth. Lateral filaments are initiated in an irregular sequence, becoming quadriverticillate. A second whorl may form below the first, particularly in tips of fertile thalli (Kylin 1928). In *Naccaria* the apical cell divides obliquely, and the first determinate filament arises from the high side of each axial cell in a 1/4 spiral. A single secondary filament is usually produced about eight segments below the first, at right angles, in the direction of the spiral (Kylin 1928). Branching in *Delisea* and *Ptilonia* is quadriverticillate. The first lateral initial is cut off from the high side of an axial cell in an alternating sequence, and develops into a major determinate or an indeterminate branch; the second is opposite the first, and forms a somewhat weaker determinate branch; the third and fourth are transverse, and contribute only cortex (Chihara & Yoshizaki 1971a; Bonin & Hawkes 1988b, c). In *Bonnemaisonia* and *Asparagopsis* the apical cell divides obliquely. Two opposite periaxial cells are produced, with the first developing into an unbranched determinate lateral, and the second into either an indeterminate branch, a modified determinate branch, or a fertile branch. Initials of cortical filaments are either ternately or quaternately branched (Kylin 1956).

In *Atractophora* the carpogonial branch is three-celled, and formed on periaxial cells of either primary or secondary whorls (Kylin 1928). Similar variability is encountered in *Naccaria naccarioides* from Australia (Womersley & Abbott 1968); however, the carpogonial branch is two-celled, borne on the basal cell of a secondary determinate filament in the type species of *Naccaria, N. wiggii*, from Atlantic Europe (Kylin 1928). In the *Delisea-Ptilonia* group, a three-celled carpogonial branch is borne alternately in successive segments on the basal cell of the second determinate filament. It either lies in the axil between two major branches (*Delisea*, Bonin & Hawkes 1988b), or laterally on a short indeterminate branch (*Ptilonia, Leptophyllis* (Bonin & Hawkes 1988c). The carpogonial branch is also three-celled in *Bonnemaisonia* and *Asparagopsis*, but it develops from the first, not the second periaxial cell, near the tip of a fertile branch (Kylin 1956).

Gonimoblast filaments ramify among extant vegetative filaments in *Naccaria naccarioides*, and additional assimilatory filaments do not seem to be produced; however, pit connections between axial cells broaden, and their nuclei enlarge in the vicinity of the carposporophyte (Hommersand, pers. obs.). Vegetative whorls resume growth in the vicinity of developing gonimoblasts, adding assimilatory filaments in *Atractophora*, and axial cells enlarge, becoming richly filled with cytoplasm (Kylin 1928). In *Naccaria wiggii* cells of the nutritive tuft contain enlarged nuclei, the carpogonium fuses with the hypogynous cell and with cells of the nutritive tuft. The supporting cell produces a branched system of secondary assimilatory filaments connected by broadened pit connections with the fusion cell and with cells of the central axis (Hommersand & Fredericq 1990). The nutritive tuft is large and highly branched in the Bonnemaisoniaceae (Chihara & Yoshizaki 1972). Before fertilization, the supporting cell and the first cell of the carpogonial branch bear ordinary cortical filaments in *Delisea* and *Ptilonia* (Chihara & Yoshizaki 1971a, 1972); however, a large pericarp is formed through secondary growth of cortical filaments associated with determinate branches (Hommersand pers. obs.). Pericarpic filaments are initiated before fertilization from the supporting cell (basal periaxial cell) and the first cell of the carpogonial branch in *Bonnemaisonia* and *Asparagopsis* (Kylin 1956); these, together with a few adjacent cells, generate a large pericarp around the carposporophyte after fertilization. These cells are linked by broadened pit connections and, ultimately, by fusion with axial cells and cells subtending the gonimoblast (Hommersand & Fredericq 1990).

In the Naccariaceae (Boillot & L'Hardy-Halos 1975, Maggs 1988), the *Delisea-Ptilonia* group (Chihara 1962, Bonin & Hawkes 1988 a, b), and *Bonnemaisonia asparagoides* (Rueness & Åsen 1982) the tetrasporophyte is a filamentous crustose plant, sometimes resembling *Hymenoclonium serpens*. *Bonnemaisonia geniculata* is intermediate, in that the carpospore first divides vertically into two daughter cells, both of which form erect and rhizoidal axes (Shevlin & Polanshek 1978). Mature tetrasporophytes raised in culture from plants obtained in

California resemble *Pleuroblepharidella*, a genus known only from tetrasporic plants at high latitudes in the western N Pacific (Wynne 1980 and included references). Bipolar germination of a carpospore produces the *Trailiella* tetrasporophytic stage of *Bonnemaisonia hamifera* from Japan (Chihara 1961) and *B. nootkana* from Pacific N America (Chihara 1965), and the *Falkenbergia* stage of *Asparagopsis* (Chihara 1961; Bonin & Hawkes 1988c).

If one considers the characters of each of the taxa just discussed, a biogeographic pattern emerges in which *Naccaria* is divided into an Indo-W Pacific cluster that includes *N. naccarioides* (Womersley & Abbott 1968) and *N. hawaiiana* (Abbott 1985a), and an Atlantic cluster, probably similar to *N. wiggii* of Atlantic Europe. *Atractophora* appears to be a relict genus restricted to Atlantic Europe. The two groups that make up the Bonnemaisoniaceae are distributed along a polar-Tethyan axis with the *Delisea-Ptilonia* group having a center of distribution in southern Australia-Tasmania-New Zealand. A few species have reached South America and extend along the islands of the West Wind Drift. *Delisea pulchra* and *D. okadai* from Japan stand sufficiently close to Southern Hemisphere species to suggest that they have undergone amphitropical distribution, reaching Japan in late Miocene or Pliocene. The counterpart of the *Delisea-Ptilonia* group in the Northern Hemisphere is the *Bonnemaisonia-Asparagopsis* group. Clearly, the most primitive member of this assemblage is *B. asparagoides* from Atlantic Europe; all of the remaining taxa probably evolved in the N Pacific. I suggest that the ancestral species of *Bonnemaisonia* arose in, or near, the Arctic Ocean in late Cretaceous, and spread to the N Pacific in late Eocene or early Oligocene, and that *B. asparagoides* is a relict of polar seas, that reached the N Atlantic in Oligocene. The *B. geniculata* and *B. hamifera* groups probably evolved at high latitudes in the N Pacific, and extended southward on both its eastern and western sides. *Asparagopsis taxiformis* evolved in parallel to *Bonnemaisonia hamifera*, probably in the NW Pacific. *A. armata* is a distinct species (Bonin & Hawkes 1988c), probably derived through the amphitropical distribution of an *A. taxiformis*-like ancestor, that reached Australia/New Zealand in the Pliocene. Subsequently, *A. taxiformis* was distributed around the world in the tropics. *Bonnemaisonia hamifera* and *Asparagopsis armata* have become widely distributed within historical times, due to man's activities.

Gigartinales

Following the merging of Cryptonemiales with Gigartinales (Kraft & Robbins 1985) the emended order is characterized solely by the interaction of the fertilized carpogonium with one or more auxiliary cells, either through the agency of connecting filaments (non-procarpial condition), or by direct fusion (procarpial condition). Pit plugs lack cap layers, but cap membranes are present (Pueschel 1989). In the following treatment I have subdivided the

Gigartinales into families or family complexes for the purpose of discussing their biogeography.

NEMATHECIAL COMPLEX (POLYIDEACEAE, RHIZOPHYLLIDACEAE, GAINIACEAE, PEYSSONNELIACEAE AND RHODOPELTIS) . This group typically possesses an isomorphic life history, and is characterized by having male and female reproductive organs (and sometimes tetrasporangia) formed in nemathecial sori generated from surface cortical cells. Spermatangial mother cells are terminal, lateral or intercalary on nemathecial filaments. Carpogonia are terminal, sometimes associated with a 3-4-celled carpogonial branch. Auxiliary cells are intercalary in unmodified nemathecial filaments, or more often, in modified, 3-celled nemathecial filaments. Connecting filaments issuing from the carpogonium are uneven in diameter, and may be different in character from those of other red algae (Hommersand & Fredericq 1990) Gonimoblasts arise from connecting filaments, either adjacent to auxiliary cells or remote from them. Tetrasporangia are cruciately to zonately divided.

This assemblage of families and genera appears to be very old (Mesozoic?), as witnessed by the wide range of vegetative types associated with a common nemathecial reproductive structure. *Polyides* is endemic to the N Atlantic; *Ochtodes* to the tropical Atlantic and Galapagos Islands; *Chondrococcus* to the Indo W-Pacific; *Contarinia* to the Mediterranean Sea and Japan; *Gainia* to Antarctica; and *Rhodopeltis* to the western central Pacific; only *Peyssonnelia* is widespread. Maggs (1990) has suggested that species of *Peyssonnelia* fall into discrete, widely separated clusters which must be evaluated if the biogeography of this genus is to be understood.

DUMONTIACEAE/GLOIOSIPHONIACEAE. The Dumontiaceae and Gloiosiphoniaceae are closely related families characterized by the production of carpogonia and auxiliary cells in special adventitious filaments that usually develop from the lower sides of inner cortical cells. The Gloiosiphoniaceae are essentially procarpial, with auxiliary cells and carpogonia borne in the same branched filament. Connecting filaments are present only in *Gloiosiphonia*. In contrast, the Dumontiaceae are non-procarpial, with carpogonia and auxiliary cells borne in separate filaments; connecting filaments are universally present. Schmitz (1883) regarded the procarpial condition as being primitive, and he (Schmitz & Hauptfleish 1896-1897) placed the Gloiosiphoniaceae at the base of his order Cryptonemiales. Kylin (1930, 1956), on the other hand, regarded the non-procarpial condition as primitive, and placed the Dumontiaceae at the base of the Cryptonemiales. This distinction has both phylogenetic and biogeographic significance.

1. Dumontiaceae

With the removal of *Acrosymphyton* (Lindstrom 1987a) the Dumontiaceae becomes a coherent, if diverse, taxonomic group. Carpogonia and auxiliary cells are borne separately in modified branched or unbranched filaments. Typically, the last three cells of a carpogonial filament are recurved, and cells 4-5 (6) behind the tip enlarge, and contain single, enlarged nuclei, as do cells adjoining the auxiliary cell. After fertilization, the carpogonium commonly divides cutting off 1-2 derivative cells, the carpogonium and its derivatives fuse with cells of the carpogonial filament, and the resulting fusion cells produce several to many connecting filaments. A connecting filament passing close to an auxiliary cell divides into a terminal and an intercalary cell. The terminal cell continues as a connecting filament, while the intercalary cell segments, and fuses with a process emanating from the auxiliary cell. The fusion product emits secondary connecting filaments and gonimoblasts, or gonimoblasts only. Gonimoblast filaments usually coalesce to form a single cluster of carposporangia (Lindstrom & Scagel 1987, Hommersand & Fredericq 1990).

The Dumontiaceae are represented in the Atlantic Ocean by three genera: *Dudresnaya*, *Dumontia*, and *Dilsea/Neodilsea*. The record of *Pikea californica* from the Isles of Scilly, England, is probably due to a recent introduction (Maggs & Guiry 1987). Lindstrom (1985) investigated the morphology and taxonomy of all genera and most species belonging to the Dumontiaceae, and presented a cladistic analysis of her results, part of which (the tribe Dumontieae) has been published (Lindstrom 1988). Lindstrom recognized three major clusters: (1) an unnamed *Dudresnaya*-group (*Dudresnaya, Thuretellopsis, Gibsmithia* and *Kraftia*); (2) the Farlowieae (*Farlowia, Pikea, Orculifilum, Leptocladia, Weeksia* and *Constantinea*); (3) the Dumontieae (*Dasyphloea, Dumontia, Cryptosiphonia, Hyalosiphonia, Masudaphycus, Neodilsea* and *Dilsea*. I will discuss each of these separately.

A. The Dudresnaya-group - The *Dudresnaya*-group is distinguished by the comparatively primitive level of organization of the vegetative apex. Thalli are either uniaxial or multiaxial. A well-defined apical cell is absent in many species, and determinate lateral branches are not ordinarily produced in regular succession. Very often, secondary lateral filaments are added adventitiously, after the primary determinate laterals have matured, and the site of initiation of indeterminate lateral branches is also variable (Robins & Kraft 1985; Kraft & Robins 1985; Lindstrom 1985). *Gibsmithia* is Indo-Pacific, and *Kraftia* is known only from southern Australia, but *Dudresnaya* is widespread. The biogeography of species characters in *Dudresnaya* appears to be almost random at the present time (Searles 1983; Robins & Kraft 1985), suggesting that several separate clusters are involved, each having its own distribution pattern. I have identified one such cluster (listed in Table 4) characterized by apicalization of its terminal meristem. *Thuretellopsis peggiana*, from Washington and British Columbia, is closely related to *Dudresnaya* (Lindstrom 1985), from which it differs primarily in its heteromorphic life history (Richardson & Dixon 1970). Both the Farlowieae and Dumontieae have reached a higher level of thallus organization, in which determinate lateral filaments are produced and differentiate, in regular succession, close to the thallus apex.

B. The Farlowieae - The Farlowieae is a coherent group. The most primitive genera (*Farlowia, Orculifilum* and *Pikea*) probably all possess a heteromorphic life history, whereas the most advanced genera (*Leptocladia, Weeksia, Constantinea*), based on both their vegetative and reproductive characters (Lindstrom 1985), have an isomorphic life history. Spore germination is of the *Naccaria*-type, in which a derivative of the first division becomes an apical initial of a prostrate filamentous or crustose system, with a marginal meristem, as in *Farlowia* (DeCew & West 1981a) or *Pikea* (Scott & Dixon 1971; Chihara 1972). Erect thalli are initiated under long day conditions, and are initially uniaxial in the Farlowieae. Initials of determinate lateral filaments divide rapidly to produce primary ascending files that branch abaxially and toward the thallus surface. Spore germination in *Constantinea* is of the *Dumontia* type. The thallus grows from a meristematic zone composed of a cluster of rapidly dividing apical initials generated under day-shortening conditions (Lindstrom 1981; Powell 1986). These branch in a coherent fashion to produce a peltate disc and upper cortex, followed by the lower cortex. Secondary cortical and medullary filaments then fill in the space created by thallus expansion (Lindstrom 1980). Cells 2-5 of the carpogonial branch may be enlarged and lobed; pit plugs enlarge after fertilization and during gonimoblast development (Lindstrom & Scagel 1987).

The Farlowieae is primarily limited to the eastern N Pacific (Pacific North America), except for *Farlowia mollis* prox., which extends to northern and central Japan (Lindstrom & Scagel 1987), and *Pikea californica* from central Japan (Chihara 1972). The latter may well have been introduced by man within historical times. The Farlowieae could have originated in the western Pacific, from an ancestor having a heteromorphic life history like that found in *Thuretellopsis peggiana* (Richardson & Dixon 1970). The distinctive branching pattern of the erect thallus, however, characterizes the taxa found in Pacific North America, from Alaska to Baja California. Since this tribe is absent in the N Atlantic, I conclude that it migrated from west to east, without interacting at any time with the Arctic Ocean.

C. The Dumontieae - In contrast to the Farlowieae, all members of the Dumontieae possess an isomorphic life history. Spore germination (where known) is of the *Dumontia*-type in which divisions produce a hemispherical cell mass within the original cell wall. Development of the thallus is either multiaxial from a cluster of initials (Dumontia), uniaxial with a distinct apical cell (*Dasyphloea, Cryptosiphonia, Hyalosiphonia*), or uniaxial only at branch tips or in juvenile condition and otherwise multiaxial (*Masudophycus, Neodilsea, Dilsea*). Branching of the determinate filaments is initially quaternate or ternate, becoming subdichotomous with periclinal divisions, followed by successive anticlinal divisions, at approximately right angles to one another. Primary determinate filaments of the type found in the Farlowieae are absent. I will break the Dumontieae into four groups for purposes of discussion: (1) *Dasyphloea*, (2) *Cryptosiphonia*, (3) *Dumontia*, (4) *Hyalosiphonia, Masudaphycus, Neodilsea, Dilsea.*

(1) *Dasyphloea insignis* from southern Australia is the only species of this assemblage occurring outside the Northern Hemisphere. As Lindstrom (1988) states, its inclusion in the Dumontieae is tenuous. Its germination pattern is unknown, and it possesses a straight rather than a spiral trichogyne, strictly zonate tetrasporangia, and several other distinctive features.

(2) *Cryptosiphonia*, from Pacific North America, is like the Dumontieae in most respects; however, division of the apical cell is oblique, and cortical filaments are produced in pairs at right angles from each successive segment (Sjöstedt 1926). Rhizoids are sparse, and secondary pit connections are absent; its germination pattern is unknown. The strongest reason, at the moment, for including *Cryptosiphonia* in the Dumontieae, is its branched auxiliary cell filaments. My opinion is that the ancestor of *Cryptosiphonia* migrated separately to the eastern Pacific without having passed into the Arctic Ocean; if so, it is a remnant of an otherwise extinct line.

(3) There are two species of *Dumontia*, *D. contorta* in the N Atlantic, Arctic and N Pacific Oceans, and *D. simplex* in the NW Pacific. *D. contorta* has an isomorphic life cycle in which tetraspores germinate directly into discoid microthalli which give rise to erect multiaxial macrothalli under short-day conditions (Rietema and Klein 1981). The erect thallus develops from a bundle of uniseriate filaments composed of flat discoid cells generated by a cluster of rapidly dividing apical initials (Rietema 1984). *D. simplex* is unbranched, while *D. contorta* produces uniaxial laterals, apparently by the divergence of axial filaments from the main thallus (Wilce & Davis 1984). By my reckoning, *Dumontia* possesses the largest number of ancestral characters of any member of the Dumontieae. There are no secondary modifications of pit connections in either vegetative or reproductive filaments, no secondary pit connections, and no extraneous cell fusions. The carpogonial and auxiliary cell filaments are short and unbranched, or branched only at the base. I suspect that *Dumontia* is a relict genus that was conserved in the Arctic Ocean, appearing in the Atlantic following the traverse of the Barent Shelf, and the broadening of the Greenland-Norwegian Sea in the Oligocene. Its extension into the N Pacific could be a recent event, since Pliocene, or may represent an older vicariant distibution in Oligocene.

(4) As Lindstrom (1988) has pointed out, *Hyalosiphonia*, *Masudaphycus*, *Neodilsea*, and *Dilsea*, share many characters in common and are undoubtedly closely related. *Hyalosiphonia* is uniaxial, and the arrangement of the determinate branches is quadriverticillate. Even so,

secondary pit connections are present and reproductive development is elaborate, and includes a large gonimoblast fusion cell, fusions of auxiliary cell branches with vegetative cells, and an emergent cystocarp (Chihara & Yoshizaki 1971b). Branches of *Masudaphycus* grow by means of apical cells, but the thallus quickly becomes multiaxial, with extensive formation of rhizoids and secondary pit connections. The life history is imperfectly known, but includes bisporic tetrasporangia, and apparent apomictic formation of carposporangia (Lindstrom 1985). *Neodilsea* and *Dilsea* are multiaxial except, perhaps, in young microthalli. The two genera are presently separated on the basis of a single character: tetrasporangia lateral in *Neodilsea* vs. intercalary in *Dilsea.*

I propose that the ancestor of the *Hyalosiphonia-Dilsea* assemblage originated in the N Pacific, and interacted with the Arctic Ocean in late Eocene or early Oligocene, whereupon it radiated new genera and species. Its nearest living relative would be *Dumontia.* According to Lindstrom's (1988) cladogram the Arctic and N Atlantic species *Dilsea integra* and *D. carnosa* are primitive in the *Dilsea/Neodilsea* complex. Her cladogram includes several reversals, including reduction and loss of a gonimoblast fusion cell, and the loss of fusions between auxiliary cell branches and vegetative cells. Her interpretation favors an early separation between N Atlantic and N Pacific taxa of *Dilsea/Neodilsea*, analogous to that proposed earlier for *Palmaria.* Alternatively, should it prove that *D. integra* and *D. carnosa* are advanced members of the *Dilsea/Neodilsea* complex, *Dilsea* could have originated in the eastern N Pacific, and extended its distribution through the Arctic Ocean in the N Atlantic since the opening of the Bering Sea in the Pliocene.

2. Gloiosiphoniaceae

The Gloiosiphoniaceae can be divided into two groups: one procarpial, lacking connecting filaments (*Thuretella, Gloeophycus, Schimmelmannia, Peleophycus*); the other both procarpial and non-procarpial, with connecting filaments (*Gloiosiphonia*). *Thuretella* (Kylin 1930) and *Gloeophycus* (Lee & Yoo 1979, Kaneko *et al.* 1980) exhibit a primitive type of vegetative growth in which the thallus is uniaxial, determinate laterals are produced irregularly in whorls of four, and indeterminate branches arise adventitiously from inner cortical cells. In both genera, the auxiliary cell is the terminal cell of a two-celled side branch, borne on the fourth cell below the carpogonium, which is recurved. Just as in many Dumontiaceae the carpogonium elongates and divides transversly, cutting off 1-2 derivatives, one of which fuses with the nearby auxiliary cell. The fusion product cuts off a single gonimoblast initial which bears the gonimoblast. *Peleophycus* (Abbott 1985b) resembles *Thuretella/Gloeophycus*, except that a carpogonial and 1-2 auxiliary cell branches are produced separately from the same inner cortical cell and lie in proximity to one another. In *Schimmelmannia* (Kylin 1930; Baardseth 1941; Acleto 1972) there are five determinate laterals per whorl. The initial of each lateral produces a leading axis, which branches abaxially and laterally, and indeterminate branches arise adventitiously from apical cells of primary determinate filaments. The auxiliary cell is the terminal cell of a basal branch of the female reproductive apparatus that bears the carpogonium; again, the carpogonium divides twice, and one of the carpogonial derivatives fuses with the nearby auxiliary cell. A fertilized carpogonial nucleus divides once or twice in a prescribed manner in members of nearly all orders of Florideophycidae, whether auxiliary cells are present or not (Hommersand & Fredericq 1990). I would argue that the tendency for carpogonial cell derivatives to fuse with a nearby cell (potential auxiliary cell) evolved first, prior to the formation of connecting filaments in Gloiosiphoniaceae and Dumontiaceae, and that connecting filaments were derived secondarily.

Vegetative development in *Gloiosiphonia* is fundamentally like that of *Schimmelmannia*, in that the primary determinate initial generates an ascending file of cells that branch laterally and abaxially, and indeterminate branches are produced through the continued growth of apical cells of primary determinate filaments (Coomans & Hommersand 1990). The female reproductive apparatus is more highly branched, and may be polycarpogonial, but is otherwise typical of Gloiosiphoniaceae. The carpogonium of *Gloiosiphonia* is subtended by a large hypogynous cell containing an enlarged nucleus (Hommersand & Fredericq 1990). The fertilized carpogonium cuts off 1-2 initials that develop directly into connecting filaments. Upon approaching an auxiliary cell a connecting filament septates, the terminal cell continues as a connecting filament, and the intercalary segment fuses with a process produced by the auxiliary cell. In *Gloiosiphonia* the diploid nucleus enters the auxiliary cell, divides, and a gonimoblast initial is cut off that produces the gonimoblast (Oltmanns 1898). I believe that the direct development of connecting filaments from the carpogonium is a derived feature, in this instance related to the elaboration of the hypogynous cell, which sustains connecting filament formation in the absence of secondary fusions.

With this information as background, I will now discuss the biogeography of the Gloiosiphoniaceae. *Thuretella* from Western Europe, and *Gloeophycus* from Hokkaido and Korea, are almost certainly related, through entities that were once present in the Arctic Ocean. *Schimmelmannia* appears to be an Atlantic genus that is now widely distributed (Acleto 1972), occurring in Europe (*S. ornata*), Cape Verde Is. (*S. bollei*), Tristan da Cunha (*S. elegans*), St. Paul I. (*S. fruenfeldii*), California (*S. plumosa*), and Peru (*S. dawsonii*). The central west coasts of North and South America are extensions of the Atlantic province in this analysis. Amphitropical distribution would have brought *Schimmelmannia* to the South Atlantic, and dispersal along the subantarctic-subtropical convergence could have carried it as far east as St. Paul I. in the South Indian Ocean (See Hommersand 1986). The occurrence of *S. plumosa* in Japan (Chihara 1972), like that of *Pikea californica*, is most likely a modern distribution, the result of man's activities. *Peleophycus* from the Hawaiian Islands (Abbott 1985b) is intermediate in some respects between the Dumontiaceae and Gloiosiphoniaceae, and its phylogenetic relationships and biogeography are uncertain. It may be a relict taxon whose distribution followed formation of the Hawaiian chain, as it extended in an arc southeast from Kamchatka.

Gloiosiphonia is especially intriguing in that the type species, *G. capillaris* of the N Atlantic, has been recorded from both the western and eastern sides of the N Pacific. De Cew *et al.* (1981) investigated the life histories of *G. capillaris* (=*G. californica*) and *G. verticillaris* from Pacific North America, and Morohoshi & Masuda (1980) studied cultured plants of '*G. capillaris*' from Japan, showing that the developmental pattern of the N Pacific species clearly differed from that of N Atlantic *G. capillaris*. N Atlantic plants generally produce deciduous determinate branches that are replaced by corticated indeterminate branches, whereas N Pacific plants develop corticated main axes and laterals directly. The cortication pattern of the two is also different (De Cew *et al.* 1981). *Gloiosiphonia* has a heteromorphic life history involving alternation of an erect gametophyte with a crustose tetrasporophyte. Maggs (1983) showed that the crustose tetrasporophyte of *G. capillaris* from Ireland (=*Cruoriopsis hauckii* Batters) has cell fusions between the hypothallial cells that lie in parallel rows. No cell fusions were seen in the crustose stage of *G. californica* studied by DeCew *et al.* These data suggest that Atlantic and Pacific species which, superficially look very much alike, are, in fact, quite different. I suggest that *Gloiosiphonia* originated in the N Pacific or Arctic Ocean in late Cretaceous or early Coenozoic, and that modern representatives of this genus entered the N Atlantic in the Neogene.

KALLYMENIACEAE. The Kallymeniaceae is distantly related to the Dumontiaceae (and Gloiosiphoniaceae), although the connection is an ancient one. The principal similarities are: carpogonial and auxiliary cell branches that originate adventitiously; connecting filaments that are uninucleate and unsegmented from the point of their departure from the carpogonial fusion cell to the point where they contact and fuse with an auxiliary cell. According to Norris (1957), the primitive condition in Kallymeniaceae is one in which carpogonia and auxiliary cells are borne in separate, identical polycarpogonial branch systems. Loss of carpogonial branches has led to monocarpogonial female and auxiliary cell branches that retain only the basal cells of carpogonial branches, sometimes called subsidiary cells. Following fertilization the supporting cell fuses with the basal cells of the carpogonial branches or subsidiary cells. Normally, connecting filaments, generated from the supporting cell of a carpogonial branch, carry the diploid nucleus to the vicinity of an auxiliary cell (non-procarpial), or the gonimoblast develops directly from the supporting cell bearing the fertilized carpogonium (procarpial). The procarpial condition is thought to be derived secondarily (Norris 1957). Genera are separated largely on the basis of vegetative characters -- principally the shape and size of medullary cells. A distinction should be made between primary medullary filaments and secondary filaments, which are often rhizoidal.

The Kallymeniaceae has two major centers of diversity: the boreal N Pacific (N Japan to California) and the SW Pacific (Australia-Tasmania-New Zealand), with secondary centers in western Europe and South Africa. I will consider only *Pugetia*, *Kallymenia*, and the *Cirrulicarpus-Meredithia* complex in this discussion. *Pugetia* is characterized by thin blades in which primary medullary cells, cortical cells and some secondary rhizoidal cells are cellular. Supporting cells and subsidiary cells are round to variously lobed and multinucleate. The carpogonial branch apparatus is polycarpogonial or monocarpogonial, and carpogonial branches are present or absent on auxiliary cell branches in different species. Gonimoblasts are small and compact. Norris (1957) regards *Pugetia* as primitive in the Kallymeniaceae. Its distribution (New Zealand, South Africa, California and the Galapagos Islands) is interesting. There is no reason why it should not occur in the N Atlantic. The species are so different from one another that I suspect that the genus is quite old.

The type species of *Kallymenia*, *K. reniformis*, is blade-like, with more or less prominent central stellate cells and long, narrow, secondary medullary filaments. The carpogonial branch apparatus is polycarpogonial, composed of rounded cells, and the auxiliary cell apparatus bears only subsidiary cells. According to Norris (1957) supporting cells and basal cells of carpogonial branches are initially multinucleate, becoming uninucleate through nuclear fusion at maturity. Only uninucleate cells were encountered by Hommersand & Fredericq (1990) in both the carpogonial and auxiliary cell apparatus in the material they investigated. It may be that all but one nucleus degenerates. Species with uninucleate auxiliary and subsidiary cells containing enlarged nuclei, as in *K. reniformis* from western Europe, occur in South Africa (Norris 1964), in the perforate species *K. westii* from the Caribbean Sea (Ganesan 1976), and *K. pertusa* from the Gulf of California (Norris & Norris 1973). Thus, the uninucleate *Kallymenia* group, containing enlarged nuclei in the carpogonial and auxiliary cell apparatus, appear to form an Atlantic cluster. In my opinion, this group originated in either South Africa or in western Europe, spread by amphitropical distribution, and also extended its range into warm seas in the Caribbean and eastern N Pacific. Species with multinucleate auxiliary and subsidiary cells occur in Australia (Womersley & Norris 1971), and in the Mediterranean Sea (Codomier 1971), and represent a Tethyan distribution pattern. Interestingly, all of the perforate *Kallymenia* species found so far, in warm waters of the Indo-W Pacific and Japan, are like *Kallymenia cribrosa* from Australia (Womersley & Norris 1971) in being monocarpogonial with large, lobed, multinucleate subsidiary cells on the carpogonial branch apparatus (Ganesan 1976).

A major problem in the biogeography of the Kallymeniaceae is to determine the relationship between the large, diverse flora in the N Pacific (particularly the cold water flora of the NW Pacific), and the equally large and diverse flora of the SW Pacific in Australia and New

Zealand. Species from the region of eastern Siberia belonging to *Beringia*, *Crossocarpus*, *Kallymeniopsis* (Perestenko 1975) and *Cirrulicarpus* (Tokida & Masaki 1956), all have coarse thalli, usually with stellate primary medullary cells, and with an extensive system of secondary rhizoidal filaments; all are monocarpogonial, with large, elongate, multinucleate subsidiary cells. Elsewhere, this combination of features is encountered in '*Cirrulicarpus*' *carolinensis* from North Carolina (Hansen 1977a), *Meredithia microphylla* from Europe (Codomier 1973, Guiry and Maggs 1985), and '*Cirrulicarpus*' australis from Australia (Hansen 1977b). *M. microphylla* and '*C*'. *carolinensis* have heteromorphic life histories in which an erect gametophyte alternates with a *Hymenoclonium*-like, prostrate filamentous crust, in which cell fusions take place between cells in the basal layer. Crusts give rise to erect thalli in '*C*'. *carolinensis* (Hansen 1977a), and to tetrasporophytes with terminal zonate tetrasporangia in *M. microphylla* (Guiry & Maggs 1985). '*Cirrulicarpus*' *australis* and *C. gmelini* have isomorphic life histories, in which erect tetrasporophytes bear irregularly cruciate terminal ('*C*'. *australis*) or intercalary (*C. gmelini*) tetrasporangia. The type species of *Cirrulicarpus*, *C. gmelini*, from Japan, is morphologically advanced, with cystocarps grouped in rings and "giant cells" that stain differentially with aniline blue, similar to those found in *Crossocarpus lamuticus* from Siberia and *Erythrophyllum delesserioides* from Pacific North America.

Based on the evidence presented here, I conclude that the '*Cirrulicarpus*' *carolinensis*-'*C*'. *australis*-*Meredithia* group originated along an axis extending from western Europe to Australia. Their nearest relatives may be two Mediterranean *Kallymenia* species, *K. requienii* and K. *feldmannii*, studied by Codomier (1974), in which cell fusions occur in the basal layer of the crustose stages; these species, however, have isomorphic life histories.

What then is the relationship of the NW pacific cluster that appears to include: *Cirrulicarpus*, *Erythrophyllum*, *Beringia*, *Crossocarpus*, *Kalymeniopsis* and, perhaps, *Hommersandia* (Hansen & Lindstrom 1984)? The key may lie in an analysis of the morphology and taxonomy of the plant known as '*Kallymenia*' *schmitzii* from the boreal N Atlantic. Should this entity prove to belong to the *Cirrulicarpus-Meredithia* complex, the distribution of the whole assemblage may be polar-Tethyan. One possibility is that the boreal group originated in the Arctic in the Cretaceous or the Paleogene, and was secondarily distributed to the N Pacific and N Atlantic Oceans. The distribution of the *Cirrulicarpus-Meredithia* complex probably involved more than one pathway.

HALYMENIACEAE

The Halymeniaceae (formerly Grateloupiaceae or Cryptonemiaceae) is characterized by carpogonia and auxiliary cells formed separately in special branched filaments called 'ampullae' (Sjöstedt 1926, Chiang 1970). Carpogonial ampullae often cease to be formed after the first few fertilizations, while connecting filaments persist, diploidizing auxiliary cells progressively as the thallus grows. In *Halymenia trigona*, for example, I have seen secondary connecting filaments emerging near diploidized auxiliary cells that were approaching ampullae close to the thallus apex, containing auxiliary cells in the process of differentiating. This strategy is similar to that seen in *Schizymenia* of the Nemastomataceae (pers. obs.), and appears to ensure the production of a large crop of carpospores from a limited number of fertilizations. According to Sjöstedt (1926) the carpogonial and auxiliary cell ampullae are homologous to the carpogonial and auxiliary cell branches of a *Dudresnaya* (Dumontiaceae). Recently, Norris (1989) equated the ampullae of Halymeniaceae (as Cryptonemiaceae) with structures he had seen in *Titanophora*, a genus currently placed in the Nemastomataceae, Gigartinales, in Kylin's (1956) system. Norris' proposal calls for the establishment of a new family, Titanophoraceae (Norris in ed), and the resurrection of the order Cryptonemiales to include the Halymeniaceae and families presently allied with the Nemastomataceae. A third possibility is that the Cryptonemiaceae is related to the Endocladiaceae and Tichocarpaceae, procarpial families that produce adventitious filaments containing both carpogonia and auxiliary cells in the same branch system. If so, the connecting filaments in the non-procarpial Halymeniaceae originated independently of connecting filaments in Dumontiaceae. Each of the three alternatives carry different implications for the origin and biogeography of the Halymeniaceae. Whatever its origin, the Halymeniaceae appears to me to be a modern family that diversified and became widely distributed primarily during the Cenozoic. My candidate for a primitive member of the Halymeniaceae would be *Dermocorynus montagnei* from Atlantic Europe, a diminutive species with auxiliary cell ampullae that are relatively unbranched, like those of *Grateloupia* (Guiry & Maggs 1982). They are unlike the highly branched auxiliary cell ampullae of *Dermocorynus occidentalis* from California (Chiang 1970), which resemble those of '*Grateloupia*' *intestinalis* from New Zealand (Kraft 1977) and Chile (Leister, pers. comm.), and members of the *Aeodes* complex found primarily in the Southern Hemisphere in Australia, New Zealand and South Africa. Except for the temperate southern hemisphere *Aeodes* group, the Halymeniaceae appear to be primarily an Indo-W Pacific assemblage that has extended its range secondarily into the east Atlantic and Caribbean regions and Pacific N America, probably by way of southern Africa. Only a part of this flora has reached Pacific North America. The modern genus *Prionitis* (Kawaguchi 1989), with centers of distribution in Japan and Pacific North America, is significantly absent in the N Atlantic. The same is true of Pacific American species of *Cryptonemia* and *Halymenia*.

GIGARTINACEAE COMPLEX (CHONDRIELLACEAE, PETROCELIDACEAE, GIGARTINACEAE, PHYLLOPHORACEAE)

This special assemblage is characterized by: multiaxial growth with abundant secondary pit connections between subcortical and medullary cells; a supporting cell bearing a curved 3-celled carpogonial branch, sometimes with the basal cell bearing a side branch; a procarpial condition in which the supporting cell serves as an auxiliary cell; gonimoblast filaments issuing inwardly from the auxiliary cell or its fusion cell and forming secondary pit connections with cortical or medullary cells, or with tissues derived secondarily from them; carposporangia in chains borne in a placenta composed of gonimoblast and gametophyte tissues or, in a few instances, with the secondary formation of an external pustule bearing tetrasporangia (Kylin 1956; Mikami 1965). Gonimoblast cells are initially uninucleate, whereas the gametophytic cells with which they fuse are multinucleate (Hommersand & Fredericq 1990). The Chondriellaceae is monotypic, and found only in the Juan Fernandez Islands; the Petrocelidaceae, with one genus, *Mastocarpus*, occurs primarily in the N Atlantic and N Pacific, while the Gigartinaceae and Phyllophoraceae are widely distributed in both the Northern and Southern Hemispheres (Guiry & Garbary 1990).

In my opinion the Gigartinaceae complex initially divided into two assemblages, one centered in the SW Pacific (southeastern Gondwanaland) and the other in the NW Pacific (NW Laurasia), during the opening of the Tethyan Ocean in the Mesozoic. The original northern group includes the families Phyllophoraceae, Petrocelidaceae and the *Chondrus* group, currently placed in the Gigartinaceae, while the original southern group includes *Gigartina*, as typified by *G. pistillata*, and *Rhodoglossum*, as typified by *R. lanceolatum*. The *Gigartina teedii* group originated in the NW Pacific, possibly as a result of a later dispersal of an entity belonging to the southern hemisphere assemblage. The principal difference between the Southern and Northern Hemisphere assemblages relates to the manner of cystocarp development after fertilization. In the southern group gametophyte cells surrounding the diplodized auxiliary cell divide and grow simultaneously with the formation and penetration of gonimoblast filaments, to form a placenta composed entirely of secondary diploid and haploid tissues. At the same time, a filamentous involucre is produced around the developing placenta that can become quite massive. Gonimoblast cells link to gametophyte placental cells by secondary pit connections, and to cells of the enveloping involucre by elongated terminal gonimoblast cells, at about the time the carposporangia begin to differentiate (pers. obs.). In the Northern Hemisphere group, gonimoblasts issuing from the auxiliary cell, ramify through the subtending cortical and medullary tissues, making secondary pit conections with them, as illustrated by Mikami (1965) for *Gymnogongrus* and *Chondrus*. Ordinary cortical and medullary cells are transformed secondarily into a nutritive tissue, in response to fertilization and the diploidization of the auxiliary cell, either by enlarging or by undergoing transverse intercalary divisions in the affected region. Kim (1976) also recognized Northern and Southern Hemisphere groups when he divided the Gigartinaceae into two genera: *Chondrus* and *Gigartina*, based on the presence and degree of compaction of the enveloping tissue (involucre) in the mature cystocarp and the relative level of immersion of the tetrasporangial sori in the cortex and medulla. For the most part, he placed northern hemisphere species in *Chondrus* and southern hemisphere species in *Gigartina*.

A. Phyllophoraceae - This family is usually distinguished by the presence of a cellular rather than a filamentous or arachnoid medulla, and of tetrasporangia borne in chains, though other characters may be important (Guiry & Garbary 1990). *Stenogramme*, *Petroglossum*, *Ozophora* and most species of *Phyllophora* possess an isomorphic life history. Most *Gymnogongrus* and '*Ahnfeltia*'-like Phyllophoraceae, and *Phyllophora traillii*, have a heteromorphic life cycle in which an erect gametophyte alternates with a crustose tetrasporophyte (DeCew & West 1981b; Masuda 1983; Maggs 1989). A few species of *Gymnogongrus* and *Phyllophora* have the tetrasporangia borne on outgrowths (pustules), formed directly on the gametophyte in place of cystocarps. These have variously been called tetrasporoblasts (Schotter 1968) or carpotetrasporophytes (Dixon & Irvine 1977).

The western N Pacific is a likely center of origin for the Phyllophoraceae, with *Gymnogongrus flabelliformis* from Japan serving as a possible archtypal taxon. This species has an especially simple cystocarp in which gonimoblast filaments become connected directly by secondary pit connections to seemingly little modified cortical cells (Mikami 1965). From the western N Pacific, species of *Gymnogongrus* may have dispersed westward to South Africa (and then to Europe?) and, by amphitropical distribution, to New Zealand. Similarly, other species could have migrated eastward to Pacific North and South America. *Stenogramme interrupta* may have a similar distribution. The blade-like forms, *Petroglossum* and *Ozophora*, appear to have originated in the eastern N Pacific (Pacific N America) and derived forms, *Schottera* and *Phyllophora sicula*, may have evolved subsequently in the N Atlantic. Some species of *Gymnogongrus* and the ancestors of *Phyllophora* could have migrated to the N Atlantic via the Arctic Ocean in Oligocene, and been distributed secondarily to antarctic and subantarctic waters, perhaps in Oligocene or Miocene. Interestingly, tetrasporoblasts are found only in Atlantic species of *Gymnogongrus* and *Phyllophora*, with one exception - *Gymnogongrus chiton*, found in Pacific N America. The distribution of this unusual character between genera suggests that it may have been transferred between genera by a vector, perhaps a bacterial plasmid.

B. Petrocelidaceae - *Mastocarpus* (referred to as *Gigartina* in papers published prior to Guiry et al.(1984) is characterized by a heteromorphic life history in which an erect gametophyte alternates with a crustose tetrasporophyte (formerly the genus *Petrocelis*) wherein the tetrasporangia are solitary and intercalary in erect filaments. *Mastocarpus* has a filamentous medulla and bears its cystocarps in papillae, as in *Gigartina*; however, the medullary cells to which gonimoblast filaments fuse are transformed into a nutritive tissue by repeated intercalary cell divisions, as in some Japanese species of *Chondrus* (Mikami 1965). Species of *Mastocarpus* in the N Atlantic and N Pacific are morphologically similar. A dioecious, southern population of *Mastocarpus stellatus*, having a heteromorphic life history with a crustose tetrasporophyte, and a monoecious northern population with a 'direct' life cycle, were identified by Guiry & West (1983). In the western N Pacific a northern group, *Mastocarpus pacificus* (Masuda *et al.* 1984), is distinct from a southern group, *Mastocarpus* sp. (Masuda *et al.* 1987); the latter crosses with *M. stellatus* from Europe (Guiry & Garbary 1990). In the eastern N Pacific (Pacific N and S America) *Mastocarpus* is represented by *M. papillatus* with both heteromorphic and 'direct' life histories, and by *M. jardinii* (West *et al.* 1983; Zupan & West 1988). I suggest that *Mastocarpus* may have evolved in the Arctic Ocean and was distributed secondarily to the N Atlantic and N Pacific, perhaps in two waves.

C. *Chondrus* group - *Chondrus* is currently represented by *C. crispus* in the N Atlantic, by an assortment of species in the NW Pacific (Japan), and one in Peru and Chile. The '*Iridaea splendens*' group and *Gigartina leptorhynchos* in Pacific North America, possibly also the '*Iridaea laminarioides*' group in Pacific South America, and '*Iridaea capensis*' and '*I. convoluta*' in South Africa, may also form part of this assemblage (Leister, pers. comm.). As in *Mastocarpus* of the Petrocelidaceae, inner cortical and medullary tissues tend to undergo intercalary divisions in response to fertilization; however, the life history, unlike that in *Mastocarpus*, involves an alternation of isomorphic generations. The course of evolution of the *Chondrus* group appears to involve the extension of intercalary divisions to vegetative tissues, sometimes coupled with the secondary formation of a weak to strong involucre around the expanding carposporophyte. In the '*Iridaea splendens*' group from Pacific N America the juvenile thallus is cylindrical and lacks intercalary cell divisions, whereas the adult is blade-like, composed of a network of interconnecting cortical and medullary filaments, in which intercalary cell divisions are a regular feature of thallus growth (Zollner 1977). *Chondrus crispus* is unusual in that intercalary divisions occur primarily in juveniles as small as 0.5 mm tall, and become less frequent in older thalli (Zollner 1977). Tetrasporangia are formed in chains in adventitious filaments that establish secondary pit connections with inner cortical and medullary cells (Kim 1976).

From a consideration of the morphological evidence, the *Chondrus* group appears to have originated in the NW Pacific, and to have spread along the Pacific Coast of North America, where blade-like taxa are presently placed in '*Iridaea*'. The South African and some South American species of '*Iridaea*' also appear to belong to the *Chondrus* group; however, the direction of their distribution, whether from west to east or east to west, remains unresolved. Most likely the ancestor of *Chondrus crispus* originated in the Arctic Ocean, or crossed it in the Paleogene, and differentiated numerous ecological forms in the North Atlantic during the Neogene. According to this interpretation, *C. crispus* is one of the more advanced species in the *Chondrus* group.

D. *Gigartina pistillata* group - The counterpart of the *Chondrus* group in the Southern Hemisphere is the *Gigartina pistillata* group. Cortical and medullary filaments only rarely undergo intercalary cell divisions. The cystocarp consists of an inner placental tissue composed of secondary cortical cells and gonimoblast filaments surrounded by an enveloping involucre. Tetrasporangia are mostly adventitious in origin, borne in short chains as in the *Chondrus* group. They are formed in sori that detach as a unit, leaving pits in *Gigartina pistillata* and its nearest relatives. The *G. pistillata* group has its center of distribution in Australia and New Zealand. Several species also occur in South Africa, including a taxon referred to *G. pistillata*. This entity is closely related to *G. disticha* from Australia, and to *G. pistillata* from western Europe. Contrary to my earlier opinion (Hommersand 1986), I now believe that the European *G. pistillata* represents a vicariant distribution of a former Atlantic-Tethyan population of *Gigartina* that became separated with the closure of the Tethyan Seaway, and that the '*G. pistillata*' of South Africa evolved in parallel to the European taxon from an Australian ancestor similar to *G. disticha*. Other more distantly related species occurring in New Zealand, South Africa, or South America and the islands of the West Wind Drift, that are presently placed either in '*Gigartina*' or '*Iridaea*', are distinguised by their distinctly different types of tetrasporangial sori (Leister, pers. comm.)

E. *Gigartina teedii* group - In this assemblage tetrasporangia are transformed from primary cortical cells rather than from secondary cortical or medullary filaments. Cystocarp placental tissue is irregular, consisting of a mixture of large and small cells surrounded by a weakly developed involucre (Hommersand & Fredericq 1990). *Gigartina tenella* from Japan possesses the simplest vegetative organization of any member belonging to the *G. teedii* group (Zollner 1977). This assemblage probably originated in the NW Pacific, and spread eastward along the Pacific Coast of North and South America and into the Atlantic Ocean. Its present center of diversity is Pacific North and South America from southeastern Alaska to south-central Chile.

NEMASTOMATACEAE COMPLEX. This assemblage contains genera that possess septate, branched connecting filaments, and auxiliary cells that correspond to ordinary vegetative cells. As in the Halymeniaceae discussed earlier, connecting filaments persist, and can interact with many auxiliary cells. The assemblage as circumscribed here includes the families Cruoriaceae, Haemeschariaceae, Nemastomataceae (Gymnophlaeaceae), Calosiphoniaceae, Titanophoraceae (Norris in ed) and the Sebdeniaceae. Only the Sebdeniaceae is known to have an isomorphic life history.

1. Cruoriaceae-Haemeschariaceae

These two crustose families have isomorphic gametophytes and tetrasporophytes. Gonimoblasts develop directly on the connecting filaments, either remote from, or in proximity to, auxiliary cells (Maggs & Guiry 1989; Wilce & Maggs 1989). Both are endemics of the N Atlantic and Arctic Oceans; they could be relict taxa.

2. Nemastomataceae

The Nemastomataceae is characterized by multiaxial growth (except *Titanophora*), absence of secondary pit connections (except the basal crust in *Schizymenia*), carpogonial branches separate from auxiliary cells, gonimoblasts developing from the connecting filament near an auxiliary cell, or directly from the auxiliary cell itself, and probably a heteromorphic life cycle, with prostrate or crustose tetrasporophytes bearing zonate tetrasporangia. Reports of an isomorphic life history and cruciate tetrasporangia may be due to confusion with *Sebdenia* (Hansen 1989). The family, or part of it, may have a N Atlantic-Tethyan distribution with *Nemastoma dichotoma*, in which gonimoblasts develop from the connecting filament (Berthold 1884), as its most primitive representative. Other species of *Nemastoma*, *Platoma* and *Predaea* might then represent more advanced Indo-W Pacific taxa, having secondary distributions in the Atlantic Ocean. Alternatively, the group may be so old that no pattern is revealed. Genera such as *Adelophyton* from west Australia with auxiliary cells on adventitious filaments (Kraft 1975b), *Titanophora* from the Indo-W Pacific with uniaxial growth, calcification of the medulla, and 'ampullae' (Norris 1989), and *Schizymenia* with a world-wide distribution in temperate waters, and with carpogonia fusing with adjacent nutritive cells (Ardré 1977), suggest an early origin for the family.

3. Calosiphoniaceae

The Calosiphoniaceae contains two heteromorphic genera, *Schmitzia* and *Calosiphonia*, in which an erect, uniaxial gametophyte alternates with a crustose tetrasporophyte (Mayhoub 1975; Maggs & Guiry 1985). In *Calosiphonia* a connecting filament approaches an auxiliary cell and septates; the intercalary segment then fuses with the auxiliary cell. The diploid nucleus divides; one of the products enters the auxiliary cell, divides again and a gonimoblast initial is cut off towards the outside, which then produces several gonimolobes (Knauss 1988). The auxiliary cells of *Schmitzia*, in contrast, differentiate in the vicinity of a growing connecting filament, and each auxiliary cell sends out a process that fuses onto a segment of the connecting filament. Gonimoblasts are generated directly from the connecting filament at points of contact with the auxiliary cell processes (Knauss 1988). Two distinctly different reproductive patterns occur in *Schmitzia*. In *S. neapolitana* and *S. hiscockiana* from western Europe, processes from the carpogonium fuse with nearby vegetative cells to produce a fusion cell, from which the connecting filaments emanate. (Early stages of connecting filament formation are similar in

Calosiphonia.) All subsequent fusions involve auxiliary cells. In the species of *Schmitzia* from the Indo-W Pacific and the western N Atlantic, connecting filaments fuse with vegetative cells at a distance from the carpogonium, and the pit connections break down within adjoining files of cortical filaments to form a branched fusion network, linked directly to the connecting filament. The nutritive system supporting connecting filaments and gonimoblasts is thus more elaborate in the Indo-W Pacific/western Atlantic cluster than in the NE Atlantic (Europe) cluster. The family as a whole exhibits an Atlantic-Tethyan distribution, with the more primitive representatives occurring in western Europe and the Mediterranean Sea. Secondary distributions from the Indo-W Pacific by way of southern Africa brought taxa carrying characters of the Indo W-Pacific cluster to the western Atlantic (Wilce & Sears pers. comm.), and perhaps to California in the eastern Pacific (K. Miller pers. comm.).

4. Sebdeniaceae

The Sebdeniaceae, containing a single genus *Sebdenia*, is characterized by an isomorphic life cycle, stellate inner cortical and medullary cells resulting from frequent formation of secondary pit connections, medullary gland cells (when present), monoecious gametophytes with simple, outwardly directed carpogonial branches and gonimoblasts, cellular branched connecting filaments that continue onward after contacting an auxiliary cell (or coenocytic, unbranched, and terminating at the auxiliary cell in *S. polydactyla*), and irregularly cruciate tetrasporangia (Hansen 1989). *Sebdenia* is world-wide in tropical to warm-temperate waters. The type species, *S. monardiana* from the Mediterranean Sea, appears to be the most primitive species in that it lacks medullary gland cells, and connecting filaments develop directly from the carpogonium, without fusing with nearby cells (Sjöstedt 1926). Most other species possess medullary gland cells, and the carpogonium fuses with the supporting cell or some other neighboring cell, before emitting connecting filaments. *Sebdenia polydactyla* from the Indo-W Pacific, and presumably also the western Atlantic, may be the most advanced species. The overall distribution fits an Atlantic-Tethyan pattern, with secondary migration from the Indo-W Pacific to the western Atlantic.

FURCELLARIACEAE/SOLIERIACEAE COMPLEX. The cluster of families that include the Furcellariaceae, Solieriaceae, Rissoellaceae, Cystocloniaceae and Hypneaceae, which Kylin (1932, pp. 76-79) treated as an evolutionary line within the Gigartinales, all possess zonately divided tetrasporangia, and an auxiliary cell that produces gonimoblasts inwardly (usually) from a single gonimoblast initial.

1. Furcellariaceae

At present, the Furcellariaceae contains three genera, *Furcellaria* (N Atlantic), *Halarachnion* (N Atlantic and NW Pacific) and *Neurocaulon* (N Atlantic & NW Pacific) characterized by: a filamentous vegetative organization in which secondary pit connections are absent; carpogonial branches that are directed inward; auxiliary cells that are derived from primary cortical cell; a tendency for the carpogonium to divide once or twice before producing connecting filaments; connecting filaments that are septate and branched, septation of the connecting filament being near an auxiliary cell; the formation of a distal portion that continues as a connecting filament, and a proximal portion that interacts with a process from the auxiliary cell; and a gonimoblast initial, cut off from the fused connecting filament segment, that gives rise inwardly to gonimoblasts (Knauss 1988; Knauss & Hommersand 1989).

(1) *Furcellaria lumbricalis* is cylindrical and dichotomously branched, whereas the rest of the genera in this assemblage are blade-like. Connecting filaments emanating from the carpogonium or its derivative cells branch profusely, but do not appear to fuse with neighboring vegetative cells. Otherwise, the female reproductive system is identical to that seen in other Furcellariaceae. *Furcellaria* is limited in its distribution to the N Atlantic, suggesting that it may have originated there and was never part of an Arctic or Pacific flora.

(2) *Halarachnion ligulatum* from the NE Atlantic (Europe) has a heteromorphic life history in which the tetrasporophyte is a discoid crust, corresponding to a plant known as '*Cruoria*' *rosea* (Maggs & Guiry 1989), whereas *H. latissimum* from the NW Pacific (Japan) is isomorphic. Vegetative and reproductive development of the two species is similar, the chief difference being that, in *Halarachnion latissimum*, the process from the connecting filament cuts off a derivative cell laterally, which elongates and then fuses with the auxiliary cell (Knauss 1988). An Arctic origin, and long separation of the two species is indicated by these observations.

(3) As in *Halarachnion ligulatum*, *Neurocaulon reniforme* from the Mediterranean Sea possesses a heteromorphic life history involving a peltate, blade-bearing gametophyte and a crustose tetrasporophyte (Codomier 1972). *Neurocaulon japonicum* from Japan is blade-like but not peltate; its systematic position needs reinvestigating.

2. *Turnerella/Opuntiella*

With three species, *Turnerella pennyi* (N Atlantic), *Turnerella mertensiana* and *Opuntiella californica* (both N Pacific), this assemblage possesses some characters in common with the Furcellariaceae, and others with the Solieriaceae. It is distinguished by the presence of prominent 'gland cells' containing a homogeneous proteinaceous material (Young 1979) not found in either of the other two families. The heteromorphic life history involves an erect, blade-like gametophyte and a crustose tetrasporophyte in *Turnerella pennyi* (South & Hooper 1972; Maggs 1983), *Turnerella mertensiana* (Kasahara 1980), and probably also in *Opuntiella californica* (West & Hommersand 1981). The female reproductive system resembles that in the Furcellariaceae. The carpogonium divides once or twice, and connecting filaments septate and link to auxiliary cells before continuing on in the same manner as in the Furcellariaceae (Knauss 1988; Hansen pers. comm.); however, unlike the Furcellariaceae, conjunctor cells and secondary pit connections are abundant between vegetative cells. The gonimoblast fusion cell in *Turnerella penneyi* is comparatively small, and similar to that seen in *Halarachnion* (Knauss 1988). It is larger and more elaborate in *Opuntiella* (Kylin 1934) and in *Turnerella mertensiana* (pers. obs.). If *Turnerella pennyi*, which appears to be structurally simpler than the other two species, is primitive for the group, then an Arctic origin, and subsequent distribution to the N Atlantic and N Pacific is indicated.

3. Solieriaceae

With the removal of the *Turnerella-Opuntiella* group, the Solieriaceae becomes an Atlantic-Tethyan and Atlantic- Indo-W Pacific group, depending on the antiquity of the taxa involved, and the direction in which they migrated. Thalli are multiaxial or uniaxial with abundant rhizoids and secondary pit connections. One or more connecting filaments arise directly from the carpogonium, with each capable of fusing with just one auxiliary cell. The diploid nucleus enters the auxiliary cell, and divides; the distal daughter nucleus is then cut off along with the gonimoblast initial. Either a large central fusion cell is formed through the fusion of inner gonimoblast cells with the gonimoblast initial and with the auxiliary cell and cells of the bearing filament, or a placenta is formed through lateral fusion between gonimoblast cells and cells of secondary nutritive filaments originating from the cortex and medulla of the gametophyte (Kraft 1975a, 1984; Gabrielson & Hommersand 1982a, b; Gabrielson & Kraft 1984).

Five genera with fusion cells have reached the N Atlantic: *Solieria* (Western Europe, West Africa, North Carolina to Brazil), *Flahaultia* (North Africa), *Anatheca* (West Africa) *Meristotheca* (Brazil) and *Eucheuma* (Florida to Brazil). *Solieria, Flahaultia* and *Anatheca* could have reached the N Atlantic by way of the Tethyan Seaway, but *Meristotheca* and *Eucheuma*, found only in the western Atlantic, are related to taxa from East Africa, and were, I believe, distributed from the Indo-W Pacific by way of southern Africa at a later date.

Except for *Placentophora* (Kraft 1975a), a distinct entity from New Zealand related to *Callophycus* (Kraft 1984), all of the genera and species having a placental cystocarp occur in the western North Atlantic and the adjacent eastern N Pacific from Alaska to Peru. They include *Agardhiella, Sarcodiotheca, Meristiella* (Gabrielson & Cheney 1987) and '*Eucheuma*' *uncinatum* (Gabrielson 1980). This complex may or may not be monophyletic, but the taxa are so diverse that I believe their ancestors were distributed to the Atlantic Ocean in the Cenozoic, probably by way of the Tethyan Seaway.

4. Rissoellaceae

The family Rissoellaceae is monotypic and endemic to the Mediterranean Sea. It is separated from the Solieriaceae by its procarpial condition, with several carpogonial branches and prominent nutritive filaments produced in the vicinity of the auxiliary cell before fertilization (Kylin 1930, 1956).

5. Cystocloniaceae/Hypneaceae

The Cystocloniaceae and Hypneaceae are segregated from the Solieriaceae primarily on being procarpic, without connecting filaments. The auxiliary cell lies just distal to the supporting cell, essentially in the same position occupied by the auxiliary cell in the non-procarpic family Solieriaceae. The fertilized carpogonium fuses directly with the auxiliary cell, and a gonimoblast initial is cut off distally as in Solieriaceae. Cortical cells distal to the auxiliary cell become modified into a nutritive tissue shortly after fertilization, followed by gonimoblast fusions to produce a central fusion cell in most genera. The Cystocloniaceae-Hypneaceae are nearly as diverse, and probably nearly as old, as the Solieriaceae. They exhibit a clear N Atlantic-Tethyan distribution pattern, with Atlantic and Indo-W Pacific genera distinct from one another (despite confusion in the literature).

(1) *Cystoclonium* is a N Atlantic genus, which may, however, include a representative species in Antarctica. The thallus is cylindrical, with apical growth initiated by an obliquely dividing apical cell (Kylin 1923). The chief distinguishing feature of *Cystoclonium* is the production of special nutritive filaments that fuse directly onto the expanding gonimoblast fusion cell as the gonimoblast radiates into the interior of the thallus (Thrainsson 1986).

(2) *Rhodophyllis*, far from being the widespread genus reported from both Atlantic and Pacific Oceans, is actually a western European endemic with one or two species. It grows by means of a wedge-shaped apical cell with two cutting faces that initiates axial cells alternately in a zig-gag pattern. Each axial cell forms one lateral and two transverse periaxial cells (three in all) that produce the blade and cortex (Thrainsson 1986). Species of *Rhodophyllis* from Australia-New Zealand have the axial cells cutting off only two periaxial cells each, as in all other investigated Cystocloniaceae. The mature gonimoblast consists of a central fusion cell and modified gonimoblast filaments that fuse with inner cells of the surrounding pericarp - a unique character.

(3) *Fimbrifolium* is superficially similar to both *Rhodophyllis* from Europe and *Craspedocarpus* from Australia. The gonimoblast initial is cut off toward the outside, and a cluster of nutritive filaments are produced below (Hansen 1980). These filaments later fuse with the central fusion cell, in a manner somewhat reminiscent of the behavior of the nutritive filaments in *Cystoclonium*. This behavior produces a cystocarp that is superficially similar to that seen in *Craspedocarpus*, but its development is quite different. *Fimbrifolium* is distributed from Massachusetts to Greenland in the NW Atlantic, and north of the Faroes in the NE Atlantic. There may be related entities in the N Pacific, most likely distributed there since Pliocene.

(4) There is a sizeable cluster of species in Australia and New Zealand that resemble *Rhodophyllis* from Europe; however, generate a special nutritive tissue toward the inside of the developing gonimoblast, with which gonimoblast filaments later fuse. In some species the nutritive tissue is well-defined and compact. Species of this type are variously placed in *Rhodophyllis*, *Craspedocarpus*, or *Calliblepharis*; all require reinvestigation. This particular type of nutritive tissue is an Indo-W Pacific character; however, a highly specialized form of this behavior is seen in *Calliblepharis ciliata* and *C. jubata* in Europe. The nutritive tissue in the European species forms an elaborate network that is larger and more complex than that seen in any Indo-W Pacific taxa (Thrainsson 1986). I believe that *Calliblepharis* originated in the Indo-W Pacific, and arrived later in the N Atlantic during the final stages in the collapse of the Tethyan Seaway (Miocene). Recently, *Craspedocarpus* has been reported from the western Atlantic, in North Carolina and Brazil (Schneider 1988). I believe that this range extension is almost certainly due to distribution around the tip of southern Africa in the Neogene.

(5) The family Hypneaceae is closely related to the Indo-W Pacific segment of the Cystocloniaceae. Apical cells have two cutting faces that divide in the same manner as in *Cystoclonium*. Each axial cell cuts off two periaxial cells at 90^0 angles. A compact basal nutritive tissue is produced at the base of the developing gonimoblast. Uninucleate cells fuse with cells in the basal nutritive tuft and also form filaments that fuse with inner cells of the pericarp (Hommersand & Fredericq 1990). The result is a basketwork of sterile and fertile tissues that seems more distinct than it really is. *Hypnea* appears to have evolved in the Indo-W Pacific, and to have been distributed rapidly around the world in the Neogene.

CAULACANTHACEAE/CUBICULOSPORACEAE. These two old families contain several divergent, widely disseminated genera (Gabrielson & Hommersand 1982a, Kraft 1973). Only the cosmopolitan genera *Caulacanthus* (and the closely related genus *Feldmannophycus*) and *Catenella* are reported from the NE Atlantic and Mediterranean Sea.

SPHAEROCOCCACEAE/PHACELOCARPACEAE. The Sphaerococcaceae with one genus, *Sphaerococcus*, is endemic to Europe and the Mediterranean Sea. Its nearest relatives are unknown; however, if the suggestion by Searles (1968) that the monocarpogonial procarpic condition of *Sphaerococcus* could represent a reduction from the polycarpogonial procarpic condition in the Phacelocarpaceae proves correct, the two families may be related, and their distributions would be N Atlantic-Tethyan. The tetrasporophytic phase of *S. coronopifolius* is *Haematocelis fissurata*, a crustose plant bearing zonate tetrasporangia (Maggs & Guiry 1982).

PLOCAMIACEAE. The Plocamiaceae is procarpic with the supporting cell functioning as an auxiliary cell. There are no cell fusions, only broadened pit connections between the auxiliary cell and cells of the gonimoblast (Hommersand & Fredericq 1990). *Plocamium* is world-wide in distribution with centers of diversity in Australia-New Zealand, South Africa and Pacific North America. There are no known close relatives. *Plocamium cartilagineum* is reported from Australia-New Zealand, Europe and Pacific North America. The European and North American representatives are especially similar in their appearance suggesting a recent dispersal, perhaps via the Arctic Ocean in the Pliocene, since this species is unknown in the western Atlantic.

Rhodymeniales

RHODYMENIACEAE. The primary distribution of the Rhodymeniaceae appears to be N Atlantic-Tethyan, especially the seemingly primitive members of the Rhodymeniaceae (including the *Fauchea-Gloioderma* group, *Halichrysis* and *Chrysymenia*). *Botryocladia*, which is closely related to *Gloiosaccion*, probably had a Tethyan origin. The genus is widely distributed in the Indo-W Pacific, Atlantic and eastern Pacific Oceans, perhaps due to migration via southern Africa in Miocene. *Rhodymenia* is widely dispersed in cool-temperate waters. The *Rhodymenia pseudopalmata* group is represented in Australia-New Zealand, South Africa, Europe and Pacific North and South America (Sparling 1957). Dispersal by currents in Miocene is indicated.

CHAMPIACEAE. The Champiaceae is represented in the N Atlantic by the cosmpopolitan species, *Champia parvula*, and by representatives of the *Lomentaria* and *Chylocladia* group. European species of *Lomentaria* are distinct suggesting a N Atlantic-Tethyan distribution. *Chylocladia* is known from South Australia, South Africa and Europe, and its sister genus, *Gastroclonium*, is known from Europe and Pacific North America, with one outlier, *G. pacificum*, from both California and Japan. The disjunct distribution of this latter species may be due to the agency of man. The *Chylocladia* group is distinguished from all other Rhodymeniales by the formation of a large, multinucleate fusion cell bearing sessile carposporangia. Often cystocarps are formed in pairs from adjacent auxiliary cells (Bliding 1928). At firts hand, the *Chylocladia* group would appear to be a N Atlantic cluster with outliers in Pacific North America (and Japan). The presence of a distinctive species of *Chylocladia* in South Australia (Reedman & Womersley 1976), however, complicates this picture, suggesting a possible Teythan origin of the group coupled with the secondary dispersal of *Chylocladia* from the austral Indo-W Pacific to western Europe by way of South Africa.

Ceramiales

The Ceramiales is the classical example of a monophyletic order in the Rhodophyta. Spore germination is characteristically bipolar with the production of an initial descending rhizoid and a primary erect axis. Carpogonial branches are almost always 4-celled, borne on a periaxial cell that cuts off or is rarely transformed into an auxiliary cell. Transfer of the diploid nucleus is effected by direct fusion between the carpogonium and auxiliary cell, or more commonly is mediated by one or more connecting cells cut off from the carpogonium. Four families are recognized within the order: Ceramiaceae, Dasyaceae, Delesseriaceae and Rhodomelaceae (Kylin 1956). Analysis of the comparative morphology and biogeography of the different tribes in each family suggests that the four families are about equally old, and contain tribes or groups of varying antiquity - some old and some more recent. Biogeographic patterns are recognizable at the tribal and group levels in the Ceramiaceae, Delesseriaceae and Rhodomelaceae, and at the generic and infrageneric levels in the Dasyaceae.

CERAMIACEAE. Three major tribes of the Ceramiaceae, Antithamnieae, Ptiloteae and Ceramieae, are united by having direct fusion of the carpogonium with the auxiliary cell. Although connecting cells are present in the Antithamnieae, they do not appear to be functional (Hommersand pers. obs.). These three tribes exhibit a polar-Tethyan distribution pattern.

1. Antithamnieae

In the Antithamnieae, the genera *Pterothamnion* (including *Platythamnion*), *Antithamnionella* and *Scagelia*, in which gland cells cut off laterally from intercalary cells, have major centers of distribution in Australia-New Zealand, Europe and the N Pacific, especially Pacific North America. Distribution to the N Pacific could have occurred by way of the Arctic Ocean. *Antithamnion*, characterized by having gland cells cut off terminally from apical initials, is temperate to tropical in its distribution. (A detailed analysis of the Antithamnieae is given by Athanasiadis 1990).

2. Ptiloteae

The Ptiloteae is characterized by a tendency for apicalization of carpogonial branches and auxiliary cells. The genus *Gymnothamnion* (tribe Gymnothamnieae, Kajimura 1989), which is bilaterally branched, uncorticated, and widely distributed from the Indo-W Pacific to western Europe, may possibly be archetypal to the Ptiloteae. The status of *Tokidaea* (Yoshida 1974, Lindstrom & Wynne 1982) from the North Pacific is similar to that of *Gymnothamnion*. The N Atlantic genus *Plumaria*, with its quadriverticillately branched corticated axes, clearly belongs in the Ptiloteae. *Ptilota*, which bears lateral branches from every segment, occurs in both the N Atlantic and N Pacific, whereas *Neoptilota*, which has one or more unbranched segments between branched segments, is restricted to the N Pacific. The Southern Hemisphere genera *Euptilota*, *Georgiella* and *Plumariopsis* (Moe & Silva 1983) are unrelated to the Ptiloteae of the Northern Hemisphere.

3. Ceramieae

Two prominent, widely dispersed genera belonging to the Ceramieae are represented in the N Atlantic: *Microcladia* and *Ceramium*. *Ceramium* is poorly understood, and too complex for consideration here. *Microcladia*, however, consists of two clusters: one with bilateral symmetry and radial branching that is distributed from the Antipodes south of New Zealand (N. Adams, pers. comm.) to South Africa, Tristan da Cunha and Pacific North America (Hommersand 1963; Stegenga 1986); the second group, which includes the type species *M. glandulosa* from the NE Atlantic (Europe), and several species from Japan, is dorsiventral (Hommersand pers. obs.). Itono (1977) transferred the Japanese species to *Herpochondria* based on a comparison with *M. coulteri* from Pacific North America, a member of the first cluster. Taken together, these two groups can be visualized as having a polar-Tethyan distribution, with secondary dispersal of the Southern Hemisphere group to the S Atlantic and E Pacific, possibly in Miocene.

4. Other tribes

Several tribes in Ceramiaceae that possess connecting cells exhibit an Atlantic-Tethyan distribution. In the Ceramiaceae, for example, the primitive genus *Gulsonia* (Wollaston & Womersley 1959) is represented by two disjunct species: *G. mediterranea* from the Mediterranean Sea, and *G. annulata* from Australia. There is a cluster of related tribes that includes the Spermothamnieae, Sphondylothamnieae, Compsothamnieae and Griffithsieae, all of which have terminalized carpogonial branches borne on shortened terminal segments, often with more than one auxilary cell. Most specialists on each of these tribes regard species that possess a complete set of whorled branchlets as being primitive (Gordon 1972; Baldock 1976; Huisman 1985), and interpret losses of branch systems as being advanced. I expect, however, that the inner whorl, derived from sterile cells in the fertile segment in the Sphondylothamnieae, will prove to be a derived character. The Spermothamnieae is clearly one of the most advanced tribes in this assemblage in that wall deposition and cell growth is diffuse (Garbary & Belliveau 1990). I regard *Spermothamnion repens* (N Atlantic) as being primitive in Spermothamnieae, *Sphondylothamnion multifidum* (NE Atlantic) as primitive in Sphondylothamnieae, and *Compsothamnion thuyoides* (NE Atlantic), with two auxiliary cells, as being primitive compared to *Pleonosporium* with one auxiliary cell in the Compsothamnieae. I agree with Baldock (1976) in his interpretation of the phylogeny of the Griffithsieae, in which *Halurus equisetifolius*, *Griffithsia setacea* [=*G. flosculosa*] and *Anotrichium barbatum* (all European taxa), are treated as being close to the hypothetical ancestor. Recently, however, Millar (1986) has discovered an archetypal form, *Baldockia verticillata* in eastern Australia, which stands at the base of the *Anotrichium* line.

DASYACEAE. The Dasyaceae, characterized by its monopodial filamentous germlings (Richardson 1981) and sympodial development of the adult axes (Rosenberg 1933), is a distinct family among the Ceramiales. The *Dasya* group possesses radial symmetry, whereas the *Heterosiphonia* group exhibits dorsiventral symmetry. The *Dasya* group appears to be an old warm-water assemblage, while the *Heterosiphonia* group has diversified extensively in subantarctic waters. *Dasya* can be subdivided into two species clusters: a *Dasya hutchinsiae* group in which carpogonial branches are borne on pseudolaterals; and a *D. baillouviana* group in which carpogonial branches are borne on normal or dwarfed indeterminate branches (Parsons 1975). Both groups are represented in the SW Pacific (Australia) and the NW Pacific (Japan); however, the ancestors of *D. baillouviana* appear to have migrated from the NW Pacific (Japan) to the NE Pacific (Pacific North America), the Caribbean Sea and thence to Europe. Only one species of *Heterosiphonia*, *H. plumosa*, is widespread in Europe (Rosenberg 1933). Its development is similar to that of the subantarctic *H. berkeleyi* group rather than to any of the tropical or subtropical species. Perhaps it was distributed to Europe by way of South Africa.

DELESSERIACEAE. The Delesseriaceae have evolved rapidly in recent times with the result that a high degree of variation is encountered in extant genera and species. I have observed only one character that appears to manifest a bipolar or polar-Tethyan distribution in members of the subfamily Delesserioideae. In the *Hypoglossum*-group and related taxa in the Southern Hemisphere, the fertilized carpogonium cuts off two connecting cells distally, whereas in the *Membranoptera* and *Phycodrys* groups in the N Atlantic and N Pacific, the fertilized carpogonium cuts off one distal and one proximal lateral connecting cell, the latter being the one that fuses with the auxiliary cell. There are many parallelisms in vegetative growth between southern and northern hemisphere taxa, and many genera have secondary distributions in both hemispheres. Genera belonging to the *Cryptopleura* group (*Hymenena*, *Acrosorium*) and the *Myriogramme* group (*Haraldiophyllum*, *Myriogramme*) may originally have had a N Atlantic-austral-Tethyan distribution; however, some species clusters appear to have been distributed secondarily to the Atlantic and eastern Pacific Oceans by way of southern Africa, probably in Miocene (Hommersand 1986).

There is an assemblage consisting of the *Nitophyllum* group (the genera *Neohypophyllum*, *Polyneurella*, and *Polyneuropsis*), and the '*Papenfussia*' group containing the genus *Pollexfenia*, with primary centers of distribution in the SE Pacific (Australia-New Zealand) and the N Pacific, that is characterized by adaxial branching of third and fourth order lateral filaments, and a single sterile group, usually coupled with a cover cell flanking the carpogonial branch (pers obs.; for references see Wynne 1983). *Nitophyllum punctatum* from western Europe is the most advanced species of the *Nitophyllum* group in that tetrasporangia become linked to neighboring vegetative cells by secondary pit connections, with all but the original nucleus degenerating (Kylin 1924).

RHODOMELACEAE. According to Hommersand (1963) the most primitive tribes in the Rhodomelaceae are the Bostrychieae, Lophothalieae and the Rhodomeleae. Parsons (1980) separated the tribe Brongniartelleae from the Lophothalieae. Taken together, these clusters of tribes exhibits a polar-austral-Tethyan distribution with the most primitive members of the Bostrychieae occurring in the SW Pacific, the Lophothalieae possibly being an Australian endemic, and with the Brongniartelleae ranging from the SW Pacific (*Brongniartella australis*) through the tropics to Europe (*B. byssoides*) and the N Pacific (*Beringiella, Schizochlaenion, Sinosiphonia*) (Parsons 1980, Wynne & Norris 1982, Tseng & Cheng 1983).

The Rhodomeleae with *Rhodomela* and *Odonthalia* are distributed between the N Atlantic and N Pacific. Masuda (1982) regards the dorsiventral, N Pacific species of *Rhodomela* as being primitive, comparing them to the dorsiventral Amansieae; however, the ontogeny of the pericentral cells is different (see below). As Kylin (1914) demonstrated, branching and pericentral cell formation can be very irregular in *Rhodomela confervoides* from Europe. I regard this unstable condition as being primitive for the tribe Rhodomeleae, and I believe that the Rhodomeleae evolved in arctic waters from a *Brongniartella*-like ancestor, and was later dispersed into both the N Atlantic and N Pacific Oceans. The origin of *Odonthalia* is not so clear.

The tribes Pterosiphonieae and Polysiphonieae are closely related and some genera appear to show a polar-austral-Tethyan distribution. The taxonomy of this group is confused, however, and biogeographic interpretations are not possible at this time.

The dorsiventral tribe Amansieae provides an excellent example of a N Atlantic-Tethyan distribution. The terete genus *Halopitys* and the compressed genus *Rytiphlaea* are probably European endemics. As I have discovered, these two genera, like all other Amansieae, are both dorsiventrally and bilaterally asymmetrical. This unique feature is related to the fact that trichoblasts are dorsal, with the first pericentral cell cut off adjacent to the trichoblast on one side of the orthostichous rows, and the second adjacent to the trichoblast on the other side, also in orthostichous rows. As a result, the thallus tip curves dorsiventrally and to one side. In the *Amansia* group, wings are formed by longitudinal divisions of the two opposite dorsal and ventral pericentral cells, and in the *Kützingia* group, they are formed from the lateral pair of pericentral cells (Kylin 1956). Species of *Amansia, Enantiocladia* and *Vidalia* , which probably originated in the Indo-W Pacific, have been distributed secondarily to the tropical western Atlantic. I suspect that *Vidalia volubilis*, from the Mediterranean Sea, is a segregate of a spirally twisted variety of *V. obtusiloba* from the Caribbean.

Saito (1967) has recognized two assemblages in the genus *Laurencia* based on the position of the tetrasporangium -- whether adaxial or abaxial on the fertile pericentral cell. The adaxial arrangement is found in the *Laurencia pinnatifida* group in the NE Atlantic and the *L. spectabilis* group in Pacific North America. All Indo-W Pacific species of *Laurencia* have abaxial tetrasporangia, and this character has been distributed to the Caribbean Sea and western Europe with *L. obtusa* and its relatives. A similar pattern may exist in the genus *Chondria* with the *C. tenuissima* group possibly being a N Atlantic assemblage, and the *C. dasyphylla* group being world-wide in distribution.

Many tribes in the Rhodomelaceae are represented by species or species clusters found in both the Indo-W Pacific and western Atlantic regions. Many that have reached the Azores, Canary Islands, western Europe and the Mediterranean Sea are listed in Table 7.

DISCUSSION

In my opinion, presently recognized families of red algae evolved within the Tethyan Ocean and adjoining polar regions of the supercontinent Pangea, beginning in the Mesozoic. This is not to say that Rhodophyta were absent in the Paleozoic or earlier eras; rather, the characters by which we recognize families today evolved since the event that separated the Permian from the Triassic. Even given the present imperfect state of our knowledge of red algal evolution, I see most phylogenetic lines converging backwards in time and space, to a point of origin within the Tethyan Ocean; or, conversely, radiating forward from such a point. According to this hypothesis, phylogenetic radiation of taxa that survive at the present time is explosive, extending in the direction of the expanding Tethyan Ocean, and not implosive from the margins of the disintegrating Pangea. This general trend in seaweed distribution continued over a period of 150 million years, and was only altered during the latter half of the Cenozoic, in response to climatic changes that led to thermal stratification of the oceanic water column, and to substantive increases in flow rates of surface currents (43-38 Ma). Thereafter, species migration by oceanic currents replaced plate tectonics and seafloor spreading as the major driving force behind speciation and species distribution (Hommersand 1986).

Biogeographic information referred to in this study is incorporated in Tables 1-7, and in the previous section on Morphology and Taxonomy of the Rhodophyta in relation to their Biogeography. I will discuss each regional affinity separately.

Biogeographic relationships between Rhodophyta of the SW Pacific (Australia-New Zealand) and the NW Pacific (Aleutian Islands to northern Japan) (Table 1)

A few assemblages are so old that their distributions seem to reflect events associated with the opening of the Tethyan Ocean during the first half of the Mesozoic. According to a thesis advocated in Audley-Charles & Hallam (1988), continental blocks rifted from Australian Gondwanaland, drifted northward and collided with parts of southeast Asia from mid-Permian to late Jurassic. Taxa separated in this manner would have evolved independently in the Northern and Southern Hemispheres. Division between southern and northern assemblages need not, however, have been brought about through plate tectonics. There may once have been a continuous distribution of related taxa along the margin of the Tethys, with vicariant patterns developing as linkages between the southern and northern regions were broken.

The Bangiaceae with two genera, *Bangia* and *Porphyra*, may well fit such a distribution pattern, depending on whether species clusters existing in Australia and New Zealand are distinct from those found in the NW Pacific. Similarly, the Gracilariaceae could have been

Table 1. Polar-Tethyan distributions between Southwest Pacific and the North Pacific (especially the NW Pacific) Oceans

Taxonomic Group	SW Pacific Australia, New Zealand	NW Pacific Aleutians, Siberia, Japan
Bangiales	*Porphyra*-in part	*Porphyra*-in part
Gracilariales	*Melanthalia* *Curdiea*	*Gracilariopsis* *Gracilaria* *Hydropuntia*
Gelidiales	*Pterocladia lucida*	Gelidiaceae (others)
Bonnemaisoniaceae	*Delisea*	*Bonnemaisonia*
Dumontiaceae complex	Dumontiaceae - in part	Gloiosiphoniaceae Endocladiaceae? Tichocarpaceae?
Tribe Dumontieae	*Dasyphloea*	*Dumontia* *Cryptosiphonia* *Hyalosiphonia/Dislsea* group
Kallymeniaceae	*Pugetia/Kallymenia* group	*Kallymeniopsis* group
Gigartinaceae complex	Gigartinaceae Chondriellaceae	Phyllophoraceae Petrocelidaceae *Chondrus* group
Furcellariaceae/Solieriaceae	Solieriaceae	Furcellariaceae *Turnerella* group
Ceramiaceae	Antithamnieae (*Pterothamnion* group) Gymnothamnieae Ceramieae-in part	Antithamnieae (*Pterothamnion* group) Ptiloteae Ceramieae-in part
Delesseriaceae	*Hypoglossum* group *Nitophyllum* group	*Membranoptera* group '*Neohypophyllum*' group '*Pollexfenia*' group
Rhodomelaceae	Bostrychieae Lophothalieae Brongniartelleae	Brongniartelleae Rhodomeleae

divided into southern and northern assemblages, with *Melanthalia* and *Curdiea* originating in the southern Tethys and *Gracilariopsis*, *Gracilaria*, and *Hydropuntia* evolving in the northern Tethys, and afterwards becoming widely distributed. In the Gelidiaceae one relict species, *Pterocladia lucida*, appears to have originated in the southern Tethys, the rest having evolved originally in the northern Tethys.

The Gigartinaceae complex, including the families Chondriellaceae, Gigartinaceae, Petrocelidaceae and Phyllophoraceae, is a major assemblage that appears to be divided into southern and northern groups. The southern group includes certain species of *Gigartina*, *Rhodoglossum* and *Iridaea* (Gigartinaceae) and, perhaps, *Chondriella* (Chondriellaceae). The northern group contains a different mixture of species currently placed in *Gigartina*, *Rhodoglossum* or *Iridaea*, as well as the genus *Chondrus* and the families Petrocelidaceae and Phyllophoraceae. The taxonomy of all of these groups is currently in a state of confusion, complicating the biogeography story.

A southern/northern dichotomy can be drawn between the Dumontiaceae and Gloiosiphoniaceae, in which the most primitive taxa belonging to the Dumontiaceae have their center of distribution in the Southern Hemisphere in the vicinity of Australia, whereas the Gloiosiphoniaceae has its center of distribution in the Northern Hemisphere in the N Pacific and N Atlantic Oceans. Earlier, I argued that the procarpial condition found in most genera belonging to the Gloiosiphoniaceae is primitive, and that connecting filaments evolved twice in this assemblage: once in the Dumontiaceae in the Southern Hemisphere; and again in *Gloiosiphonia* in the Northern Hemisphere. Should the procarpial families Endocladiaceae and Tichocarpaceae ultimately prove to be related to the Gloiosiphoniaceae, their N Pacific distribution could, perhaps, further strengthen this argument. The story is further complicated by the fact that the Dumontiaceae has evolved and proliferated tribes, genera and species in the Northern Hemisphere, the Dumontieae having evolved primarily in the boreal N Pacific with one genus (*Dumontia*) in the N Atlantic, and the Farlowieae in Pacific North America. This distribution represents a second separation of taxa within Dumontiaceae that took place at a later time, probably in the late Cretaceous or early Paleogene.

There exists a south-north distribution pattern that appears to have followed a diagonal pathway extending from the austral Tethys to the Arctic Ocean (probably by way of the Turgai Strait and West Siberian Sea, Vakhrameev 1987), and from there to the N Pacific, beginning in Cretaceous and ending in Paleocene. Distributions that appear to have followed this pathway are also represented in the N Atlantic by members that appear to be primitive, or are intermediate between taxa of the SW Pacific and those of the NW Pacific. Groups that appear to have a polar-austral-Tethyan distribution include the Bonnemaisoniaceae, Kallymeniaceae, Furcellariaceae/Solieriaceae and the tribes of Ceramiales listed in Table 1. *Bonnemaisonia asparagoides* from the N Atlantic, with its heteromorphic life history and terminal cystocarps, is intermediate between the austral *Delisea* group and the *Bonnemaisonia* species of the N Pacific. *Meredithia microphylla* and '*Cirrulicarpus*' *carolinensis* (North Atlantic), with heteromorphic life histories, are probably primitive with respect to this character, compared to '*Cirrulicarpus*' *australis* (Australia) and *Cirrulicarpus gmelinii* (NW Pacific) which have isomorphic life histories. The Furcellariaceae and the *Turnerella* group (boreal), in which the gonimoblast develops directly from the fused connecting filament, stands in contrast to the Solieriaceae (primarily austral), in which the gonimoblast develops directly from the auxiliary cell, one per connecting filament. A polar tethyan pathway can also be invoked to account for the distribution of the Antithamnieae, Ptiloteae, and Ceramieae (Ceramiaceae), the *Hypoglossum* group (primarily austral), the *Membranoptera* group (primarily boreal), components of the *Nitophyllum* group (Delesseriaceae), and certain Rhodomelaceae linked through the tribe Brongniartelleae. A south/north longitudinal distribution cannot be ruled out, and should be borne in mind when considering alternative possibilities for the biogeography of any of these taxa.

Distribution of boreal taxa between the North Atlantic and the North Pacific Oceans (Table 2)

Lindstrom (1987b) discussed the biogeographic relationships of possible sister groups belonging to nine families of Rhodophyta that are distributed between the N Pacific and N Atlantic Oceans. She argues that cladistic analysis can provide biogeographic tracks suggesting the origins and migrations of species across the northern oceans, and tentatively concludes that the large pool of N Pacific endemics, that already existed prior to the innundation of the Bering Land Bridge, may have provided the seed stock for what appears to be related species, that may have evolved in the Arctic and cool temperate North Atlantic during the last 3 million years. Wilce (1990), on the other hand, concludes that very few species originating in the N Pacific appear to have reached the N Atlantic since Pliocene, despite the fact that the direction of the prevailing current is from south to north through the Bering Strait. In contrast, numerous introductions of Arctic and North Atlantic species into the N Pacific have certainly occurred, especially along the eastern side of the Bering Strait in Alaska (Lindstrom pers. comm.) If there has indeed been a strong bias against migration from the N Pacific to the N Atlantic in the last four million years, the reason may be that species adapted to a moderate climate, like that of the N Pacific, do not convert readily to a stressful environment, like that of the Arctic Ocean, or the N Atlantic during Pleistocene. The opposite may be true of species adapted to severe environments, when confronted with the opportunity of migrating into a region having a moderate climate.

In this paper I am concerned with possible evolutionary and biogeographic relationships between N Pacific and N Atlantic taxa, that were established much earlier than the formation of the modern Bering Strait at ca. 3.7 Ma. Admittedly, it is very difficult to distinguish between older and more recent separations. The central event in the evolution of the Arctic Ocean was the establishment of a plate boundary between Asia and North America associated with the rifting of the Chukota-Alaska-North Canadian margin. Details of this process are controversial;

Table 2. Distributions of boreal taxa between the North Atlantic and North Pacific Oceans. Possible Pliocene and recent distributions from N Pacific to N Atlantic are placed in brackets

Taxonomic group	N Pacific	N Atlantic
Bangiales	*Bangia atropurpurea* *Porphyra*	*Bangia atropurpurea* *Porphyra*
Acrochaetiales	(B-phycoerythrin) *Acrochaetium* Rhodophysemataceae	(B-phycoerythrin) *Acrochaetium* Rhodophysemataceae
Acrochaetiales	(R-phycoerythrin) *Colaconema* *Rhodochorton*	(R-phycoerythrin) *Colaconema* *Rhodochorton*
Palmariales	*Palmaria mollis* [*Devaleraea ramentacea*?] *Halosaccion*	*Palmaria palmata* *Devaleraea ramentacea*
Acrochaetiales	*Nemalion*	*Nemalion helminthoides*
Ahnfeltiales	*Ahnfeltia fastigiata* [*Ahnfeltia plicata*]	*Ahnfeltia plicata*
Bonnemaisoniales	*Bonnemaisonia geniculata* (*Polyblepharidiella*) *B. hamifera* group *Asparagopsis*	*Bonnemaisonia asparagoides*
Dumontieae	*Dumontia contorta* *Dumontia simplex* Hyalosiphonia Masudaphycus *Neodilsea/Dilsea*	*Dumontia contorta* *Dilsea integra*? *Dilsea carnosa*?
Gloiosiphoniaceae	*Gloeophycus* *Gloiosiphonia*	*Thuretella* *Schimmelmania* *Gloiosiphonia capillaris*
Kallymeniaceae	*Kallymeniopsis* group	'*Kallymenia*' *schmitzii*?
Phyllophoraceae	*Gymnogongrus*-in part *Petroglossum*?	*Gymnogongrus*-in part *Phyllophora*
Petrocelidaceae	*Mastocarpus*	*Mastocarpus stellatus*
Gigartinaceae	*Chondrus* (Japan) '*Iridaea*' *splendens* group *Gigartina leptorhynchos*	*Chondrus crispus*

Table 2 cont. on following page

Table 2 cont.

Nemastomataceae	*Platoma?* *Schizymenia*	*Platoma bairdii* *Schizymenia dubyi*
Furcellariaceae	*Halarachnion latissimum* *Neurocaulon?*	*Halarachnion ligulatum* *Neurocaulon reniforme*
Turnerella group	*Opuntiella californica* *Turnerella mertensiana*	 *Turnerella pennyi*
Antithamnieae	*Platythamnion* *Scagelia pylaesii* *Antithamnionella floccosa*	*Pterothamnion plumula* *Scagelia pylaesii* [*Antithamnionella floccosa*?]
Ptiloteae	 *Ptilota* *Neoptilota*	*Plumaria* *Ptilota plumosa* [*Ptilota serrata*]
Ceramieae	*Microcladia* (Japan)	*Microcladia glandulosa*
Delesseriaceae	*Pantoneura* *Membranoptera* *Phycodrys* *Erythroglossum/Polyneura*	*Pantoneura baerii* *Membranoptera alata* *Phycodrys rubens* *Erythroglossum/Polyneura*
Brongniartelleae	*Beringiella* *Schizochlaenion* *Sinosiphonia*	*Brongniartella*
Rhodomeleae	*Neorhodomela* *Rhodomela* *Odonthalia*	 *Rhodomela* [*Odonthalia dentata*?]

however, a boundary is thought to have been established by 95 Ma (Barron *et al.* 1981; Barron 1987). It is not certain that compression followed immediately, or only later, perhaps as late as Eocene (Srivastava pers. comm). It is conceivable that a traverse existed or was formed periodically, either through the Bering Strait, or across the Chukota-Laurasian boundary in the Cretaceous and Paleogene. If so, the Arctic Ocean functioned during this period as an extension of he N Pacific, and taxa evolving in either region would be part of the ame flora. Ultimately, the Arctic Ocean became isolated from the N Pacific with the establishment of a continuous Bering Land Bridge. Even so, isolation of the Arctic Ocean was never complete, since passageways to the N Pacific and the West Siberian (Asia) and Cannonball (North America) inland seas remained open for long periods of time. Such passageways would have functioned as an overflow mechanism permitting latitudinal movements in response to temperature changes.

While the Arctic Ocean has probably functioned as a passageway for the migration of taxa across the north polar region, it may have served more directly as a center for the evolution of structural and physiological adaptations to cool temperatures and long photoperiods. Such adaptations may have required a heteromorphic life history involving alternation between an erect gametophyte and a perennial crustose tetrasporophyte. One or more representatives of nearly every phylogenetic line having a boreal distribution pattern exhibits photoperiodic behavior and possesses a heteromorphic life history.

Taxa that evolved in the Arctic Ocean and adjoining seas, that had generated adaptive life histories in response to a cool climate and long photoperiods, may have migrated to cool regions of the N Pacific. The NW Pacific (Aleutian Islands to Japan), which appears to have received the greater part of the early Arctic flora, would have cooled more rapidly than the NE Pacific (Pacific North America), as a result of prevailing patterns of ocean circulation. Temperatures are known to have dropped from 14^{0}C in Eocene to 3^{0}C in Oligocene in the N Pacific (Frakes 1979). The present flora of the NW Pacific extending from 40^{0}N to 55^{0}N latitude, is distinctly different from the Indo-W Pacific flora to the south, although elements of the two tend to interdigitate.

The mid-Atlantic Ridge system propagated northwards in the Cenozoic, ultimately leading to connection between the N Atlantic and Arctic Oceans. My impression is that this process was completed around anomaly 13 in Oligocene with the opening of a passageway across the Barents Shelf (Dowdeswell 1988). The result would have been to bring Arctic taxa into the North Atlantic. Many of the floristic elements in the N Atlantic that are related to algae of the N Pacific appear to me to be structurally less complex and, therefore, more primitive than their counterparts in the N Pacific. I believe that the majority are relicts of a previous Arctic flora that evolved little and may have suffered major extinctions since entering the N Atlantic. Examples include: *Palmaria palmata*, *Bonnemaisonia asparagoides*, *Thuretella schousboei*, *Gloiosiphonia capillaris*, *Dumontia contorta*, *Gymnogongrus* spp, *Mastocarpus stellatus*, *Chondrus crispus*, *Phyllophora* spp., *Schizymenia dubyi*, *Halarachnion ligulatum*, *Turnerella pennyi*, *Microcladia glandulosa*, *Plumaria elegans*, *Ptilota plumosa* and *P. serrata*, *Pantoneura baerii*, *Membranoptera alata*?, *Phycodrys rubens*, and *Rhodomela confervoides*. Rates of morphological change and speciation appear to have been lower in the N Atlantic than in the N Pacific during the Neogene.

An eastward distribution from the NW Pacific (Japan) to the NE Pacific (Pacific North America) and the N Atlantic (Table 3)

Several families, genera and species clusters, present on both sides of the N Pacific that also occur in the N Atlantic, apparently originated in the western N Pacific (China, Japan) and spread eastward. A possible pathway would have been alongside the Bering Land Bridge in

Table 3. Distributions between temperate regions of the western Pacific (China, Japan), the eastern Pacific (Pacific North and South America), the western Atlantic (Atlantic North and South America) and the eastern N Atlantic (Europe and North Africa)

Taxonomic group	Pacific	Atlantic
Liagoraceae	(in part)	in part
Galaxauraceae	*Pseudogloiophloea* group	*Pseudogloiophloea* group
	(*Scinaia okamurae*)	(*Scinaia halliae, verae*)
	(*Scinaia confusa*)	(*Scinaia furcellata*)
	Scinaia japonica	*Scinaia johnstonii*?
	Scinaia johnstonii	*Scinaia complanata*
	Scinaia hormoides	*Scinaia hormoides*
	Scinaia cottonii	*Scinaia cottonii*?
	Scinaia latifrons	
Gracilariaceae	*Gracilariopsis*	*Gracilariopsis*
	Gracilaria textorii group	*Gracilaria textorii* group
	Gracilaria verrucosa group?	*Gracilaria verrucosa* group?
Phyllophoraceae	*Gymnogongrus*-in part	*Gymnogongrus*-in part?
	Stenogramme	*Stenogramme*
	Petroglossum	*Phyllophora sicula*?
	Ozophora	*Schottera nicaeensis*?
Gigartinaceae	*Gigartina teedii* group	*Gigartina teedii* group
Dasyaceae	*Dasya baillouviana* group?	*Dasya baillouviana* group

late Cretaceous or early Cenozoic before global cooling, and the presumed introduction of Arctic taxa into the western N Pacific had taken place. Possible candidates for such a distribution pattern include: several species clusters in *Scinaia*, *Gracilariopsis* and certain species of *Gracilaria*, segments of the Phyllophoraceae, part of which may have migrated to the N Atlantic through the Arctic Ocean, the *Gigartina teedii* group (Gigartinaceae) and the *Dasya baillouviana* group (Dasyaceae). Other taxa having widespread distributions in the northern hemisphere are possibilities.

A North Atlantic-Tethyan distribution (Table 4)

A large component of the N Atlantic red seaweed flora has special affinities with the algae of the Indo-W Pacific, especially the temperate algae of the SE Pacific. These taxa all belong to separate phylogenetic lines from those that show a polar-Tethyan distribution. I believe that all originated comparatively recently, having, for the most part, evolved and radiated toward the end of the Cretaceous and during the Cenozoic. This is somewhat surprising, since many taxa in this floristic assemblage have long been regarded as being primitive, both from morphological and phylogenetic standpoints. There is no reason, however, why thalli that are relatively simple structurally and reproductively might not undergo a phase of rapid evolution and geographic expansion, especially following mass extinctions, such as those that occurred at the Cretaceous-Tertiary boundary. The most primitive elements in each of the phylogenetic lines belonging to this broad association are generally conserved in the N Atlantic, probably because the Indo-W Pacific region is larger, and provided a more stable environment over a longer period of time for speciation and adaptive radiation to have occurred. I would argue that taxa belonging to this floristic assemblage evolved along a gradient extending from western Europe to Australia, and that the present location of primitive vs. advanced forms is not necessarily an indication of the center of origin of any particular taxon.

Assemblages listed in Table 4 include: (1) the Nemastomataceae complex consisting of the Cruoriaceae, Hemichariaceae, Calosiphoniaceae, Nemastomataceae, Sebdeniaceae,

Table 4. Atlantic austro-Tethyan distribution. (Distributions between the austral Indo-W Pacific and NE Atlantic Oceans. *Indicates extended distribution to the western Atlantic. **Indicates extended distributions to both the western Atlantic and eastern Pacific Oceans

Taxonomic group	Austral Indo-W Pacific	Northeast Atlantic
Liagoraceae	*Helminthora australis*	*Helminthora divaricata*
		*Helminthora stricta***
Dumontiaceae	*Dudresnaya australis*	*Dudresnaya verticillata*
		Dudresnaya crassa?
		Dudresnaya georgiana
Kallymeniaceae	'*Cirrulicarpus*' *australis*	*Meridithia microphylla*
		'*Cirrulicarpus*' *carolinensis*
	Kallymenia	*Kallymenia* (Mediterranean)
Gigartinaceae	*Gigartina disticha*	*Gigartina pistillata*
Nemastomataceae		*Nemastoma dichotoma*
	Platoma	*Platoma cyclocolpa**
Calosiphoniaceae		*Calosiphonia vermicularis*
		Schmitzia neapolitana
		Schmitzia hiscockiana
	'*Schmitzia*'	
Sebedeniaceae	*Sebdenia polydactyla*	*Sebdenia monardiana*
		*Sebdenia****
Solieriaceae	*Solieria robusta*	*Solieria chordalis*
		*Solieria filiformis**
		Flahaultia
		*Agardhiella**
		*Sarcodiotheca***
Cystocloniaceae/Hypneaceae		*Rhodophyllis divaricata*
		*Cystoclonium purpureum**
		*Fimbrifolium dichotomum**
	Calliblepharis group	*Calliblepharis ciliata*
		Calliblepharis jubata
	Hypnea	[*Hypnea musciformis*]
Rhodymeniaceae	*Fauchea*/*Gloioderma* group	*Fauchea*/*Gloioderma* group
	Botryocladia group	*Botryocladia****
Lomentariaceae	*Lomentaria*	*Lomentaria articulata*
		Lomentaria clavellosa
Crouanieae	*Gulsonia annulata*	*Gulsonia mediterranea*
	Crouania	*Crouania annulata**
Spermothamnieae		*Spermothamnion repens**

Table 4 cont. on following page

Table 4 cont.

Sphondylothamnieae	*Medeiothamnion*	*Sphondylothamnion multifidum*
Griffithsieae		*Halurus equisetifolius*
		Griffithsia flosculosa
	Anotrichium	*Anotrichium barbatum**
	Griffithsia	*Griffithsia corallinoides*
Delesseriaceae	*Hymenena*	*Cryptopleura***
	Myriogramme group	*Myriogramme bonnemaisonii*
	Nitophyllum group	*Nitophyllum punctatum*
Amansieae		*Halopitys incurvus*
		Rytiphlaea tinctoria
	Amansia	
	Vidalia obtusiloba	[*Vidalia volubilis*]
Chondrieae	*C. dasyphylla* group	[*Chondria dasyphylla*]
		C. tenuissima group
Laurencieae		*Laurencia pinnatifida/*
		L. spectabilis group**
	Laurencia obtusa group	[*Laurencia obtusa*]

Titanophoraceae (Norris in ed.), (2) the Halymeniaceae, (3) a cluster composed of the Furcellariaceae, Solieriaceae, Cystocloniaceae, Hypneaceae, Sarcodiaceae? and Wurdemanniaceae?, (4) a series of derived taxa, that evolved later in the vicinity of Australia, including the Cubiculosporaceae, Dicranemaceae, Acrotylaceae, Mychodeaceae, and Mychodiellaceae; (5) one or two groups in the Rhodymeniales and, (6) several tribes or groups in the Ceramiales, most of which are now widely distributed. Tribes in the Ceramiaceae include the Crouanieae (in part), and the Spermothamnieae, Sphondylothamnieae and Griffithsieae. Included among the Delesseriaceae are components of the *Cryptopleura*-group, the *Myriogramme*-group and the *Nitophyllum*-group; and in the Rhodomelaceae, elements of the Chondrieae/Laurencieae and the Amansieae.

Distributions between the SW Pacific (Australia-New Zealand), South Africa and the NE Atlantic (Europe), sometimes with extensions to the western Atlantic and eastern Pacific (Table 5)

The coast of southern Africa may have served as a gateway or way station for the distribution of species which originated in south Australia-Tasmania or New Zealand, and were subsequently distributed to western Europe and Pacific North and South America. It is also

Table 5. Distributions between SW Pacific (Australia-New Zealand), South Africa, and the NE Atlantic (Europe). *Indicates presence in the western Atlantic. **Indicates presence in the eastern Pacific (Pacific America). ***Indicates presence in both the western Atlantic and the eastern Pacific

Taxonomic group	SW Pacific (Australia/NZ)	NE Atlantic/South Africa
Kallymeniaceae	*Pugetia*	*Pugetia*** (absent in Europe)
	Kallymenia	*Kallymenia reniformis* group ***
	Thamnophyllis	*Thamnophyllis*
Gigartinaceae	*Gigartina disticha*	*Gigartina pistillata*
Cryptonemiaceae	*Grateloupia 'filicina'*	*Grateloupia filicina****
		Grateloupia dichotoma
		Dermocorynus montagnei
	Grateloupia 'doryphora'	*Grateloupia doryphora****
Plocamiaceae	*Plocamium cartilagineum*	*Plocamium cartilagineum***
		P. glomeratum (S Africa)
Rhodymeniaceae	*Rhodymenia*	*Rhodymenia pseudopalmata group****
Champiaceae	*Chylocladia grandis*	*Chylocladia verticillata*
		Chylocladia capensis
		*Gastroclonium***
Antithamnieae	*Pterothamnion*	*Pterothamnion***
	Antithamnionella	*Antithamnionella****
Ceramieae	*Ceramium rubrum group*	*Ceramium rubrum* group***
	'Microcladia'	*'Microcladia'* (except Europe)
Spermothamnieae	*Spermothamnion*	*Spermothamnion**
	Tiffaniella	*Tiffaniella****
Sphondylothamnieae	*Bornetia*	*Bornetia***
Compsothamnieae		*Compsothamnion*
	Pleonosporium	*Pleonosporium****
Callithamnieae	*Callithamnion*	*Callithamnion****
		Aristothamnion (S Africa)
Delesseriaceae	*Platysiphonia*	*Platysiphonia miniata*
		*Platysiphonia****
	Hypoglossum	*Hypoglossum hypoglossoides***
		*Hypoglossum****
	Apoglossum	*Apoglossum ruscifolium***

Table 5 cont. on following page

Table 5 cont.

	Hymenena/Acrosorium	*Myriogramme minuta* *Acrosorium*** *Hymenena***
Dasyaceae	*Dasya scoparia*	*Dasya hutchinsiae*? *Dasya californica***
	Heterosiphonia berkeleyi group	*Heterosiphonia*? (S Africa) *Heterosiphonia plumosa*
Bostrychieae	*Bostrychia harveyi*	*Bostrychia harveyi* (S Africa) *Bostrychia scorpioides* (Europe)
Polysiphonieae	*Pterosiphonia* group *Aphanocladia*	*Pterosiphonia*** *Aphanocladia stichidiosa* (Europe) *Tayloriella* (S Africa)
	Polysiphonia-in part	*Polysiphonia*-in part
Streblocladieae	*Streblocladia*	*Streblocladia***

possible that some of these were first distributed from Australia to Europe via the Tethyan Seaway. In most instances the more primitive taxa appear to be in Australia-New Zealand, and the more advanced ones in South Africa and the Atlantic or eastern Pacific Oceans. Species having this distribution pattern divide into two groups: those comming from West and South Australia that are part of the Indian Ocean flora, and those coming from Tasmania, South Australia and New Zealand, that are part of the SW Pacific flora. Earlier, I postulated that the latter group came via a passageway between East and West Antarctica, and that the former came by way of the South Equatorial Current and East Africa in Miocene (Hommersand 1986). The warm-temperate Indian Ocean cluster could have migrated around the Cape during warm periods, and the cool-temperate Pacific cluster during cooler periods in Miocene. Ultimately, a cool temperate boundary was established at the Cape of Good Hope, and along the west coast of South Africa and Namibia, with the emergence of the Benguela current around 6 Ma, blocking the further dispersal of the most recent arrivals from West Australia.

Distributions between the N Atlantic, the SE Atlantic and islands of the Scotia Sea and West Wind Drift (Table 6)

There are several taxa, which almost certainly originated in the Northern Hemisphere, that are prominent components of the antiboreal floras of southern South America, the Falkland Islands, Antarctica, islands of the Scotia Sea and islands of the West Wind Drift. The most prominent and widely distributed of these is *Palmaria*, with two species: *P. decipiens* and *P.*

Table 6. Distributions between the boreal North Atlantic and the antiboreal South Atlantic (South America, Antarctica, Islands of the Scotia Sea, and Islands of the West Wind Drift)

Taxonomic group	Boreal North Atlantic	Antiboreal South Atlantic
Bangiales	*Porphyra umbilicalis* group	*Porphyra columbina* group
Acrochaetiales	*Acrochaetium* *Audouinella* *spetsbergensis/concrescens*	*Acrochaetium* *Audouinella* *spetsbergensis/concrescens*
	Colaconema daviesii *Colaconema botryocarpa* group *Rhodochorton purpureum*	*Colaconema daviesii* *Colaconema botryocarpa* group *Rhodochorton purpureum*
Palmariales	*Palmaria palmata*	*Palmaria decipiens* *Palmaria georgica*
Hildenbrandiales	*Hildenbrandia*?	*Hildenbrandia lecannellieri*?
Ahnfeltiales	*Ahnfeltia plicata*	*Ahnfeltia plicata* *Ahnfeltia elongata*
Phyllophoraceae	*Gymnogongrus*? *Phyllophora*	*Gymnogongrus*? *Phyllophora*
Cystocloniaceae	*Cystoclonium purpureum*	*Cystoclonium obtusangulum*
Plocamiaceae	*Plocamium cartilagineum*?	*Plocamium cartilagineum*?
Rhodymeniaceae	*Rhodymenia*?	*Rhodymenia*?
Antithamnieae	*Pterothamnion plumula*?	*Pterothamnion plumula*?
Callithamnieae	*Callithamnion granulatum*	*Callithamnion gaudichaudii*
Delesseriaceae	*Pantoneura bairii* *Delesseria sanguinea* *Phycodrys rubens*	*Pantoneura plocamioides* *Delesseria lancifolia*? *Phycodrys quercifolia* *Phycodrys*

georgica, both of which are thought to be closely related to *Palmaria palmata* from the N Atlantic (Ricker 1987). *Palmaria* is unusual in that it is widely distributed in both antarctic and subantarctic waters. Other genera recorded from the N Atlantic and Antarctica inlude: *Porphyra*, certain Acrochaetiaceae, *Kallymenia*, *Callophyllis*, *Plocamium*, *Rhodophyllis*, *Cystoclonium*, *Phyllophora*, *Pantoneura*, *Delesseria* and *Phycodrys*. Additional genera recorded from both the N Atlantic and southern South America or the islands of the West Wind Drift include: *Ahnfeltia*, *Rhodophyllis*, *Gymnogongrus*, *Gigartina*, *Pterothamnion*, *Callithamnion*, *Ceramium*, *Myriogramme*, *Heterosiphonia* and *Polysiphonia*.

Many of the species belonging to these genera are antiboreal in origin, and are not related to boreal taxa. Those that originated in the N Atlantic must have undergone bipolar distribution from north to south at a time of global climatic change. For those species that reached Antarctica, this must have taken place before formation of the West Wind Drift as a permanent feature of the southern Ocean. A north-south bipolar distribution may have occurred during an early period of global cooling in Oligocene, or later in mid Miocene (16-14 Ma), at a time when subsidence of the Iceland Ridge linked the North Polar Sea, as a heat sink, to the North Atlantic Ocean, and an ice cap formed over East Antarctica (Schniter 1980).

Algae having an Indo-W Pacific-Caribbean-N Atlantic distribution (Table 7)

Many species present in the Caribbean Sea (or in West Africa) are identical to or closely related to species found in warm-temperate regions of the N Atlantic or Mediterranean Sea. Some of these represent taxa that have their center of origin in the Indo-W Pacific and were subsequently distributed to the Atlantic Ocean by a southern route around South Africa. Some may have reached the NW Atlantic (Europe) by amphitropical distribution. Others were distributed to the Caribbean Sea by way of the South Equatorial and Guiana currents (Hommersand 1986). Transport of Caribbean algae to the NE Atlantic is probably related to the history of the Gulf Stream in late Miocene and Pliocene. Many Indo-W Pacific species that reached the western Atlantic (Caribbean Sea) have not extended their range to the NE Atlantic and Mediterranean Sea.

Table 7. List of species distributed between the Indo-W Pacific, the western Atlantic (Caribbean Sea) and the NE Atlantic (western Europe, North Africa and the Mediterranean Sea)

Galaxaura oblongata	*Gelidium pusillum*	*Pterocladia capillacea*
Gelidiella tenuissima	*Peyssonnelia* spp.	*Hildenbrandia* spp.
Gracilaria bursa-pastoris?	*Gracilaria cervicornis*?	*Caulacanthus ustulatus*
Catenella caespitosa	*Platoma cyclocolpa*	*Predaea feldmannii*
Hypnea musciformis	*Grateloupia filicina*?	*Halymenia trigona*
Champia parvula	*Antithamniella elegans*	*Gymnothamnion elegans*
Wragelia argus	*Wrangelia penicillata*	*Ceramium diaphanum*
Ceramium flaccidum	*Ceramium strictrum*	*Ceramium tenuissimum*
Spyridia filamentosa	*Ptilothamnion pluma*	*Pleonosporium caribaeum*
Monosporus pedicellatus	*Anotrichium tenue*	*Heterosiphonia crispella*
Taenioma nanum	*Platysiphonia miniata*	*Cottoniella filamentosa*
Caloglossa leprieurii	*Hypoglossum hypoglossoides*	*Acrosorium uncinatum*
Bostrychia radicans	*Chondria dasyphylla*	*Herposiphonia secunda*
Lophosiphonia spp.	*Polysiphonia* spp.	

CONCLUSIONS

The results given in Tables 1, 2 and 4 are particularly interesting. Families and genera listed in Tables 1 and 2 are distributed primarily between the SW Pacific and NW Pacific, either longitudinally along a south-north axis, or diagonally along a polar-Tethyan axis with the more primitive taxa usually residing in the Southern Hemisphere. In contrast, families and genera recorded in Table 4 are distributed from west to east along a N Atlantic-Tethyan axis with the primitive members more often occurring in the N Atlantic.

I have constructed tentative phylogenetic trees for taxa belonging to each of the two patterns, and find that there is little overlap between them at the family level, and none at the level of major phylogenetic lines (orders or superfamilies). An exception is the Ceramiales in which only the most primitive tribes (according to my evaluation) exhibit a south-north distribution (Table 1). What could be the cause of this large-scale biogeographic separation of rhodophytan taxa?

At the beginning of the discussion section, I expressed the belief that present-day red algal families evolved in the expanding Tethyan Ocean beginning in the Mezozoic. This was the world of the thermosphere, in which ocean temperatures were relatively uniform over a broad latitudinal expanse, with only the polar regions being significantly cooler. Gradual cooling at the poles began in the Cretaceous, and extended into the Paleogene at an accelerated rate from 53 to 40 Ma (Kennett 1980), but a dramatic change in world climate, from the conditions of the thermosphere to those of the psychrosphere, did not occur until the end of the Eocene around 38 Ma (Benson *et al.* 1984).

The world of the circumglobal thermosphere was very different from that of the later psychrosphere. Currents were not as strong and upwelling less likely, with the result that average nutrient levels, like average temperatures, were more uniform. Under these conditions, taxa presently found in temperate waters should have been widely distributed, either as a result of plate tectonics and seafloor spreading, or through dispersal by currents at opportune times. The cataclysmic event that gave rise to the Cenozoic, and the 'climatic deterioration' that ushered in the Neogene, changed the character of the world's oceans altogether. Energy became canalized with the intensification of ocean currents and the establishment of stable temperature- and density-dependent convergencies. Biotic interactions were intensified with the evolution of new groups of algal herbivores, especially in the tropics. Some older phylogenetic lines either became extinct or had their distributions limited to temperate regions beyond the range of the most predaceous herbivores.

Whole new phylogenetic lines appear to have evolved in response to the changing climatic conditions along the N Atlantic-Tethyan axis (Table 4). Families and genera presently composing the Indo-W Pacific flora, form a wedge that displaced taxa of the old thermosphere, and became the algae of the new 'tropics'. Many extended their range around southern Africa into the tropical N Atlantic. A close inspection of tropical assemblages reveals that each has developed one or more strategies for coping with biotic interactions, particularly herbivory. These include: heteromorphic life cycles with long-lived filamentous or crustose stages coupled with ephemeral gametophytic stages (Liagoraceae, Galaxauraceae [*Scinaia* assemblage], Dumontiaceae-in part, Calosiphoniaceae, Nemastomataceae); a system of connecting filaments for continued diploidization of auxiliary cells and production of carposporangia from a few fertilizations (Nemastomataceae, Sebdeniaceae, Halymeniaceae); a flattened crustose or peltate habit (Peyssonneliaceae, Solieriaceae-in part, Rhodymeniaceae-in part); a bushy habit suited for resistance to herbivory in the intertidal or subtidal fringe (Gracilariaceae-in part, Hypneaceae-in part); an ephemeral life cycle and fugitive behavior (Ceramiaceae, Rhodomelaceae-in part); and lastly, the production of secondary metabolites that ward off herbivores or regulate their behavior (many families, but especially the Rhodomelaceae).

It is the newer families with a N Atlantic-Tethyan distribution that appear to have taken greatest advantage of the changing conditions of the Cenozoic. The older polar austral-Tethyan groups have largely been displaced, or have reinvaded the tropics in deeper water. The picture is complex, but the overall evolutionary and biogeographic pattern is clear.

REFERENCES

Abbott IA (1985a) Vegetative and reproductive morphology in *Reticulocaulis* gen. nov. and *Naccaria hawaiiana* sp. nov. (Rhodophyta, Naccariaceae). J Phycol 21: 554-561

Abbott IA (1985b) *Peleophycus multiprocarpium* gen. et sp. nov. (Gloiosiphoniaceae, Rhodophyta). Pac Sci 38: 324-332

Acleto OC (1972) Structure and reproduction of *Schimmelmannia dawsonii* sp. nov. (Rhodophyceae, Cryptonemiales). Phycologia 11: 5-9

Anathasiadas, A (1990) Evolutionary biogeography of the North Atlantic Antithamnioid algae (Rhodophyta, Ceramiales). In: Garbary DJ, South GR (eds) Evolutionary biogeography of the marine algae of the North Atlantic. Springer Verlag, Berlin, p 219-240

Ardre F (1977) Sur le cycle du *Schizymenia dubyi* (Chauvin ex Duby) J. Agardh (Nemastomacée, Gigartinale). Rev Algol NS 12: 73-86

Audley-Charles MG, Hallam A (Eds) (1988) Gondwana and Tethys. Oxford University Press, Oxford

Baardseth E (1941) The marine algae of Tristan da Cunha. Results Norwegian Sc Exp Tristan da Cunha 1937-1938. No. 9. Oslo, p 1-174

Baldock RN (1976) The Griffithsieae group of the Ceramiaceae (Rhodophyta) and its southern Australian representatives. Austr J Bot 24: 509-593

Barron EJ (1987) Cretaceous plate tectonic reconstructions. Palaeogr Palaeoclimatol Palaeocol 59: 3-29

Barron EJ, Harrison CGA, Sloan JL, Hay W (1981) Paleogeography, 180 millon years ago to the present. Eclogae Geol Helv 74: 443-470

Benson RH, Chapman RE, Deck LT (1984) Paleoceanic events and deep sea ostracods. Science 224: 1334-1336
Berthold G (1884) Cryptonemiaceen In: Flora und Fauna des Gulfes von Neapel und der Angrendzenden Meeres-Abschnitte. W. Engelmann, Leipzig
Bliding C (1928) Studien über die Florideenordnung Rhodymeniales. Lunds Univ Årsskr, NF Avd 2 24(3): 1-74
Boillot A, L'Hardy-Halos MTh (1975) Observations en culture d'une Rhodophycee Bonnemaisoniale: le Naccaria wiggii (Turner) Endlicher. Bull Soc Phycol France 20: 30-35
Bonin DR, Hawkes MW (1988a) Systematics and life histories of New Zealand Bonnemaisoniaceae (Bonnemaisoniales, Rhodophyta): I. The genus *Asparagopsis*. New Zealand J Bot 26: 577-590
Bonin DR, Hawkes MW (1988b) Systematics and life histories of New Zealand Bonnemaisoniaceae (Bonnemaisoniales, Rhodophyta): II. The genus *Delisea*. New Zealand J Bot 26: 619-632
Bonin DR, Hawkes MW (1988c) Systematics and life histories of New Zealand Bonnemaisoniaceae (Bonnemaisoniales, Rhodophyta): III. The genus *Ptilonia*. New Zealand J Bot 26: 633-644
Chiang YM (1970) Morphological studies of red algae of the family Cryptonemiaceae. Univ Calif Publ Bot 58: 1-83
Chihara M (1961) Life cycle of the bonnemaisoniaceous algae in Japan (1). Sci Rep Tokyo Kyoiku Daigaku Sect B 10: 121-153
Chihara M (1962) Life cycle of the bonnemaisoniaceous algae in Japan (2). Sci Rep Tokyo Kyoiku Daigaku Sect B 11: 27-53
Chihara M (1965) Germination of the carpospores of *Bonnemaisonia nootkana*, with special reference to the life cycle. Phycologia 5: 71-79
Chihara M (1972) Germination of carpospores of *Pikea californica* and *Schimmelmannia plumosa* as found in Japan, with special reference to their life histories. Mém Soc Bot Fr 1972: 313-322
Chihara M, Yoshizaki M (1971a) Morphological study of the thallus, structure and reproductive organs of *Delisea fimbriata* (Florideophycidae, Rhodophyta). Phycologia 10: 149-154
Chihara M, Yoshizaki M (1971b) The thallus structure and the repoductive system of *Hyalosiphonia caespitosa* (Cryptonemiales, Rhodophyta). Bot Mag Tokyo 84: 319-325
Chihara M, Yoshizaki M (1972) Bonnemaisoniaceae: Their gonimoblast development, life history and systematics. In: Abbott IA & Kurogi M (eds) Contributions to the systematics of benthic marine algae of the North Pacific. Japanese Society of Phycology, Kobe, Japan, p. 243-252
Codomier L (1971) Recherches sur les *Kallymenia* (Cryptonemiales, Kallymeniacées) I. Les espèces Méditerranèennes. Vie Milieu 22: 1-54
Codomier L (1972) Le cycle du *Neurocaulon reniforme* (P. et R.) Zanardini (Rhodophycée, Gigartinale). Soc Bot Fr Mém 1972: 293-310.
Codomier L (1973) Sur le développement des spores et la formation du thalle rampant de *Kallymenia microphylla* J. Ag. (Rhodophyceae, Cryptonemiales). Giorn Bot Ital 107:269-280
Codomier L (1974) Recherches sur les *Kallymenia* (Cryptonémiales, Kallymeniacées). II. Developpement des spores et morphogenése. Vie Milieu 24: 369-388
Cole K, Conway E (1980) Studies in the Bangiaceae: reproductive modes. Bot Mar 23: 545-553
Cole KM, Hymes BJ, Sheath RG (1983) Karyotypes and reproductive seasonality of the genus *Bangia* (Rhodophyta). Protoplasma 102: 253-279
Coomans RG, Hommersand MH (1990) Vegetative growth and organization. In: Cole KM & Sheath RG (eds) Biology of the red algae. Cambridge Univ. Press, Cambridge (In press)
DeCew TC, West JA (1981a) Investigations on the life histories of three *Farlowia* species (Rhodophyta: Cryptonemiales, Dumontiaceae) from Pacific North America. Phycologia 20: 342- 351
DeCew TC, West JA (1981b) Life histories in the Phyllophoraceae (Rhodophyta:Gigartinales) from the Pacific coast of North America. I. *Gymnogongrus linearis* and *G. leptophyllus*. J Phycol 17: 240-250

DeCew TC, West JA, Ganesan EK (1981) The life histories and developmental morphology of two species of *Gloiosiphonia* (Rhodophyta: Cryptonemiales, Gloiosiphoniaceae from the Pacific Coast of North America. Phycologia 20: 415-423
Dixon PS, Irvine LM (1977) Seaweeds of the British Isles. Vol.1. Rhodophyta. Part 1. Introduction, Nemaliales, Gigartinales. British Museum (Nat. Hist.), London
Dowdeswell EK (1988) Cenozoic Panarctic Tectonic events. In: Harland WB, Dowdeswell EK (eds) Geological evolution of the Barent's Shelf region. Graham, Trodman, London Dordrecht Burton, p 33-46
Dring MJ (1988) Photocontrol and development of algae. Ann Rev Plant Phys 39: 157-174
Frakes LA (1979) Climate throughout geologic time. Elsevier, Amsterdam
Fredericq S, Hommersand MH (1989) Proposal of the Gracilariales ord. nov. (Rhodophyta) based on an analysis of the reproductive development of *Gracilaria verrucosa*. J Phycol 25: 213-227
Fredericq S, Hommersand MH (1990) Diagnoses and key to the genera of the Gracilariaceae. Proc XIIIth Int Seaweed Symp Vancouver, British Columbia (In press)
Gabrielson PW (1980) Morphological studies of selected northern hemisphere genera in the family Solieriacee, Gigartinales (Rhodophyta). Unpublished PhD dissertation, University of North Carolina, Chapel Hill
Gabrielson PW, Cheney DP (1987) Morphology and taxonomy of *Meristiella* gen. nov. (Solieriaceae, Rhodophyta). J Phycol 23: 481-493
Gabrielson PW, Garbary DJ, (1987) A cladistic analysis of Rhodophyta: Florideophycidean orders. Br Phycol J 22: 125-138
Gabrielson PW, Hommersand MH (1982a) The Atlantic species of *Solieria* (Gigartinales, Rhodophyta): their morphology, distribution and affinities. J Phycol 18: 31-45
Gabrielson PW, Hommersand MH (1982b) The morphology of *Agardhiella subulata* representing the Agardhielleae, a new tribe in the Solieriaceae (Gigartinales, Rhodophyta). J Phycol 18: 46-58
Gabrielson PW, Kraft GT (1984) The marine algae of Lord Howe Island (N.S.W.): the family Solieriaceae (Gigartinales, Rhodophyta). Brunonia 7: 217-251
Ganesan EK (1976) On *Kallymenia westii* sp. nov. (Rhodophyta, Cryptonemiales) from the Caribbean Sea. Bol Inst Oceanogr Oriente 15: 169-175
Garbary DJ (1978) On the phylogenetic relationships of the Acrochaetiaceae (Rhodophyta). Br Phycol J 13: 247-254
Garbary DJ, Belliveau DJ (1990) Diffuse growth, a new pattern of cell wall deposition for Rhodophyta. Phycologia 29: 98-102
Garbary DJ, Gabrielson PW (1987) Acrochaetiales (Rhodophyta) taxonomy and evolution. Cryptogamie Algol 8: 241-52
Glazer AN, West JA, Chan C (1982) Phycoerythrins as chemotaxonomic markers in red algae: a survey. Biochem Syst Ecol 10: 203-215
Gordon EM (1972) Comparative morphology and taxonomy of the Wrangelieae, Sphondylothamnieae, and Spermothamnieae (Ceramiaceae, Rhodophyta). Aust J Bot Suppl 4: 1-180
Guiry MD (1974) A preliminary consideration of the taxonomic position of *Palmaria palmata* (Linnaeus) Stackhouse = *Rhodymenia palmata* (Linnaeus) Greville. J Mar Biol Assoc U.K. 54: 509-528
Guiry MD, Garbary DG (1990) A preliminary phylogenetic analysis of the Phyllophoraceae, Gigartinaceae and Petrocelidaceae (Rhodophyta) in the North Atlantic and North Pacific. In: Garbary DJ, South GR (eds), Evolutionary biogeography of the marine algae of the North Atlantic Ocean. Springer Verlag, Berlin, p 265-290
Guiry MD, Maggs CA (1982) The morphology and life history of *Dermocorynus montagnei* Crouan frat. (Halymeniaceae; Rhodophyta) from Ireland. Br Phycol J 17: 215-228
Guiry MD, Maggs CA (1985) Reproduction and life history of *Meredithia microphylla* (J. Ag.) J. Ag. (Kallymeniaceae, Rhodophyta) from Ireland. Giorn Bot Ital 118: 105-125
Guiry MD, West JA (1983) Life history and hybridization studies on *Gigartina stellata* and *Petrocelis cruenta* (Rhodophyta) in the North Atlantic. J Phycol 19: 474-494
Guiry MD, West JA, Kim DH, Masuda M (1984). Reinstatement of the genus *Mastocarpus* Kützing (Rhodophyta). Taxon 33:53-63

Hansen GI (1977a) *Cirrulicarpus carolinensis*, a new species in the Kallymeniaceae (Rhodophyta). Occ Pap Farlow Herb 12:1-22
Hansen GI (1977b) A comparison of the species of *Cirrulicarpus* (Kallymeniaceae, Rhodophyta). Occ Pap Farlow Herb 12: 23-34
Hansen GI (1980) A morphological study of *Fimbrifolium*, a new genus in the Cystocloniaceae (Gigartinales, Rhodophyta). J Phycol 16: 207-217
Hansen GI (1989) *Schizymenia dawsonii* and its relation to the genus *Sebdenia* (Sebdeniaceae, Rhodophyta). Taxon 38: 54-59
Hansen GI, Lindstrom SC (1984) A morphological study of *Hommersandia maximicarpa* gen. et sp. nov. (Kallymeniaceae, Rhodophyta) from the North Pacific. J Phycol 20: 476-488
Hommersand MH (1963) The morphology and classification of some Ceramiaceae and Rhodomelaceae. Univ Calif Publ Bot 35: 165-366
Hommersand MH (1986) The biogeography of the South African marine red algae: a model. Bot Mar 24: 257-270
Hommersand MH, Fredericq S (1988) An investigation of cystocarp development in *Gelidium pteridifolium* with a revised description of the Gelidiales (Rhodophyta). Phycologia 27: 254-272
Hommersand MH, Fredericq S (1990) Sexual reproduction and cystocarp development. In: Cole KM & Sheath RG (eds) Biology of the red algae. Cambridge Univ. Press, Cambridge (In press)
Huisman JH (1985) *Rhipidothmnion secundum* gen. et sp. nov. and *Spermothamnion miniatum* sp. nov. (Ceramiaceae, Rhodophyta) from eastern Australia. Phycologia 24: 55-66
Itono H (1977) Studies on the Ceramiaceous algae (Rhodophyta) from the southern part of Japan. Biblioth Phycol 35: 1-499
Joosten AMT, Hoek van den C (1986) World-wide relationships between red seaweed floras: a multivariate approach. Bot Mar 24: 195-214
Kajimura M (1989) Gymnothamnieae trib. nov. (Ceramiaceae, Rhodophyta) Bot Mar 32: 121-130
Kaneko T, Matsuyama K, Yamada I (1980) On *Gloeophycus koreanum* I.K. Lee & Yoo (Rhodophyta, Gloiosiphoniaceae) in Hokkaido. Jap J Phycol 28: 97-104
Kapraun DF, Lemus AJ (1987) Field and culture studies of *Porphyra spiralis* var. *amplifolia* Oliveira Filho et Coll (Bangiales, Rhodophyta) from Isla de Margarita, Venezuela. Bot Mar 30: 483-490
Kasahara K (1980) On the life history of *Turnerella mertensiana* (Rhodophyta, Gigartinales). Bot Mag, Tokyo 93: 117-123
Kawaguchi S (1989) The genus *Prionitis* (Halymeniaceae, Rhodophyta) in Japan. J Fac Sci, Hokkaido Univ Ser V (Bot) 14: 193-257
Kennett JP (1982) Marine Geology. Prentice Hall Inc., Englewood Cliffs, New Jersey
Kim DH (DH) 1976 A study of the development of cystocarps and tetrasporangial sori in Gigartinaceae. Nova Hedwigia 27: 1-145
Knauss ME (1988) Reproductive development and systematics of some primitive families of Gigartinales (Rhodophyta). Unpubl PhD Dissertation, University of North Carolina, Chapel Hill
Knauss ME, Hommersand MH (1989) Vegetative and reproductive development of *Halarachnion ligulatum* (Gigartinales, Rhodophyta). Br Phycol J 24: 39-52
Kraft GT (1973) The morphology of *Cubiculosporum koronicarpis* gen. et sp. nov., representing a new family in the Gigartinales (Rhodophyta). Amer J Bot 60: 872-882
Kraft GT (1975a) Morphology of *Placentophora* (Solieriaceae, Gigartinales, Rhodophyta), a new genus based on *Sarcodiotheca colensoi* from New Zealand J Phycol 11: 399-410
Kraft GT (1975b) Consideration of the order Cryptonemiales and the families Nemastomaceae and Furcellariaceae (Gigartinales, Rhodophyta) in light of the morphology of *Adelophyton corneum* (J. Agardh) gen. et comb. nov. from Southren Australia. Br Phycol J 1: 279-290
Kraft GT (1977) The morphology of *Grateloupia intestinalis* from New Zealand, with some thoughts on generic criteria within the family Cryptonemiaceae (Rhodophyta). Phycologia 16: 43-51
Kraft GT (1984) Taxonomic and morphological studies of tropical and subtropical species of *Callophycus* (Solieriaceae, Rhodophyta). Phycologia 23: 5-71

Kraft GT (1989) *Cylindrotaxis rotundus* gen. et sp. nov. and its generic relationship within the Liagoraceae (Nemaliales, Rhodophyta). Phycologia 28: 275-304

Kraft GT, Robins PA (1985) Is the order Cryptonemiales defensible? Phycologia 24: 67-77

Kylin H (1914) Studien über die Entwicklungsgeschichte von *Rhodomela virgata* Kjellm. Sv Bot Tidskr 8: 33-69

Kylin H (1923) Studien über die Entwicklungsgeschichte der Florideen. Kungl Svenska Vidensk Handl 63(11): 1-139

Kylin H (1924) Studien über die Delesseriaceen. Lunds Univ. Årsskr., NF Avd 2, 20(6): 1-111

Kylin H (1928) Entwicklungsgeschichtliche Florideenstudien. Lunds Univ Årsskr, NF, Avd 2, 26(4): 1-127

Kylin H (1930) Uber die Entwicklungsgeschichte der Florideen. Lunds Univ Årsskr, NF Avd 2, 28(8): 1-88

Kylin H (1932) Die Florideenordnung Gigartinales. Lunds Univ Årsskr, NF Avd 2, 28(8): 1-88

Kylin H (1934) Uber die systematische Stellung der Gattungen *Opuntiella* und *Turnerella*. Kungl Fysiogr S_llsk I Lund Forhandl 4: 1-6

Kylin H (1956) Die Gattungen der Rhodophyceen. Gleerups, Lund

Lee YP, Lee IK (1988) Contribution to the generic classification of the Rhodochortaceae (Rhodophyta, Nemaliales). Bot Mar 31: 119-131

Lee IK, Yoo SA (1979) *Gloeophycus koreanum* gen. et sp. nov. (Rhodophyta, Gloiosiphoniaceae) from Korea. Phycologia 18: 347-354

Lindstrom SC (1981) Female reproductive structures and strategy in a red alga, *Constantinea rosa-marina* (Gmelin) Postels et Ruprecht (Dumontiaceae, Cryptonemiales). Jap J Phycol 29: 251-257

Lindstrom SC (1985) A phylogenetic systematics study of the red algal family Dumontiaceae. Unpublished PhD thesis, Univ of British Columbia

Lindstrom SC (1987a) Acrosymphytaceae, a new family in the order Gigartinales sensu lato (Rhodophyta). Taxon 36: 50-53

Lindstrom, S (1987b) Possible sister groups and phylogenetic relationships among selected North Pacific and North Atlantic Rhodophyta. Helgol Meersunters 41: 245-260

Lindstrom SC (1988) The Dumontieae, a resurrected tribe of red algae (Dumontiaceae, Rhodophyta) Phycologia 27: 89-102

Lindstrom SC, Scagel RF (1987) The marine algae of British Columbia, northern Washington and southeast Alaska: division Rhodophyta (red algae), class Rhodophyceae, order Gigartinales, family Dumontiaceae, with an introduction to the order Gigartinales. Can J Bot 65: 2202-2232

Lindstrom SC, South GR (1989) Evidence of species relationships in the Palmariaceae (Palmariales, Rhodophyta) based on starch gel electrophoresis. Crypt Bot 1: 32-41

Lindstrom SC, Wynne MJ (1982) *Tokidaea chilkatensis* sp. nov. and *T. serrata* (Wynne) comb. nov. (Ceramiaceae, Rhodophyta) from Alaska. Syesis 14: 33-43

Maggs CA (1983) A phenological study of the epiflora of two maerl beds in Galway Bay. Unpubl PhD Dissertation, University College, Belfast

Maggs CA (1988) Interspecific life history variability in the Florideophycidae (Rhodophyta). Bot Mar 31: 465-490

Maggs CA (1989) *Erythrodermis allenii* Batters in the life history of *Phyllophora traillii* Holmes ex Batters (Phyllophoraceae, Rhodophyta). Phycologia 28: 305-317

Maggs CA (1990) Distribution and evolution of con-coralline crustose red algae in the North Atlantic. In: Garbary DJ, South GR (eds), Evolutionary biogeography of the marine algae of the North Atlantic. Springer Verlag, Berlin, p 241-264

Maggs CA, Guiry MD (1982) The life history of *Haematocelis fissurata* Crouan frat. (Rhodophyta: Sphaerococcaceae). Br Phycol J 17: 235

Maggs CA, Guiry MD (1985) Life history and reproduction of *Schmitzia hiscockiana* sp. nov. (Rhodophyta, Gigartinales) from the British Isles. Phycologia 24: 297-310

Maggs CA, Guiry MD (1987) An Atlantic population of *Pikea californica* (Dumontiaceae, Rhodophyta). J Phycol 23: 170-176

Maggs CA, Guiry MD (1989) A re-evaluation of the crustose red algal genus *Cruoria* and the family Cruoriaceae. Br Phycol J 24: 253-269

Maggs CA, Pueschel CM (1989) Morphology and development of *Ahnfeltia plicata* (Rhodophyta): proposal of the Ahnfeltiales ord. nov. J Phycol 25: 333-351

Maggs CA, McLachlan JL, Saunders GW (1989) Infrageneric taxonomy of *Ahnfeltia* (Ahnfeltiales, Rhodophyta). J Phycol 25: 351-368

Magne F (1989) Classification et phylogénie des Rhodophycées. Cryptogamie Algol 10: 101-115

Masuda M (1982) A systematic study of the tribe Rhodomeleae (Rhodomelaceae, Rhodophyta) J Fac Sc, Hokkaido Univ Ser V, 12: 209-400

Masuda M (1983) The life history of *Ahnfeltia concinna* J. Agardh (Rhodophyta, Gigartinales) from Japan. Jap J Phycol 31: 180-189

Masuda M, West JA, Kurogi M (1987) Life history studies in culture of a *Mastocarpus* species (Rhodophyta) from central Japan. J Fac Sc Hokkaido Univ, Ser V (Bot) 14: 11-38

Masuda M, West JA, Ohno Y, Kurogi M (1984) Comparative reproductive patterns in culture of different *Gigartina* subgenus *Mastocarpus* and *Petrocelis* populations from northern Japan. Bot Mag, Tokyo 97: 107-125

Mayhoub H (1975) Nouvelles observations sur le cycle de développement de *Calosiphonia vermicularis* (J. Agardh) Schmitz (Rhodophycée, Gigartinale). CR Acad Sc Paris D 280: 2441-2443

Meer JP van der, Todd ER (1980) The life history of *Palmaria palmata* in culture. A new type for the Rhodophyta. Can J Bot 58: 1250-1256

Mikami H (1965) A systematic study of the Phyllophoraceae and Gigartinaceae from Japan and its vicinity. Sci Papers Inst Algol Res Fac Sci, Hokkaido Univ 5: 181-285

Millar AJK (1986) *Baldockia verticillata* (Griffithsieae, Ceramiales), a new red algal genus and species from eastern Australia. Phycologia 25: 87-97

Moe RL, Silva PC (1983) Morphological and taxonomic studies on Antarctic Ceramiaceae (Rhodophyceae) III *Georgiella* and *Plumariopsis* (tribe Ptiloteae). Br Phycol J 18: 275-298

Morohoshi H, Masuda M (1980) The life history of *Gloiosiphonia capillaris* (Hudson) Carmichael, Cryptonemiales. Jap J Phycol 28: 81-92

Norris RE (1957) Morphological studies on the Kallymeniaceae. Univ Calif Publ Bot 28: 251-334

Norris RE (1964) The morphology and taxonomy of South African Kallymeniaceae. Bot Mar 7: 90-129

Norris RE (1989) Studies on sexual reproduction in *Titanophora weberae* and possible phylogenetic implications in the red algae. J Phycol 25(suppl): 6

Norris RE, Norris JN (1973) *Kallymenia pertusa* (Rhodophyceae, Cryptonemiales) from the Gulf of California. Phycologia 12: 71-74

Oltmanns F (1898) Zur Entwicklungsgeschichtliche der Florideen. Bot Ztg 56: 99-140

Parsons MJ (1975) Morphology and taxonomy of the Dasyaceae and the Lophothalieae (Rhodomelaceae) of the Rhodophyta. Austr J Bot 23: 549-713

Parsons MJ (1980) The morphology and taxonomy of *Brongniartella* Bory sensu Kylin (Rhodomelaceae, Rhodophyta). Phycologia 19: 273-295

Perestenko LP (1975). Red algae of the far eastern seas of the U.S.S.R. foliose Cryptonemialean algae (Cryptonemiales, Rhodophyta). Bot Zh USSR 60: 1676-1689

Powell J (1986) A short day photoperiodic response in *Constantinea subulifera*. Am Zool 26:4 79-487

Pueschel CM (1987) Absence of cap membranes as a characteristic of some red algal orders. J Phycol 23: 150-156

Pueschel CM (1989) An expanded survey of the ultrastructure of red algal pit plugs. J Phycol 25: 625-636

Pueschel CM, Cole KM (1982) Rhodophycean pit plugs: an ultrastructural survey with taxonomic implications. Am J Bot 69: 703-720

Pueschel CM, Magne F (1987) Pit plugs and other ultrastructural features of systematic value in *Rhodochaete parvula* (Rhodophyta, Rhodochaetales). Cryptogamie Algol 8: 201-9

Reedman DJ, Womersley HBS (1976) Southern Australian species of *Champia* and *Chylocladia* (Rhodymeniales: Rhodophyta). Trans R Soc S Aust 100: 75-104

Richardson JP (1981) Persistence and developemnt of *Dasya baillouviana* (Gmelin) Montagne (Rhodophyceae, Dasyaceae) in North Carolina. Phycologia 20: 385-391

Richardson N, Dixon PS (1970) Culture studies on *Thuretellopsis peggiana* Kylin. J. Phycol. 6: 154-159

Ricker RW (1987) Taxonomy and biogeography of Macquarie Island Seaweeds. British Museum (Nat Hist), London

Rietema H (1984) Development of erect thalli from basal crusts in *Dumontia contorta* (Gmel.) Rupr. (Rhodophyta, Cryptonemiales). Bot Mar 27: 29-36

Rietema H, Klein AWO (1981) Environmental control of the life cycle of *Dumontia contorta* (Rhodophyta) kept in culture. Mar Ecol Prog Ser 4: 23-29

Robins PA & Kraft GT (1985) Morphology of the type and Australian species of *Dudresnaya* (Dumontiaceae, Rhodophyta). Phycologia 24: 1-34

Rosenberg T (1933) Studien über Rhodomelaceen und Dasyaceen. Akad Abh Lund: 1-87

Rueness J, Asen PA (1982) Field and culture observations on the life history of *Bonnemaisonia asparagoides* (Woodw.) C. Ag. (Rhodophyta) from Norway. Bot Mar 25: 577-587

Saito Y (1967) Studies on Japanese species of *Laurencia*, with special reference to the comparative morphology. Mem Fac Fish Hokkaido Univ 15: 1-81

Saunders GW (1987) Investigation and taxonomic considerations of the red algal genus *Rhodophysema* Batters and a discussion of the Acrochaetiales-Palmariales complex of the Florideophycidae. Unpubl MS thesis, Acadia University

Saunders GW, McLachlan JL (1990) Taxonomic considerations of the genus *Rhodophysema* and the Rhodophysemataceae fam. nov. (Rhodophyta, Florideophycidae). Proc Nova Scotia Inst Sci 39: 19-26

Schneider CW (1988) *Craspedocarous humilis* sp. nov. (Cystocloniaceae, Gigartinales) from North Carolina, and a reappraisal of the genus. Phycologia 27: 1-9

Schmitz F, Hauptfleisch P (1896-97) Rhodophyceae. In: Engler A & Prantl K (eds) Die Natürlichen Pflanzenfamilien, vol 1(2): 298-306

Schnitter D (1980) Global paleoceanography and its deep water linkage to antarctic glaciation. Earth Sci Rev 16: 1-193

Scott J, Broadwater S (1989) Ultrastructure of vegetative organization and cell division in the freshwater red alga *Compsopogon*. Protoplasma 152: 112-122

Scott J, Broadwater S (1990) Cell division. In: Cole KM, Sheath RG (eds), The biology of the red algae. Cambridge University Press, Cambridge

Scott JL, Dixon PS (1971) The life history of *Pikea californica* Harv. J Phycol 7:295-300

Scott J, Thomas J, Saunders B (1988) Primary pit connections in *Compsopogon coeruleus* (Balbis) Montagne (Compsopogonales, Rhodophyta). Phycologia 27: 327-333

Searles RB (1968) Morphological studies of red algae of the order Gigartinales. Univ California Publs Bot 43: 1-100

Searles RB (1983) Vegetative and reproductive morphology of *Dudresnaya georgiana* sp. nov. (Rhodophyta, Dumontiaceae). Phycologia 22: 309-316

Sheath RD, Cole KM (1984) Systematics of *Bangia* (Rhodophyta) in North America). I. Biogeographic trends in morphology. Phycologia 23: 383-396

Shevlin DE, Polanshek AR (1978) Life history of *Bonnemaisonia geniculata* (Rhodophyta): a laboratory and field study. J Phycol 14: 282-289

Sjöstedt LG (1926) Floridean studies. Lunds Univ Årsskr, NF, Avd 2, 22(4):1-95

South GR, Hooper RG (1972) The life history of *Turnerella pennyi* (Harv.) Schmitz. Br Phycol J 7: 221-233

Sparling SR (1957) The structure and reproduction of some members of the Rhodymeniaceae. Univ California Publs Bot 29: 319-396

Stegenga H (1985) The marine Acrochaetiaceae (Rhodophyta) of southern Africa. S Afr J Bot 51: 291-330

Stegenga H (1986) The Ceramiaceae (excl. Ceramium) (Rhodophyta) of the South West Cape Province, South Africa. Bibliotheca Phycologica 74: 1-141

Svedelius N (1924) On the discontinuous geographical distribution of some tropical and subtropical marine algae. Arkiv für Bot 19(3): 1-70

Thrainsson SA (1986) Morphological studies on some members of the family Cystocloniaceae. Unpubl MA thesis, University of North Carolina, Chapel Hill

Tokida J, Masaki T (1956). Studies on the reproductive organs of red algae. II. On *Erythrophyllum gmelini* (Grun.) Yendo. Bull Fac Fish Hokkaido Univ 7: 63-71

Tseng CK, Cheng PL (1983) *Sinosiphonia*, a new red algal genus of the Rhodomelaceae from North China. Chin J Oceanol Limnol 1: 352-358

Vakhrameev (1987) Cretaceous paleogeography of the U.S.S.R. Palaeogr, Palaeoclimatol, Palaeocol 59: 57-67

West JA, Hommersand MH (1981) Rhodophyta: Life histories. In: Lobban CS & Wynne MJ (eds), The Biology of Seaweeds. University of California Press, Berkeley, p 133-193

West JA, Guiry MD, Masuda M (1983). Further investigations on the genetic affinities and life history patterns of the red alga *Gigartina*. In: Tseng CK (ed.) Proceedings of the joint China-US Phycology Symposium 1981. Science Press, Beijing, p 137-166

Wilce RT (1990) Role of the Arctic Ocean as a bridge between the Atlantic and Pacific Oceans: fact and hypothesis. In: Garbary DJ, South GR (eds), Evolutionary biogeography of the marine algae of the North Atlantic. Springer-Verlag, Berlin, p 323-348

Wilce RT, Davis AN (1984) Development of *Dumontia contorta* (Dumontiaceae, Cryptonemiales) compared with that of other higher red algae. J Phycol 20: 336-351

Wilce RT, Maggs CA (1989) Reinstatement of the genus *Haemescharia* (Rhodophyta, Haemeschariaceae fam. nov.) for *H. polygona* and *H. hennedyi* comb. nov. (=*Petrocelis hennedyi*). Can J Bot 67: 1465-1479

Wollaston EM, Womersley HBS (1959) The structure and reproduction of *Gulsonia annulata* (Rhodophyta). Pacif Sci 12: 55-62

Womersley HBS, Abbott IA (1968) Structure and reproduction of *Neoardissonea* Kylin (Rhodophyta-Naccariaceae). J Phycol 4: 173-177

Womersley HBS, Norris RE (1971) The morphology and taxonomy of Australian Kallymeniaceae (Rhodophyta). Austr J Bot suppl 2: 1-62

Wynne MJ (1980) *Pleuroblepharidella* nom. nov. (Bonnemaisoniaceae, Rhodophyceae) proposed for *Pleuroblepharis* Wynne. Taxon 29: 325-326

Wynne MJ (1983) The current status of the genera in the Delesseriaceae (Rhodophyta). Bot Mar 26: 437-450

Wynne MJ, Norris RE (1982) *Schizochloenion* gen. nov. (Rhodomelaceae, Ceramiales), a new red algal genus from the north-eastern North Pacific. Phycologia 21: 288-298

Young DN (1979) Fine structure of the 'gland cells' of the red alga, *Opuntiella californica* (Solieriaceae, Gigartinales). Phycologia 18: 288-295

Yoshida T (1974) Sur un genre nouveau, *Tokidaea* (Ceramiaceae, Rhodophyta) du nord du Japon. Bull Mus Nat Hist Nat 3(9): 161-170

Zollner J (1977) Variation in the developmental vegetative anatomy among the Gigartinaceae (Gigartinales, Rhodophyta). Unpubl. MA thesis, University of North Carolina, Chapel Hill

Zupan JR, West JA (1988) Geographic variation in the life history of *Mastocarpus papillatus* (Rhodophyta). J. Phycol 24: 223-229

TAXONOMIC INDEX

SUBJECT INDEX

NATO ASI Series G

Vol. 1: **Numerical Taxonomy.** Edited by J. Felsenstein. 644 pages. 1983. (out of print)

Vol. 2: **Immunotoxicology.** Edited by P.W. Mullen. 161 pages. 1984.

Vol. 3: **In Vitro Effects of Mineral Dusts.** Edited by E.G. Beck and J. Bignon. 548 pages. 1985.

Vol. 4: **Environmental Impact Assessment, Technology Assessment, and Risk Analysis.** Edited by V.T. Covello, J.L. Mumpower, P.J.M. Stallen, and V.R.R. Uppuluri. 1068 pages. 1985.

Vol. 5: **Genetic Differentiation and Dispersal in Plants.** Edited by P. Jacquard, G. Heim, and J. Antonovics. 452 pages. 1985.

Vol. 6: **Chemistry of Multiphase Atmospheric Systems.** Edited by W. Jaeschke. 773 pages. 1986.

Vol. 7: **The Role of Freshwater Outflow in Coastal Marine Ecosystems.** Edited by S. Skreslet. 453 pages. 1986.

Vol. 8: **Stratospheric Ozone Reduction, Solar Ultraviolet Radiation and Plant Life.** Edited by R.C. Worrest and M.M. Caldwell. 374 pages. 1986.

Vol. 9: **Strategies and Advanced Techniques for Marine Pollution Studies: Mediterranean Sea.** Edited by C.S. Giam and H.J.-M. Dou. 475 pages. 1986.

Vol. 10: **Urban Runoff Pollution.** Edited by H.C. Torno, J. Marsalek, and M. Desbordes. 893 pages. 1986.

Vol. 11: **Pest Control: Operations and Systems Analysis in Fruit Fly Management.** Edited by M. Mangel, J.R. Carey, and R.E. Plant. 465 pages. 1986.

Vol. 12: **Mediterranean Marine Avifauna: Population Studies and Conservation.** Edited by MEDMARAVIS and X. Monbailliu. 535 pages. 1986.

Vol. 13: **Taxonomy of Porifera from the N.E. Atlantic and Mediterranean Sea.** Edited by J. Vacelet and N. Boury-Esnault. 332 pages. 1987.

Vol. 14: **Developments in Numerical Ecology.** Edited by P. Legendre and L. Legendre. 585 pages. 1987.

Vol. 15: **Plant Response to Stress. Functional Analysis in Mediterranean Ecosystems.** Edited by J.D. Tenhunen, F.M. Catarino, O.L. Lange, and W.C. Oechel. 668 pages. 1987.

Vol. 16: **Effects of Atmospheric Pollutants on Forests, Wetlands and Agricultural Ecosystems.** Edited by T.C. Hutchinson and K.M. Meema. 652 pages. 1987.

Vol. 17: **Intelligence and Evolutionary Biology.** Edited by H.J. Jerison and I. Jerison. 481 pages. 1988.

Vol. 18: **Safety Assurance for Environmental Introductions of Genetically-Engineered Organisms.** Edited by J. Fiksel and V.T. Covello. 282 pages. 1988.

Vol. 19: **Environmental Stress in Plants. Biochemical and Physiological Mechanisms.** Edited by J.H. Cherry. 369 pages. 1989.

Vol. 20: **Behavioural Mechanisms of Food Selection.** Edited by R.N. Hughes. 886 pages. 1990.

Vol. 21: **Health Related Effects of Phyllosilicates.** Edited by J. Bignon. 462 pages. 1990.

Vol. 22: **Evolutionary Biogeography of the Marine Algae of the North Atlantic.** Edited by D.J. Garbary and G.R. South. 441 pages. 1990.

MIX
Papier aus verantwortungsvollen Quellen
Paper from responsible sources
FSC® C105338

If you have any concerns about our products,
you can contact us on
ProductSafety@springernature.com

In case Publisher is established outside the EU,
the EU authorized representative is:
Springer Nature Customer Service Center GmbH
Europaplatz 3, 69115 Heidelberg, Germany

Printed by Libri Plureos GmbH
in Hamburg, Germany